Supercritical Technology Applied to Food, Pharmaceutical and Chemical Industries

Supercritical Technology Applied to Food, Pharmaceutical and Chemical Industries

Editors

Maria Angela A. Meireles
Ádina L. Santana
Grazielle Náthia-Neves

Basel • Beijing • Wuhan • Barcelona • Belgrade • Novi Sad • Cluj • Manchester

Editors
Maria Angela A. Meireles
University of Campinas
(Unicamp)
Campinas
Brazil

Ádina L. Santana
Kansas State University
Manhattan
United States

Grazielle Náthia-Neves
Research Group for Bioactives–
Analysis and Application
National Food Institute
Technical University of
Denmark
Lyngby
Denmark

Editorial Office
MDPI AG
Grosspeteranlage 5
4052 Basel, Switzerland

This is a reprint of articles from the Special Issue published online in the open access journal *Processes* (ISSN 2227-9717) (available at: https://www.mdpi.com/journal/processes/special_issues/Supercritical_Technology).

For citation purposes, cite each article independently as indicated on the article page online and as indicated below:

Lastname, A.A.; Lastname, B.B. Article Title. *Journal Name* **Year**, *Volume Number*, Page Range.

ISBN 978-3-7258-1409-1 (Hbk)
ISBN 978-3-7258-1410-7 (PDF)
doi.org/10.3390/books978-3-7258-1410-7

Contents

About the Editors

Maria Angela A. Meireles

Dr. Maria Angela A. Meireles is a retired professor of Food Engineering at the University of Campinas (Unicamp), Brazil, where she began working in 1983 as an Assistant Professor. She holds a PhD in Chemical Engineering from Iowa State University, United States of America (1982), and both an MSc and a BS degree in Food Engineering from Unicamp (1979 and 1977, respectively). Dr. Meireles has published 230 research papers in peer-reviewed journals and has held more than 520 presentations at scientific conferences. Dr. Meireles has supervised approximately 50 PhD dissertations, 30 MSc theses, and 40 undergraduate research projects. Her research is in the field of the production of extracts from aromatic, medicinal, and spice plants using supercritical fluid extraction and conventional techniques, such as steam distillation and green solvent extraction, including the determination of process parameters, process integration and optimization, extraction fractionation, and techno and economic analysis. She has coordinated scientific exchange projects between Unicamp and European universities in France, Germany, the Netherlands, and Spain. Nationally, she coordinated a project (SuperNat) that involved six Brazilian institutions (Unicamp, UFPA, UFRN, UEM, UFSC, and IAC) and a German university (TUHH). She belongs to the editorial boards of *The Journal of Supercritical Fluids* and the *Journal of Food Processing Engineering* (Blackwell-Publishing).

Ádina L. Santana

Dr Ádina L. Santana is currently a Postdoctoral Researcher at the Grain Science and Industry Department at Kansas State University (United States of America). She obtained her PhD in Food Engineering from the University of Campinas (Unicamp), Brazil (2017), on the use of supercritical technology for the measurement of phase behavior and the reutilization of agricultural waste via extraction, encapsulation, and economic evaluation. She previously worked as a Postdoctoral Researcher at the Department of Food Science and Nutrition at Unicamp (2018) where she worked with supercritical technology integrated with bioprocessing to select polyphenols and alkaloids. In 2019, Dr. Santana joined the Food Science and Technology Department at the University of Nebraska–Lincoln (United States of America), before joining Kansas State University (2021), where she is currently working with the valorization of phenolic compounds in sorghum. Over the years, Dr. Santana has trained undergraduate and graduate students, co-edited two books (*Supercritical Fluid Biorefining*, Springer, 2020, and *Supercritical Antisolvent Precipitation* Process, Springer, 2019), published approximately 40 papers in peer-reviewed journals and 28 book chapters, and performed more than 100 reviews for 45 peer-reviewed journals.

Grazielle Náthia-Neves

Dr Grazielle Náthia-Neves earned her MSc in Food Engineering (2016) at the University of Campinas, Brazil, on the expertise of biotechnology and bioengineering for the reutilization of food waste via anaerobic fermentation, and she completed her PhD in Food Engineering (2019) at the same institution on the use of green technologies for the extraction of bioactive compounds with the use of supercritical carbon dioxide and pressurized liquids for the utilization of undervalued colorants. Previously, she served as a Postdoctoral Researcher at the University of Valladolid (Spain).

Preface

The current demands for sustainability have made consumers concerned about the origin of the products generated in different industries. For this reason, green technologies are increasingly required and optimized for industrial processing. In the last few decades, supercritical technology has emerged as an environmentally friendly unit operation of industry that answers such sustainability demands via the development of efficient and low-cost processes, the generation of products of commercial value, and a reduction in waste emissions.

The scope of the Special Issue "Supercritical Technology Applied to Food, Pharmaceutical and Chemical Industries" is multidisciplinary and includes one editorial and fifteen outstanding papers covering advances in the theory and practice of processing strategies that involve supercritical-based technology for the extraction of bioactive compounds (saponins, carotenoids, essential oil, phenolic compounds, and iridoids), particle formation, the pretreatment of materials for medical and colorant application, the sterilization of foods, and the transesterification of fats for biofuel production. Moreover, the theory items studied in terms of thermodynamics, mathematical modeling, and process simulation for scale-up and economic analysis provide useful information for optimal process performance and cost reduction.

For the following Special Issue, the target audience comprises, but is not limited to, researchers, students, technicians, and those with a general curiosity about green chemistry and industrial processing.

Maria Angela A. Meireles, Ádina L. Santana, and Grazielle Náthia-Neves
Editors

Editorial

Supercritical Technology Applied to Food, Pharmaceutical, and Chemical Industries

Ádina L. Santana

Grain Science and Industry Department, Kansas State University, 1301 N Mid Campus Drive, Manhattan, KS 66506, USA; adina@ksu.edu

1. Introduction

Everyday life has caused consumers to feel genuine concern about the origin of the products they consume. For this reason, green technologies are required in industrial processing to ensure the development of high-quality products. Supercritical technology is a green methodology that includes multiple types of high-pressure processes that employ substances in conditions next to or above the critical point [1–3].

Supercritical technology has emerged as an environmentally friendly and efficient alternative for use in the preparation of multiple varieties of matrix for the extraction [4–6], fractionation [7,8], and purification of molecules [9]; the transformation of molecules via chemical reactions [10]; particle formation [11]; impregnation [12]; drying [13]; and sterilization [14]. In this Special Issue, fifteen outstanding manuscripts covering novel insights into the theory and practice of supercritical fluid-based processes are published. For more information on this Special Issue, readers are strongly encouraged to visit the website: https://www.mdpi.com/journal/processes/special_issues/Supercritical_Technology (accessed on 6 April 2024).

2. Review Manuscripts

The thermodynamic background [15] of the gas–lattice model and its potential to describe processes at a supercritical state was reviewed by Tovbin [16]. Supercritical fluids possess applications as refrigerant fluids. The optimization of heat transfer using supercritical fluids has been studied via the use of the gas–lattice approach [17,18].

The use of supercritical technology to valorize corn byproducts was reviewed by Santana and Meireles [19], who proposed the use of a novel process according to the biorefinery approach [20–23]. The proposed biorefining method consisted of integrating traditional dry-grinding, performed in an industrial setting, with the supercritical carbon dioxide (SC-CO_2)-based extraction of corn-dried distiller's grains with solubles (DDGS) to obtain an oil that was rich in the carotenoids known as lutein and zeaxanthin. This was followed by the use of pressurized liquid to extract phenolic acids from the semi-defatted corn DDGS, and by the concentration of the extract into precipitated particles.

3. Research Manuscript: Particle Formation Techniques

Particle formation using supercritical technology offers advantages like the control of particle size and morphology, high encapsulation efficiency, and the low degradation of molecules [11,24,25]. In this Issue, Tirado and coworkers [26] modelled the supercritical fluid extraction of emulsions process (SFEE) of a ternary CO_2/ethyl acetate/water system in order to design equipment that could meet industrial requirements regarding the permitted quantities of residual organic materials in the leaving streams. In SFEE, SC-CO_2 is used to extract the organic phase from an organic phase/water (O/W) emulsion in which the target molecule and its coating material have been previously solubilized. After solvent removal, both compounds precipitate, generating particles that are suspended in the water phase with the aid of a surfactant [27].

Citation: Santana, Á.L. Supercritical Technology Applied to Food, Pharmaceutical, and Chemical Industries. *Processes* 2024, 12, 861. https://doi.org/10.3390/pr12050861

Received: 7 April 2024
Revised: 20 April 2024
Accepted: 22 April 2024
Published: 25 April 2024

4. Research Manuscript: Chemical Reactions

Supercritical transesterification is attractive in comparison with conventional transesterification as a method with which to produce biodiesel since it requires little time and no catalyst [28]. In this Issue, García-Morales and coworkers [29] investigated the potential of alcohols at supercritical state in terms of the transformation of waste beef tallow into fatty acid alkyl esters or biodiesel. We found conversion rates higher than 90% at 335–390 °C for supercritical iso-butanol and at 360 °C for supercritical ethanol.

5. Research Manuscript: Sterilization of Foods

The sterilization of bacteria with SC-CO_2 emerged as a method because of the mild temperatures used in comparison to conventional techniques [30,31]. In this Issue, Dacal-Gutiérrez and coworkers [32] observed that the inactivation of *Clostridium* spores in low-moisture honey is not effective when using SC-CO_2 at temperatures lower than 70 °C, but that the use of carbon dioxide, modified with cinnamon essential oil, significantly reduced the presence of spores at 60 °C.

6. Research Manuscripts: Removal of Undesirable Compounds

Chiu and coworkers [33] observed that SC-CO_2 has good biocompatibility in the decellularization of porcine hide for the reconstruction of an abdominal wall that had been injured by hernia. The decellularization of tissues with SC-CO_2 is a pretreatment protocol that is used to remove undesirable tissue and molecules (protein and lipids) for biomedical applications [34].

Náthia-Neves and coworkers [35] studied the extraction of colorants from unripe genipap defatted with SC-CO_2. SC-CO_2 extraction worked as a pretreatment, ensuring the plant material was suitable for the subsequent ultrasound-assisted extraction of genipin and geniposide with water and ethanol mixtures [36].

7. Research Manuscripts: Extraction of Bioactive Compounds

Supercritical technology is used for the extraction of multiple bioactive compounds, including phenolic compounds [37], carotenoids [38], phytosterols [39], and cannabinoids [40].

Qamar and coworkers [41] employed the half-fractional factorial design to select the best conditions for the SC-CO_2 extraction of cannabis flowers and found that the optimal conditions were 45 °C, 250 bar, and 180 min.

Boumghar and coworkers [42] selected the Box–Behnken experimental design to optimize the supercritical fluid extraction of decarboxylated cannabis flower. The authors used decarboxylation to pretreat the raw material in order to increase the affinity of cannabinoids to CO_2. The optimal conditions for the extraction of cannabinoids were 55 °C, 235 bar, 2 h, and a CO_2 flow rate of 15 g/min.

Popescu and coworkers [43] investigated the potential of oil seeds as modifier to CO_2 in order to increase the recovery of carotenoids from tomato slices. After supercritical fluid extraction, two products were obtained: a solid, oleoresin rich in lycopene, and an oil fraction rich in other carotenes and linolenic acid.

Duong and coworkers [44] optimized the recovery of saponins from *Hedera nepalensis* leaves with the use of pressurized liquid extraction. The Box–Behnken design was adopted by the authors to select the extraction time, solvent used, and temperature. The extracts showed antimicrobial activity by inhibiting the growth of three types of bacteria.

Santana and coworkers [45] investigated the effects of the post-acidification of pressurized liquid extracts of sorghum on the concentration of phenolic compounds. The authors observed that acidification considerably improved the concentration of 3-deoxyanthocyanidins and cyanidin, but decreased the concentration of other phenolics, including taxifolin, quercetin, and chlorogenic acid. Sorghum (*Sorghum bicolor* L.) is the fifth most-produced cereal worldwide and is a source of diverse classes of phenolic compounds, including tannins, benzoic- and cinnamic acids, 3-deoxyanthocyanidins, and flavonols [46,47].

8. Research Manuscripts: Modeling, Simulation, and Economic Evaluation

Bushnaq et al. [48] proposed a three-step process in order to enhance the yield of sugars from date palm. This was based on (1) the freeze-drying of dates, (2) supercritical fluid extraction with CO_2 modified with water, and (3) the spray-drying of a supercritical-based extract. After simulation, the authors reported that a highest rate of sugar recovery can be reached at a CO_2–water ratio of 0.07, CO_2 flow rate of 31,000 kg/h, 65 °C, and 308 bar. SC-CO_2 extraction is useful as a pretreatment technique in the extraction of sugars, allowing manufacturers to remove undesirable compounds of raw material [49].

Before implementing a process in the market, it is important to have knowledge of the economic feasibility of the process. Economic evaluations consider the components involved, the costs of processing, the final products, as well as the economic fluctuations that affect the price of inputs [50–52]. Cruz Sánchez and coworkers [53] extracted lavender flowers with SC-CO_2 at 60 °C and 180 and 250 bar, simulated the process for the capacities of 20 L, 50 L, and 100 L. They observed that the cost of manufacturing was lower at 50 and 100 L, and that the price of equipment was the item that most affected the return on equity. The return on equity is a parameter that indicates a process's profitability [54].

Best and coworkers [55] investigated the economic profitability of the extraction of *Mauritia flexuosa* pulp using two scenarios: (a) conventional extraction and (b) conventional extraction integrated with SC-CO_2 extraction. They concluded that scenario (b) was the most feasible economically, since it was enabled researchers to obtain two types of products—namely, an oil rich in carotenoids, and an extract with high phenolic content.

9. Conclusions

The results obtained from the research published in this Special Issue support the industrial use of supercritical technology via the application of antioxidant extracts in food, pharmaceutical industries, and the medical sector, as well as the conversion of underused fat into value-added fuels. Additionally, the theoretical aspects explored in this Issue, with explorations into thermodynamics, mathematical modeling, and economic evaluation, provide useful information for the optimization of processes and reduction of costs.

Conflicts of Interest: The author declares no conflicts of interest.

References

1. Brunner, G. Applications of Supercritical Fluids. *Annu. Rev. Chem. Biomol. Eng.* **2010**, *1*, 321–342. [CrossRef]
2. Brunner, G. *Gas Extraction: An Introduction to Fundamentals of Supercritical Fluids and the Application to Separation Processes*, 4th ed.; Steinkopff-Verlag Heidelberg: New York, NY, USA, 1994.
3. Prasad, S.K.; Sangwai, J.S.; Byun, H.-S. A review of the supercritical CO_2 fluid applications for improved oil and gas production and associated carbon storage. *J. CO_2 Util.* **2023**, *72*, 102479. [CrossRef]
4. Pimentel-Moral, S.; Borrás-Linares, I.; Lozano-Sánchez, J.; Arráez-Román, D.; Martínez-Férez, A.; Segura-Carretero, A. Supercritical CO_2 extraction of bioactive compounds from *Hibiscus sabdariffa*. *J. Supercrit. Fluids* **2019**, *147*, 213–221. [CrossRef]
5. Priyanka; Khanam, S. Influence of operating parameters on supercritical fluid extraction of essential oil from turmeric root. *J. Clean. Prod.* **2018**, *188*, 816–824. [CrossRef]
6. Kuvendziev, S.; Lisichkov, K.; Zeković, Z.; Marinkovski, M.; Musliu, Z.H. Supercritical fluid extraction of fish oil from common carp (*Cyprinus carpio* L.) tissues. *J. Supercrit. Fluids* **2018**, *133*, 528–534. [CrossRef]
7. Shukla, A.; Naik, S.N.; Goud, V.V.; Das, C. Supercritical CO_2 extraction and online fractionation of dry ginger for production of high-quality volatile oil and gingerols enriched oleoresin. *Ind. Crops Prod.* **2019**, *130*, 352–362. [CrossRef]
8. Reverchon, E.; De Marco, I. Supercritical fluid extraction and fractionation of natural matter. *J. Supercrit. Fluids* **2006**, *38*, 146–166. [CrossRef]
9. Gallo-Molina, A.C.; Castro-Vargas, H.I.; Garzón-Méndez, W.F.; Martínez Ramírez, J.A.; Rivera Monroy, Z.J.; King, J.W.; Parada-Alfonso, F. Extraction, isolation and purification of tetrahydrocannabinol from the *Cannabis sativa* L. plant using supercritical fluid extraction and solid phase extraction. *J. Supercrit. Fluids* **2019**, *146*, 208–216. [CrossRef]
10. Knez, Ž. Enzymatic reactions in dense gases. *J. Supercrit. Fluids* **2009**, *47*, 357–372. [CrossRef]
11. Palazzo, I.; Campardelli, R.; Scognamiglio, M.; Reverchon, E. Zein/luteolin microparticles formation using a supercritical fluids assisted technique. *Powder Technol.* **2019**, *356*, 899–908. [CrossRef]

12. Liu, X.; Jia, J.; Duan, S.; Zhou, X.; Xiang, A.; Lian, Z.; Ge, F. Zein/MCM-41 Nanocomposite Film Incorporated with Cinnamon Essential Oil Loaded by Modified Supercritical CO_2 Impregnation for Long-Term Antibacterial Packaging. *Pharmaceutics* **2020**, *12*, 169. [CrossRef] [PubMed]
13. Şahin, İ.; Özbakır, Y.; İnönü, Z.; Ulker, Z.; Erkey, C. Kinetics of Supercritical Drying of Gels. *Gels* **2018**, *4*, 3. [CrossRef] [PubMed]
14. Scognamiglio, F.; Blanchy, M.; Borgogna, M.; Travan, A.; Donati, I.; Bosmans, J.W.A.M.; Foulc, M.P.; Bouvy, N.D.; Paoletti, S.; Marsich, E. Effects of supercritical carbon dioxide sterilization on polysaccharidic membranes for surgical applications. *Carbohydr. Polym.* **2017**, *173*, 482–488. [CrossRef] [PubMed]
15. Tovbin, Y.K. Possibilities of the Molecular Modeling of Kinetic Processes under Supercritical Conditions. *Russ. J. Phys. Chem. A* **2021**, *95*, 429–444. [CrossRef]
16. Tovbin, Y.K. Molecular Modeling of Supercritical Processes and the Lattice—Gas Model. *Processes* **2023**, *11*, 2541. [CrossRef]
17. Yang, Z.; Luo, X.; Chen, W.; Chyu, M.K. Mitigation effects of Body-Centered Cubic Lattices on the heat transfer deterioration of supercritical CO_2. *Appl. Therm. Eng.* **2021**, *183*, 116085. [CrossRef]
18. Shi, X.; Yang, Z.; Chen, W.; Chyu, M.K. Investigation of the effect of lattice structure on the fluid flow and heat transfer of supercritical CO_2 in tubes. *Appl. Therm. Eng.* **2022**, *207*, 118132. [CrossRef]
19. Santana, Á.L.; Meireles, M.A.A. Valorization of Cereal Byproducts with Supercritical Technology: The Case of Corn. *Processes* **2023**, *11*, 289. [CrossRef]
20. Herrero, M.; Ibañez, E. Green extraction processes, biorefineries and sustainability: Recovery of high added-value products from natural sources. *J. Supercrit. Fluids* **2018**, *134*, 252–259. [CrossRef]
21. Santana, Á.L.; Santos, D.T.; Meireles, M.A.A. Perspectives on small-scale integrated biorefineries using supercritical CO_2 as a green solvent. *Curr. Opin. Green Sustain. Chem.* **2019**, *18*, 1–12. [CrossRef]
22. Abecassis, J.; de Vries, H.; Rouau, X. New perspective for biorefining cereals. *Biofuels Bioprod. Biorefining* **2014**, *8*, 462–474. [CrossRef]
23. Chatzifragkou, A.; Charalampopoulos, D. 3-Distiller's dried grains with solubles (DDGS) and intermediate products as starting materials in biorefinery strategies. In *Sustainable Recovery and Reutilization of Cereal Processing By-Products*; Galanakis, C.M., Ed.; Woodhead: Cambridge, UK, 2018; pp. 63–86. ISBN 978-0-08-102162-0.
24. Rosa, M.T.M.G.; Alvarez, V.H.; Albarelli, J.Q.; Santos, D.T.; Meireles, M.A.A.; Saldaña, M.D.A. Supercritical anti-solvent process as an alternative technology for vitamin complex encapsulation using zein as wall material: Technical-economic evaluation. *J. Supercrit. Fluids* **2020**, *159*, 104499. [CrossRef]
25. Baldino, L.; Adami, R.; Reverchon, E. Concentration of Ruta graveolens active compounds using SC-CO_2 extraction coupled with fractional separation. *J. Supercrit. Fluids* **2018**, *131*, 82–86. [CrossRef]
26. Tirado, D.F.; Cabañas, A.; Calvo, L. Modelling and Scaling-Up of a Supercritical Fluid Extraction of Emulsions Process. *Processes* **2023**, *11*, 1063. [CrossRef]
27. Prieto, C.; Calvo, L. Supercritical fluid extraction of emulsions to nanoencapsulate vitamin E in polycaprolactone. *J. Supercrit. Fluids* **2017**, *119*, 274–282. [CrossRef]
28. Tobar, M.; Núñez, G.A. Supercritical transesterification of microalgae triglycerides for biodiesel production: Effect of alcohol type and co-solvent. *J. Supercrit. Fluids* **2018**, *137*, 50–56. [CrossRef]
29. García-Morales, R.; Verónico-Sánchez, F.J.; Zúñiga-Moreno, A.; González-Vargas, O.A.; Ramírez-Jiménez, E.; Elizalde-Solis, O. Fatty Acid Alkyl Ester Production by One-Step Supercritical Transesterification of Beef Tallow by Using Ethanol, Iso-Butanol, and 1-Butanol. *Processes* **2023**, *11*, 742. [CrossRef]
30. Warambourg, V.; Mouahid, A.; Crampon, C.; Galinier, A.; Claeys-Bruno, M.; Badens, E. Supercritical CO_2 sterilization under low temperature and pressure conditions. *J. Supercrit. Fluids* **2023**, *203*, 106084. [CrossRef]
31. Martín-Muñoz, D.; Tirado, D.F.; Calvo, L. Inactivation of Legionella in aqueous media by high-pressure carbon dioxide. *J. Supercrit. Fluids* **2022**, *180*, 105431. [CrossRef]
32. Dacal-Gutiérrez, A.; Tirado, D.F.; Calvo, L. Inactivation of Clostridium Spores in Honey with Supercritical CO_2 and in Combination with Essential Oils. *Processes* **2022**, *10*, 2232. [CrossRef]
33. Chiu, Y.-L.; Lin, Y.-N.; Chen, Y.-J.; Periasamy, S.; Yen, K.-C.; Hsieh, D.-J. Efficacy of Supercritical Fluid Decellularized Porcine Acellular Dermal Matrix in the Post-Repair of Full-Thickness Abdominal Wall Defects in the Rabbit Hernia Model. *Processes* **2022**, *10*, 2588. [CrossRef]
34. Chou, P.R.; Lin, Y.N.; Wu, S.H.; Lin, S.D.; Srinivasan, P.; Hsieh, D.J.; Huang, S.H. Supercritical Carbon Dioxide-decellularized Porcine Acellular Dermal Matrix combined with Autologous Adipose-derived Stem Cells: Its Role in Accelerated Diabetic Wound Healing. *Int. J. Med. Sci.* **2020**, *17*, 354–367. [CrossRef]
35. Náthia-Neves, G.; Santana, Á.L.; Viganó, J.; Martínez, J.; Meireles, M.A.A. Ultrasound-Assisted Extraction of Semi-Defatted Unripe Genipap (*Genipa americana* L.): Selective Conditions for the Recovery of Natural Colorants. *Processes* **2021**, *9*, 1435. [CrossRef]
36. Náthia-Neves, G.; Vardanega, R.; Meireles, M.A.A. Extraction of natural blue colorant from Genipa americana L. using green technologies: Techno-economic evaluation. *Food Bioprod. Process.* **2019**, *114*, 132–143. [CrossRef]
37. Garmus, T.T.; Paviani, L.C.; Queiroga, C.L.; Cabral, F.A. Extraction of phenolic compounds from pepper-rosmarin (*Lippia sidoides* Cham.) leaves by sequential extraction in fixed bed extractor using supercritical CO_2, ethanol and water as solvents. *J. Supercrit. Fluids* **2015**, *99*, 68–75. [CrossRef]

38. Torres, R.A.C.; Santana, Á.L.; Santos, D.T.; Albarelli, J.Q.; Meireles, M.A.A. A novel process for CO_2 purification and recycling based on subcritical adsorption in oat bran. *J. CO_2 Util.* **2019**, *34*, 362–374. [CrossRef]
39. Alvarez-Henao, M.V.; Cardona, L.; Hincapié, S.; Londoño-Londoño, J.; Jimenez-Cartagena, C. Supercritical fluid extraction of phytosterols from sugarcane bagasse: Evaluation of extraction parameters. *J. Supercrit. Fluids* **2022**, *179*, 105427. [CrossRef]
40. Qamar, S.; Torres, Y.J.M.; Parekh, H.S.; Falconer, J.R. Effects of Ethanol on the Supercritical Carbon Dioxide Extraction of Cannabinoids from Near Equimolar (THC and CBD Balanced) Cannabis Flower. *Separations* **2021**, *8*, 154. [CrossRef]
41. Qamar, S.; Torres, Y.J.M.; Parekh, H.S.; Falconer, J.R. Fractional Factorial Design Study for the Extraction of Cannabinoids from CBD-Dominant Cannabis Flowers by Supercritical Carbon Dioxide. *Processes* **2022**, *10*, 93. [CrossRef]
42. Boumghar, H.; Sarrazin, M.; Banquy, X.; Boffito, D.C.; Patience, G.S.; Boumghar, Y. Optimization of Supercritical Carbon Dioxide Fluid Extraction of Medicinal Cannabis from Quebec. *Processes* **2023**, *11*, 1953. [CrossRef]
43. Popescu, M.; Iancu, P.; Plesu, V.; Bildea, C.S. Carotenoids Recovery Enhancement by Supercritical CO_2 Extraction from Tomato Using Seed Oils as Modifiers. *Processes* **2022**, *10*, 2656. [CrossRef]
44. Duong, H.T.; Trieu, L.H.; Linh, D.T.T.; Duy, L.X.; Thao, L.Q.; Van Minh, L.; Hiep, N.T.; Khoi, N.M. Optimization of Subcritical Fluid Extraction for Total Saponins from Hedera nepalensis Leaves Using Response Surface Methodology and Evaluation of Its Potential Antimicrobial Activity. *Processes* **2022**, *10*, 1268. [CrossRef]
45. Santana, Á.L.; Peterson, J.; Perumal, R.; Hu, C.; Sang, S.; Siliveru, K.; Smolensky, D. Post Acid Treatment on Pressurized Liquid Extracts of Sorghum (*Sorghum bicolor* L. Moench) Grain and Plant Material Improves Quantification and Identification of 3-Deoxyanthocyanidins. *Processes* **2023**, *11*, 2079. [CrossRef]
46. Lee, H.-S.; Santana, Á.L.; Peterson, J.; Yucel, U.; Perumal, R.; De Leon, J.; Lee, S.-H.; Smolensky, D. Anti-Adipogenic Activity of High-Phenolic Sorghum Brans in Pre-Adipocytes. *Nutrients* **2022**, *14*, 1493. [CrossRef]
47. Pontieri, P.; Pepe, G.; Campiglia, P.; Mercial, F.; Basilicata, M.G.; Smolensky, D.; Calcagnile, M.; Troisi, J.; Romano, R.; Del Giudice, F.; et al. Comparison of Content in Phenolic Compounds and Antioxidant Capacity in Grains of White, Red, and Black Sorghum Varieties Grown in the Mediterranean Area. *ACS Food Sci. Technol.* **2021**, *1*, 1109–1119. [CrossRef]
48. Bushnaq, H.; Krishnamoorthy, R.; Abu-Zahra, M.; Hasan, S.W.; Taher, H.; Alomar, S.Y.; Ahmad, N.; Banat, F. Supercritical Technology-Based Date Sugar Powder Production: Process Modeling and Simulation. *Processes* **2022**, *10*, 257. [CrossRef]
49. Arumugham, T.; AlYammahi, J.; Rambabu, K.; Hassan, S.W.; Banat, F. Supercritical CO_2 pretreatment of date fruit biomass for enhanced recovery of fruit sugars. *Sustain. Energy Technol. Assess.* **2022**, *52*, 102231. [CrossRef]
50. Caffrey, K.R.; Veal, M.W.; Chinn, M.S. The farm to biorefinery continuum: A techno-economic and LCA analysis of ethanol production from sweet sorghum juice. *Agric. Syst.* **2014**, *130*, 55–66. [CrossRef]
51. Espada, J.J.; Pérez-Antolín, D.; Vicente, G.; Bautista, L.F.; Morales, V.; Rodríguez, R. Environmental and techno-economic evaluation of β-carotene production from *Dunaliella salina*. A biorefinery approach. *Biofuels Bioprod. Biorefining* **2019**, *14*, 43–54. [CrossRef]
52. Gwee, Y.L.; Yusup, S.; Tan, R.R.; Yiin, C.L. Techno-economic and life-cycle assessment of volatile oil extracted from *Aquilaria sinensis* using supercritical carbon dioxide. *J. CO_2 Util.* **2020**, *38*, 158–167. [CrossRef]
53. Cruz-Sánchez, E.; García-Vargas, J.M.; Gracia, I.; Rodriguez, J.F.; García, M.T. Supercritical CO_2 extraction of lavender flower with antioxidant activity: Laboratory to a large scale optimization process. *J. Taiwan Inst. Chem. Eng.* **2024**, *157*, 105404. [CrossRef]
54. Cruz Sánchez, E.; García-Vargas, J.M.; Gracia, I.; Rodríguez, J.F.; García, M.T. Pilot-Plant-Scale Extraction of Antioxidant Compounds from Lavender: Experimental Data and Methodology for an Economic Assessment. *Processes* **2022**, *10*, 2708. [CrossRef]
55. Best, I.; Cartagena-Gonzales, Z.; Arana-Copa, O.; Olivera-Montenegro, L.; Zabot, G. Production of Oil and Phenolic-Rich Extracts from *Mauritia flexuosa* L.f. Using Sequential Supercritical and Conventional Solvent Extraction: Experimental and Economic Evaluation. *Processes* **2022**, *10*, 459. [CrossRef]

 *processes*

Article

Molecular Modeling of Supercritical Processes and the Lattice—Gas Model

Yuri Konstantinovich Tovbin

Kurnakov Institute of General and Inorganic Chemistry, Russian Academy of Sciences, Moscow 119991, Russia; tovbinyk@mail.ru

Abstract: The existing possibilities for modeling the kinetics of supercritical processes at the molecular level are considered from the point of view that the Second Law of thermodynamics must be fulfilled. The only approach that ensures the fulfillment of the Second Law of thermodynamics is the molecular theory based on the discrete–continuous lattice gas model. Expressions for the rates of the elementary stage on its basis give a self-consistent description of the equilibrium states of the mixtures under consideration. The common usage today of ideal kinetic models in SC processes in modeling industrial chemistry contradicts the non-ideal equation of states. The used molecular theory is the theory of absolute reaction rates for non-ideal reaction systems, which takes into account intermolecular interactions that change the effective activation energies of elementary stages. This allows the theory to describe the rates of elementary stages of chemical transformations and molecular transport at arbitrary temperatures and reagent densities in different phases. The application of this theory in a wide range of state parameters (pressure and temperature) is considered when calculating the rates of elementary bimolecular reactions and dissipative coefficients under supercritical conditions. Generalized dependencies are calculated within the framework of the law of the corresponding states for the coefficients of compressibility, shear viscosity, and thermal conductivity of pure substances, and for the coefficients of compressibility, self- and mutual diffusion, and shear viscosity of binary mixtures. The effect of density and temperature on the rates of elementary stages under supercritical conditions has been demonstrated for a reaction's effective energies of activation, diffusion and share viscosity coefficients, and equilibrium constants of adsorption. Differences between models with effective parameters and the prospects for developing them by allowing for differences in size and contributions from the vibrational motions of components are described.

Keywords: non-ideal reaction systems; supercritical conditions; lattice gas model; theory of the absolute rate of a reaction

Citation: Tovbin, Y.K. Molecular Modeling of Supercritical Processes and the Lattice—Gas Model. *Processes* **2023**, *11*, 2541. https://doi.org/10.3390/pr11092541

Academic Editors: Maria Angela A. Meireles, Ádina L. Santana and Grazielle Nathia Neves

Received: 20 May 2023
Revised: 27 July 2023
Accepted: 7 August 2023
Published: 24 August 2023

1. Introduction

Transition to supercritical (SC) conditions of a gas mixture are connected with the increase in temperature and pressure in a system [1–4]. Processes in supercritical fluids (SCFs) are allocated in a separate area of research and practical applications owing to their physicochemical properties. Many physical properties of SCFs (density, viscosity, and speed of diffusion) are intermediate between the properties of a liquid and gas. SCFs combine properties of gases at high pressures (low viscosity, high diffusion factor) and liquids (high dissolving ability); they possess fast mass transfer carried out thanks to their low viscosity and high factor of diffusion. Further, SCFs possess very small interphase tension, low viscosity and the high factor of diffusion allowing SCF to get into porous environments easier in comparison with liquids, high sensitivity of dissolving ability of SCF to pressure or temperature change, and simplicity of division SCF and the substances dissolved in them at pressure dump.

All these properties have allowed the development of high-pressure technologies involving sub- and supercritical liquids, and have allowed the possibility of receiving products with special features or projecting new processes that are viable and harmless to the environment. Now, SCF processes are an important part of industrial chemistry. Some achievements in this direction are reflected in reviews [5–19].

In supercritical (SC) conditions, a wide range of various technological processes is currently implemented. Among them are homogeneous chemical reactions in bulk phases and processes for creating new materials, including nanoparticles, heterogeneous catalytic reactions, physicochemical processes in porous media, chromatography, extraction, and many others. The presence of fluids in the SC system can change the nature of the implementation of chemical reactions in comparison with their flow under normal conditions (i.e., at a temperature below the critical one and a pressure of the order of one atmosphere). This change is due to the increased density of the SCF compared to the density of the gas phase and the rapid dissipation of heat in it. Density changes in the components in reaction systems lead to shifts in all chemical equilibria and allow the occurrence of processes that are unlikely under normal conditions. This is the basis for the search for new ways to implement physical and chemical processes that allow the development of new, more environmentally friendly industries. In the existing set of SCF processes, their various exploiting is possible as an environment (for inert fluids), solvents (for associated fluids) or reagents, and for all these manifestations of SCFs, it is desirable to have a common approach for modeling their implementation, since it affects their fundamental physico-chemical features. The overall flow of the processes can be controlled by changing the pressure in the system. The increase always leads to an increase in the total density of matter in the system. In SCF processes, the temperature also rises. These two factors can influence volumetric and surface processes in different ways. Thus, for adsorption inside porous materials and at open surfaces, the increase in temperature always reduces the adsorption, and adsorption increases with raising pressure.

For the practical realization of technological processes, a search for optimum modes that are carried out by means of modeling methods is required. Modeling questions in different technologies are connected with the necessity of the account of scale transition from process studying in vitro to technological reactors.

In the general case, the same methods that have been developed earlier for other, different technologies are applied to modeling SCF processes. The general equations of nonequilibrium thermodynamics on the macro-level [20–22] and the molecular theory on the molecular level [23,24] are usually used for the description of technological processes. Traditionally these methods share the description of the kinetic and equilibrium processes concerning the molecular and above-molecular levels.

The most widespread equations on the above-molecular level are hydrodynamic equations. The equations of a hydrodynamic mode of flows in gases and liquids describing describe the dynamics of the system in terms of concepts the theory of continuous media concerning such equations, and in the more general case, the equations include the following types of the kinetic equations: (1) hydrodynamic Navier–Stokes equations or analogs for complex molecular systems [25–30]; (2) chemical kinetics equations within the limits of ideal models (the law of mass action) which operate only with one or partial functions of distributions (or a concentration of reagents) [31–35]; (3) classical thermodynamics equations for the simplification of calculations of nucleation and coagulation processes [36–40]; and (4) thermodynamics equations of irreversible processes which include points (1)–(3) [20,21], containing all types of molecular mechanisms of transport processes in addition to convective flow.

Chemical kinetics equations within the limits of ideal models, i.e., using the law of mass action, comprise information only about a concentration of reagents. These equations also are included in the equations of the hydrodynamic level at any size of volume of the system. The equations of classical thermodynamics are often used for calculations of nucleation processes to simplify or eliminate the calculation of stages of condensation and desorption of single molecules to a formed phase. The new phase (drop) is described through functions of exceeded free energy (through an interface tension).

At the hydrodynamic modeling level, there is a very large number of specific algorithms for specific processes [41,42]. Among them, we can mention the finite element method [43,44], which is focused on calculations of systems with complex geometric configurations and irregular physical structures. In the finite element method, the problem of finding the function is replaced by the problem of finding a finite number of its approximate values at specific points. This may explain the view of the finite element method as a grid method designed for solving microlevel problems, for which the model of the object is defined by a system of differential equations in partial derivatives with given boundary conditions.

The existing statistical physics methods of modeling the molecular level are the following: the molecular dynamics (MD) method [45–49], the kinetic Monte Carlo method [50–55], Brownian (or Langevin) dynamics [56,57], the Boltzmann equation (for gases in a continuum) [58,59], the Boltzmann discrete equation [60–63], the lattice-gas model (LGM) [64–66], microscopic hydrodynamics [67,68], and the lattice automata method [69–73]. More detail on different methods of modeling nonequilibrium processes, ranges of intervals of time, and areas of their application is given in Appendix 8 of the monograph [68]. The majority of technological SCF processes are described by means of gas and hydrodynamics equations, and local chemical kinetics equations. For examples of these, see various works on kinetics for SCF extraction [74–90].

For many technological processes in SCFs, the modeling problem is reduced to calculations of molecular distributions of components in different phases in almost equilibrium conditions. In such situations, one can use only thermodynamic models for the equations of state that allow the calculation of factors of interphase distribution of components in processes of solubility or extraction [91–96]. So, the Peng–Robinson equation of state [97] and the Mukhopadhyaya and Rao mixing law [98] were actively used in the modeling of solubility in the following research: the phase diagram of the system "CO_2—diethylene glycol monoethyl ether (ethylcarbitol)" [99], the phase equilibrium of the propylene glycol–propane/butane system and the solubility of propylene glycol in supercritical propane–butane mixture [100], the solubility of ammonium palmitate in supercritical carbon dioxide [101], the solubility of bio-diesel fuel components (methyl esters of stearic and palmitic acid) in supercritical carbon dioxide [102], and in many other systems.

On the other hand, different equations of states can be used in one work. In [103] 14 equations of states were used; similarly, 13 equations of states were considered in [104].

Modeling questions in all technologies are connected with the transition scale from process studying in vitro to technological reactors. Often, for technological problems, one uses rough or simplified models for a targeted outcome, but a broader range of information can be attained with more correct models based on kinetic models of processes, thermodynamic models, and equations of states of non-ideal systems.

All possible approaches have been used for describing SCF processes. Equilibrium characteristics are described by the equations on concentration (or density) reagents whereas nonequilibrium processes are described by the kinetic equations on changes of these concentration in time. Depending on the intensity of molecular mixing, the considered equations can relate to local volumes or to the system as a whole. In such modeling technological processes, situations arise when two levels of models are used simultaneously: models of non-ideal systems for describing the equilibrium properties of SCFs and kinetic models based on the law of mass action, i.e., models for ideal systems, which does not agree with the second law of thermodynamics [22–24]. (1) According to the second law of thermodynamics, there should be unified mathematical models that describe both the relaxation stage of the displacement of the nonequilibrium state of the system to the direction of its equilibrium and the limiting equilibrium state itself during long periods. However, it is obvious that kinetic models based on the law of mass action cannot, in principle, transform during long periods into equations of state for non-ideal systems. To describe such a transition, it is necessary to have kinetic models and their mathematical equations for non-ideal reaction systems.

At the base of all modeling, there are local equilibrium models. The state of the theory and methods of calculation of SCFs are reflected in the collection of works [105]. These reviews deal with the modern theory of critical phenomena, methods to correlate near critical experimental results, and approaches to understanding the behavior of near critical fluids from microscopic theory. However, the question about the connection between kinetic models and equilibrium equations of state in non-ideal systems has not been discussed.

The alternative is approaches in molecular-kinetic theory that are constructed based on the so-called lattice-gas model (LGM) [68,106]. The advantage of the LGM is that this method is the only one of the above methods that provides self-consistency in the description of the rates of stages of non-ideal reaction systems with their equilibrium state in accordance with the second law of thermodynamics. Also, this approach gives a uniform method of the description of three-aggregate systems.

The purpose of this review is to present the possibility of modeling the physicochemical processes occurring at the molecular level in the LGM for the SC phases.

Expressions for the rates of elementary stages in ideal and non-ideal systems are presented in Section 2. These expressions are discussed in Section 3 from the point of view of the fulfillment of the second law of thermodynamics. Also discussed is the area of thermodynamic parameters near the critical point, in which it is inappropriate to carry out technological processes. Sections 4–6 present the possibilities of modeling kinetic SCF processes using the theory of non-ideal reaction systems based on the LGM: Section 4 outlines a model of the effective pair potential (Section 4); Section 5 describes the influence of SCFs on equilibrium and kinetic characteristics (5); and Section 6 describes the LGM and dissipative coefficients (6). Extensions of the LGM are indicated in Section 7. Section 8 gives the conclusions.

2. Molecular Level

2.1. Ideal Systems

The law of mass action defines the following expression for the rate of the bimolecular stage [21,22,31–35]:

$$U_{ij} = k_{ij}n_i n_j, \, k_{ij} = k_{ij}{}^0 exp(-\beta E_{ij}), \tag{1}$$

where k_{ij} is the rate constant of elementary reaction $i + j \rightarrow products$; n_i is the concentration of molecules, measured as the number of i-type molecules in a unit volume; $k_{ij}{}^0$ is the rate constant pre-exponential factor; E_{ij} is the reaction's energy of activation between i and j reagents; $\beta = 1/k_B T$, k_B is the Boltzmann constant, and T is the temperature. In Expression (1), for a heterogeneous process, the area which does not change during the process is expressed in terms of the concentration of adsorbed particles θ_i [35], which determines the fraction of the surface occupied by component i: $U_{ij} = k_{ij}\theta_i\theta_j$ (the product $n_i n_j$ is replaced by the product $\theta_i\theta_j$). To calculate the rate constants, the theory of absolute reaction rates is used [31]. This theory expresses the rate constants of the elementary steps with the partition functions of the reactants and the activated complex (AC) of the stage. Equation (1) assumes that the stage of chemical transformation is slow, the particles move completely independently, and an equilibrium distribution of components is realized in space. This means two important things: there is no intermolecular interaction in the system and there are no diffusion inhibitions at the micro- and macro-levels.

It also follows from Expression (1) that an increase in temperature exponentially increases the rate of the stage. The effect of SCF molecules as reagents is manifested only through the factor n_i. However, the rate of the stage slows down with the increased pressure, even if the SCF component is inert. This pressure effect is due to the filling of the SCF of the volume of the system and a decrease in the probability of approaching the reagents, as well as the fact that, at high molecular densities, the formation of different associates around each reagent is inevitable.

Equation (1) is written for the elementary stage of the chemical transformation. When modeling real systems, it should be taken into account that a chemical process usually consists of several elementary stages, depending on the mechanism of a chemical reaction and on the transport stages of molecular transfer. The rate of each stage is affected by the temperature and concentrations of the components, which complicates the description of the gross SCF process. Almost any chemical process consists of several elementary stages determined by the mechanism of chemical transformation. To model SCF processes in continuous or batch devices, it is important to know the dissipative transfer coefficients.

As the concentration of SCF increases, the influence of intermolecular interactions increases, which affects the nature of all kinetic processes (chemical reactions, transfer coefficients, adsorption, and catalytic processes). Intermolecular interactions in dense gases lead to deviations from the law of mass action and it is necessary to turn to the theory of non-ideal reaction systems. The influence of pressure increase on the rate of the catalytic process of ammonia synthesis was first demonstrated in [107] (see also [35]). An increase in pressure to 300 atmospheres led to a change in all effective rate constants of elementary stages.

The main problem in the theory of non-ideal reaction systems is to take into account the influence of the environment on the rate of the elementary stage [32,33,68,106,108–110]. Reagent molecules in the dense phase are constantly surrounded by their neighbors and their intermolecular interactions change the potential surface of the elementary stage. In general, this changes the activation energies of the reaction for each local environment of the reactants.

2.2. Non-Ideal Systems and the Lattice-Gas Model

The theory of non-ideal reaction systems is also based on the theory of absolute reaction rates, which exploits the concept of an activated complex (AC) of an elementary stage. In this case, the AC is a particle with its own interparticle interaction potentials. The spatial distribution of all components of the reaction system (reagents, AC, and SCF) is described within the framework of the lattice-gas model (LGM) on discrete-continuum scales [66,68,106] in the quasi-chemical approximation (QCA). The QCA reflects the effects of direct correlations between all interacting particles. It should be recalled that the LGM allows one to describe the entire range of dimensionless particle densities from zero to one (in mole fractions), which allows it to be used to solve all problems of CSF processes. The choice of QCA as the basic approximation for calculating the equilibrium and kinetic characteristics is associated with full agreement with the second law of thermodynamics—this approximation provides a self-consistent description of the rates of elementary stages and the equilibrium state of the entire system. Below, for simplicity, we restrict ourselves to the equations of the bimolecular stage and take into account the interaction of nearest neighbors.

The entire volume of the system under consideration is divided into separate cells in the LGM. The cell size is chosen on the order of the average particle size so that it can be considered that this cell is occupied or free (vacant). The cell occupancy state is fixed by the index i, where $1 \leq i \leq s - 1$ (s is the number of system components) and the symbol $i = s$ refers to free cells. Above, the notation for the fraction of cells occupied by particles of type i was introduced as θ_i (or its number density). Then, the normalization condition will be written as $\sum_{j=1}^{s} \theta_i = 1$ and the value $\theta = \sum_{j=1}^{s-1} \theta_i$ is the complete occupancy of a lattice system by all i components of the system, $1 \leq i \leq s - 1$. The quantity is the fraction of free sites (recall that vacancies are not thermodynamic characteristics). The ratio $x_i = \theta_i / \theta$ is the mole fraction of component i.

The lattice structure is characterized by the number of nearest neighbors z. Between particles in neighboring cells, lateral interactions are taken into account; the energy parameter of this interaction between a pair of particles ij is denoted by ε_{ij}. The analogous parameter of any particle with a neighboring vacancy is equal to zero. The probability of finding a pair of particles ij at neighboring nodes is characterized by the value θ_{ij}—this is the pair distribution function of particles. Such functions are needed to describe the probability that reactants i and j will meet in dense phases so that a chemical reaction can occur.

The rate of a bimolecular reaction of the Langmuir–Hinshelwood type $U_{fg}{}^{AB}$ is written in the theory of non-ideal reaction systems [66,106] as the following expression (subscripts f and g indicate the numbers of sites where reagents A and B are located):

$$U_{fg}^{AB} = k_{fg}^{AB} \exp(-\beta \varepsilon_{fg}) \theta_{fg}^{AB} \Lambda_{fg}^{AB},$$

(2)

where Λ_{fg}^{AB} is the function of imperfections, defined as

$$\Lambda_{fg}^{AB} = \prod_{h \in (z(f)-1)} S_{fh}^{A} \prod_{h \in (z(g)-1)} S_{gh}^{B}, \quad S_{fh}^{A} = \sum_{j=1}^{s} t_{fh}^{Aj} \exp(\beta(\varepsilon_{Aj}^{*} - \varepsilon_{Aj}))$$

(3)

The subscript h refers to the nearest neighbors of site f or g. The neighboring sites g or f themselves are not included in the set of values of the index h; the function S_{gh}^{A} is defined similarly to the function S_{fh}^{A} (indices A and f change to indices B and g). Here the symbol ε^{*}_{ij} represents the interaction parameter of the AC reaction for an i-type particle with a neighboring particle of type j.

The function $t_{fh}^{ij} = \theta_{fh}^{ij}/\theta_{f}^{i}$ is the conditional probability of finding particles j next to particles i. Here, the numbers of neighboring sites (subscripts) are introduced only to indicate differences in positions on the lattice of the reactants: $\theta_{fh}^{ij} = \theta_{ij}$ and $t_{fh}^{ij} = t_{ij}$. In non-ideal systems, $\theta_{ij} \neq \theta_i \theta_j$, which corresponds to the correlated distribution of components in space. The case of equality $\theta_{ij} = \theta_i \theta_j$ corresponds to the chaotic distribution of components, which is typical for ideal systems (see Formula (1)).

Intermolecular interactions change the probability of encounters of reagents (the factor θ_{AB} instead of the product $\theta_A \theta_B$) and the heights of activation barriers through the functions $\Lambda_{fg}{}^{AB}$. Functions S_i take into account the influence of neighbors on the magnitude of activation barriers through the difference in interaction parameters due to the influence of the neighboring particle j (via $\delta \varepsilon_{ij} = \varepsilon^{*}_{ij} - \varepsilon_{ij}$). The exponential factor with $\beta \varepsilon_{fg}^{AB}$ in Formula (2) is necessary for the transition at low system densities from Formula (4) to Expression (1) for the law of mass action, as for an ideal system [66].

Equations (2) and (3) contain the pair function θ_{ij} that characterizes the probability that two particles i and j can be on neighboring sites. The calculation of pair functions θ_{ij} in non-ideal reaction systems is always carried out in some particular approximation because the problem cannot be solved exactly [64–66]. In this case, the so-called quasi-chemical approximation (QCA) is used [23,64–66]. Historically, it was the first approximation in which the effects of direct correlations between interacting molecules were taken into account. There, each pair of neighboring molecules is considered independent of other molecules in the system. The function θ_{ij} depends on the interaction energy of molecules and concentrations of components.

The pairwise distribution functions are found from the solution of the system of algebraic equations in the QCA together with the normalization condition:

$$\theta_{ij}\theta_{ss} = \theta_{is}\theta_{sj} \exp(-\beta \varepsilon_{ij}), \quad \sum_{j=1}^{s} \theta_{ij} = \theta_i.$$

(4)

It follows from Equations (4) that i and j particles attract one another when $\varepsilon_{ij} > 0$; analogously, i and j particles repulse one another when $\varepsilon_{ij} < 0$.

When calculating the rates of reactions on a homogeneous surface, it is necessary to first find the surface concentrations of the reagents, which are determined from the QCA Equation (5) of adsorption in the presence of SCF molecules:

$$a_i P_i = \left(\frac{\theta_i}{\theta_s}\right) S_i{}^z, \quad S_i = 1 + \sum_{j=1}^{s-1} x_{ij} t_{ij},$$

(5)

where P_i represents the partial pressures in the gas phase ($1 \leq i \leq s - 1$); θ_i is the degree of filling the surface with particles i; $a_i = a_i{}^0\, exp(\beta Q_i)$, $a_i{}^0$ is the pre-exponential of Henry's constant; Q_i is the binding energy of particle i with the surface; and $x_{ij} = \exp(-\beta \varepsilon_{ij}) - 1$.

From the terms for the rates of individual stages, the right-hand sides of the kinetic equations of the simulated processes (specified in Appendix A) are formed.

The rates of two-site stages have the form of Equations (2) and (3), and the rates of single-site stages in QCA have the form of Equation (A6) in Appendix A.

3. Thermodynamics and Kinetics

The developed theory of non-ideal reactionary systems [66,106] answers a number of requirements regarding its connection with chemical thermodynamics:

1. The second law of thermodynamics and connection between models of equilibrium and kinetics;
2. A self-consistence of equilibrium and kinetics in ideal systems;
3. A self-consistence of equilibrium and kinetics in non-ideal systems;
4. The equations of a state for non-ideal systems and their connection with kinetic models;
5. Why it is impossible to use factors of activity for the AC in kinetic models;
6. Thermodynamic parameters of the critical area and the requirement of technologies.

3.1. The Second Law of Thermodynamics and Connection between Models of Equilibrium and Kinetics

Clausius' formulation of the second law of thermodynamics contained a way to consider the transfer of nonmechanical energy in the first law of thermodynamics. This formulation is rather complex to understand because it simultaneously contains both the *process* of development toward equilibrium in a closed system and the limiting equilibrium *state* [25,111,112]. The mathematical formulation of the combined equation of thermodynamics is $dU \leq TdS - PdV$, where U is the internal energy, P is pressure, and V is the volume of the system; the sign of the equality corresponds to the equilibrium. A quantitative measure of the process considered in the second law of thermodynamics is entropy S which characterizes the thermal motion of matter. The entropy in the time-limited equilibrium state of a system does not depend on the transition to the equilibrium state and it is maximal with respect to all other states.

Clausius' statements about the limiting states of equilibrium systems with extreme properties of entropy were adopted as a basis in thermodynamics by Gibbs. This allowed him to deal with various states without specifying transitions between them. Gibbs's mathematical formulation of the combined equation of thermodynamics ($dU = TdS - PdV$) divided Clausius' second law of thermodynamics into two parts: equilibrium and nonequilibrium.

Dividing the second law of thermodynamics into two parts [112–114], Gibbs omitted the requirements, contained in bases of thermodynamics, about a self-consistence description of reaction rates and an equilibrium condition [115]. The existing law of mass action [21,34] was highlighted, although it is not valid in non-ideal systems that comprise the majority of real processes. The requirement of self-consistence did not become obligatory for the kinetic theory of non-ideal reactionary systems (including all of Prigogine's works [22]). This requirement is not present either for transport processes where it is absolutely necessary for the calculation of dissipative factors [20,23].

The requirement for the self-consistency of the description of the rates of stages in a non-ideal reaction system and its equilibrium state was introduced only in Ref. [115], as a necessary refinement of the second law of thermodynamics for modeling all kinetic processes.

3.2. Self-Consistence of Equilibrium and Kinetics in Ideal Systems

Irreversible processes take place until either a stationary state or equilibrium is established (excluding the possibility of realization of periodic processes). If several irreversible processes are superimposed and the final state attained corresponds to equilibrium, then in certain cases it is possible to obtain general conditions for the coefficients that describe irreversible processes, without the application of the thermodynamics of irreversible processes.

In the general case of more complex elementary stages of chemical reactions in the gas phase, the law of mass action, which was empirically established by Guldberg and Waage (1867), is used to describe reaction rates. For reversible reactions of a general form, we can write $\sum_i \nu_i [A_i] \underset{k_2}{\overset{k_1}{\rightleftharpoons}} \sum_j \nu_j [A_j]$, where the symbols A_i and A_j in brackets denote different reacting particles, and the values of ν_i and ν_j are equal to the negative and positive values of the stoichiometric coefficient (the sign of the coefficient is determined by their location: on the left or right side of the equation). The constants k_1 and k_2 are the reaction rate constants in the forward and backward directions. Numerically, they are equal to the reaction rate at single values of the concentration of each of their reagents in the forward direction.

The rate of the considered reaction within the framework of the law of mass action [21,34] will be written as

$$U = k_1 \prod_i n_i{}^{\nu_i} - k_2 \prod_j n_j{}^{\nu_j} \tag{6}$$

In the equilibrium state, the rate is zero, $w = 0$, and it follows from (6) that the rate constants in the forward and backward directions are related to each other in the form

$$k_1/k_2 = \prod_j n_j{}^{\nu_j} / \prod_i n_i{}^{\nu_i} = K \tag{7}$$

where $K = k_1/k_2$ is the equilibrium constant of the stage.

Let us consider the simplest process of adsorption–desorption of gas phase molecules without dissociation. The rate of desorption of particles A from the occupied areas of the surface will be written as $U_A = K_A \theta_A$, where K_A is the desorption rate constant and θ_A is the fraction of the occupied surface. The adsorption rate on free surface areas (V is the symbol of vacancies) will be written as $U_V = K_V P \theta_V$, where K_V is the adsorption rate constant, P is the pressure in the gas phase, and θ_V is the fraction of the free surface ($\theta_V = 1 - \theta_A$).

At equilibrium $U_A = U_V$, it follows that $K_A \theta_A = K_V P \theta_V$, or $\theta_A = K_V P (1 - \theta_A)/KA$ or $\theta_A = K_V P/(K_A + K_V P)$, and $\theta_A = a_1 P/(1 + a_1 P)$, $a_1 = K_V/K_A$ is the adsorption coefficient without dissociation. If the adsorption process proceeds with the dissociation of gas phase molecules, then the rates of desorption and adsorption will be rewritten as $U_{AA} = K_{AA} \theta_{AA} = K_{AA}(\theta_A)^2$ and $U_{VV} = K_{VV} P \theta_{VV} = K_{VV} P (1 - \theta_A)^2$. Hence, $\theta_A = a_2 P^{1/2}/(1 + a_2 P^{1/2})$ is the Langmuir isotherm of dissociating molecules and $a_2 = (K_{VV}/K_{AA})^{1/2}$ is the adsorption coefficient in the presence of dissociation.

In both cases, one can write down one fractionally rational function $\theta_A = y/(1 + y)$, where $y = y_1 = a_1 P$ for adsorption without dissociation and $y = y_2 = a_2 P^{1/2}$ in the presence of dissociation. It follows that in the coordinates $\theta_A = \theta_A(y)$ both dependencies behave equivalently. The coincidence of both dependences $\theta_A = \theta_A(y)$ means that the equilibrium adsorption of dissociating molecules does not depend on what occurs first: dissociation of molecules in the gas phase or their adsorption. Both of these equilibrium dependences are obtained from the condition that the second law of thermodynamics according to Clausius is satisfied—from equalizing the velocities of the oppositely directed velocities of the stages. As a result, the four different rates of elementary stages give one equilibrium concentration dependence.

This result is obtained within the framework of the law of mass action, which is fulfilled for an ideal gas mixture and dilute solutions. The following expression for the chemical potential can be written $\mu_i = \mu_i{}^0 + kT \ln(n_i)$, where $\mu_i{}^0$ and n_i are the chemical potential of the standard state and the molar volume concentration of component i. However, in kinetic models for non-ideal systems it is impossible to use activities or activity coefficients [115] as natural thermodynamic replacements of concentrations (see below).

3.3. Self-Consistence of Equilibrium and Kinetics in Non-Ideal Systems

The theory on the basis of the LGM provides a self-consistent description of an equilibrium condition of a reactionary mixture and rates of elementary stages. This means that by equalizing rates of oppositely directed processes of any stage, the equation of equilibrium distribution of components [66,106] can be derived. These self-consistent processes are only those where local correlations, such as QCA, are taken into account.

The above statement can be formally explained in the following way. The expression for the equilibrium constant K (7) of the two-molecular stage A + B $\leftrightarrow$ C + D can be considered as a product of two independent monomolecular processes A $\leftrightarrow$ C and B $\leftrightarrow$ D. That is, global equilibrium does not depend on the way it is realized in the system—it goes through different equilibrium reactions or coincides with a condition of equilibrium of the environment as a whole. In this case, the theory of non-ideal reactionary systems guarantees that the second law of thermodynamics is realized as it happens in the theory of ideal reactionary systems.

Exactly the same situation with obtaining equilibrium dependences $\theta_A = \theta_A(y)$ in non-ideal reaction systems will be considered for adsorption and desorption processes with ($m = 2$) and without ($m = 1$) dissociation. Let us consider the equalities of the rates in both directions, $U_A = U_V$ and $U_{AA} = U_{VV}$, expressed by Formulas (2) and (3) in the QCA.

To illustrate the logic of the second law of thermodynamics for a non-ideal system, let us again (as for the ideal system) consider the process of adsorption of dissociating A_2 molecules in the form of two routes: the dissociation of A_2 molecules into atoms A in the bulk phase followed by their adsorption or the adsorption of A_2 molecules followed by the dissociation process. Both routes are described by the rates of stages in the forward (adsorption) and reverse (desorption) directions. At equilibrium, adsorption isotherms for A_2 molecules and A atoms should be obtained. Moreover, the degree of surface filling θ_A must be the same regardless of the route, i.e., the final equilibrium state does not depend on the way the equilibrium is reached. This fact reflects the concept of Clausius twice: the equilibrium state itself follows from the equality of the rates of the stages and for different routes, it receives a single mathematical dependence for the degree of surface filling on pressure.

As a visual illustration, we present Figure 1, which shows the concentration factors of the rates for the four indicated stages and the equilibrium values θ_A corresponding to the equalities of the rates of the stages for chemisorption ($\varepsilon_{AA} < 0$), without dissociation (Figure 1a) and in the presence of dissociation (Figure 1b). Here, $V_i = U_i/K_i$, $V_{ii} = U_{ii}/K_{ii}$. Each field consists of three curves for the logarithm of the concentration factors for rates of non-dissociative (Figure 1a) and dissociative desorption (Figure 1b), adsorption (curve 1), desorption (curve 2), and the logarithm of the isotherm of adsorption (curve 3, right y-axis). As in ideal systems (Section 3.2), both curves 3 are identical for non-ideal systems.

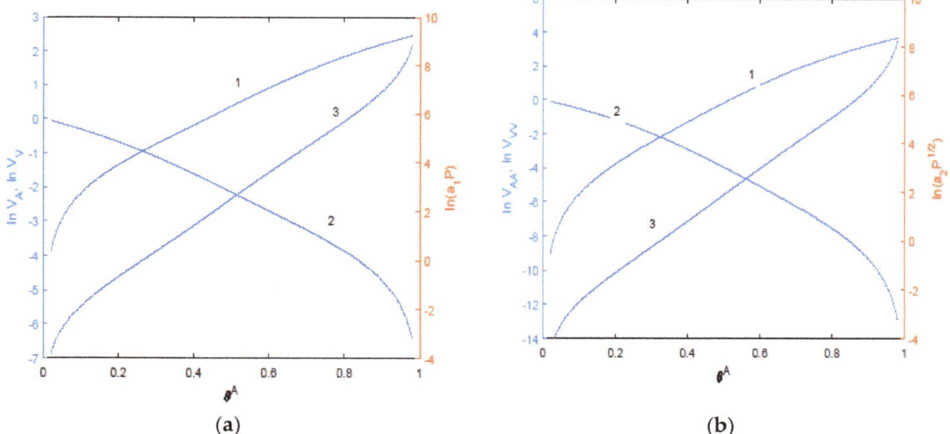

Figure 1. Concentration factors (V_i and V_{ii}) in rates of non-dissociative (**a**) and dissociative (**b**) adsorption (curves 1) and desorption (curves 2) for a non-ideal chemisorption ($\varepsilon_{AA} < 0$) system ($V_i = U_i/K_i$, $V_{ii} = U_{ii}/K_{ii}$). Curves 3 (right y-axis) correspond to the equilibrium isotherm.

A rigorous mathematical proof of the self-consistency of the stage rates and the equilibrium distribution of the components of a non-ideal system in QCA is given in [66,106].

Other approaches, where the distribution of particles is not correlated, do not provide a self-consistent description of the kinetics and equilibrium stages (see more detail in [66,106]). In particular, all one-particle approaches without the effects of correlation are not self-consistent expressions for rates of two-molecular stages processes. These are average field approximation, chaotic approximation, and approximation of functional density.

3.4. The Equations of a State for Non-Ideal Systems and Their Connection with Kinetic Models

In the work of Fisher M.E. [116], it was shown that the van-der-Waals equation for non-ideal gases corresponds in LGM to the approach of an average field. All other constructions for phenomenological equations of a state are derived in the same style [103,104,117,118]; this leads to the obvious contradiction when modeling SCF processes: various phenomenological equations of a state reflect the non-ideality of a reactionary system, whereas the equations for kinetic models do not reflect the non-ideality of systems.

All kinetic models for SCF processes are based on an ideal model with the law of mass action. This breaks the concept of the self-consistent description for the majority of SCF processes. In order to avoid this controversy, it is necessary to use the equations for non-ideal systems in the LGM. Then, all features of the movement of molecules in SCF conditions can be reflected through internal statistical sums of molecules, taking into account the effects of correlations and preservation of the self-consistency description of kinetics and equilibrium.

3.5. Why It Is Impossible to Use Factors of Activity for AC in Kinetic Models

For non-ideal reaction systems, all expressions for the elementary stage rates are interpreted within TARR [31]. So, instead of using Equation (1), the rate of a bimolecular stage is written as

$$U_{ij} = k_{ij}^* n_i n_j, \tag{8}$$

where k_{ij}^* are the rate constants of the elementary stage, written as

$$k_{ij}^* = k_{ij}{}^0 \frac{\alpha_i \alpha_j}{\alpha_{ij}^*} exp(-\beta E_{ij}) = k_{ij}\alpha_i\alpha_j/\alpha_{ij}^*, \tag{9}$$

where k_{ij}^* is a pre-exponential factor of rate constants; E_{ij} is the energy of activation of the reaction i + j → products; k_{ij} is the rate constant in the ideal reaction system (1); α_i is the activity coefficient of i-type reagents; and α_{ij}^* denotes the activity coefficients of ACs. Calculating the activity coefficients for reagents and for ACs requires the use of the well-known theory of non-ideal solutions [119–122].

However, for kinetic models, the above method leads to errors. This is because of the calculation of the activities in a non-ideal mixture with the required averaging on all configurations of all components. The analysis of the applicability of TARR to the condensed phases in the form of Equation (8) has shown [68,115,123] that the basic condition of TARR on the equilibrium presence between the AC and surrounding molecules is violated during the elementary process.

This means that the elementary stage is realized at fixed positions of all neighbors since the relaxation time of each neighbor's environment is much longer than the time of the AC formation. Therefore, an introduction of the concept for AC activities deforms the basic TARR statement. As a result, the use of Equations (8) and (9) instead of Equations (2) and (3) can lead to appreciable differences in the values of reaction rates, especially at intermediate densities.

Figure 2 illustrates the qualitative difference in concentration dependences of reaction rates for different environment relaxations in isothermal conditions (the curves describe the desorption system of type CO-Pt).

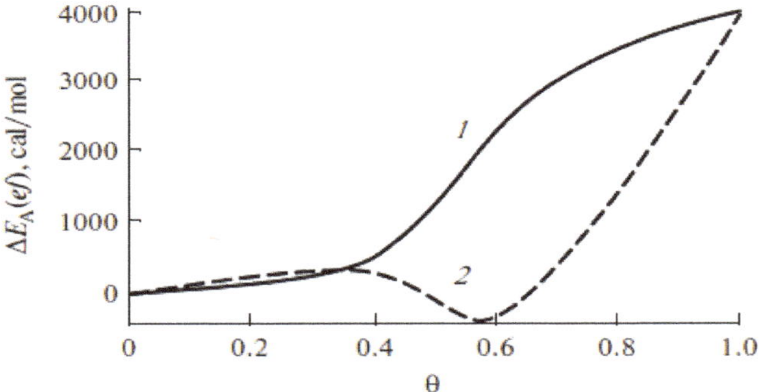

Figure 2. Concentration dependence of $E_A(ef)$ for monomolecular desorption at 300 K calculated in the case of (1) fast and (2) slow elementary stage [123].

The effective activation energy of desorption ($exp(-\beta E_A(ef)) = U_A/(\theta_A K_A{}^0)$), and accordingly the reaction rate, increases monotonically with increasing surface coverage when calculated by Equation (A6) for the unimolecular stage with the function S_{fh}^A (3) (this stage is fast without taking into account environment relaxation). When calculated by the Formula (9) for the unimolecular stage [123], the effective activation energy for desorption under conditions of equilibrium relaxation of the environment (this stage is slow) varies nonmonotonically. The nonmonotonic behavior of the desorption rate contradicts the physics of the process, especially in view of the fact that, at short distances, chemisorbed species repel each other. In general, in non-ideal systems, the nonmonotonic behavior of the reaction rate is possible only if the potential is attractive [66,106,124]. Furthermore, the negative effective activation energy of the sole desorption reaction does not have a physical sense. The behavior of curve 2 is due to an incorrect procedure for averaging over all configurations of all neighbors with respect to the AC. This drastically changes Equation (3), in which the functions S_{fh}^A have a different relationship with the concentrations and functions t_{ij}. As a result of this averaging procedure, the nature of the effective interaction between particles changes: instead of repulsion, their effective attraction appears (which distorts the physical nature of the system). When the density of the system changes, the particles behave as in the case of a first-order phase transition with a nonmonotonic concentration dependence of the rate of the desorption stage.

The above comparison shows how much the effect of the environment relaxation influences the rate of a unimolecular process. Therefore, the issue of the correctness of using one or another method for averaging the contribution from the neighbors plays a fundamental role in the dynamics of the elementary stages. So, the incorrect use in TARR representations about AC activity (as in Equation (9)) excludes formal extension of kinetic models on non-ideal reactionary systems via a conception of "activity" instead of a concentration.

3.6. Thermodynamic Parameters of the Critical Area and the Requirement of Technologies

It should be remembered that for successful realization of chemical reactions in SC conditions, it is necessary to displace on temperature upwards from the critical point of phase transitions of the first order both in solid and in liquid phases. Otherwise, near the critical point, processes of alignment of density are slowed down (or diffusion transport stages are at a loss), and this complicates a current of chemical multiphase processes [125–128]. The reason for such a delay is the general thermodynamic relations connected with the equality to a zero in the critical point derivative of chemical potential μ_i on concentration components i in the vicinity of the coordinates of the critical point (P_{cr}, T_{cr}): $(\partial \mu_i / \partial n_i)_{P_{cr}, T_{cr}} = 0$ and $(\partial^2 \mu_i / \partial n_i^2)_{P_{cr}, T_{cr}} = 0$. This peculiarity to the same extent concerns both the average field and the QCA approach in the LGM. It must be valid in any calculation method.

Attention has been paid to the same circumstance in [129] in the analysis of conditions for performing an experiment by definition of the parameters of the "coil–globule" conformational transition for the polymeric chains dissolved in SC–CO$_2$. On the basis of the gas-dynamic model of the impulse jet expansion of a van der Waals gas, a strategy experiment on the determination of the parameters of the "coil–globule" transition of the polymer chain in SC carbon dioxide was developed. To use the condition of constant isochoric heat capacity outside the near-critical point in modeling, it is necessary to determine the structure of the near-critical region.

For such purposes, the authors use techniques [130] based on representations about thermal stability [131] and increases in fluctuations with near-critical areas [132]. Taking into account the isentropic flow of the process and the behavior of CO$_2$ in the near-critical region, the conditions of expansion corresponding to the model was determined. An experimental design (geometry and dimensions of the basic elements of the installation, and the duration of the impulse valve) was developed. Possible variants of the experiment and its data processing were discussed.

The area of thermodynamic parameters that is necessary to exclude from the area of search for carrying out the experiment is specified in Figure 3 (solid lines). This area is limited by curves C-SC—the line of local minima of stability, C-MSC—the line of maxima of fluctuations, and line SC-MSC—the line of a supercritical isotherm [129], where C is the critical point, SC is the super-critical point, and MSC is the point of the maximum of fluctuations on the supercritical diagram.

Taking into account the developments of the authors of [129], Figure 3 shows the general scheme reflecting the combined effects (dashed lines MSC-A and SC-B) of slowing down mass transfer processes near the critical region (lower dashed line AB) and effects with maximum fluctuations (solid lines from [129]) of thermodynamic parameters.

Thus, from the point of view of the implementation of optimal modes of technological processes, the proximity of thermodynamic parameters to the parameters of the critical point is not appropriate. Search for them is required at such a distance from the critical point in order to avoid both the slowdown of the process near the critical point (P_{cr}, T_{cr}) and the presence of large fluctuations of parameters on the supercritical isotherm, to ensure the stability of obtaining the target product.

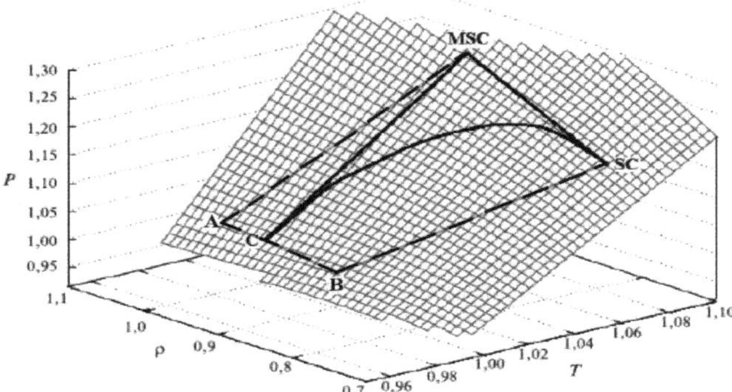

Figure 3. A supercritical area on the phase diagram of the van-der-Waals substances (A-MSC-SC-B-A) for which thermodynamics parameters are not desirable for technological processes.

4. Model of the Effective Pair Potential

4.1. Internal Motions of Particles

The initial interpretation of the LGM refers to a rigid lattice with fixed parameters of the lateral interaction. These conditions limit the possibility of using the LGM to describe experimental data even in equilibrium [119–122]. In order to expand the potential possibilities of the LGM in interpreting different systems, the LGM equations [133] take into account the motions of the center of mass of molecules inside the cells. This led to integral equations [134], similar to the theory of liquid [135–138], but not without violating the condition of a single filling of each cell.

As a result, the QCA continuum was formulated [133], which made it possible to use traditional ideas about lattice models, including the concepts of the excluded (or accessible) cell volume for the center of mass for molecules, the softness (or deformability) of the lattice structure, and the vibrational motions of molecules. Extensions of models of internal motions of molecules in LGM are associated with different degrees of freedom of molecules in the condensed phase. The free volume of the cell is associated with the translational motion of the center of mass of the molecule. Vibrational motions are always realized in bound ensembles of molecules. The softness of the lattice structure is formed due to the average displacement of molecules relative to each other. Traditionally, the concepts of translational and oscillatory motion refer to a rarefied gas and a solid body. In the SCF system for dense gases and liquids, these concepts have a conventional meaning, since molecules are constantly interacting with each other.

"Excluded" volume [139]. The movement of a selected particle in a dense fluid phase is hindered by neighboring molecules. They block part of the space and it becomes inaccessible for the movement of the selected particles. If the lattice structure is free, then the center of mass of the particle can be located at any point inside the cell. If the lattice structure is filled, then the center of mass of the particle can shift inside the cell only near its center.

The available volume for the movement of the center of mass in this case is equal to the value $V(\theta \sim 1) = \kappa^3$, where the value κ is the root-mean-square displacement of the molecule from the center; it can be estimated from the theory of harmonic motions in a solid for temperatures above the Debye temperature. That is, the value of κ is found from the parameters of the paired Lennard–Jones interaction potential [139]. In the general case, for any fluid densities θ, the available volume can be estimated as the ratio $V(\theta)/V(\theta = 0) = L(\theta)^3$, where $L(\theta) = t_{AV} + \kappa \, t_{AA}$, where the function t_{AA}, characterizing the conditional probability of finding two neighboring molecules A, is defined in Section 2.2. More accurate estimates of the available volume $V(\theta)/V(\theta = 0)$ are possible if a geometric model [139] is used that refines different positions of neighboring molecules [139].

Lattice softness. The softness of the lattice structure means that the average distance between fluid particles determines the average size of the lattice parameter λ. This value is found from the condition of minimum free energy of the system [140]—it is uniquely related to the parameters of potentials of interparticle interactions without introducing additional parameters. (In a number of situations, one can use the virial theorem to find λ [141,142].) For a gas, this λ quantity is related to the properties of a dimer, while, in a solid, all neighbors are taken into account. An increase in fluid density leads to a decrease in the lattice parameter. This fact correlates well with experimental and molecular dynamics data.

Molecular vibrations. Vibrations of molecules affect the average values of intermolecular interactions and all equilibrium distributions. Small deviations of molecules relative to the average size of the value λ lead [143–145] to a temperature dependence of the lateral interaction parameter of the type $\varepsilon = \varepsilon_0(1 - uT)$, where the function u reflects the vibrational motion of molecules. Previously, this form of dependence of the interaction parameter was considered as a convenient fitting function [119–121,146–148]. This approach [133] makes it possible to express the function u in terms of potential functions without introducing additional parameters.

The listed molecular properties of the effective pair parameter of interparticle interaction preserve the technique of calculating traditional lattice models and greatly simplify the consideration of continuum displacements of particles inside the cell. This reduces the calculation time in the LGM compared to using the technique of the integral equation by two to three orders of magnitude.

4.2. Vapor–Liquid Systems

One of the most popular potentials for describing vapor–liquid systems is the Lennard–Jones potential with parameters: σ (hard sphere diameter) and ε_0 (isolated dimer potential well depth) [23]. The values of these parameters are actively used in calculations for any density using the Monte Carlo and molecular dynamics methods. They were determined from experimental data for low-density gases (described up to the second virial coefficient). With increasing fluid density, it is necessary to add triple interactions [23,137,149,150]. Therefore, to use the effective pair potential ($\varepsilon_{\mathrm{ef}}$) of the Lennard–Jones type in the LGM for any densities [151–153], the following function was used:

$$\varepsilon_{AA}(r) = 4\varepsilon_{ef}\left(\left(\frac{\sigma}{r}\right)^{12} - \left(\frac{\sigma}{r}\right)^{6}\right), \varepsilon_{ef} = \varepsilon_0(1 - d_{tr}\Delta_{1r}t_{AA})(1 - uT), \tag{10}$$

Equation (10) reflects the dependence of the effective pair potential on temperature (u) and on triple interactions (d_{tr}) for nearest neighbors (Δ_{1r}—Kronecker symbol) in the form $d_{tr} = 0.2(z - 1)\varepsilon_3/\varepsilon_0$ [66,154], where ε_3 is the triple interaction parameter. The t_{AA} function reflects the presence of a third particle A nearby (it is defined in Section 2.2). For simplicity, it is assumed that the contributions from concentration and temperature are taken into account separately. For a quantitative description of experimental systems, it is possible to involve contributions from several coordination spheres. A similar structure to Equation (10) for effective pair potentials is also preserved for multicomponent mixtures [155].

As an example, the influence of the considered parameters in Figure 4a describes the concentration dependence of the compressibility factor $Z = p/n\kappa T$, where $n^* = n\sigma^3$ is the reduced number density, for argon in the volume ($\sigma = 0.34$ nm and $\varepsilon_0 = 119$K [23]). The comparison was made for the virial expansion (curves 2–4) [156] and for the LGM [151] at $T = 162$ K [157]. Figure 4a shows that at least five terms of the virial expansion are required to agree with the experimental data and with the LGM.

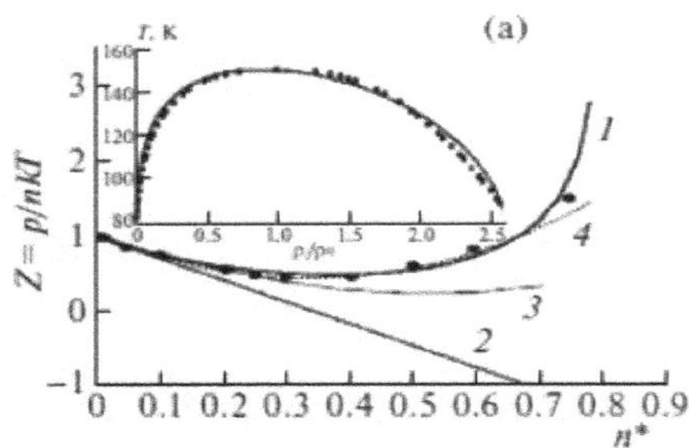
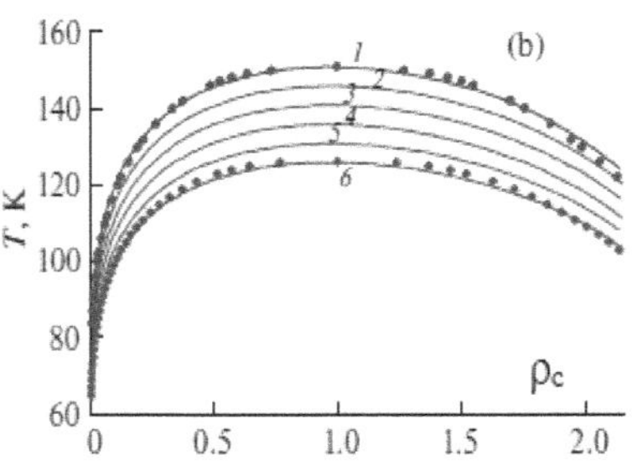

Figure 4. (**a**) Concentration dependences of the argon compressibility factor in [151]. Experimental values at 162 K [157] (dots); LGM calculations (1); calculations using the virial equations in [156]; (2) with regard to the second virial coefficient; (3) with regard to the second and third virial coefficients; (4) with regard to the second–fourth virial coefficients. The inset shows the phase diagram of argon. Symbols represent experimental values from [158]; the solid line represents calculations in [155] for $d_{tr} = 0.15$, $u = 0.00075$. (**b**) Phase diagrams of Ar, N_2, and an Ar–N_2 mixture at different compositions with nitrogen mole fractions $x_{N2} = $ (1) 0, (2) 0.2, (3) 0.4, (4) 0.6, (5) 0.8, (6) 1.0. Symbols represent experimental values from [159]; solid lines represent calculations in [160].

The stratification curve for argon with potential (10) was calculated taking into account the contribution of the calibration function (inset in Figure 4a), which is necessary to describe the critical region (see below). The density of argon is given in the reduced form ϱ/ϱ_c, where ϱ_c is the density at the critical point. Throughout the region, there is agreement between the experimental data [158] and the calculation by the LGM, including the value of the critical parameter β equal to 0.37.

The same approach was used to calculate the stratification curves for the Ar-N_2 binary mixture [159]. Curves for different compositions of the mole fraction of nitrogen are shown in Figure 4b.

For other gases, the molecular theory based on the LGM gives the same satisfactory agreement with experiments [161]. This is due to a certain extent to the fact that the considered gases do not have specific interactions and obey the law of the corresponding states. For these, a generalized compressibility coefficient was constructed [23,26,156] with the coordinates Z, "reduced" pressure, which is the same for different substances at the correspondingly introduced "reduced" temperatures (all quantities are normalized to their critical parameters). Such a generalized experimental dependence [23,26,156] was described in the LGM with an accuracy of about 3–4% (see Figure 5). These calculations reflect the main properties of Z for different temperatures up to pressures of 1000 atmospheres [152]. If the components of the mixtures do not violate the conditions for using the law of the corresponding states (as in Figure 4b), then the generalized dependences Z can be used for estimates and for mixtures [26]. Curves for different compositions of the mole fraction of nitrogen are shown in Figure 4b.

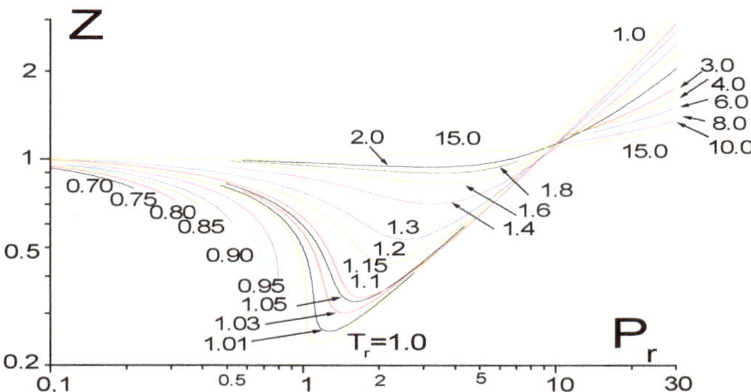

Figure 5. Generalized compressibility factors of dense gases that obey the law of corresponding states [161].

Figure 5 includes the critical region. In works [134,162–164] (based on the ideas of scaling theory [165–167]), it was proposed to introduce a calibration function as an approximate method for calculating the equilibrium distributions of components in the LGM near the critical point. This approach separates the contributions in the equation for local isotherms from the short-range potential and from large-scale fluctuations. The last contributions are taken into account by the calibration function. For more details, see [68,134,162–164].

Potential functions (10) were also used to calculate the equilibrium characteristics of mixtures at high pressures in the bulk phase [168]. Here, the potential of the LGM to describe the stratification of ammonia–nitrogen gases is shown in Figure 6. The field on the left shows the experimental measurements (abscissa axis—percentage). In the right field, the same points are marked with symbols (abscissa axis—mole fractions) for the same temperatures.

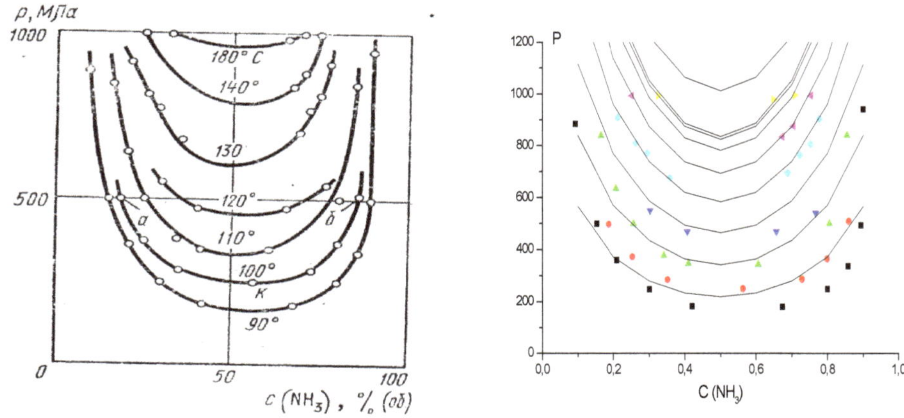

Figure 6. Dependences of a system's pressure on the concentration of ammonia in a binary ammonia–nitrogen mixture at different temperatures. Experimental values from [168] are on the left; calculated values are on the right. Temperature values on the right lines from bottom to top: 90, 110, 120, 130, 140, 170, and two lines around 180, and 210°.

The reason for the deviation of the experimental value from the theoretical one is that the model equations use the rules for combining parameters of the LD-type potential for molecules obeying the law of the corresponding states. This calculation agrees qualitatively with the experiment. The calculation uses the usual concept of the law of corresponding states for all components of the mixture, although for ammonia this assumption is conditional. Nevertheless, the use of LGM provides qualitatively acceptable results, although the degree of deviation of ammonia from the law of corresponding states increases with increasing pressure.

5. Influence of SCFs on Equilibrium and Kinetic Characteristics

While analyzing the specifics of the effect of SCFs on a variety of technological processes, the following three circumstances should be taken into account: (1) an increase in the role of lateral interactions with increasing pressure; (2) shifts in equilibrium concentrations with increasing temperature, which can lead in reaction systems to the implementation of stages that are unlikely for low temperatures (thermal dissociation of water sharply increases the value of the ion product) [2,169]; (3) under SC conditions, the states of the materials of the reaction system change (for example, the state of coke on catalysts or the properties of polymer matrices in membranes change). Nevertheless, the main question remains about changing the rates of the stages that form the gross processes, relating to points two and three. Its decision determines the accuracy of the description of the entire gross process.

The thermodynamic and structural characteristics of condensed systems are described using the methods of statistical thermodynamics based on the knowledge of the potential functions of interparticle interaction. SCF systems are created using CO_2 molecules, low molecular weight alkanes, alcohols, freons, water, etc. Such solvents are multicomponent mixtures of low molecular weight substances. To model SC systems, it is necessary to be able to calculate the bulk properties of SCFs and their contacts with the surfaces of solids (non-porous and porous). For polymer matrices, the dissolution of SCF molecules through open surfaces is possible. All these properties of SCF systems can be taken into account within the framework of the unified LGM technique [66,106] discussed above.

5.1. Effect of SCFs on the Characteristics of Adsorption Processes

It was noted above that an increase in pressure always leads to an increase in the density of the mixture components, and this causes a shift in the equilibrium conditions and changes the rates of elementary stages [1,3,170]. This fact affects all surface processes: catalytic, adsorption, membrane, etc. Depending on the composition of the gas mixture, we will conditionally divide practical situations into two cases. The first case refers to a gas mixture with strongly adsorbing particles. For these, an increase in pressure increases the surface concentration of reactants or adsorbed particles.

The second case refers to gas mixtures with weakly adsorbing or inert molecules. In this case, an increase in pressure also increases the near-surface number of molecules, which can affect the competition for filling part of the surface between strongly and weakly adsorbing molecules. This fact is important if the amount of weakly adsorbing molecules in the volume exceeds the proportion of strongly adsorbing molecules. Such competition makes it possible, in principle, to more accurately control the course of the surface process. Many polymeric systems are of this type; their interaction with supercritical carbon dioxide (SC–CO_2) has been extensively studied [171,172]. CO_2 molecules are a solvent for some polymers, combining many important technological factors such as environmental friendliness, low cost, ease of removal from the polymer, incombustibility, etc.

In paper [173], the processes of chemisorption and physical adsorption were analyzed depending on the degree of surface coverage in the presence in the system of both the main components for the studied surface processes and the influence of the presence of SCF components, which, as a rule, are weakly adsorbing components. Methods for modeling surface processes are given in [66,68].

As an example, typical isotherm curves of the adsorption of component A are shown [173] for a rise in the SCF pressure. As the SCF pressure increases, the surface coverage by component A decreases (see Figure 7); thus, component A is displaced from the adsorbent surface.

There is the displacement of the adsorbed molecules of component A from the adsorbent surface upon an increase in the pressure of the SCF for fixed surface coverages $\theta_A = 0.05$ (1), 0.5 (2), and 0.85 (3). A comparison of Figure 6a,b leads to the conclusion that, on the strongly binding surface, component A is displaced from adsorption sites at higher SCF pressures more slowly and over a broader range of pressures of component SCF than for the weakly binding surface. The same sort of curves will apply for chemisorption—values of partial pressure change (the range of pressure of the basic component A at chemisorption and physical adsorption differs by approximately ≈ 4–6 orders of magnitude) and the course of concentration dependences remains similar.

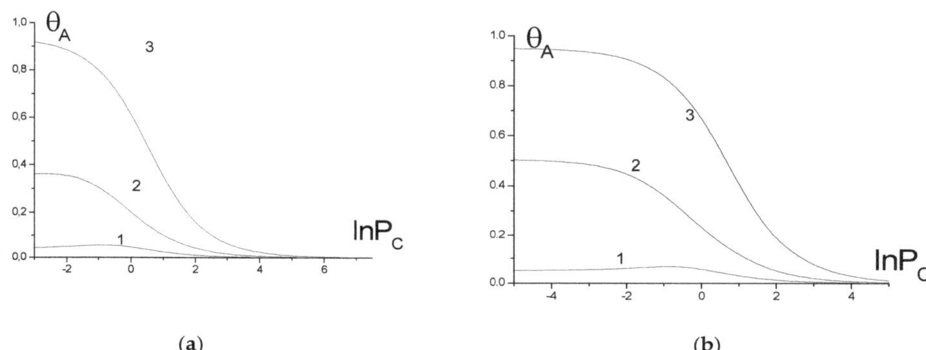

(a)   (b)

Figure 7. Dependence of adsorption of the basic component A in binary mixture of component A and component SCF for rising pressure of SCF for fixed concentration component A corresponding to $\theta_A = 0.05$ (1), 0.5 (2), and 0.85 (3) for weak (**a**) and strong (**b**) adsorption [173].

In work [173] the influence of SCFs on the stratification characteristics of the adsorbed particles A is investigated. With adding the component C, isotherms of component A change, and, hence, the phase diagram changes. Additionally, it was found that at a great enough pressure of an inert component C, there is a weakening of the connection between components A. (Here, the two-dimensional situation reflects the real three-dimensional process of SCF interaction with polymeric matrices).

5.2. Effect of an SCF on the Concentration Dependence of the Rate of a Reaction

Relation (2) defines the effective activation energy $E_{AB}{}^{ef}$ for a bimolecular reaction [170] as

$$W_{12} = U_{AB}/(k_{AB}{}^{0}\theta_A\theta_B) = \exp(-\beta E_{AB}{}^{ef}) \tag{11}$$

where $E_{AB}{}^{ef} = E_{AB} + \varepsilon_{AB} - \beta^{-1}\ln(\theta_{AB}/\theta_A\theta_B) - (z-1)\ln(S_A\,S_B)$.

At low concentrations of reagents, $E_{AB}{}^{ef} = E_{AB}$. At high fluid densities, when the proportion of reagents A and B is small, and the proportion of SCF is large ($\theta_C >> \theta_A + \theta_B$), then

$$E_{AB}{}^{ef} \approx E_{AB} - (z-1)(\delta\varepsilon_{AC} + \delta\varepsilon_{BC}) = E_{AB} + (z-1)(1-\alpha)(\varepsilon_{AC} + \varepsilon_{BC}).$$

In this expression, the equivalence $\alpha_{ij} = \alpha$ is used for both reagents.

Formula (11) determines the effect of the interaction of the SCF with reagents on the value of $E_{AB}{}^{ef}$. If $\alpha < 1$, then the presence of the SCF associates around the reagents increases $E_{AB}{}^{ef}$. If $\alpha > 1$, then the SCF associates decrease $E_{AB}{}^{ef}$. The difference in values of $\Delta E_{AB}(\theta) = E_{AB}{}^{ef} - E_{AB}$ indicates the effect of the SCF on the activation energy of the stage.

In that case, when the SCF plays the role of a solvent, then the influence of intermolecular interactions manifests itself in a change in $E_{AB}{}^{ef}$ with varying pressure and temperature. In the case of a large value of $E_{AB}{}^{ef}$, the influence of the SCF is small if the rate of the chemical reaction itself remains in the region of the kinetic regime. However, due to the addition of a large amount of inert SCF molecules, with an increase in the pressure in the system, the nature of the flow of the bimolecular reaction can change and move from the kinetic regime to the diffusion regime.

A detailed analysis of the ratios between $E_{ij}{}^{ef}$ for chemical reactions and transport stages was carried out in [170]. The contributions of the total density of the system and the role of intermolecular interactions at different temperatures of the SCF of the system were discussed. The theory of non-ideal reaction systems allows the SCF process to separate into failure contributions from pressure and temperature. In situations where the contribution of temperature predominates, the differences between the values of $E_{ij}{}^{ef}$ and E_{ij} are small, so Equation (1) can be used. If the contribution from pressure is predominant, then the differences between $E_{ij}{}^{ef}$ and E_{ij} limit the region of the SCF for concentrations where the mixture can be considered an ideal one.

Such a molecular interpretation distinguishes the chemical features of reaction systems at high pressures from the collective properties of multicomponent systems. Thus, for water molecules, an increase in temperature of more than ~10 degrees above the critical temperature leads to a decrease in density and water remains as a dense vapor up to very high pressures (almost 10 P_{cr}) [174].

For the case when the SCF is a reactant, the dependence of the effective reaction energy is shown in Figure 8. A three-component system (A, B, C) with the chemical reaction A + B (here $\theta_A = \theta_B$) is considered, in which the reaction between molecules A and C is also possible (C is the SCF component). The proportion of component C varies and it is demonstrated how the effective activation energy $E_{AC}(ef)$ changes with respect to the energy E_{AC} of the same reaction in an ideal system as a function of the total amount of substance $\theta = \theta_A + \theta_B + \theta_C$.

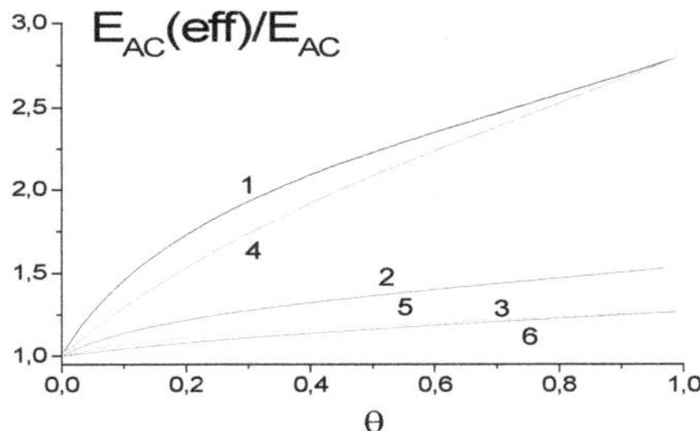

Figure 8. Dependences E_{AC} on the degree of occupancy θ at $\varepsilon_{CC} = 0.38$ (energy, kcal/mol), $\varepsilon_{AA} = 4\varepsilon_{CC}$, and $\varepsilon_{BB} = 2.4\varepsilon_{CC}$ for $T = 1.1T_c$; $E_{12} = (1)$ 3, (2) 10, (3) 20; and $T = 2.5T_c$, $E_{12} = (4)$ 3, (5) 10, (6) 20. Curves correspond to reagent concentrations $x_A = x_B = 0.01$-mole fractions.

Here, the SCF manifests itself through lateral interactions and as one reagent; therefore, in general, the role of the SCF component increases, although only lateral interactions make a smaller contribution compared to the contribution from the main chemical reaction A + B (for these, the relative contribution of lateral interactions reaches values of 4–5). Obviously, an increase in the activation energy of the parallel stage A + C reduces the contribution from lateral interactions (curves 3 and 6).

The theory of reaction rates in condensed phases [66,68] demonstrates that, under supercritical conditions, intermolecular interactions cannot be ignored even for an inert supercritical component, which may exert a significant effect on the rates of elementary processes.

The curves shown in Figure 8 indicate the characteristic intervals of possible deviations from the law of mass action in the typical diapasons of supercritical processes in the gas phase, in the temperature and activation energy, which differs from a process in an ordinary gas atmosphere at 1 atm. An increase in the density of the supercritical component decreases the probability of the reactants meeting one another and the SCF–reactant lateral interactions, though comparatively weak, stabilize the initial states and reduce the rates of the reactions. The latter circumstance is significant for most reactive reactants, such as ozone [2,175]. The calculations were carried out for the entire SCF density range, so it can readily be seen from the calculated curves whether it is possible to neglect the interactions between the SCF molecules under the given conditions. (If the lateral interactions can be neglected, then $E_{AB}(ef) = E_{AB}$ at any θ_C.)

In many cases, such a state may arise with an increase in the density of inert SCF ($\theta \rightarrow 1$), when the reagent transport stage begins to limit the processes, especially at low concentrations of the main chemical reagents. This problem, however, does not occur if the SCF component is one of the reactants. In this case, the bimolecular stage A + C transforms into a quasi-monomolecular one. In general, when analyzing the role of the SCF, one should take into account the possibility of changing the mechanism of the process under study and the state of the accompanying materials, in addition to traditional ideas about the increase in the rate of stages with increasing temperature and its decrease with increasing pressure, together with a decrease in the self-diffusion coefficient and an increase in the viscosity coefficient.

A comprehensive analysis of the entire system as a whole is required because a change in one of the thermodynamic parameters (pressure or temperature) can worsen the implementation of elementary stages. Using the concepts of specific mechanisms of gross processes, the kinetic equations of non-ideal reaction systems make it possible to relate the characteristics of the SCF processes to similar processes under normal conditions.

5.3. Effect of an SCF on the Concentration Dependence of the Dissipative Coefficients

The theory of non-ideal reaction systems provides ways to consider transport characteristics. The simplest characteristic of molecular transport in mixtures is the self-diffusion coefficient of component i ($1 \leq i \leq (s - 1)$), which characterizes the thermal motion of type i molecules under equilibrium conditions. In practice, the self-diffusion coefficient is usually associated with the motion of an isotopic label locally introduced into some region of the system and the temporal distribution of the label over the rest of the solution is monitored. For labeled type i molecules, we have the following expression for the local partial self-diffusion coefficient [68,176]:

$$D_i^* = z_{fg}^* \, U_{iV} / \theta_f^{\,i}, \tag{12}$$

where z_{fg}^* is the number of possible hops to nearest-neighbor sites g for the fth cell along the direction in which the label moves. The expression for w_{iV} is given by Formula (2). This refers to the bimolecular hop of molecule i, $i + V \rightarrow V + i$, in which the first "reactant" is a moving type i molecule, and the second "reactant" is the vacancy into which the molecule i is transferred. The activation energy of this process is $E_{iV} = 0$ for a bulk phase and $E_{iV} > 0$ for the surface migration step. With the growth of the full density of a system, the fraction of free volume decreases and the factor of self-diffusion of any particle decreases.

The apparent activation energy of the self-diffusion of component i is written in the form of Formula (11):

$$E_{iV}(ef) = E_{iV} - \beta^{-1}\ln(\theta_{iV}/\theta_i\theta_V) - (z - 1)\ln(S_i \, S_V), \tag{13}$$

For high SCF coverages, it follows from Formula (13) that $\Delta E_{iV} = E_{iV}(ef) - E_{iV} = (z - 1)(1 - 2\alpha)\varepsilon_{AC}$. For $\alpha = 0.5$, $\Delta E_{iV} = 0$ and the decrease in the self-diffusion coefficient is only due to the decrease in the free volume fraction.

Figure 8 plots the concentration dependences of the self-diffusion coefficient of the main component A (D_A) in the SCF bulk ($E_{iV} = 0$). The effect of the density of the supercritical component was studied as a function of the density θ_C varying between zero and $1 - \theta_A$ for two main component coverages, $\theta_A = 0.01$ and 0.1.

In these calculations, we fixed the mole fraction of the main component A relative to the supercritical component C; accordingly, all curves begin at $\theta = 0$. The curves descend as the total density increases. This is due to the decrease in the free volume, which is necessary for molecular transport. The self-diffusion coefficients of the SCF in the bulk are similar to those of the main component. The behavior of the D_C curves is similar to the behavior of the curves shown in Figure 9. The concentration dependence of the self-diffusion coefficients depends strongly on the nature of the supercritical component. The stronger the interaction between SCF molecules, the greater the extent to which diffusion slows down with an increasing SCF concentration. The self-diffusion coefficient decreases as the temperature is raised.

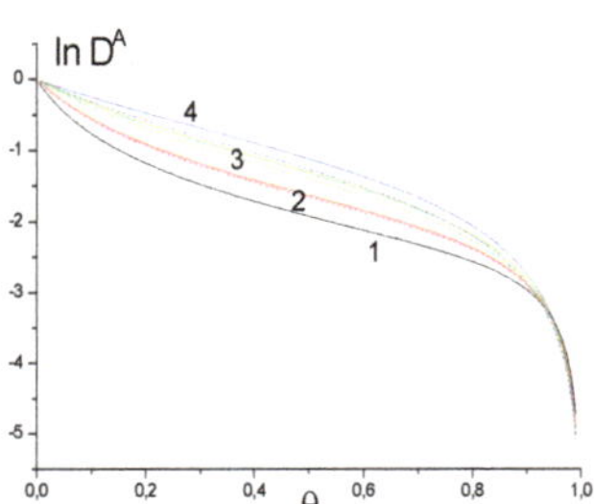

Figure 9. Self-diffusion coefficient of component A (D_A) as a function of $\theta = (\theta_A + \theta_C)$ at $\gamma_A = \varepsilon_{AA}/\varepsilon_{CC}$ = (1) 1, (2) 1.6, (3) 3.2, and (4) 6.4 for $\tau = 1.15$. Solid lines: $\theta_A = 0.01$; dotted lines: $\theta_A = 0.1$.

Shear viscosity. Another important transport characteristic is the shear viscosity. Knowledge of this characteristic is necessary for calculating flow velocities in various reactors. Like the self-diffusion coefficient, viscosity is expressed in terms of the thermal velocities of molecules. The local shear viscosity η_{fg} for spherical molecules of comparable sizes is expressed as follows [25]:

$$\eta = \left[\sum_{j=1}^{s-1} x_j \left(\eta_j\right)^{-1}\right]^{-1}, \eta j = \theta j / U_{iV}, x_i = \theta_i/\theta, \theta = \sum i = 1^{s-1}\theta_i, \tag{14}$$

where x_i is the mole fraction of component i and θ is the total coverage of the system.

For pure components, it follows from this expression that η depends on temperature as $T^{1/2}$ and depends linearly on the density. For high densities, we have an exponential temperature dependence, as in Eyring's conventional model [31]. Equation (14) allows viscosity to be calculated for any composition of a multicomponent mixture.

Figure 10 plots viscosity versus the total mixture density θ. In these calculations, we fixed the mole fraction of the main component A relative to the supercritical component C; accordingly, all curves begin at $\theta = 0$. With this method of expressing the amount of the main component A, the difference between the viscosities at $\theta_A = 0.01$ and 0.1 is smaller than in the case of the fixed amounts of component A. The concentration dependences of viscosity are normalized to the viscosity of component A in a rarefied gas. The calculations were carried out at a fixed ε_{AA} value for component A and a decreasing ε_0 value or an increasing $\gamma_A = \varepsilon_{AA}/\varepsilon_{CC}$ ratio. An increase in γ_A leads to an increase in the energy of the lateral interactions between the main component A and the SCF. As a result, the viscosity of the system decreases. Therefore, by changing the SCF, it is possible to vary the viscosity of the system in a fairly wide range. This range is temperature-dependent: it widens with increasing temperature.

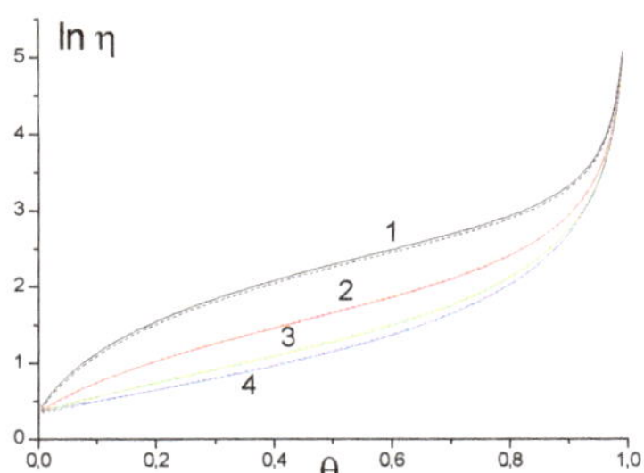

Figure 10. Viscosity as a function of θ at γ_A = (1) 1, (2) 1.6, (3) 3.2, and (4) 6.4 for τ =1.15. Solid lines: θ_A = 0.01; dotted lines: θ_A = 0.1.

The calculated data indicate that the self-diffusion coefficients and viscosity characterizing the transport properties of the entire supercritical system depend on the density of the supercritical component to a lesser extent than the reaction rates. This is due to the fact that the thermal velocities of molecular migration, which determine both of the transport coefficients, depend on temperature much less strongly than chemical reactions.

The above results in Section 5 correlate well with the known similar relationships for the same steps occurring at atmospheric pressure. At the same time, they were obtained by an analysis of the processes in a very wide SCF density range. The validity of the above inferences throughout the pressure and temperature ranges typical for supercritical processes suggests that this good correlation is a more general point than a simple corollary of the law of mass action.

6. LGM and Dissipative Coefficients

Real technological SCF processes are implemented in reactors and for their modeling it is necessary to know not only the equations for describing chemical reactions, but also the general thermophysical flows of momentum, energy, and mass transfer. As noted in Appendix A, the dissipative coefficients are directly related to the elementary rates of transport processes; therefore, the same model potentials of intermolecular interaction are used for their calculation, both for chemical reactions and equilibrium distributions (according to Section 3.1). Therefore, the effective pair potential (10) was used to calculate all dissipative coefficients. These calculations were carried out both for individual components (coefficients of self-diffusion, viscosity, and thermal conductivity) and for binary mixtures (coefficients of mutual diffusion and viscosity). These calculations refer to the group properties of simple fluids, which reflect the general patterns for many molecules that obey the law of corresponding states. There are no strong specific bonds for molecules of this kind. Comparisons were made with the so-called generalized diagrams, which were actively used earlier in technological calculations. Specifying the properties of molecules, such diagrams correspond to an accuracy of the order of 3–5%.

Calculations of shear viscosity coefficients for specific molecules and their generalized normalized values of the coefficients are shown in Figure 11. The accuracy of these calculations is about 5%.

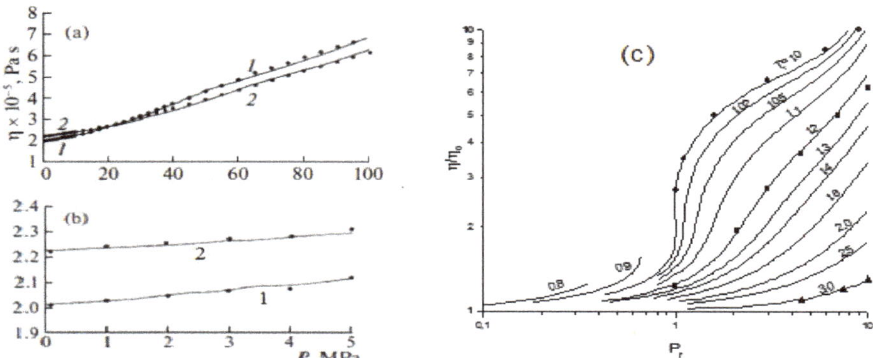

Figure 11. (**a**) Dependences of O_2 shear viscosity η on pressure P at different gas temperatures: (1) 289, (2) 328 K. Dots are experimental values from [177]. (**b**) Analogous dependences for low densities. (**c**) Generalized diagram of reduced shear viscosity η/η_0, depending on reduced pressure P_r at different reduced temperatures Tr. Dots are experimental values from [23,26].

Another important dissipative coefficient is the coefficient of thermal conductivity. For dense gases and liquids, this coefficient is calculated based on the simulation of two channels of heat transfer: the diffusion mechanism of transfer of atoms and/or molecules in space through a selected plane (see Appendix A), and the mechanism of particle collision among themselves, when no movement occurs through a selected plane of particles from different half-spaces (this mechanism was first introduced for rarefied gases by Enskog [178]). Calculations of the thermal conductivity coefficient are demonstrated for the bulk phase [158,177] (Figure 12a) and for its generalized dependence [26] (Figure 12b).

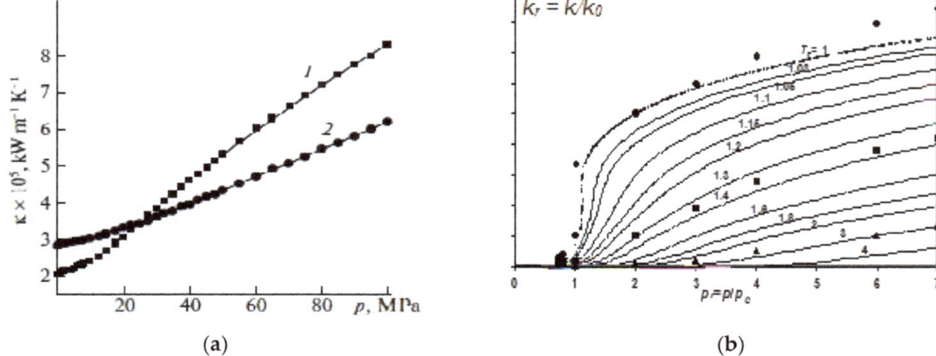

Figure 12. (**a**) Heat capacity coefficients of argon at (1) 273 and (2) 400 K. Dots represent experimental values [158,177]; lines represent calculations [152]. (**b**) Experimental data on the generalized heat capacity coefficient in [26]. Dots represent experimental values; lines represent calculations.

For multicomponent mixtures, the self-diffusion and mutual diffusion coefficients are necessary for modeling mass transfer processes. These coefficients are shown in Figure 13a,b. The self-diffusion coefficient characterizes the motion of a labeled particle in a one-component system under the condition of a complete equilibrium distribution of particles within the system. The coefficient of mutual diffusion characterizes the nonequilibrium of a binary system with respect to its composition. Experimental data on the indicated mass transfer coefficients are presented in [26] with the help of weighted average correlations, which make it possible to display both types of coefficients in wide ranges of temperature and pressure changes. In Figure 13b, the interdiffusion coefficients refer both to individual gases (argon and nitrogen) and to their mixtures.

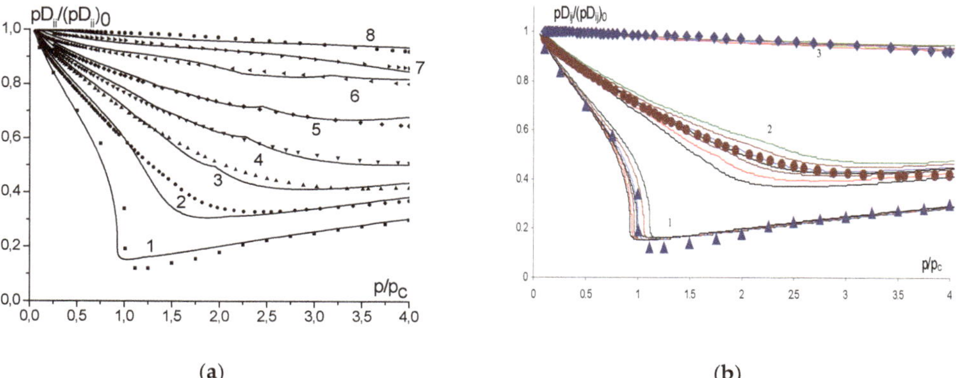

(a)   (b)

Figure 13. (**a**) Generalized diagrams of the self-diffusion coefficient for the $p/p_C < 4$ range of pressures. Lines represent calculations in [160]; dots represent experimental values from [26] at τ = (1) 1, (2) 1.1, (3) 1.2, (4) 1.3, (5) 1.4, (6) 1.6, (7) 2, and (8) 3. (**b**) Generalized diagrams of the mutual diffusion coefficients of an Ar–N$_2$ mixture with x_{N_2} = 0, 0.2, 0.4, 0.5, 0.6, 0.8, and 1.0-mole fractions at τ = (1) 1, (2) 1.3, and (3) 3 in the $p/pc < 4$ range of pressures. Series of curves calculated for each τ value are displayed from the bottom up as the concentration of nitrogen grows [160]. Dots represent experimental data recalculated according to the rules in [26].

Weighted average correlations over gas densities were actively used for approximate estimates of various characteristics. For shear viscosity coefficients of binary mixtures, the approach of Dean and Steele was popular (cited in [26]). Based on experimental data for mixtures of non-polar dense gases (light hydrocarbons, and hydrocarbons with inert gases and air components), an approximation expression was proposed that is almost linear in the logarithmic scale. The numerical values of this approximate formula for the experimental data are shown in Figure 14 by thin lines.

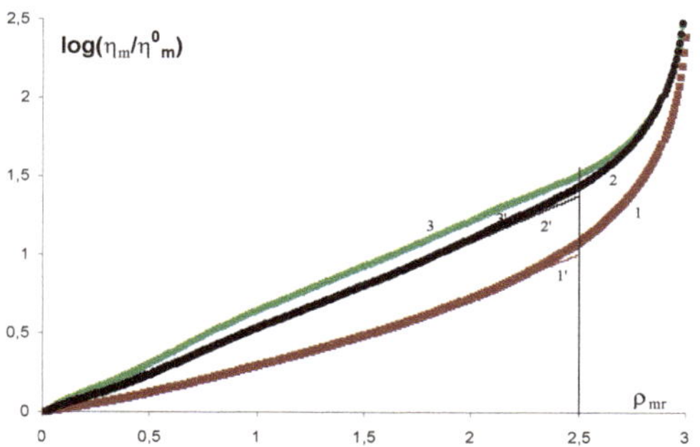

Figure 14. Logarithmic dependences η_m / η_m^0 on reduced density ϱ_{rm}, obtained for a binary Ar–N$_2$ mixture at different reduced temperatures τ: (1) 0.75, (2) 1, and (3) 3; thick lines represent the calculated values from [160]; thin lines represent experimental values from [26].

Figure 14 compares the dependences η_m / η_{m0} (where the denominator η_{m0} is a normalization factor to the properties of a rarefied gas) on the reduced density ρ_{rm}, obtained for the Ar $-$ N$_2$ binary mixture for three values of τ according to this empirical formula and to the LGM (thick lines are the calculation of the LGM [160]). Here, the reduced temperature

$\tau = T/T_{cr} = 0.75, 1,$ and 3, where T_{cr} is the critical temperature reflecting the change in the value as a function of the composition of the mixture.

7. Extension of the Models

This section briefly mentions the directions where approaches based on the LGM have been developed. Above, the components of the mixture were assumed to be approximately commensurate in size in order to singly occupy a cell of the lattice structure, and a spherically symmetric pair potential was used for the model interaction potential. These restrictions were lifted in subsequent works when nonspherical effective potentials were considered, jointly taking into account electrostatic contributions with the Lennard–Jones potential. Also, the kinetic equations are discussed below in the LGM, which are necessary for describing three-aggregate states.

7.1. Nonspherical Potential Functions

Refusal to use spherically symmetric effective pair functions (10) has been considered in the framework of lattice models since the beginning of their development [119–122] (see also [179–181]). This direction of work made it possible to move to a more accurate accounting of the size of molecules and their shape, which can differ greatly from spherical ones. This includes models of interactions with local atom–atom potential functions for individual functional groups of molecules, as well as their simplified description, which involves the use of energy contact interaction parameters.

As an example of the development of such approaches, a model was considered in [182] that makes it possible to take into account the differences between hard spheres of particles and an integer number of cells. Thus, one of the main SCF components, the CO_2 molecule, has noticeable differences in shape from a spherical one. The ratio of its long and short axes is 1.38. As an example, Figure 15 shows the compressibility factor $\chi = PV/(PV)_s$ of CO_2 molecules [183], where the product of pressure and specific volume $(PV)_s$ refers to normal conditions. The calculations were carried out for different potential models: curves 1–3 are given for a single-site model; curve 4 reflects the difference in the shape of the hard sphere of the molecule. (For comparison, curve 2 is given for the strong contribution of the triple interaction).

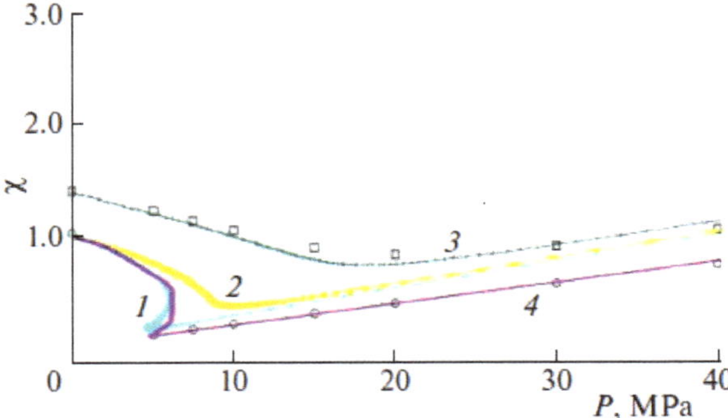

Figure 15. Comparison of experimental [184] (dots) and calculated lines of compressibility factors for CO_2 molecules; calculations were made for (1, 2, 4) $T = 273$ and (3) 373 K.

The discussed modification of the potential is better than the experimental curve [184]. (Figure 15 presents the compressibility factor described by the theory with accuracies of up to ~4%). Curves 1 and 4 for a temperature of 272 K reflect the separation of CO_2 molecules ($T_{cr} = 304$ K). Calculations of the separation curve of CO_2 molecules demonstrate a number of differences depending on the interaction potential; their consideration depends on the goal of modeling the processes with their participation.

The [182] approach also reflects the possibility of taking into account different energy contact interactions in nonspherical molecules. Such potential modifications can be proposed for other SCF components of mixtures and for the reactant molecules themselves. The refinement of differences in the shape of molecules from spherical is important for the calculation of all dissipative coefficients. When modeling processes inside porous materials and on open surfaces, the role of molecular asymmetry increases, since the spatial distribution of molecules is additionally affected by the surface potential of the solid. The nonsphericity of molecules is important in the analysis of short-range order and in the transition to the appearance of long-range order.

7.2. Water Molecules

Water is a component that is actively used in SCF processes. To expand the possibility of using LGM approaches with different potential functions, in [185,186] a potential was considered with a combination of the contributions of the dipole potential and the Lennard–Jones potential and simultaneous generalization according to a more general statistical description: direct correlations in the QCA were taken into account instead of the average field description of interacting molecules [187,188]. Previously, it was known [32,33,108–110] that the use of only electrostatic interaction is not enough to describe the thermodynamic characteristics of water.

The joint allowance for the dipole potential and the Lennard–Jones potential made it possible to improve the quantitative description of the experimental data [189,190]. In particular, it was found that the LGM describes the stratification curve of water with the same accuracy as for argon. The microscopic nature of the LGM makes it possible to obtain a description of dipole systems without resorting to the macroscopic concept of permittivity. The same approach can be applied to many other polar liquids. On the other hand, the LGM [185,186] used in [191] makes it possible to describe the experimental data on the self-diffusion of water molecules [190].

In this work, we studied the temperature behavior of the water self-diffusion coefficient. This process can be written in the form of the bimolecular reaction $(H_2O)_f + V_g = V_f + (H_2O)_g$, in which the numbers of the neighboring sites f and g are used as indices and the symbol Vg denotes a vacant site with the number g, i.e., the diffusion of water molecules is a special case of a bimolecular reaction occurring on the f and g sites.

The theory allows the determination of the water self-diffusion coefficient and the study of its temperature dependence (Figure 16). Figure 16 presents the water self-diffusion coefficient described by the theory with accuracies of up to ~8%. Our calculations were performed under the condition that the considered jump of a molecule to the neighboring vacancy occurs much more rapidly than the local rearrangement of the overall environment. According to the Frenkel hypothesis [192], the transition to the thermal motion of molecules is associated with the clearly correct hypothesis of a weak change in the state of neighbors when the considered molecule jumps. Assuming [31] an equilibrium distribution of its neighbors would contradict the logic of the process: for a selected molecule to move, it is necessary that its analogous neighbors are quickly attuned to the transition state of this step. Many jumps of the neighbors are required for one jump of the considered molecule to occur.

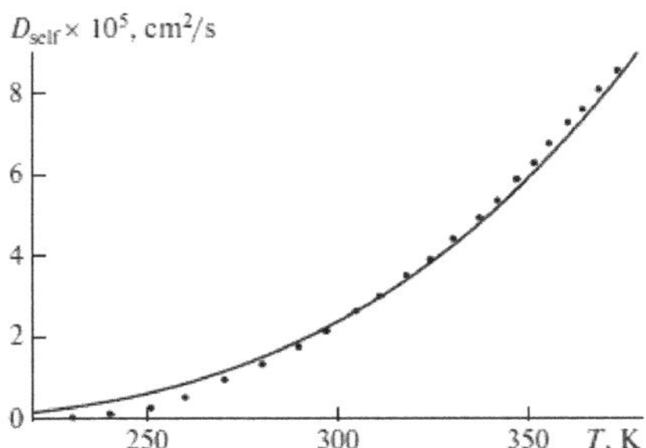

Figure 16. Comparison of the temperature dependence of the water self-diffusion coefficient. Dots represent the experimental values in [190]; lines represent the calculations performed in the work [191].

By comparing the constructed theoretical curve with our experiment's results [185], we found the optimum value of parameter α for the water molecule energy parameters from [193]: $\alpha = 0.65$. (Note that the same experiment can also be approximated by molecular dynamics [194].) The obtained good agreement testifies to the adequacy of the lattice model and the validity of its assumptions. The α value above demonstrates the relatively low activation energy that a water molecule must overcome during a jump in the course of this reaction.

7.3. Kinetic Equations

The mathematical formulation of the second law of thermodynamics is written as the inequality $dS \geq dQ/T$, where Q is the amount of heat [20–22,111]. To calculate the change in entropy during the transition processes between different initial and final states, it is necessary to know the state of the system at two points (the equal sign refers to reversible processes). The transition between initial and final states is described by kinetic equations, which determine their key importance in modeling any technological processes [112].

Currently, kinetic theory at the atomic and molecular level in the lattice-gas model (LGM) [66,68] can be used in almost the entire time range, from the characteristic times of atomic vibrations to macroscopic, including equilibration times. The theory considers the full set of elementary processes of movements of molecules and their chemical reactions occurring in the system on the set of lattice sites. To construct the general structure of the kinetic equations of the lattice model, we assume that the lattice sites are not equivalent to each other. The nature of the inhomogeneity of the lattice sites is considered known and constant over time. From a physical standpoint, the nonuniform distribution of particles is due to both the interaction between the particles of the system and the possible additional influence of external fields or interactions. The formulation of the problem under consideration allows covering from a unified point of view a wide range of issues related to the nonuniform distribution of particles at the gas–solid interface: the spatial distribution of the particles on a uniform surface, the presence of ordering in them, changes in the distribution of atoms of the solid adsorbed particles along the normal to the interface, and the distribution of particles laterally interacting on the nonuniform surfaces and, correspondingly, in any three-dimensional volume of the porous body. General ideas about the construction of the kinetic equations in the LGM are described below in Appendix A.

The LGM provides a unified approach to modeling equilibrium and kinetic processes in three states of aggregation. It uses a single set of energy parameters to describe different stages of a multi-stage gross process both in kinetics and in the equilibrium state.

Many practically important circumstances, such as the presence of several heterogeneous phases in systems, the multistage nature of the overall process, the presence of external fields, etc., lead to the complication of the structure of the kinetic equations. Particularly important in this regard are the ratios of the relaxation times of the reactants in nonequilibrium states, which must be taken into account when modeling using kinetic equations.

All these kinetic equations should be built based on the theory of non-ideal reaction systems [66,106], taking into account all possible options for changing the potential functions of intermolecular interaction. Accordingly, equations were adapted to describe the processes in complex porous materials [68], in which phase and supercritical states are realized. For these, their own equations of state and equations for dissipative coefficients were also constructed. Such developments led to the creation of microscopic hydrodynamics [68]. This describes the processes of porous systems, inside which chemical and/or catalytic reactions take place in the fields of pore walls, in the presence of vapor and liquid phases. In the same way, the LGM methods made it possible to develop approaches for the formation of new phases [115] taking into account the microscopic description of their interfaces (such processes are often implemented in SCF). New kinetic equations have also been developed that reflect the consideration discussed above for differences in molecular sizes [195] and the possibility of the presence of charged particles in reaction systems, including electron transfer [196,197].

At present, the kinetic equations in the MGM have reached the general level of describing all three-aggregate states from a unified position [115,196,198], starting from the elementary stages of changing the position of particles in space and/or elementary chemical transformations. This makes it possible to simulate many processes that are implemented in SCFs, associated with a change in the phase state of materials and chemical processes occurring in the SCF conditions. As an example, we can note the development of the fundamentals of the process of swelling of polymer matrices [195]. In this work, a molecular model for the sorption of low molecular weight molecules, accompanied by a change in the volume of the polymer, is proposed. The kinetic equations make it possible to relate changes in the local densities of the sorbate and fragments of polymer chains to each other, although it should be noted that there is, up to now, no complete theory of the swelling of polymers in SCFs. However, the analysis [170,173] of separated stages shows that SCFs can rather strongly affect the state of the adsorbents and catalysts that can qualitatively be observed in the studies of the combined effects of SCF–CO_2 and ozone on coke-like deposits on a surface of Pt–Re/γ-Al_2O_3 bimetallic reforming catalysts [199].

8. Conclusions

This review presents two levels of discussion on the modeling of SCF processes. The traditional level reflects the specific modeling methods used to describe the physicochemical characteristics of the systems under study, the rates of chemical reactions, and dissipative coefficients. In this regard, LGM-based approaches are presented as universal modeling methods for three-aggregate states of systems. A new level of discussion of the applicability of the LGM is associated with the important circumstance of agreement with the second law of thermodynamics when the models for calculating the rates of stages and the equilibrium distribution of components must be consistent with each other. Otherwise, the models will belong to different classes and their molecular parameters will not be preserved when discussing different modes of the process, which makes it difficult to model gross processes. It should be recalled that the historically known expressions for chemical equilibrium according to Guldberg–Waage's mass action law [34,35] and Langmuir's adsorption isotherms [200] (Section 3.2) were first obtained based on the Clausius second law of thermodynamics.

In modeling real technological processes using existing methods discussed in the Section 1, situations arise when two levels of models are used simultaneously: models of non-ideal systems for describing the equilibrium properties of SCF and kinetic models based on the law of mass action, i.e., models for ideal systems, which does not agree with the second law of thermodynamics.

This review demonstrates the great potential of the LGM for modeling a wide range of SCF processes, consisting of different elementary stages in the overall processes of chemical transformations and transport stages, which ensure self-consistency of the description of the kinetics of these processes with equilibrium states in non-ideal reaction systems.

The first comparisons of different methods of molecular modeling (integral equations, molecular dynamics methods, Monte Carlo, and LGM) already showed [201] that the LGM is in no way inferior in calculating all thermodynamic characteristics (although it does not give a continuum distribution of molecules inside the cell). It is difficult to give a full account of the practical applications of the LGM (although some of this material is presented in this review). The development of the statistical methods of the LGM [66,106] made it possible to significantly enhance its capabilities, especially for highly inhomogeneous systems (these are problems for phase boundaries, adsorption, and absorption). The introduction of the modifications discussed in this review, taking into account the internal motions of molecules inside the cells [133,134,139], formulated a unified approach to describing the molecular distribution in three-aggregate systems [115]. New developments in the LGM make it possible to perform calculations with more complex multiparticle potentials. Today, joint calculations are available for these using quantum-chemical methods for determining the energy of intermolecular interactions [66,106,154,202], including those with the Coulomb potential [197,203]. The statistical substantiation of initially purely empirical potential functions of the type (10) obtained in the LGM allows one to operate with molecular potentials, similar to all other methods of statistical physics. This makes it possible to search for the parameters of potential functions from the solution of inverse problems by describing the experimental data more quickly and more accurately due to the significant gain in time when using the LGM methods. In particular, the parameters of the "excluded" volume fraction, the cell size, and the vibrational contribution of molecules introduced into the LGM operate only on the properties of potential functions and do not have additional parameters.

Mathematical models developed in the LGM correspond to all times of realization of processes, including their relaxation stages in all three-aggregate systems and reaching equilibrium distributions according to the second law of thermodynamics. In terms of the computing speed, LGM is at least 2–4 orders of magnitude faster than the molecular dynamics method and its accuracy is not inferior to it.

It should be noted that, up to now, modeling based on the LGM has been carried out for the simplest molecular mixtures; however, its extensions make it possible to switch to more complex systems that have more complex phase diagrams and distinct physico-chemical properties, including those containing both upper and lower critical points, as well as systems in which there are diagrams without top and right borders. Models for such equilibrium characteristics have been discussed earlier in the literature [118–122] and recent developments in the MGM (Section 7) make it possible to generalize them to the corresponding kinetic models, which will be consistent with the second law of thermodynamics.

Modern approaches based on the LGM [66,68,106,115] can be applied, as well as the thermodynamic approach [114,204], for any phases and from interfaces. They are actively used to simulate many practically important processes in a wide range of pressure and temperature changes. They describe the distribution of components of heterogeneous systems and their phase diagrams, to model the chemical synthesis reactions, and so on. With their help, one can calculate the processes of adsorption, catalysis, and growth of crystals, and the processes of transport of molecules through various polydisperse materials (porous bodies, membranes, and thin films), as well as consider the rheological properties of molecular and other aspects of physical and chemical mechanics, etc.

An important advantage of the LGM is the extension of the theory of chemical transformations to non-ideal reaction systems, which takes into account the influence of interparticle interactions and preserves the effects of correlation between interacting particles. This makes it possible to eliminate the existing contradiction in the modeling of SC processes, when the equations of state describe non-ideal mixtures, and the kinetic models are based on the law of mass action for ideal systems or the dissipative coefficients are built on the basis of Boltzmann-type equations for rarefied gases. The kinetic equations in the LGM give a self-consistent calculation of the equations of state and rates of elementary stages, which satisfies the second law of thermodynamics. This fact is of strategic importance for the organization of the correct modeling of SC processes.

Modeling methods with the help of LGM make it possible to (1) assess the processes of absorption of SCF molecules by polymer matrices, (2) describe the processes of growth of a new phase (nanosized particles in CSF) and analysis of their size dependences, (3) analyze the effect of concentrations of SCF molecules, taking into account their influence on shifts in equilibrium and changes in the rates of stages in bulk phases, (4) study the factors in porous materials that affect the processes of chemical kinetics, and the growth of a new phase inside pores and in near-wall regions under supercritical conditions, and (5) assess the role of critical regions inside porous materials on SCF transfer processes of molecules and their chemical transformations, etc.

Thus, for almost all modeling parameters, the LGM has the most significant potential for describing SC states of fluids and processes with their participation in comparison with all other existing methods of statistical physics in calculating both equilibrium and nonequilibrium characteristics. This provides the basis for using the LGM approach for practical applications to various technological SCF processes.

Funding: This research received no external funding.

Data Availability Statement: The results of research on this topic are reflected in the list of references.

Acknowledgments: Preparing the manuscript for publication was performed as part of a State Contract for the Institute of General and Inorganic Chemistry, Russian Academy of Sciences, in the field of basic scientific research, 2023.

Conflicts of Interest: The author declares no conflict of interest.

Nomenclature

a_{i0}	pre-exponential factor of the Henry constant for molecules of type i
d_{tr}	triple interaction parameter associated with its energy ε_3
$D_i{}^*$	partial self-diffusion coefficient for molecules of type i
E_{ij}	reaction's energy of activation between i and j reagents
$E_A(ef)$	effective activation energy of desorption
k_B	Boltzmann constant
k_{ij}	rate constant of elementary reaction $i + j \rightarrow$ products
$k_{ij}{}^0$	rate constant pre-exponential factor for elementary reaction $i + j \rightarrow products$
k_1 and k_2	reaction rate constants in the forward and backward directions
K	equilibrium constant of the stage
M	number of sites in the system
n_i	concentration of i-type molecules
Q_i	energy of i-particle bonding with the surface
Q	amount of heat
Q_s	statistical sum of the system
P	pressure
P_i	partial pressure of i-type molecules
$P(\{\gamma_f{}^i\},\tau)$	probability of finding the system at the time τ in a state $\{\gamma_f{}^i\}$. For the sake of brevity, this state is denoted as $\{I\} \equiv \{\gamma_f{}^i\}$
s	number of occupation states of any cell or site
S	entropy
S_m	molecular property in flow
T	temperature
$t_{fh}{}^{ij} = \theta_{fh}{}^{ij} / \theta_f{}^j$	function of the conventional probability of j particles being near i particles (fh represents the numbers of sites containing these particles)
U	internal energy
U_{ij}	rate of an elementary stage of a bimolecular reaction $i + j \rightarrow$ products

$U_f^j(\alpha)$	rate of the elementary single-site stage $i \leftrightarrow b$ with number α in the site f
$U_{fg}^{ij}(\alpha)$	rate of the elementary two- site stage $i + j_\alpha \leftrightarrow b + d_\alpha$ with number α in two sites fg
V	volume of the system
u	contribution from the vibrational motion of molecules to energy parameters
$W_\alpha(\{I\} \to \{II\})$	probability of the elementary process α which resulted at time τ in the transfer of the system from the initial state $\{I\}$ to the final state $\{II\}$
$x_i = \theta_i/\theta$	mole fraction of component i among all molecules of the mixture.
z	nearest neighbors of any site or cell
$z_{fg}{}^*$	the number of possible hops to nearest-neighbor sites g for the fth cell along the direction in which the label moves
Z	compressibility factor
α	number of stages in the total process
α_i	activity coefficient of i-type reagents
$\alpha_{ij}{}^*$	denotes the activity coefficients of ACs
$\alpha_{ij} = \varepsilon_{ij}{}^*/\varepsilon_{ij},$ for simplicity	is used for both reagents
$\alpha_{ij} = \alpha$	
γ_f^j	variable determined the occupation state of site with number f ($1 \leq f \leq M$) by particle of type i ($1 \leq i \leq s$)
ε_{ij}	parameter of this interaction between ij pairs of neighboring particles
$\varepsilon_{ij}{}^*$	interaction parameter for reaction AC using i-type particles and neighboring j-type particles
η	shear viscosity coefficient
κ	heat capacity coefficient
κ_D	mean square displacement of particles in a solid in the harmonic approximation
λ	average cell size
Λ_{fg}^{AB}	non-ideality function for the two-site stage
Λ_f^i	non-ideality function for a one-site stage
μ_i	chemical potential of component i
μ_i^0	chemical potential of the standard state for component i
ρ	mean free path of a particle
θ_i	concentration of particles type i in the (surface or bulk)) system
$\theta = \sum_{I=1}^{s-1} \theta_i$	complete occupancy of a lattice system by all i components of the system, $1 \leq i \leq s-1$
θ_{fh}^{ij}	probability of two particles i and j being on nearest neighboring sites f and h (for homogeneous system θ_{ij}, is the pair particle distribution function)
ν_i	stoichiometric coefficient

Appendix A. Kinetics Equation in the LGM

A change in the state of a condensed phase, which we are modeling in the LGM by a lattice structure at a molecular level, is associated with a change in the state of its individual sites as a result of the realization of elementary processes. Fixation of the molecule in the center of the cell corresponds to the state of its occupation. Mathematically, this event is described by γ_f^j, where f is the cell number, $1 \leq f \leq M$, subscript i denotes the state of occupation of the cell with number f, $1 \leq f \leq s$, and s is the number of the states of cell occupation including a vacancy (M is the number of sites) [66]. For the two components of the lattice systems (s = 2) of any site of the lattice structure corresponding to the one-component system for which i = A or V (vacancies). If the site f contains an adsorbed particle A, then $\gamma_f^A = 1$ and $\gamma_f^V = 0$; if the cell is free, then there is a vacancy, so $\gamma_f^A = 0$ and $\gamma_f^V = 1$. The random variables γ_f^j are subject to the following relations:

$$\sum_{i=1}^{s} \gamma_f^i = 1, \text{ and } \gamma_f^i \gamma_f^j = \Delta_{ij} \gamma_f^i,$$ where Δ_{ij} is the Kronecker symbol, which means that any site is unavoidably occupied by one, but only one, particle.

The state of the sites changes at the expense of two main types of elementary processes, namely, the migration of the particles and their participation in chemical transformations. In the first case, the particles change their position in space while remaining unchanged in a chemical respect. In the second case, the particles transform into their other chemical compounds without changing their position. The description of the process when a particle simultaneously changes its coordinates and partly changes its chemical nature by entering into the composition of a complex or an intermediate does not virtually differ from the description of the two basic types of processes.

We denote by $\{\gamma_f{}^i\} = \gamma_1{}^i, \gamma_2{}^j, \ldots, \gamma_M{}^n$ the complete set (or full list) of values $\gamma_f{}^i$ of all lattice sites, which uniquely determine the complete configuration of the locations of the particles on the lattice at time τ, and, by $P(\{\gamma_f{}^i\},\tau)$, the probability of finding the system at this time in a state $\{\gamma_f{}^i\}$. For the sake of brevity, this state is denoted as $\{I\} \equiv \{\gamma_f{}^i\}$. Let the common studied process consist of many stages and through α we denote the number of elementary stages in the process. The master equation for the evolution of the full distribution function of the system in a state $\{I\}$, due to the implementation of the elementary processes α in condensed phases, has the following form (the so-called Glauber-type equation) [66,163,205,206]:

$$\frac{d}{d\tau}P(\{I\},\tau) = \sum_{\alpha,\{II\}} [W_\alpha(\{II\} \to \{I\})P(\{II\},\tau) - W_\alpha(\{I\} \to \{II\})P(\{I\},\tau)], \qquad (A1)$$

where $W\alpha(\{I\} \to \{II\})$ is the probability of the elementary process α (the probability of transition via channel α), which results at time τ in the transfer of the system from the initial state $\{I\}$ to the final state $\{II\}$. In Equation (A1) the sum is taken over the different types of direct processes (index α) and all reversed processes $\{II\}$, in the state of occupation of each site in the system changes.

If the elementary process occurs at one site, the lists of states of occupation of the sites of the system $\{I\}$ and $\{II\}$ differ only for this site. Single-site processes are processes associated with changes in the internal degrees of freedom of the particle, the adsorption and desorption of non-dissociating molecules, and the reaction by the impact mechanism. If the elementary process occurs in two neighboring lattice sites, then lists of the states $\{I\}$ and $\{II\}$ differ in the conditions of occupation of these two sites. Two-site processes are exchange reactions, adsorption, desorption of the dissociating molecules, migration processes by the vacancy and exchange mechanisms, etc. The sum of the states $\{II\}$ corresponds to the change in states of occupation for all lattice sites. The interconnection of the states $\{I\}$ and $\{II\}$ depends on the mechanism of the process that defines a set of elementary stages α.

Equation (A1) is written in the Markov approximation for which it is assumed that the relaxation processes of the internal degrees of freedom of all particles are faster than the process of changes of the state of occupation of different sites of the lattice system. The transition probabilities W_α are subject to the condition of detailed balance:

$$W_\alpha(\{I\} \to \{II\})\, exp(-\beta H(\{I\})) = W_\alpha(\{II\} \to \{I\})\, exp(-\beta H(\{II\})), \qquad (A2)$$

where $H(\{I\})$ is the total energy of the lattice system in the state $\{I\}$. In equilibrium, $P(\{\gamma_f{}^j\}, \tau \to \infty) = exp(-\beta H(\{\gamma_f{}^j\}))/Q_s$; here, Q_s is the statistical sum of the system [66]. Expressions for $W_\alpha(\{I\} \to \{II\})$ are constructed with all the molecular features of the system taken into account [68].

The large dimension of system (A1) does not allow it to be used to study the dynamics of macroscopic systems by direct integration, so the kinetic equations are based on the functions of distributions of a lower order through which the distribution functions of high order are closed. To this end, instead of the full distribution function $P(\{\gamma_f{}^j\},\tau)$, the evolution of the system is described using a shortened method of defining it by time distribution function (correlators) determined by

$$\theta_{f_1 \ldots f_m}^{i_1 \ldots i_m}(\tau) = <\gamma_{f_1}^{i_1} \ldots \gamma_{f_m}^{i_m}> = \sum_{i_1=1}^{s} \ldots \sum_{i_N=1}^{s} \prod_{n=1}^{m} \gamma_{f_n}^{i_n} P(\{\gamma_f^i\},\tau), \qquad (A3)$$

The closed system of equations for the first ($\theta_f^i = <\gamma_f^i>$) and second ($\theta_{fg}^{ij} = <\gamma_f^i \gamma_g^j>$) correlations in the general form can be written as

$$\frac{d}{dt}\theta_f^i = \sum_\alpha \left[U_f^b(\alpha) - U_f^i(\alpha) \right] + \sum_h \sum_j \sum_\alpha \left[U_{fh}^{bd}(\alpha) - U_{fh}^{ij}(\alpha) \right], \qquad (A4)$$

$$\begin{aligned}
\frac{d}{dt}\theta_{fg}^{ij} &= \sum_\alpha \left[U_{fg}^{bd}(\alpha) - U_{fg}^{ij}(\alpha) \right] + P_{fg}^{ij} + P_{gf}^{ji}, \\
P_{fg}^{ij} &= \sum_\alpha \left[U_{fg}^{(b)j}(\alpha) - U_{fg}^{(i)j}(\alpha) \right] + \sum_h \sum_m \sum_\alpha \left[U_{hfg}^{(cb)j}(\alpha) - U_{hfg}^{(mi)j}(\alpha) \right]
\end{aligned} \qquad (A5)$$

where $U_f^i(\alpha)$ is the rate of the elementary single-site processes i $\leftrightarrow$ b (here $h \in z_f$), $U_{fg}^{ij}(\alpha)$, and α is the rate of the elementary two-site processes $i + j_\alpha \leftrightarrow b + d_\alpha$ ($h \in z$) on the nearest sites; the second term in P_{fg}^{ij} describes the stage $i + m \leftrightarrow b + c$ on neighboring sites f and h (and the term of P_{gf}^{ji} describes the stages on sites g and h and similar stages on sites f and h) [66,106]. All the rates of the elementary stages $U_f^i(\alpha)$ and $U_{fg}^{ij}(\alpha)$ are calculated in the framework of the theory of absolute reaction rates for non-ideal reaction systems written in the quasi-chemical approximation of the interparticle interaction. The rates of two-site stages $U_{fg}^{ij}(\alpha)$ have the form of Equations (2) and (3), and the rates of single-site stages $U_f^i(\alpha)$ in the QCA are expressed as

$$U_f^i(\alpha) = k_f^i(\alpha)\theta_f^i \Lambda_f^i(\alpha), \Lambda_f^i(\alpha) = \prod_{h \in z(f)} S_{fh}^i(\alpha), S_{fh}^i(\alpha) = \sum_{j=1}^{s} t_{fh}^{ij} \exp[\beta(\varepsilon_{ij}^*(\alpha) - \varepsilon_{ij})] \qquad (A6)$$

where $\Lambda_f^i(\alpha)$ is the non-ideality function. The functions t_{fh}^{ij} are defined in Section 2.2. For Equations (A4) and (A5) we have the normalizing ratios (see also Section 2.2), which are executed at any time.

Equations (A4) and (A5) describe elementary processes at the micro level. To move to macroscopic transport equations, two levels are distinguished: local, which preserves the written equations of chemical kinetics for a non-ideal reaction system, and macroscopic, which describes the transport of molecules and their properties on the hydrodynamic distance and time scales, which makes it possible to construct dissipative coefficients for non-ideal systems. Transfer processes in the LGM correspond to the stage of displacement of molecules in space through free sites (or vacancies), both for gas and for liquids and solids.

From the kinetic theory, it follows [23–26] that the transport coefficients characterize flows for small deviations from the equilibrium state of the system. We shall characterize the state of the fluid in a bulk by concentration θ and temperature T. The equilibrium distribution of the particles relative to each other will be calculated in the quasichemical approximation taking into account the direct correlations between the interacting particles. (Recall that the simpler mean-field approximation does not account for the correlation effects and does not provide a self-consistent description of the equilibrium distribution of the particles and the rates of elementary processes, so it cannot be used [66,106]).

To calculate the kinetic coefficients as usual [23,207,208], we select a plane in space 0 and consider the particle fluxes and the momentum and energy transferred by them. We use the concept of the average speed of moving particles w. We draw two planes parallel to plane 0 (with $x = 0$) at distances $x = \pm\rho$, where ρ is the mean free path of a particle; then, the properties of the particles S_m in these planes can be written as $S_m(x = +\rho) = S_m(x = 0) \pm \rho dS_m/dx$, where the symbol S_m means the concentration, the momentum (in the direction y, for example) or the energy of the particles moving along the axis X. The flow of quantity S_m through plane 0 consists of two oppositely directed movements of particles from the planes $x = \pm\rho$.

Two channels of transfer of momentum and energy operate in dense fluids. The first is connected with the movement of the particles, as in the rarefied phase, and the other is determined by collisions between particles. The particle in question may not cross plane 0 if its path is blocked by other particles in sites on plane 0 or a particle in close proximity on the other side of plane 0 prevents it from crossing a given plane. Both cases are not considered by the elementary kinetic theory of gases and the kinetic theory of condensed systems must be used [66,106].

In this way, the transmission of the property S_m through the selected plane is calculated, where S_m is modified by the following variables: (1) the number of molecules—for calculation of self-diffusion coefficient D_i^* and mass transfer coefficients D_{ij}, (2) the number of impulses—for the calculation of the shear η and bulk viscosity ξ, and (3) the amount of energy—for the calculation of the thermal conductivity κ. There are two channels of transfer of the property S_m: $\eta = \eta_1 + \eta_2$, $\kappa = \kappa_1 + \kappa_2$: (1) the transfer of molecules via a separated plane—the calculation of the coefficients D_i^*, D_{ij}, η_1, and κ_1; (2) transfer of the property (momentum and energy) through collisions—the calculation of the coefficients η_2, ζ, and κ_2 [68]. The results of calculating the dissipative coefficients are also shown in Figures 8–13.

References

1. Savage, P.E.; Gopalan, S.; Mizan, T.I.; Martino, C.J.; Brock, E.E. Reactions at supercritical conditions: Applications and fundamentals. *AIChE J.* **1995**, *41*, 1723–1778. [CrossRef]
2. Galkin, A.A.; Lunin, V.V. Subcritical and supercritical water: A universal medium for chemical reactions. *Russ. Chem. Rev.* **2005**, *74*, 21–35. [CrossRef]
3. Zalepugin, D.Y.; Tilkunova, N.A.; Chernyshova, I.V.; Polyakov, V.S. Development of Technologies based on Supercritical Fluids. *Sverhkriticheskie Flyuidy Teor. Prakt.* **2006**, *1*, 27–51.
4. Bogdan, V.I.; Koklin, A.E.; Kazansky, V.B. Regeneration of Deactivated Palladium Catalysts of Selec-tive Acetylene Hydrogenation be the Supercritical CO_2. *Sverhkriticheskie Flyuidy Teor. Prakt.* **2006**, *1*, 5–12.
5. Knez, Ž.; Markočič, E.; Leitgeb, M.; Primožič, M.; Hrnčič, M.K.; Škerget, M. Industrial applications of supercritical fluids: A review. *Energy* **2014**, *77*, 235–243. [CrossRef]
6. Gandhi, K.; Arora, S.; Kumar, A. Industrial applications of supercritical fluid extraction: A review. *Int. J. Chem. Stud.* **2017**, *5*, 336–340.
7. Mukhopadhyay, M. *Natural Extracts Using Supercritical Carbon Dioxide*; CRC Press: Boca Raton, FL, USA, 2000. [CrossRef]
8. Gopaliya, P.; Kamble, P.R.; Kamble, R.; Chauhan, C.S. A Review Article on Supercritical Fluid Chroma-tography. *Int. J. Pharma Res. Rev.* **2014**, *3*, 59–66.
9. Martinez, A.C.; Meireles, M.A.A. Application of Supercritical Fluids in the Conservation of Bioactive Compounds: A Review. *Food Public Health* **2020**, *10*, 26–34.
10. Bhardwaj, L.; Sharma, P.K.; Visht, S.; Garg, V.K.; Kumar, N. A review on methodology and application of supercritical fluid technology in pharmaceutical industry. *Pharm. Sin.* **2010**, *1*, 183–194.
11. Sapkale, G.N.; Patil, S.M.; Surwase, U.S.; Bhatbhage, P.K. Supercritical Fluid Extraction. *Int. J. Chem. Sci.* **2010**, *8*, 729–743.
12. Gumerov, F.M.; Gabitov, F.R.; Gazizov, R.A.; Bilalov, T.R.; Yarullin, R.S. Future Trends of Sub- and Supercritical Fluids Appli-cation in Biodiezel Fuel Production. *Sverhkriticheskie Flyuidy Teor. Prakt.* **2006**, *1*, 66–76.
13. Gumerov, F.M.; Sabirzyanov, A.N.; Gumerova, G.I. *Sub- and Supercritical Fluids in Processes of Polymer Processing*; Fan: Kazan, Russia, 2007.
14. Myasoedov, B.F.; Kulyako, Y.M.; Shadrin, A.Y.; Samsonov, M.D. Supercritical Fluid Extraction of Radionuclides. *Sverhkriticheskie Flyuidy Teor. Prakt.* **2007**, *2*, 5–24.
15. Ahmad, T.; Masoodi, F.A.; Rather, S.A.; Wani, S.M.; Gull, A. Supercritical Fluid Extraction: A Review. *J. Biol. Chem. Chron.* **2019**, *5*, 114–122. [CrossRef]
16. Parhi, R.; Suresh, P. Supercritical Fluid Technology: A Review. *J. Adv. Pharm. Sci. Technol.* **2013**, *1*, 13–36. [CrossRef]
17. Zhou, J.; Gullón, B.; Wang, M.; Gullón, P.; Lorenzo, J.M.; Barba, F.J. The Application of Supercritical Fluids Technology to Recover Healthy Valuable Compounds from Marine and Agricultural Food Processing By-Products: A Review. *Processes* **2021**, *9*, 357. [CrossRef]
18. Aymonier, C.; Loppinet-Serani, A.; Reverón, H.; Garrabos, Y.; Cansell, F. Review of supercritical fluids in inorganic materials science. *J. Supercrit. Fluids* **2006**, *38*, 242–251. [CrossRef]
19. Manjare, S.D.; Dhingra, K. Supercritical fluids in separation and purification: A review. *Mater. Sci. Energy Technol.* **2019**, *2*, 463–484. [CrossRef]
20. De Groot, S.R.; Mazur, P. *Nonequilibrium Thermodynamics*; North-Holland: Amsterdam, The Netherlands, 1962.
21. Haase, R. *Thermodynamik der Irreversible Processe*; Dr. Dierrich Steinkopff Verlag: Darmstadt, Germany, 1963.
22. Prigogine, I.; Kondepudi, D. *Modern Thermodynamics: From Heat Engines to Dissipative Structures*; Wiley: New York, NY, USA, 1998.
23. Hirschfelder, J.O.; Curtiss, C.F.; Bird, R.B. *Molecular Theory of Gases and Liquids*; Wiley: New York, NY, USA, 1954.

24. Balescu, R. *Equilibrium and Nonequilibrium Statistical Mechanics*; Wiley-Interscience Publication Jonh Wiley and Sons: Hoboken, NJ, USA, 1975.
25. Landau, L.D.; Lifshitz, E.M. *Course of Theoretical Physics*; Fluid Mechanics; Pergamon: New York, NY, USA, 1987; Volume 6.
26. Bird, R.; Stewart, W.E.; Lightfoot, E.N. *Transport Phenomena*; Wiley: New York, NY, USA, 1960.
27. Collins, R. *Fluid Flow through Porous Materials*; Wiley: New York, NY, USA, 1964.
28. Sheydegger, A.E. *The Physics of Flow through Porous Media*, 3rd ed.; University of Toronto Press: Toronto, ON, Canada, 1974.
29. Nigmatulin, R.I. *Fundamentals of Mechanics of Heterogeneous Media*; Nauka: Moscow, Russia, 1973.
30. Nicholaevsky, V.N. *Mechanics of Porous and Fractured Media*; Nedra: Moscow, Russia, 1984.
31. Glasstone, S.; Laidler, K.J.; Eyring, H. *The Theory of Rate Processes: The Kinetics of Chemical Reactions, Viscosity, Diffusion, and Electrochemical Phenomena*; Van Nostrand: New York, NY, USA, 1941.
32. Entelis, S.G.; Tiger, R.P. *Reaction Kinetics in the Liquid Phase*; Khimiya: Moscow, Russia, 1973.
33. Moelwyn-Hughes, E.A. *The Chemical Statics and Kinetics of Solutions*; Academic Press: London, UK; New York, NY, USA, 1971.
34. Benson, S.W. *The Foundations of Chemical Kinetics*; McGrow–Hill: New York, NY, USA, 1960.
35. Kiperman, S.L. *Foundations of Chemical Kinetics in Heterogeneous Catalysis*; Khimiya: Moscow, Russia, 1979.
36. Volmer, M.; Flood, H. Tröpfchenbildung in Dämpfen. *Z. Phys. Chem.* **1934**, *170A*, 273–285. [CrossRef]
37. Volmer, M.; Vollmer, M. *Kinetics of New Phase Formation*; Plenum: New York, NY, USA, 1983.
38. Fuchs, N.A. *Mechanic of Aerosols*; Khimiya: Moscow, Russia, 1959.
39. Erkey, C.; Türk, M. Modeling of particle formation in supercritical fluids (SCF). *Supercrit. Fluid Sci. Technol.* **2021**, *8*, 239–259. [CrossRef]
40. Helfgen, B.; Türk, M.; Schaber, K. Hydrodynamic and aerosol modelling of the rapid expansion of supercritical solutions (RESS-process). *J. Supercrit. Fluids* **2003**, *26*, 225–242. [CrossRef]
41. Fletcher, K. *Methods in Computational Fluid Dynamics*; Wiley: New York, NY, USA, 1991; Volumes 1 and 2.
42. Hoffman, K.A.; Steve, T.C. *Computation Fluid Dynamics*; Engineering Education System: Wichite, KS, USA, 2000; Volumes 1–3.
43. Strang, G.; Fix, G. *Theory of Finite Element Method*; Wiley: New York, NY, USA, 1977.
44. Zinkevych, O.; Morgan, K. *Finite Elements and Approximation*; Wiley: New York, NY, USA, 1986.
45. Tovbin, Y.K. (Ed.) *Method of Molecular Dynamics in Physical Chemistry*; Nauka: Moscow, Russia, 1996.
46. Allen, M.P.; Tildesley, D.J. *Computer Simulation of Liquids*; Claredon Press: Oxford, UK, 2002.
47. Haile, J.M. *Molecular Dynamics Simulation: ElementaryMethods*; Wiley: New York, NY, USA, 1992.
48. Coccotti, G.; Hoover, W.G. (Eds.) *Molecular Dynamics Simulation of Statistical Mechanics Systems*; North-Holland: Amsterdam, The Netherlands, 1986; 610p.
49. Evans, D.J.; Morriss, G.P. *Statistical Mechanics of Nonequilibrium Liquids*, 2nd ed.; Cambridge University Press: Cambridge, UK, 2008.
50. Binder, K. (Ed.) *Monte Carlo Methods in Statistical Physics*; Springer: Berlin/Heidelberg, Germany; New York, NY, USA, 1979.
51. Nicolson, D.; Parsonage, N.G. *Computer Simulation and The Statistical Mechanics of Adsorption*; Academic Press: New York, NY, USA, 1982.
52. Allen, M.P. Introduction to Monte Carlo simulations. In *Observation, Prediction and Simulation of Phase Transitions in Complex Fluids*; Baus, M., Rull, L.F., Ryckaert, J.-P., Eds.; Kluwer Academic Publishers: Boston, MA, USA, 1995; p. 339.
53. Jorgensen, W.L. Monte Carlo simulations for liquids. In *Encyclopedia of Computational Chemistry*; Schleyer, P.V.R., Ed.; Wiley: New York, NY, USA, 1998; p. 1754.
54. Jorgensen, W.L.; Tirado-Rives, J. Monte Carlo vs. Molecular Dynamics for Conformational Sampling. *J. Phys. Chem.* **1996**, *100*, 14508–14513. [CrossRef]
55. Binder, K.; Landau, D.P. Capillary condensation in the lattice gas model: A Monte Carlo study. *J. Chem. Phys.* **1992**, *96*, 1444–1454. [CrossRef]
56. Lemak, A.S.; Balabaev, N.K. A Comparison Between Collisional Dynamics and Brownian Dynamics. *Mol. Simul.* **1995**, *15*, 223–231. [CrossRef]
57. Lemak, A.S.; Balabaev, N.K. Molecular dynamics simulation of a polymer chain in solution by collisional dynamics method. *J. Comput. Chem.* **1996**, *17*, 1685–1695. [CrossRef]
58. Bird, G.A. *Molecular Gas Dynamics*; Oxford University Press: Oxford, UK, 1976.
59. Lebowitz, J.K.; Montroll, E.W. (Eds.) *Nonequilibrium Phenomena I. The Boltzman Equation. Studies in Statistical Mechanics*; North-Holland Publishing Company: Amsterdam, The Netherlands; New York, NY, USA; Oxford, UK, 1983.
60. Gardiner, C.W. *Handbook of Stochastic Methods (for Physics, Chemistry and Natural Science)*, 2nd ed.; Haken, H., Ed.; Springer Series in Synergetics; Springer: Berlin/Heidelberg, Germany; New York, NY, USA; Tokyo, Japan, 1985; Volume 13.
61. Succi, S. *The Lattice Boltzmann Equation for Fluid Dynamics and Beyond*; Oxford University Press: Oxford, UK, 2001.
62. Mohamad, A.A. *Lattice Boltzmann Method: Fundamentals and Engineering. Applications with Computer Codes*; Springer: Berlin/Heidelberg, Germany, 2011.
63. Timm, K.; Kusumaatmaja, H.; Kuzmin, A.; Shardt, O.; Silva, G.; Viggen, E. *The Lattice Boltzmann Method: Principles and Practice*; Springer: Berlin/Heidelberg, Germany, 2016.
64. Hill, T.L. *Statistical Mechanics. Principles and Selected Applications*; McGraw–Hill: New York, NY, USA, 1956.
65. Huang, K. *Statistical Mechanics*; Wiley: New York, NY, USA, 1963.

66. Tovbin, Y.K. *Theory of Physicochemical Processes at the Gas–Solid Interface*; CRC: Boca Raton, FL, USA, 1991.

67. Tovbin, Y.K. Molecular Approach to Micro-dynamics: Transfer of Molecules in Narrow Pores. *Russ. J. Phys. Chem. A* **2002**, *76*, 64–70.

68. Tovbin, Y.K. *Molecular Theory of Adsorption in Porous Solids*; CRC: Boca Raton, FL, USA, 2017.

69. Wolfram, S. Cellular Automata. *Los Alamos Sci.* **1983**, *9*, 2.

70. Toffoli, T.; Margolus, N. *Cellular Automata Machines*; The MIT Press: Cambridge, MA, USA, 1987.

71. Wolfram, S. *A New Kind of Science*; Wolfram Media: Champaign, IL, USA, 2002.

72. Kier, L.B.; Seybold, P.G.; Cheng, C.-K. *Cellular Automata Modeling of Chemical Systems*; Springer: Dordrecht, The Netherlands, 2005; 175p.

73. Wolf-Gladrow, D.A. *Lattice-Gas Cellular Automata and Lattice Boltzmann Models: An Introduction*; Springer: Berlin/Heidelberg, Germany, 2000.

74. Goto, M.; Roy, B.C.; Kodama, A.; Hirose, T. Modeling Supercritical Fluid Extraction Process Involving Solute-Solid Interaction. *J. Chem. Eng. Jpn.* **1998**, *31*, 171–177. [CrossRef]

75. Oliveira, E.L.G.; Silvestre, A.J.D.; Silva, C.M. Review of kinetic models for supercritical fluid extraction. *Chem. Eng. Res. Des.* **2011**, *89*, 1104–1117. [CrossRef]

76. Al-Jabari, M. Kinetic models of supercritical fluid extraction. *J. Sep. Sci.* **2002**, *25*, 477–489. [CrossRef]

77. Sovová, H. Rate of the vegetable oil extraction with supercritical CO_2—I. Modelling of extraction curves. *Chem. Eng. Sci.* **1994**, *49*, 409–414. [CrossRef]

78. Sovová, H. Mathematical model for supercritical fluid extraction of natural products and extraction curve evaluation. *J. Supercrit. Fluids* **2005**, *33*, 35–52. [CrossRef]

79. Sovová, H. Steps of supercritical fluid extraction of natural products and their characteristic times. *J. Supercrit. Fluids* **2012**, *66*, 73–79. [CrossRef]

80. Rai, A.; Punase, K.D.; Mohanty, B.; Bhargava, R. Evaluation of models for supercritical fluid extraction. *Int. J. Heat Mass Transf.* **2014**, *72*, 274–287. [CrossRef]

81. Promraksa, A.; Siripatana, C.; Rakmak, N.; Chusri, N. Modeling of Supercritical CO_2 Extraction of Palm Oil and Tocopherols Based on Volumetric Axial Dispersion. *J. Supercrit. Fluids* **2020**, *166*, 105021. [CrossRef]

82. Roodpeyma, M.; Street, C.; Guigard, S.E.; Stiver, W.H. A hydrodynamic model of a continuous supercritical fluid extraction system for the treatment of oil contaminated solids. *Sep. Sci. Technol.* **2017**, *53*, 44–60. [CrossRef]

83. Garcia, E.C.C.; Rabi, J.A. Lattice-Boltzmann Simulation of Supercritical Fluid Extraction of Essential Oil from Gorse: Influence of Process Parameters on Yields. In Proceedings of the 14th WSEAS International Conference on Mathematics and Computers in Biology and Chemistry, Baltimore, MD, USA, 17–19 September 2013; pp. 62–67, ISBN 978-960-474-333-9.

84. Duba, K.S.; Fiori, L. Supercritical fluid extraction of vegetable oils: Different approach to modeling the mass transfer kinetics. *Chem. Eng. Trans.* **2015**, *43*, 1051–1056. [CrossRef]

85. Markom, M.; Hassim, N.; Hasan, M.; Daud, W.R.W. Modeling of supercritical fluid extraction by enhancement factor of cosolvent mixtures. *Sep. Sci. Technol.* **2020**, *56*, 1290–1302. [CrossRef]

86. Gadkari, P.V.; Balaraman, M. Mass transfer and kinetic modelling of supercritical CO_2 extraction of fresh tea leaves (*Camellia sinensis* L.). *Braz. J. Chem. Eng.* **2017**, *34*, 799–810. [CrossRef]

87. Dimić, I.; Pezo, L.; Rakić, D.; Teslić, N.; Zeković, Z.; Pavlić, B. Supercritical Fluid Extraction Kinetics of Cherry Seed Oil: Kinetics Modeling and ANN Optimization. *Foods* **2021**, *10*, 1513. [CrossRef]

88. Cabeza, A.; Sobrón, F.; García-Serna, J.; Cocero, M. Simulation of the supercritical CO_2 extraction from natural matrices in packed bed columns: User-friendly simulator tool using Excel. *J. Supercrit. Fluids* **2016**, *116*, 198–208. [CrossRef]

89. Amani, M.; Ardestani, N.S.; Honarvar, B. Experimental Optimization and Modeling of Supercritical Fluid Extraction of Oil from *Pinus gerardiana*. *Chem. Eng. Technol.* **2021**, *44*, 578–588. [CrossRef]

90. Bushnaq, H.; Krishnamoorthy, R.; Abu-Zahra, M.; Hasan, S.W.; Taher, H.; Alomar, S.Y.; Ahmad, N.; Banat, F. Supercritical Tech-nology-Based Date Sugar Powder Production: Process Modeling and Simula-tion. *Processes* **2022**, *10*, 257. [CrossRef]

91. Wilhelmsen, Ø.; Aasen, A.; Skaugen, G.; Aursand, P.; Austegard, A.; Aursand, E.; Gjennestad, M.A.; Lund, H.; Linga, G.; Hammer, M. Thermodynamic Modeling with Equations of State: Present Challenges with Established Methods. *Ind. Eng. Chem. Res.* **2017**, *56*, 3503–3515. [CrossRef]

92. Salmani, H.J.; Karkhanechi, H.; Moradi, M.R.; Matsuyama, H. Thermodynamic modeling of binary mixtures of ethylenediamine with water, methanol, ethanol, and 2-propanol by association theory. *RSC Adv.* **2022**, *12*, 32415–32428. [CrossRef]

93. Congedo, P.M.; Rodio, M.G.; Tryoen, J.; Abgrall, R. *Reliable and Robust Thermodynamic Model for Liquid-Vapor Mixture*; [Re-search Report] RR-8439, INRIA. hal-00922816; HAL: Lyon, France, 2013.

94. Alanazi, A.; Bawazeer, S.; Ali, M.; Keshavarz, A.; Hoteit, H. Thermodynamic modeling of hydrogen–water systems with gas impurity at various conditions using cubic and PC-SAFT equations of state. *Energy Convers. Manag. X* **2022**, *15*, 100257. [CrossRef]

95. Matos, I.Q.; Varandas, J.S.; Santos, J.P. Thermodynamic Modeling of Azeotropic Mixtures with [EMIM][TfO] with Cubic-Plus-Association and Cubic EOSs. *Braz. J. Chem. Eng.* **2018**, *35*, 363–372. [CrossRef]

96. Yeoh, H.S.; Chong, G.H.; Azahan, N.M.; Rahman, R.A.; Choong, T.S.Y. Solubility Measurement Method and Mathematical Modeling in Supercritical Fluids. *Eng. J.* **2013**, *17*, 67–78. [CrossRef]

97. Peng, D.Y.; Robinson, D.B. A New Two-Constant Equation of State. *Ind. Eng. Chem. Fundam.* **1976**, *15*, 59–64. [CrossRef]

98. Mukhopadhyay, M.; Rao, G.V.R. Thermodynamic modeling for supercritical fluid process design. *Ind. Eng. Chem. Res.* **1993**, *32*, 922–930. [CrossRef]

99. Bilalov, T.R.; Zavjalova, N.B.; Gumerov, F.M. Phase Diagram of the Supercritical Carbon Dioxide–Ethylcarbitol System. *Sverhkriticheskie Flyuidy Teor. Prakt.* **2019**, *14*, 27–33. [CrossRef]

100. Bilalov, T.R.; Gumerov, F.M.; Khairutdinov, V.F.; Khabriev, I.S.; Gabitov, F.R.; Zaripov, Z.I.; Ga-niev, A.A.; Mazanov, C.V. Phase Equilibrium of the Binary System Propylene Glycol—Propane/Butane. *Sverhkriticheskie Flyuidy Teor. Prakt.* **2020**, *15*, 79–86. [CrossRef]

101. Zakharov, A.A.; Bilalov, T.R.; Gumerov, F.M. Solubility of Ammonium Palmitate in Supercritical Carbon Dioxide. *Sverhkriticheskie Flyuidy Teor. Prakt.* **2015**, *12*, 60–70. [CrossRef]

102. Gumerov, F.M.; Gabitov, F.R.; Gazizov, R.A.; Bilalov, T.R.; Yakushev, I.A. Determination of Phase Equilibria Parameters in Binary Systems Containing Components of Biodiezel Fuel and Supercritical Carbon Dioxde. *Sverhkriticheskie Flyuidy Teor. Prakt.* **2006**, *1*, 89–100.

103. Bazaev, A.R.; Karabekova, B.K.; Abdurashidova, A.A. p, ρ, T, and x dependences for supercritical water-aliphatic alcohol mixtures. *Sverhkriticheskie Flyuidy Teor. Prakt.* **2013**, *8*, 11–38. [CrossRef]

104. Durakovic, G.; Skaugen, G. Analysis of Thermodynamic Models for Simulation and Optimisation of Organic Rankine Cycles. *Energies* **2019**, *12*, 3307. [CrossRef]

105. Bruno, T.J.; Ely, J.F. (Eds.) *Supercritical Fluid Technology: Reviews in Modem Theory and Applications*; CRC Press Taylor & Francis Group: Boca Raton, FL, USA, 1991.

106. Tovbin, Y. Lattice-gas model in kinetic theory of gas-solid interface processes. *Prog. Surf. Sci.* **1990**, *34*, 1–235. [CrossRef]

107. Temkin, M.I. Kinetics of ammonia synthesis at high pressures. *Zhurnal Fiz. Khimii* **1950**, *24*, 1312–1323.

108. Marcus, R.A. Chemical and Electrochemical Electron-Transfer Theory. *Annu. Rev. Phys. Chem.* **1964**, *15*, 155–196. [CrossRef]

109. Dogonadze, R.R.; Kuznetsov, A.M. Kinetics of heterogeneous chemical reactions in solutions. *Itogi Nauki Tekh. Ser. Kinet. Katal.* **1978**, *5*.

110. Kuznetsov, A.M.; Ulstrup, J. *Electron Transfer in Chemistry and Biology*; John Wiley & Sons, Ltd.: Chichester, UK, 1999; 350p.

111. Clausius, R. *Mechanical Theory of Heat*; John van Voorst: London, UK, 1867.

112. Tovbin, Y.K. Second Law of Thermodynamics, Gibbs' Thermodynamics, and Relaxation Times of Thermodynamic Parameters. *Russ. J. Phys. Chem. A* **2021**, *95*, 637–658. [CrossRef]

113. Gibbs, J.W. On the Equilibrium of Heterogeneous Substances. *Trans. Conn. Acad. Arts Sci.* **1878**, *16*, 441–458. [CrossRef]

114. Gibbs, J.W. *The Collected Works of J. W. Gibbs, in 2 Volumes*; Longmans Green: New York, NY, USA, 1928; Volume 1.

115. Tovbin, Y.K. *Small Systems and Fundamentals of Thermodynamics*; CRC Press: Boca Raton, FL, USA, 2019.

116. Fisher, M.E. *The Nature of Critical Points*; Lectures in Theoretical Physics; University of Colorado Press: Boulder, CO, USA, 1965; Volume VII.

117. Novikov, I.I. *Equation of States of Gas and Liquids*; Nauka: Moscow, Russia, 1965.

118. Walas, S.M. *Phase Equilibria in Chemical Engineering*; The C.W. Nofsinger Company Butterworth Publisher: Boston, MA, USA; London, UK; Wellington, New Zealand; Durban, South Africa; Toronto, ON, Canada, 1985.

119. Guggenheim, E.A. *Mixtures*; Clarendon: Oxford, UK, 1952.

120. Barker, J.A. Cooperative Orientation Effects in Solutions. *J. Chem. Phys.* **1952**, *20*, 1526–1532. [CrossRef]

121. Prigogine, I.P. *The Molecular Theory of Solutions*; Interscience: Amsterdam, The Netherlands; New York, NY, USA, 1957.

122. Smirnova, N.A. *The Molecular Theory of Solutions*; Khimiya: Leningrad, Russia, 1987.

123. Tovbin, Y.K.; Titov, S.V. Role of local environment relaxation in calculating the reaction rates for nonideal reaction systems. *Sverhkriticheskie Flyuidy Teor. Prakt. B* **2011**, *6*, 35–48. [CrossRef]

124. Tovbin, Y.K. Theory of adsorption—Desorption kinetics on flat heterogeneous surfaces. In *Equiliblia and Dynamics of Gas Adsorption on Heterogeneous Solid Surfaces*; Rudzinski, W., Steele, W.A., Zgrablich, G., Eds.; Elsevier: Amsterdam, The Netherlands, 1997; pp. 201–284.

125. Lifshits, I.M. To the theory of real solutions. *Zh. Eksp. Teor. Fiz.* **1939**, *9*, 481–499.

126. Krichevskii, I.R. *Phase Equilibria at High Pressures*; Goskhimizdat: Moscow, Russia, 1963.

127. Borovskii, I.B.; Gurov, K.P.; Marchukova, Y.E.; Ugaste, Y.E. *Interdiffusion Processes in Alloys*; Gurov, K.P., Ed.; Nauka: Moscow, Russia, 1973.

128. Gurov, K.P.; Kartashkin, B.A.; Ugaste, Y.E. *Mutual Diffusion in Multicomponent Metal Alloys*; Nauka: Moscow, Russia, 1981.

129. Lazarev, A.V.; Tatarenko, P.A.; Tatarenko, K.A. Gas-Dynamic Model of the Expansion of a Pulse Jet of Supercritical Carbon Dioxide: The Strategy of the Experiment. *Sverhkriticheskie Flyuidy Teor. Prakt.* **2017**, *12*, 3–13. [CrossRef]

130. Nikolaev, P.N. The singular points and phase diagram of the supercritical region of a substance. *Mosc. Univ. Phys. Bull.* **2014**, *69*, 146–151. [CrossRef]

131. Semenchenko, V.K. *Selected Chapters in Theoretical Physics*; EDUCATION: Moscow, Russia, 1966.

132. Nishikawa, K.; Kusano, K.; Arai, A.A.; Morita, T. Density fluctuation of a van der Waals fluid in supercritical state. *J. Chem. Phys.* **2003**, *118*, 1341–1346. [CrossRef]

133. Tovbin, Y.K. Molecular Aspects of Lattice Models of Liquid and Adsorption Systems. *Russ. J. Phys. Chem. A* **1995**, *69*, 105–113.

134. Tovbin, Y.K. Modern State of the Lattice- Theory of Adsorption. *Russ. J. Phys. Chem. A* **1998**, *72*, 675–683.

135. Bogolyubov, N.N. *Problems of Dynamic Theory in Statistical Physics*; Gostekhizdat: Moscow, Russia, 1946.

136. Fisher, I.Z. *Statistical Theory of Liquids*; Chicago University: Chicago, IL, USA, 1964.
137. Croxton, C.A. *Liquid State Physics–A Statistical Mechanical Introduction*; Cambridge University Press: Cambridge, UK, 1974.
138. Martynov, G.A. *Classical Static Physics*; Fluid Theory; Intellekt: Dolgoprudnyi, Russia, 2011.
139. Tovbin, Y.K.; Senyavin, M.M.; Zhidkova, L.K. Modified cell theory of fluids. *Russ. J. Phys. Chem. A* **1999**, *73*, 245–253.
140. Ono, S.; Kondo, S. *Molecular Theory of Surface Tension in Liquids, Handbuch der Physik*; Springer: Berlin/Heidelberg, Germany, 1960.
141. Zagrebnov, V.A.; Fedyanin, B.K. Spin-phonon interaction in the ising model. *Theor. Math. Phys.* **1972**, *10*, 84–93. [CrossRef]
142. Plakida, N.M. The method of two-time Green's functions in the theory of anharmonic crystals. In *Statistical Physics and Quantum Field Theory*; Nauka: Moscow, Russia, 1973; pp. 205–240.
143. Batalin, O.Y.; Tovbin, Y.K.; Fedyanin, V.K. Equilibrium properties of a liquid in a modified lattice model. *Zhurnal Fiz. Khimii* **1979**, *53*, 3020–3023.
144. Fedyanin, V.K. Thermodynamics of a Lattice System of Particles of Different Sizes with Contact Are-as of Different Types. In *Theoretical Methods for Describing the Properties of Solutions*; Interschool Collection of Scientific Works: Ivanovo, Russia, 1987; pp. 40–44.
145. Tovbin, Y.K. Allowing for Intermolecular Vibrations in the Thermodynamic Functions of a Liquid Inert Gas. *Russ. J. Phys. Chem. A* **2019**, *93*, 603–613. [CrossRef]
146. Barker, J.A. *Lattice Theories of the Liquid State*; Pergamon Press: Oxford, UK, 1963.
147. Shakhparonov, M.I. *Introduction to the Molecular Theory of Solutions*; GITTL: Moscow, Russia, 1956.
148. Morachevskii, A.G.; Smirnova, N.A.; Piotrovskaya, E.M.; Kuranov, G.L.; Balashova, I.M.; Pukinskiy, I.B.; Alekseeva, M.V.; Viktoriv, A.I. *Thermodynamics of Liquid-Vapour Equilibrium*; Morachevskii, A.G., Ed.; Khimiya: Leningrad, Russia, 1989.
149. Kaplan, I.G. *Introduction to the Theory of Molecular Interactions*; Nauka: Moscow, Russia, 1982; 312p.
150. Kiselev, A.V.; Poshkus, D.P.; Yashin, Y.I. *Molecular Basis of Adsorption Chromatography*; Khimiya: Moscow, Russia, 1986; 269p.
151. Egorov, B.V.; Komarov, V.N.; Markachev, Y.E.; Tovbin, Y.K. Concentration dependence of viscosity under conditions of its clustering. *Russ. J. Phys. Chem. A* **2000**, *74*, 778–783.
152. Tovbin, Y.K.; Komarov, V.N. Calculation of Compressibility and Viscosity of Non-ideal Gases with if Framework of the Lattice Model. *Russ. J. Phys. Chem. A* **2001**, *75*, 490–495.
153. Komarov, V.N.; Tovbin, Y.K. Self-Consistent Calculation of the Compressibility and Viscosity of Dense Gases in the Lattice-Gas Model. *High Temp.* **2003**, *41*, 181–188. [CrossRef]
154. Tovbin, Y.K. Many-Particle Interactions in Equilibrium Theories of Adsorption and Absorption. *Zhurnal Fiz. Khimii* **1987**, *61*, 2711–2716.
155. Tovbin, Y.K.; Komarov, V.N. Calculation of the compressibility coefficient of a mixture of dense gases. *Russ. J. Phys. Chem. A* **2005**, *79*, 1807–1813.
156. Reid, R.C.; Sherwood, T.K. *The Properties of Gases and Liquids. (The Restimation and Correlation)*; MeGrav-Hill Boch Company: New York, NY, USA; San Francisco, CA, USA; Toronto, ON, Canada; London, UK; Sydney, Austria, 1966.
157. Sengers, J.M.H.L.; Klein, M.; Gallagher, J. *Pressure—Volume Temperature Relationships of Gases—Virial Coefficients*, 3rd ed.; American Institute of Physics Handbook; American Institute of Physics: New York, NY, USA, 1972; 2364p.
158. Rabinovich, V.A.; Vasserman, A.A.; Nedostup, V.I.; Veksler, L.S. *Thermophysical Properties of Neon, Argon, Krypton, and Xenon*; Standartgiz: Moscow, Russia, 1976.
159. Crain, E.W.; Santag, R.E. The P-V-T Behavior of nitrogen, argon and their mixtures. *Adv. Cryog. Eng.* **1966**, *11*, 379.
160. Komarov, V.N.; Rabinovich, A.B.; Tovbin, Y.K. Calculation of concentration dependences of the transport characteristics of binary mixtures of dense gases. *High Temp.* **2007**, *45*, 463–472. [CrossRef]
161. Tovbin, Y.K. Lattice gas model in the molecular-statistical theory of equilibrium systems. *Russ. J. Phys. Chem. A* **2005**, *79*, 1903–1920.
162. Tovbin, Y.K. Calculation of Adsorption Characteristics in the "Quasi-Point" Approximation Based on the Lattice Gas Model. *Russ. J. Phys. Chem. A* **1998**, *72*, 2053–2058.
163. Tovbin, Y.K.; Rabinovich, A.B.; Votyakov, E.V. Calibration functions in approximate methods for calculating the equilibrium of adsorption characteristics. *Russ. Chem. Bull.* **2002**, *51*, 1667–1674. [CrossRef]
164. Tovbin, Y.K.; Rabinovich, A.B. Phase Diagrams of Adsorption Systems and Calibration Functions in the Lattice-Gas Model. *Langmuir* **2004**, *20*, 6041–6051. [CrossRef]
165. Patashinskii, A.Z.; Pokrovskii, V.L. *Fluctuation Theory of Phase Transitions*; Nauka: Moscow, Russia, 1975.
166. Stanley, H.E. *Introduction to Phase Transitions and Critical Phenomena*; Clarendon: Oxford, UK, 1971.
167. Ma, S.-K. *Modern Theory of Critical Phenomena*; W.A.Benjamin, Inc.: London, UK, 1976.
168. Tsiklis, D.S. *Dense Gases*; Khimiya: Moscow, Russia, 1977.
169. Kruse, A.; Dinjus, E. Hot compressed water as reaction medium and reactant: Properties and synthesis reactions. *J. Supercrit. Fluids* **2007**, *39*, 362–380. [CrossRef]
170. Rabinovich, A.B.; Tovbin, Y.K. Supercritical fluid effect on the rates of elementary bimolecular reactions. *Kinet. Catal.* **2011**, *52*, 471–479. [CrossRef]
171. Cooper, A.I. Polymer synthesis and processing using supercritical carbon dioxide. *J. Mater. Chem.* **2000**, *10*, 207–234. [CrossRef]
172. McHugh, M.A.; Krukonis, V.J. *Supercritical Fluid Extraction: Principles and Practice*; Butter-Worth Publishers: Stoneham, MA, USA, 1994.

173. Rabinovich, A.B.; Tovbin, Y.K. Effect of a supercritical fluid on the characteristics of sorption processes. *Russ. Chem. Bull.* **2010**, *59*, 1–6. [CrossRef]
174. Franck, E.U. Physicochemical Properties of Supercritical Solvents (Invited Lecture). *Berichte Bunsengesellschaft Physikalische Chemie* **1984**, *88*, 820–825. [CrossRef]
175. Lemenovskii, D.A.; Yurin, S.A.; Timofeev, V.V.; Popov, V.K.; Bagratashvili, V.N.; Gorbaty, Y.E.; Brusova, G.P.; Lunin, V.V. Reactions of Ozone with Organic Substrates in Supercritical Carbon Dioxide. *Sverhkriticheskie Flyuidy Teor. Prakt.* **2007**, *2*, 30–42.
176. Tovbin, Y.K. Molecular grounds of the calculation of equilibrium and transport characteristics of inert gases and liquids in complex narrow-pore systems. *Russ. Chem. Bull.* **2003**, *52*, 869–881. [CrossRef]
177. Anisimov, M.A.; Rabinovich, V.A.; Sychev, V.V. *Thermodynamics of Critical State*; Energoatomizdat: Moscow, Russia, 1990.
178. Chapman, S.; Cowling, T. *The Mathematical Theory of Nonequilibrium Gases*; Cambridge University Press: Cambridge, UK, 1953.
179. DiMarzio, E.A. Statistics of Orientation Effects in Linear Polymer Molecules. *J. Chem. Phys.* **1961**, *35*, 658–669. [CrossRef]
180. Chandrasekhar, S. *Liquid Crystals*; Cambridge University: Cambridge, UK, 1977.
181. Bazarov, I.P.; Gevorkyan, E.V. *Statistical Theory of Solid and Liquid Crystals*; Moscow State University: Moscow, Russia, 1983.
182. Tovbin, Y.K. Refinement of taking into account molecule sizes in the lattice gas model. *Russ. J. Phys. Chem. A* **2012**, *86*, 705–708. [CrossRef]
183. Tovbin, Y.K. Possibilities of the Molecular Modeling of Kinetic Processes under Supercritical Conditions. *Russ. J. Phys. Chem. A* **2021**, *95*, 429–444. [CrossRef]
184. Vukalovich, M.P.; Altunin, V.V. *Thermophysical Properties of Carbon Dioxide*; Atomizdat: Moscow, Russia, 1965.
185. Titov, S.V.; Tovbin, Y.K. Lattice model of a polar liquid. *Russ. Chem. Bull.* **2011**, *60*, 11–19. [CrossRef]
186. Titov, S.V.; Tovbin, Y.K. A molecular model of water based on the lattice gas model. *Russ. J. Phys. Chem. A* **2011**, *85*, 194–201. [CrossRef]
187. Bell, G.M. Statistical mechanics of water: Lattice model with directed bonding. *J. Phys. C Solid State Phys.* **1972**, *5*, 889–905. [CrossRef]
188. Bell, G.M.; Salt, D.W. Three-dimensional lattice model for water/ice system. *J. Chem. Soc. Faraday Trans.* **1976**, *72*, 76–86. [CrossRef]
189. Franks, F. (Ed.) *Water: A Comprehensive Treatise*; Plenum: New York, NY, USA; London, UK, 1972; Volume 1.
190. Eisenberg, D.; Kautsman, V. *Structure and Properties of Water*; Gidrometeoizdat: Leningrad, Russia, 1975.
191. Tovbin, Y.K.; Titov, S.V. Role of local environment relaxation in calculating the rates of elementary processes in vapor-liquid systems. *Russ. J. Phys. Chem. A* **2013**, *87*, 185–190. [CrossRef]
192. Frenkel, Y.I. *Kinetic Theory of Liquids*; Oxford University: London, UK, 1946.
193. Angell, C.A. *Water: A Comprehensive Treatise*; Franks, F., Ed.; Plenum: New York, NY, USA, 1978; Volume 7, p. 23.
194. Malenkov, G.; Tytik, D.; Zheligovskaya, E. Structural and dynamic heterogeneity of computer simulated water: Ordinary, supercooled, stretched and compressed. *J. Mol. Liq.* **2003**, *106*, 179–198. [CrossRef]
195. Tovbin, Y.K. Kinetic equations for processes of local rearrangement of molecular systems. *Russ. J. Phys. Chem. B* **2011**, *5*, 256–270. [CrossRef]
196. Tovbin, Y.K. Taking environment into account in the theory of liquid-phase reaction rates with electron transfer in the discrete solvent model. *Russ. J. Phys. Chem. A* **2011**, *85*, 238–244. [CrossRef]
197. Tovbin, Y.K. Kinetic equation for the processes of local reorganization of molecular systems with charged species. *Russ. J. Phys. Chem. B* **2012**, *6*, 716–729. [CrossRef]
198. Tovbin, Y.K. Local equations of state in nonequilibrium heterogeneous physicochemical systems. *Russ. J. Phys. Chem. A* **2017**, *91*, 403–424. [CrossRef]
199. Gaydamaka, S.N.; Timofeev, V.V.; Guryev, Y.V.; Lemenovskiy, D.A.; Brusova, G.P.; Parenago, O.O.; Bagratashvili, V.N.; Lunin, V.V. Processing of coked Pt-Re/γ-Al$_2$O$_3$ catalysts with high-concentration ozone dissolved in supercritical carbon dioxide. *Sverhkriticheskie Flyuidy Teor. Prakt.* **2010**, *5*, 76–91. [CrossRef]
200. Adamson, A. *The Physical Chemistry of Surfaces*; Wiley: New York, NY, USA, 1976.
201. Barker, J.A.; Henderson, D. What is "liquid"? Understanding the states of matter. *Rev. Mod. Phys.* **1976**, *48*, 587–671. [CrossRef]
202. Tovbin, Y.K. The problem of a self-consistent description of the equilibrium distribution of particles in three states of aggregation. *Russ. J. Phys. Chem. A* **2006**, *80*, 1554–1566. [CrossRef]
203. Tovbin, Y.K. A Theory of Liquid-Phase Reaction Rates Including Coulomb Terms in the Lattice Gas Model. *Russ. J. Phys. Chem. A* **1996**, *70*, 1655–1660.
204. Prigogine, I.; Defay, R. *Chemical Thermodynamics*; Longmans Green and Co.: London, UK, 1954.
205. Glauber, R.J. Time-Dependent Statistics of the Ising Model. *J. Math. Phys.* **1963**, *4*, 294–307. [CrossRef]
206. Tovbin, Y.K. Kinetics and equilibrium in ordered systems. *Dokl. AN SSSR* **1984**, *277*, 917–921.
207. Tovbin, Y.K. Concentration Dependence of the Transfer Coefficients of Molecules in Mesopores in the Capillary-Condensation Region. *Russ. J. Phys. Chem. A* **1998**, *72*, 1298–1303.
208. Reif, F. *Statistical Physics. Berkeley Physics Course*; Mcgraw–Hill Book Company: New York, NY, USA, 1965; Volume 5.

processes

Article

Post Acid Treatment on Pressurized Liquid Extracts of Sorghum (*Sorghum bicolor* L. Moench) Grain and Plant Material Improves Quantification and Identification of 3-Deoxyanthocyanidins

Ádina L. Santana [1,†], Jaymi Peterson [2,†], Ramasamy Perumal [3], Changling Hu [4], Shengmin Sang [4], Kaliramesh Siliveru [1] and Dmitriy Smolensky [2,*]

[1] Grain Science and Industry Department, Kansas State University, 1301 N Mid Campus Drive, Manhattan, KS 66503, USA

[2] Grain Quality and Structure Research Unit, Agricultural Research Service, U.S. Department of Agriculture, Manhattan, KS 66502, USA

[3] Agricultural Research Center, Kansas State University, Hays, KS 67601, USA

[4] Nutrition Research Institute Building, North Carolina Agricultural and Technical State University, 150 N Research Campus Dr, Kannapolis, NC 28081, USA

[*] Correspondence: dmitriy.smolensky@usda.gov

[†] These authors contributed equally to this work.

Citation: Santana, Á.L.; Peterson, J.; Perumal, R.; Hu, C.; Sang, S.; Siliveru, K.; Smolensky, D. Post Acid Treatment on Pressurized Liquid Extracts of Sorghum (*Sorghum bicolor* L. Moench) Grain and Plant Material Improves Quantification and Identification of 3-Deoxyanthocyanidins. *Processes* **2023**, *11*, 2079. https://doi.org/10.3390/pr11072079

Academic Editor: Alina Pyka-Pająk

Received: 8 June 2023
Revised: 4 July 2023
Accepted: 5 July 2023
Published: 12 July 2023

Abstract: Sorghum is a unique natural food source of 3-deoxyanthocyanidins (3-DA) polyphenols. This work evaluated the effect of acidification on sorghum extracts post pressurized liquid extraction (PLE) and its ability to increase the identification and quantification of 3-DA. The sorghum genotypes included Sumac and PI570366 (bran only) and SC991 (leaf and leaf sheath tissue). The acidification of the PLE extracts was carried out with methanol–HCl solutions at various concentrations (0, 0.5, 1, 2, and 4%, v/v). Changes in color were determined using $L^*a^*b^*$. The overall phenolic composition was estimated with the total phenolic content and the DPPH free radical scavenging assays. Quantitative and qualitative chromatographic methods determined the phenolic profile. Color analysis showed that the redness and color saturation increased after acidification. No statistical difference was found in the total phenolic content of the acidified extracts, except for SC991, which was increased. There were no differences in the antioxidant capacity following acidification in all samples. For chromatographic analysis, luteolinidin was predominant in the extracts and the 3-DA content increased after acid treatment. However, some flavonoid and phenolic acid concentrations decreased following acid treatment, including taxifolin, quercetin, and chlorogenic acid. Interestingly, 0.5% v/v HCl acidification was sufficient to increase the color, allow the detection of 5-methoxyluteolinidin, and to increase luteolinidin and 7-methoxyapigenidin by at least twofold.

Keywords: accelerated solvent extraction equipment; natural colorants; hydrochloric acid treatment; sorghum polyphenols; thin-layer chromatography; HPLC; 3-deoxyanthocyanidins; LC-MS

1. Introduction

In 1998, the Dionex Corp. [1] patented a bench-scale automated piece of apparatus that uses pressurized liquid extraction (PLE) principles. This apparatus and methodology, designated as accelerated solvent extraction (ASE), extracts analytes from a solid or semi-solid matrix with the use of compressed liquids below the subcritical state. Cardenas-Toro et al. [2] noted that the subcritical state of water happens at temperatures ranging between around 150 and 374 °C and at pressures higher than its vapor saturation pressure. When the liquid solvent is hot and pressurized, its dielectric constant decreases and contributes to an increase in the solubility of target compounds in the matrix to the solvent, resulting in rapid extraction in comparison with conventional extraction methods including Soxhlet, maceration, and reflux extraction.

Although PLE can be used for high throughput, there are some limitations to this method. For instance, most phenolic compounds are thermal sensitive and, therefore, extraction temperatures should be studied to evaluate the optimal solubility while minimizing thermal degradation [3]. Additionally, the vessels, valves, and fittings in the ASE are easily corroded by the presence of strong acids and alkali, limiting the use of solvents. The ASE machine is also limited by its inability to pressurize higher than 10 MPa.

Sorghum (*Sorghum bicolor* L. Moench) grain contains a variety of phenolic compounds, including phenolic acids (ferulic and caffeic acids), flavonoids (apigenin, luteolin, and naringenin), and 3-deoxyanthocyanidins (3-DA), namely, luteolinidin, apigenidin, and 7-methoxyapigenidin. The extraction of phenolics from sorghum has been studied worldwide using water [4] and solvent solutions containing acetone [5], ethanol [6,7], and methanol [8–10].

Despite the use of solvents that are generally recognized as safe (GRAS) for sorghum extraction methods, some phenolics are not easily extracted in various solvents. To increase extraction efficiency, acidified methanol at 1% hydrochloric acid (HCl) has been extensively used [8,9]. Hydrolysis can cleave ester bonds to release phenolics and is primarily performed through acid, alkali, or enzyme treatments [11,12]. In acidic medium, there is an increase in hydrogen ions, allowing the protonation of phenolics. This assists the release of phenolics bound to macromolecules (i.e., proteins, lignin, cellulose), subsequently increasing the solubility of hydrophobic compounds [13,14].

Solutions containing HCl are highly corrosive. In academic studies, solutions with low concentrations of HCl were shown to be efficient for their goals. Wizi et al. [14] observed that treating samples with 1% HCl (v/v) increased the colorants luteolinidin and apigenidin by approximately 17% in Chinese sorghum husks extracted by microwave and ultrasonification. Paunović et al. [15] showed that acidification with 5% HCl in an aqueous solution increased the total phenolic content in barley extracts by approximately 200% in comparison with the non-acidified extracts in 30% ethanol (v/v). Ju and Howard observed that the post acidification with HCl at 0.1% (v/v) of PLE extracts of red grape skin enhanced the detection of total anthocyanins similar to acetic acid at 7% (v/v) [16].

For this reason, this work investigates the effect of post acid treatments on phenolic composition in sorghum extracts derived from PLE technology. The goal was to determine whether post-extraction acidification would enhance the phenolic content and increase the amount of detectable phenolics, namely, 3-DA. The advantages of acidification post extraction are the ability to use PLE and obtain high-throughput extractions while limiting the amount of HCl waste generated, leading to greener chemistry when extracting sorghum phenolics.

2. Materials and Methods

2.1. Raw Material and Chemicals

Commercial Sumac bran was purchased from NU Life Markets (Scott City, KS, USA). The sorghum cultivar PI570366 was grown in Puerta Vallarta, Mexico, by Kansas State University (Hays, KS, USA), during the 2017 growing season and stored at −20 °C. The grains were decorticated in-house to produce the bran. Sorghum leaf and leaf sheaths from SC991, a high phenolic genotype, were collected during the 2019 season in Hays (KS, USA), and milled with the aid of a UDY mill (Fort Collins, CO, USA) using a 0.5 mm screen. Glacial acetic acid, acetonitrile, hydrochloric acid, methanol, chloroform, deionized water, formic acid, and ethanol were ordered from Fisher Scientific (Waltham, MA, USA).

The standards used in HPLC analysis include 7-methoxyapigenidin (Cayman, Ann Harbor, MI, USA), taxifolin (HWI, Ruelzheim, Germany), 4-hydroxybenzoic acid, syringic acid, protocatechuic acid, ferulic acid, quercetin, naringenin (Sigma-Aldrich, Darmstadt, Germany), catechin (TCI, Portland, OR, USA), kaempferol, caffeic acid, p-coumaric acid, eriodictyol, luteolin, apigenin, and the chloride salts of luteolinidin, apigenidin, cyanidin, pelargonidin, malvidin, and peonidin (Indofine, Hillsborough, NJ, USA).

Gallic acid hydrate (TCI, Portland, OR, USA), Folin-Ciocalteu (Sigma-Aldrich, Darmstadt, Germany) and sodium carbonate (Fisher Scientific, Waltham, MA, USA) were used for the TPC analysis. The 2,2-diphenyl-1-picrylhydrazyl (DPPH) and Trolox (Sigma-Aldrich, Darmstadt, Germany) were used for antioxidant potential analysis.

2.2. Pressurized Liquid Extraction (PLE)

The extraction was carried out in Accelerated Solvent Extraction 350 (Dionex, Thermo Fisher, Pittsburgh, PA, USA, Figure 1) apparatus. Then, 1 g of raw material was mixed with approximately 6 g of borosilicate glass beads (3 mm i.d., Chemglass Life Sciences, Vineland, NJ, USA) and inserted into a 10 mL stainless steel extraction vessel (Thermo Fisher, Pittsburgh, PA, USA) packed with cellulose filters at both ends.

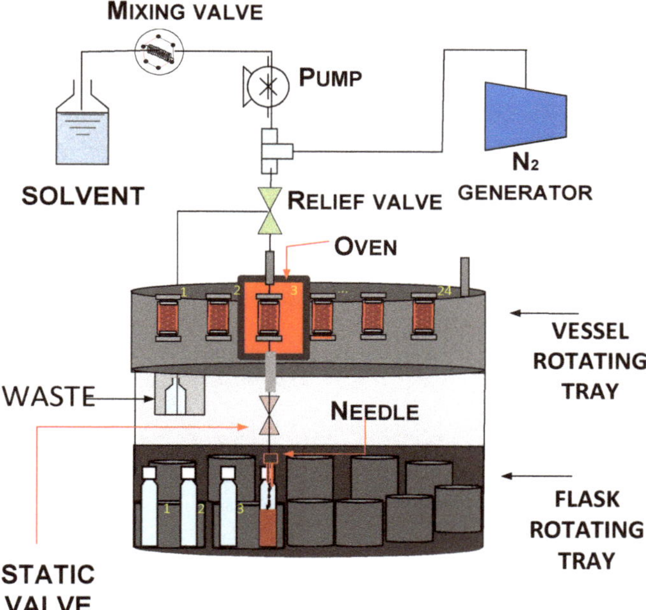

Figure 1. Schematic diagram showing the extraction process using ASE equipment (adapted from Thermo Fisher [17]).

Once the oven reached the set-up temperature, the nitrogen-driven pump pressurized the vessel with fresh solvent at 10.34 MPa, starting the extraction process. Two cycles of 5 min each were selected, i.e., the number of times consisting of vessel heating followed by pressurization with the solvent. The rinse volume was 100% of the vessel volume. Once the rinse cycles finished, a nitrogen stream (99% pure, Calypso DS, F-DGSi Nitrogen Generator, France) flushed the line throughout the vessel for 1 min. Afterwards, the vessel was depressurized, and the PLE finished.

The solvent used was ethanol:water (70:30, *v/v*) based on prior research suggesting that it is an optimal GRAS solvent for sorghum phenolic extraction [6]. At first, three levels of temperature (80 °C, 100 °C, and 120 °C) were used to select to the optimal temperature to extract phenolics for further acidification. The total extraction time was approximately 20 min. The quantity of solvent spent per experiment varied around 25–30 mL. Such variation is justified by the nature of the raw material and the quantity of solvent remaining in the line before the purge phase. Afterwards, the volume of extracts was adjusted to 30 mL to assure equal volume for all samples followed by the TPC assay (see Figure 2). Two biological replicates were conducted for each temperature tested.

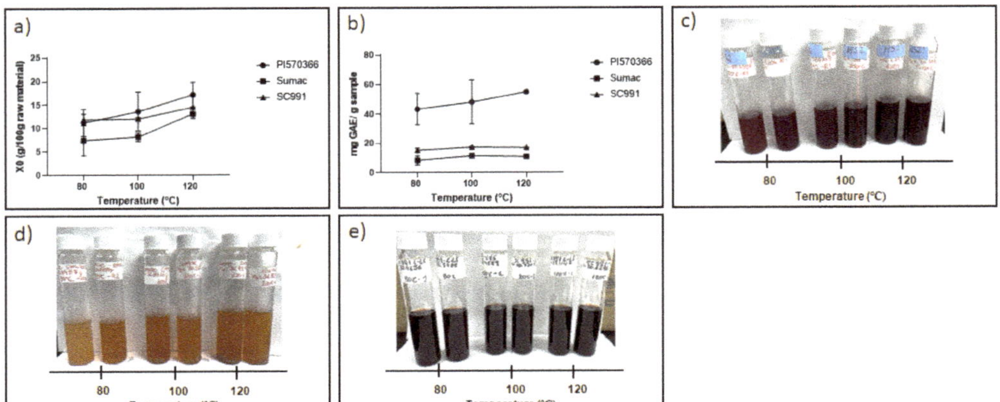

Figure 2. PLE extracts of sorghum with ethanol 70%: the effect of temperature on (**a**) global yield, (**b**) total phenolic content, and the visual representation of extracts (**c**) PI570366, (**d**) Sumac bran, and (**e**) SC991 leaf tissue.

2.2.1. Global Yield of Soluble Extract

The extracts were centrifuged at 4 °C, 3500× g, for 10 min (Thermo Fisher, Pittsburgh, PA, USA) to remove any precipitate. The precipitate was discarded, and the supernatant collected and dried with the use of a Rocket Synergy 2 Evaporation System (Thermo Fisher, Pittsburgh, PA, USA) at 60 °C for approximately 2 h. The global yield, or mass of the dried supernatant, was calculated according to Equation (1).

$$X_0 = \left(\frac{m_{EXT}}{m_{RM}} \right) \times 100 \tag{1}$$

where X_0 is the global yield, expressed as g soluble extract per 100 g of raw material, m_{EXT} (g) is the mass of dried extract, and m_{RM} (g) is the mass of raw material used in PLE, i.e., 1 g.

2.2.2. Acidification of PLE Extracts

Since HCl is highly corrosive, it cannot be incorporated into solvents for ASE equipment or any other equipment that is sensitive to strong alkali or acids [17]. In this case, the acidification of extracts with HCl solutions at small volumes was studied to optimize the release of phenolic species to improve the sample preparation of sorghum extracts for analytical measurements.

Dried PLE extracts solubilized with 10 mL of pure methanol served as the control. For acid treatments, extracts were solubilized in 10 mL of methanol–HCl solutions at the following percentages: 0.5%, 1%, 2%, and 4% (v/v), and stored at −20 °C for at least 48 h. The acid-treated samples were then dried under nitrogen. All samples, including the untreated dried samples, were resuspended in the solvent used for analysis (90% methanol, 10% ethanol). The samples were homogenized and sonicated with the aid of an ultrasonic bath (ULTRAsonik, Simi Valley, CA, USA) for 20 min. Afterwards, the samples were centrifuged at 4 °C, 3500× g for 10 min (Thermo Fisher, Pittsburgh, PA, USA) to remove any precipitate. The supernatants were used for analysis. Three biological replicates of each acidification treatment were studied.

2.3. Characterization of Extracts

2.3.1. Color Analysis

The effect of acidification in the coloration of the extracts was evaluated using a MiniScan® EZ 4500 portable spectrophotometer (45°/0° geometry, HunterLab, Reston, VA,

USA). Due to poor transmittance, SC991 plant tissue extracts and PI570366 bran extracts were diluted in pure ethanol at 1:10, *v/v* for color analysis (see Figure 3).

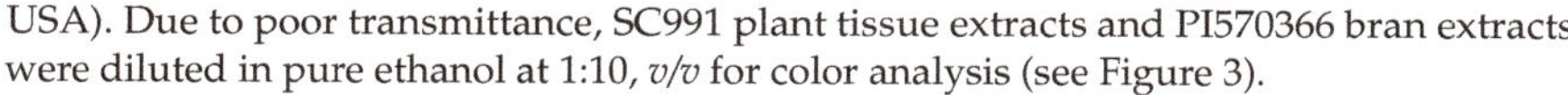

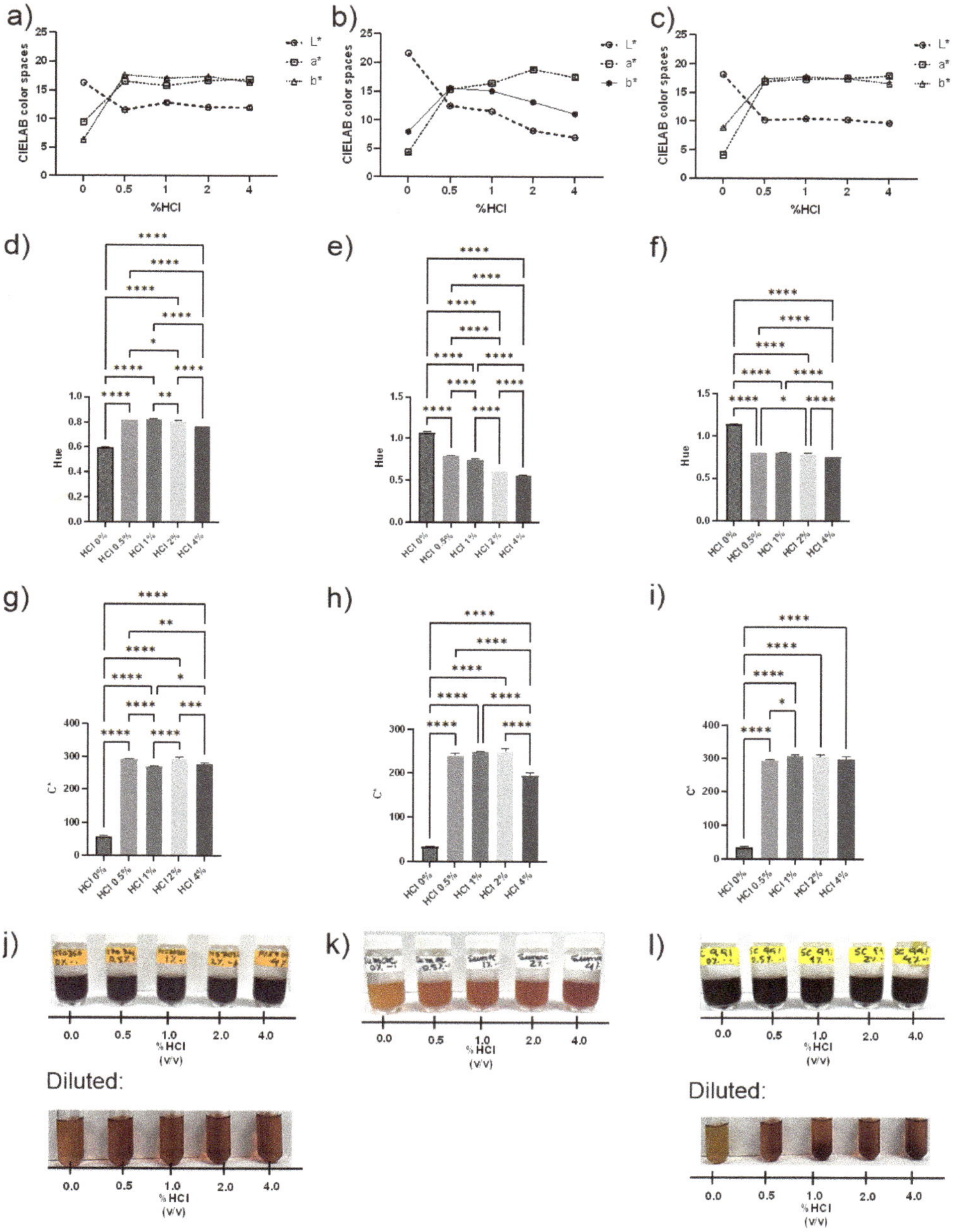

Figure 3. CIEBLAB color values for (**a**) PI570366; (**b**) Sumac; (**c**) SC991 leaf and leaf sheaths. Hue values for (**d**) PI570366; (**e**) Sumac; (**f**) SC991 and chroma values for (**g**) PI570366; (**h**) Sumac; (**i**) SC991 and overall changes in physical appearance for (**j**) PI570366; (**k**) Sumac; (**l**) SC991 with increasing acid concentration. The results are represented as mean $\pm$ standard deviation of three independent acidified extracts post extraction. Significance: * $p \leq 0.05$, ** $p \leq 0.01$, *** $p \leq 0.001$, *** $p \leq 0.0001$.

The Commission Internationale de l'Eclairage (CIE) system was used to model L^*, a^*, and b^* color coordinates. The coordinate a^* represents the color trend of red (positive) or green (negative), whereas the b^* coordinate represents the color trend between yellow (positive) and blue (negative). The chroma (C^*) represents the color saturation, and the hue angle (H^*) represents the relative intensity of redness and yellowness, where $0°$ denotes red, $90°$ yellow, $180°$ for green, and $270°$ for blue. The chroma (C^*) and hue (H^*) values were calculated using a^* and b^* values, as mentioned in Equations (2) and (3), respectively.

$$C^* = \sqrt{(a*)^2 + (b*)^2} \tag{2}$$

$$H^* = arctan\left(\frac{b*}{a*}\right) \tag{3}$$

2.3.2. Total Phenolic Content Assay

The method previously described by Herald et al. [18] was used to determine the total phenolic content (TPC) using the Folin–Ciocalteu (FC) method. The FC working reagent was prepared by diluting FC with deionized (DI) water at a 1:1 (v/v) ratio. The samples were diluted with ethanol (70% v/v) at a 1:20 ratio. To a 96-well plate, samples or standards (25 µL) were combined with 75 µL DI water and 25 µL FC working reagent. The plate was allowed to incubate at room temperature for 6 min before 100 µL of 7.5% (w/v) Na_2CO_3 was added. The plate was then heat sealed and left to incubate in the dark for 90 min before reading the absorbance at 765 nm. The absorbance was read using a BioTek Synergy 2 multi-detection plate reader (Winooski, VT, USA). The standard curve was prepared by dissolving gallic acid (1 mg/mL) in 70% ethanol and ranged in concentration from 0 to 400 µg/L. Sample absorbances were calculated using the standard curve and expressed as milligram gallic acid equivalents (GAE) per gram of raw material.

2.3.3. DPPH Antioxidant Capacity Assay

Antioxidant capacity was determined using the 2,2-diphenyl-1-picrylhydrazyl (DPPH*) radical method [19]. Briefly, 50 µL of samples were combined with 10 mL of DPPH reagent (0.1 mM in 70% ethanol). The samples and standards were briefly vortexed before incubating in the dark for 30 min. The samples were filtered using a 0.45 µm nylon syringe filter prior to reading absorbance at 517 nm. Absorbance was measured using a UV–Vis spectrophotometer (Shanghai Metash Instruments Co., Ltd., Shanghai, China). The Trolox standard curve in 70% ethanol was plotted (0–100 mg/mL). The results of triplicates were expressed as milligram Trolox equivalents (TE) per gram of raw material.

2.3.4. High-Performance Liquid Chromatography

The 1260 Infinity (Agilent, Santa Clara, CA, USA) HPLC apparatus analyzed the individual phenolics in the samples, based on the protocol of Irakli et al. [20] with adaptations described by Lee and coworkers [21]. The stationary phase consisted of a Kinetex® C18 column (150 mm × 4.6 mm, 100 Å, 2.6 µm, Phenomenex, Torrance, CA, USA) connected to a guard column (SecurityGuard™, Phenomenex, Torrance, CA, USA). The column temperature was set up at 30 °C. The mobile phases consisted of acetonitrile (A), methanol (B), and acidified water at 0.5% glacial acetic acid, v/v (C). The elution gradient of A:B:C ($v/v/v$) at 1 mL/min consisted of an initial composition of 5:5:90, followed by 0–5 min of 8:8:84, 5–15 min of 10:10:80, 15–25 min of 25:0:75, 25–35 min of 30:0:70, and 35–45 min of 60:0:40. A post time of 5 min with the initial gradient was used to stabilize the baseline for further runs.

Catechin, eriodictyol, taxifolin, 4-hydroxybenzoic-, protocatechuic- and syringic acids, quercetin, luteolin, and naringenin were investigated at 280 nm. Chlorogenic, caffeic, p-coumaric, and ferulic acids, rutin, apigenin, and kaempferol were investigated at 320 nm.

Luteolinidin, cyanidin, peonidin, malvidin, pelargonidin, apigenidin, and 7-methoxyapigenidin were investigated at 510 nm. The results were expressed as µg/g raw material.

In Sumac and PI570366, apigenidin was the second most prominent 3-DA in the extracts. However, in the acidified SC991 extracts, apigenidin could not be detected by HPLC-DAD using the current protocol, because of peak overlap with 5-methoxyluteolinidin, which was identified by LC-MS. For the calculation of 5-methoxyluteolinidin, we used the luteolinidin calibration curve multiplied by the molecular weight correction factor of 0.93, as described by Speranza and coworkers [22].

Representative chromatograms at 280 nm, 320 nm, and 510 nm, as well as information about the HPLC method validation (precision and accuracy) are available in the Supplementary Materials.

For all studied compounds, the sorghum extracts spiked with phenolics presented linear behavior with a correlation factor (R^2) of 0.998 or better (Table S2.1), which is consistent with the previous findings [23,24]. For one concentration studied, the repeatability and accuracy for the detection of 3-deoxyanthocyanidins, flavonoids, and phenolic acids for the genotypes studied (Table S2.2) were comparable with other validated HPLC methods [23–25].

2.3.5. Thin-Layer Chromatography (TLC)

A qualitative evaluation of the presence of 3-DA and the effect of acidification on the extracts was carried out with silica gel–glass TLC plates (10 × 20 cm, 60G F_{254}, Merck, Darmstadt, Germany) as a stationary phase. Stock solutions of apigenidin, luteolinidin, and 7-methoxyapigenidin at 0.09 mg/mL, in methanol:ethanol (90:10, v/v) were prepared. One milliliter of non-acidified and acidified extracts at 1% and 4% HCl were dried under nitrogen (99%, Matteson, Manhattan, KS, USA) and resuspended in 1 mL methanol:ethanol (90:10, v/v).

Two microliters of standards and four microliters of extracts were injected into the TLC plates with the use of a chromatographic syringe (Hamilton, Reno, NV, USA). One hundred milliliters of the mobile phase (chloroform:ethanol:glacial acetic acid:methanol, 95:5:1:10, $v/v/v$) were inserted into a 20 × 20 cm closed glass chamber and, subsequently, the TLC plates were eluted with samples and standards injected. After elution, the plates were recorded under visible and under UV regions with a Cole-Parmer viewing cabinet (Vernon Hills, IL, USA) equipped with a UV lamp at 366 nm. The theoretically present 3-DA in the extracts was identified via the quality of the bands eluted, and the retention factor (R_F, cm/cm) calculations of the separated zones, i.e., the R_F values of the compounds in the extracts, were compared to the R_F values of the standards.

2.3.6. LC-MS Identification of 5-Methoxy Luteolinidin

LC-MS analysis was carried out with a Thermo Q Exactive mass spectrometer coupled with a Vanquish LC system (Thermo Scientific, San Jose, CA, USA) incorporated with an electrospray ionization (ESI) for UHPLC-HRMS/MS analysis. Chromatographic separation was performed using an Accucore Gemini NX-C18 column (50 mm × 2.1 mm; i.d. 2.6 µm, Thermo Scientific, San Jose, CA, USA). The binary mobile phase system consisted of water with 0.1% formic acid (FA) (phase A) and acetonitrile (ACN) with 0.1% FA (phase B). The flow rate was 0.2 mL/min, and the injection volume was 3 µL. The column was eluted with a gradient program (2% B from 0 to 1 min, 2% to 45% B from 1 to 2 min, 45% to 80% B from 2 to 6 min, 80% to 100% B from 6 to 6.5 min, and held at 100% for 1 min, then re-equilibrated with 2% B for another 1 min). The mass conditions were optimized using a mixture of apigenidin and luteolinidin. The negative ion polarity mode was set for an ESI ion source with the voltage on the ESI interface maintained at 2.65 kV. Nitrogen gas was used as the sheath gas at a flow rate of 20 AU and the auxiliary gas at 10 AU. The collision-induced dissociation (CID) was conducted with an isolation width of 1.5 Da and a normalized collision energy of 20, 40, and 60 for MS/MS analysis.

2.4. Statistical Analysis

The one-way analysis of variance (ANOVA) with post hoc Tukey's test was used to evaluate the statistical differences. GraphPad Prism version 9.5.1 was used. Data are represented as mean $\pm$ standard deviation of three independent experiments, unless otherwise stated.

3. Results and Discussion

3.1. Selection of Temperature for Obtaining Phenolic Compounds via PLE

Increasing the temperature increased the global yield of Sumac bran, PI570366 bran, and the SC991 leaf and leaf sheaths (Figure 2a). A higher yield of crude extract was not associated with a higher concentration of target compounds, but the association of pressurized solvent with the temperature may allow the selection of other classes of compounds, or the use of temperatures higher than 100 °C may allow the formation of degradation products in the extracts, including hydroxymethylfurfural [26]. Since there was no statistical difference in TPC for the temperatures used (Figure 2b), 100 °C was selected to generate high phenolic extracts for subsequent acid hydrolysis. The TPC detected in PI570366 at 100 °C was 0 mgGAE/g, which is comparable to 45 mgGAE/g found in Chinese whole sorghum grain extracted with subcritical water at 150 °C [4]. For the replicates reproduced at 100 °C, the global yield values were 8.20 ± 0.94 g/100 g for Sumac, 13.65 ± 4.16 g/100 g for PI570366, and 12.06 ± 0.28 g/100 g for SC991 (Figure 2a).

3.2. Acid Hydrolysis of PLE Extracts

3.2.1. Color Analysis

The color, hue, and chroma results of the acidified extracts post PLE extraction are represented in Figure 3. In our extracts, the redness of extracts represented by a^* coordinate increased after acidification, while the lightness (L^*) and yellowness (represented by b^*) decreased (Figure 3a–c).

For all extracts, the hue angle ranged between 0 and 1, which is an interval expected for red color, as highlighted by the a^* values. Interestingly, increasing the HCl concentration had an inverse effect on hue in Sumac and SC991. The hue angles in the extracts were lower than the hue 31–59° range previously reported in high-tannin black sorghum extracts [27]. The color saturation represented by chroma (C^*) increased significantly after acidification in all extracts (Figure 3g–i). Our values differed from those found by Wizi and coworkers [14], who dyed wool fabrics with sorghum leaf sheath extracts acidified with HCl at 1%, and found L^* of 26.4–31.5°, $a^* = 15.5–23.7$, $b^* = 13.1–19.1$, H = 38.9–40°, and $C^* = 26.4–31.5$. Our values also differed from those found in the ethanolic extracts of leaf sheaths harvested from Benin: $L^* = 28.1°$, $a^* = 5.6$, $b^* = -0.4$ [28].

3.2.2. Effect of Acidification on Total Phenolic Content

The highest TPC detected in PI570366 was 27.56 mgGAE/g. In Sumac, the TPC ranged from 17.73 to 26.46 mgGAE/g. This was within the range of 12.03–124.85 mgGAE/g reported by Herald and coworkers [18], who detected polyphenols in non-tannin and high-tannin genotypes [18]. No statistical differences were detected in TPC between non-acidified and acidified extracts, except for SC991 between 0 and 0.5% HCl, which increased from 10.32 mg GAE/g to 15.17 mg GAE/g (Figure 4c). Kayodé and coworkers reported much higher concentrations of 103–411 mg GAE/g TPC in their acidified extracts at 1% HCl (v/v) obtained from leaf sheaths cultivated from an unknown genotype in Benin [28]. Such differences may be attributed to genotypes, location, and plant growing conditions [29].

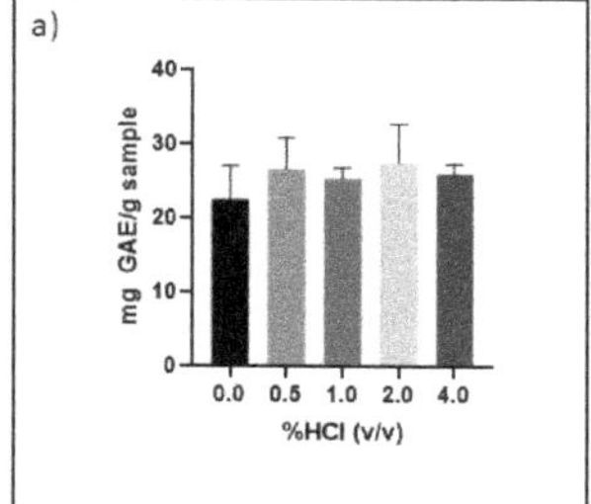 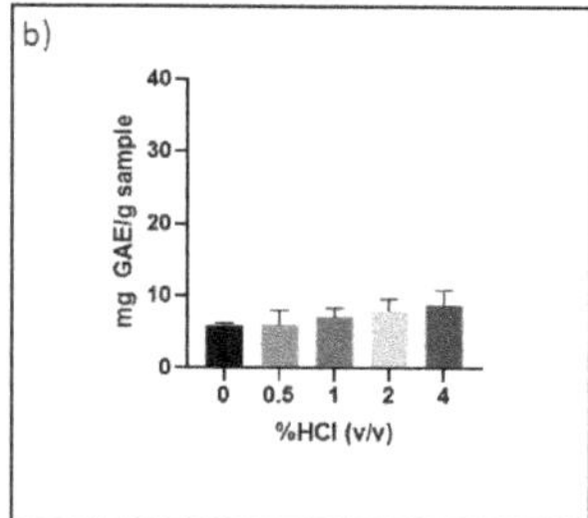 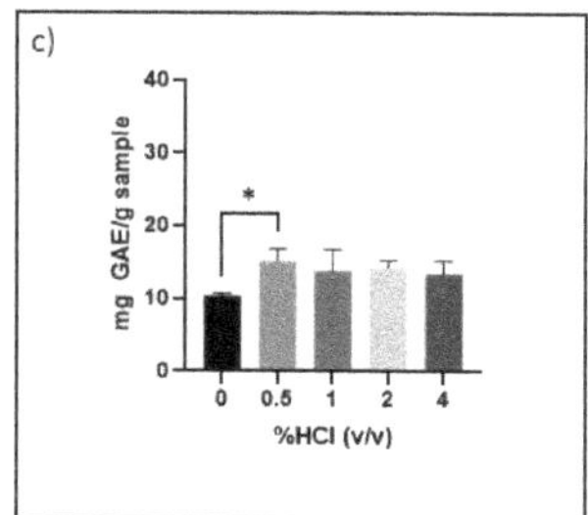

Figure 4. Total phenolic content of (**a**) PI570366, (**b**) Sumac, and (**c**) SC991 at various acid levels. The data represent mean ± standard deviations of three independent experiments. Significance: * $p \leq 0.05$.

3.2.3. Effect of Acidification on DPPH Antioxidant Capacity

The antioxidant capacity of non-acidified and acidified extracts is represented in Figure 5. Similar to TPC, there were no statistical differences detected after HCl acidification. Interestingly, the phenolic compounds' potential against free radicals is dependent on pH, i.e., at acidic conditions, the phenolics are deprotonated, leading to the H-atom transfer and consequently decreasing the oxidative stress [30]. The variety with the highest DPPH radical scavenging capacity was the PI570366 bran extract, which peaked 1968.24 mg TE/g at 0% HCl. The Sumac bran extracts ranged from 348.93 to 467.59 mg TE/g. This is in agreement with the 35.47–742.63 mg TE/g detected in the bran of tannin and non-tannin lines of sorghum [18], and higher than the 1.65–2.11 mgTE/g detected in the bran of non-tannin white and red sorghum varieties [22]. The SC991 leaves and leaf sheaths ranged from 551.85 mg TE/g to 645.27 mg TE/g. No significant effect of acidification on the antioxidant capacity was found.

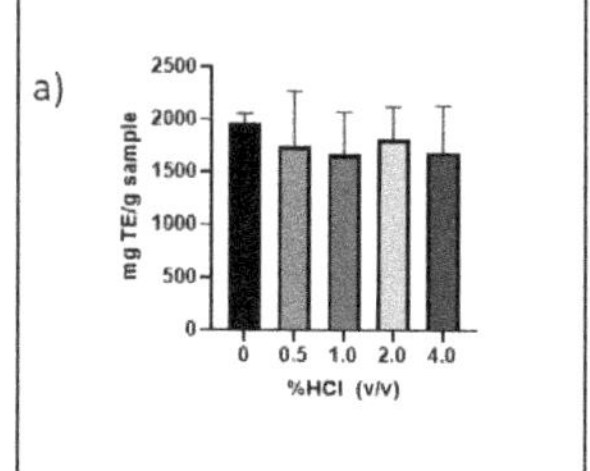 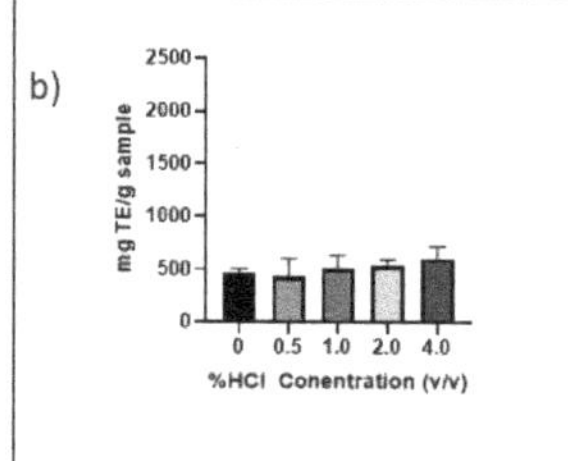 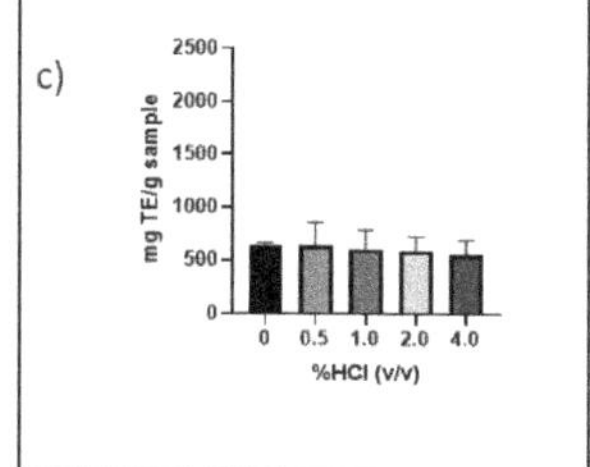

Figure 5. DPPH radical absorbance capacity for of (**a**) PI570366, (**b**) Sumac, and (**c**) SC991 at various acid levels.

3.2.4. Phenolic Profile Detected by HPLC

Kaempferol, syringic acid, rutin, pelargonidin, malvidin, and peonidin were not identified in any of the extracts. In the PLE extracts, the phenolic acids and some flavonoids were found mostly in the control solvent, i.e., methanol (Table 1A–C). Such reduced detection in phenolic species after acidification may be justified by the (A) binding of phenolics to the plant matrix—for instance, p-coumaric acid and ferulic acid are easily released after the alkaline hydrolysis of lignocellulosic materials, since alkaline hydrolysis is suitable to dissolve lignin [31]; (B) degradation of phenolics after interaction with HCl [32]; and (C) precipitation of phenolics because of the reduced solubility. In addition, many phenolics are extracted as glycosides. For instance, flavonoids are generally present in C-glycoside and O-glycoside forms. The linkage between flavonoids and sugar moieties can occur through an OH group to form O-glycosides or through carbon–carbon bonds to form C-glycosides [33]. In sorghum bran genotypes, the glycosides of luteolinidin, apigenidin, kaempferol, luteolin, and apigenin were detected previously by LC-MS [27,34,35].

Table 1. (**A**) Phenolic compounds quantified (µg/g raw material) in acidified PLE extracts obtained from Sumac bran extract. (**B**) Phenolic compounds quantified (µg/g raw material) in acidified PLE extracts obtained from PI570366 bran extract. (**C**) Phenolic compounds quantified (µg/g raw material) in acidified PLE extracts obtained from SC991 leaf and leaf sheath.

(A)

Subclass	Compound	λ nm	% HCl				
			0%	*0.50%*	*1%*	*2%*	*4%*
3-deoxyanthocyanidin	Luteolinidin	510	60.63 ± 7.88 [b]	101.56 ± 14.09 [ab]	107.89 ± 20.65 [ab]	102.71 ± 30.73 [ab]	132.30 ± 21.94 [a]
	Apigenidin		7.62 ± 0.30 [b]	14.20 ± 1.77 [a]	14.56 ± 0.22 [a]	15.19 ± 2.05 [a]	17.14 ± 1.85 [a]
Benzoic acid	Protocatechuic acid	280	15.96 ± 0.53 [b]	26.36 ± 8.71 [a]	17.27 ± 3.93 [b]	21.19 ± 2.70 [ab]	15.63 ± 1.97 [b]
Flavanol	Catechin	280	67.48 ± 2.32 [a]	7.50 ± 2.53 [b]	4.04 ± 0.91 [b]	12.16 ± 4.07 [b]	3.88 ± 2.79 [b]
Flavanonol	Taxifolin	280	261.32 ± 11.06 [a]	61.22 ± 15.93 [b]	76.39 ± 88.72 [b]	54.61 ± 20.57 [b]	88.14 ± 101.36 [b]
Flavanone	Eriodictyol	280	10.71 ± 0.47 [a]	20.92 ± 13.21 [a]	58.89 ± 49.82 [a]	78.46 ± 35.40 [a]	86.02 ± 99.28 [a]
	Naringenin		0 [#b]	12.20 ± 5.18 [a]	20.12 ± 13.88 [a]	24.12 ± 6.17 [a]	27.67 ± 27.26 [a]
Flavonol	Quercetin	280	26.09 ± 0.56 [a]	13.72 ± 2.28 [a]	12.35 ± 4.63 [a]	11.46 ± 5.19 [a]	11.45 ± 14.01 [a]
Cinnamic acid	Chlorogenic acid	320	14.09 ± 0.24 [a]	0 [#b]	0 [#b]	0 [#b]	0 [#b]
	Caffeic acid		9.25 ± 0.13 [a]	6.16 ± 0.37 [b]	6.09 ± 1.31 [b]	5.38 ± 0.09 [b]	0 [#c]
	p-Coumaric acid		2.53 ± 0.49 [a]	2.06 ± 1.59 [a]	4.23 ± 3.25 [a]	3.53 ± 1.13 [a]	0.78 ± 0.84 [a]
	Ferulic acid		6.28 ± 0.35 [a]	6.59 ± 0.52 [a]	6.94 ± 3.59 [a]	5.87 ± 0.81 [a]	5.10 ± 0.09 [a]
Flavone	Apigenin	320	5.18 ± 0.21 [a]	5.41 ± 0.10 [a]	5.42 ± 0.18 [a]	5.55 ± 0.20 [a]	7.74 ± 4.19 [a]
	Luteolin		27.87 ± 1.26 [b]	43.24 ± 3.00 [ab]	40.12 ± 6.25 [ab]	41.03 ± 4.73 [ab]	46.83 ± 10.19 [a]
Anthocyanidin	Cyanidin	510	0 [#b]	35.95 ± 0.36 [ab]	43.51 ± 13.25 [a]	41.32 ± 4.79 [ab]	60.95 ± 30.81 [a]

(B)

Subclass	Compound	λ nm	% HCl				
			0%	*0.50%*	*1%*	*2%*	*4%*
3-deoxyanthocyanidin	Luteolinidin	510	613.87 ± 83.52 [b]	1067.40 ± 117.92 [a]	1113.65 ± 96.63 [a]	1197.85 ± 82.93 [a]	1156.78 ± 221.78 [a]
	Apigenidin		85.77 ± 11.17 [b]	161.81 ± 33.35 [ab]	169.41 ± 22.10 [a]	142.17 ± 46.67 [ab]	167.34 ± 20.75 [a]
	7-methoxyapigenidin		129.79 ± 4.70 [b]	672.37 ± 66.07 [a]	744.85 ± 30.73 [a]	835.84 ± 110.99 [a]	726.24 ± 75.45 [a]
Benzoic acid	Protocatechuic acid	280	7.16 ± 0.38 [b]	18.84 ± 1.40 [a]	18.69 ± 1.05 [a]	18.22 ± 4.58 [a]	16.39 ± 5.25 [a]
	4-hydroxybenzoic acid		13.05 ± 3.16 [ab]	21.14 ± 3.69 [a]	18.59 ± 2.36 [a]	17.04 ± 12.36 [a]	0 [#b]

Table 1. *Cont.*

(B)

Subclass	Compound	λ nm	% HCl				
			0%	0.50%	1%	2%	4%
Flavanol	Catechin	280	41.62 ± 5.97 [b]	129.34 ± 21.54 [ab]	81.94 ± 23.07 [b]	121.70 ± 51.67 [b]	308.40 ± 141.11 [a]
Flavanone	Eriodictyol	280	12.11 ± 1.42 [b]	75.40 ± 16.50 [a]	41.43 ± 15.80 [ab]	58.82 ± 27.78 [ab]	81.11 ± 10.45 [a]
	Naringenin		0 [#c]	21.30 ± 4.64 [ab]	17.23 ± 3.40 [b]	24.79 ± 7.47 [ab]	32.07 ± 2.91 [a]
Flavone	Luteolin	280	32.22 ± 6.50 [b]	53.17 ± 6.29 [ab]	49.19 ± 3.91 [ab]	56.70 ± 10 [a]	48.16 ± 4.28 [ab]
	Apigenin		5.08 ± 0.23 [a]	5.87 ± 0.67 [a]	5.70 ± 0.07 [a]	5.92 ± 0.18 [a]	5.82 ± 0.64 [a]
Cinnamic acid	Chlorogenic acid	320	30.05 ± 5.26 [a]	11.42 ± 1.08 [b]	11.15 ± 0.95 [b]	12.13 ± 0.37 [b]	12.61 ± 1.28 [b]
	p-Coumaric acid		2.19 ± 1.35 [ab]	2.93 ± 0.81 [a]	0 [#b]	0.94 ± 1.92 [ab]	0 [#b]
	Ferulic acid		7.19 ± 0.99 [a]	12.23 ± 1.93 [a]	8.48 ± 1.34 [a]	8.60 ± 8.89 [a]	7.26 ± 1.09 [a]
Anthocyanidin	Cyanidin	510	0 [#b]	67.10 ± 17.20 [ab]	92.53 ± 11.72 [ab]	139.79 ± 67.41 [ab]	165.35 ± 108.58 [a]

(C)

Subclass	Compound	λ nm	% HCl				
			0%	0.50%	1%	2%	4%
3-deoxyanthocyanidin	Luteolinidin	510	295.43 ± 20.10 [a]	1499.37 ± 508.47 [a]	1486.01 ± 396.13 [a]	1514.00 ± 710.02 [a]	1595.43 ± 692.57 [a]
	Apigenidin		40.29 ± 7.70 [a]	NQ	NQ	NQ	NQ
	7-methoxyapigenidin		123.13 ± 17.81 [b]	1170.42 ± 313.44 [a]	1515.29 ± 499.65 [a]	1517.42 ± 358.27 [a]	1640.87 ± 284.82 [a]
	5-methoxyluteolinidin		0 [#b]	1733.41 ± 110.50 [a]	1861.12 ± 95.90 [a]	1881.75 ± 355.67 [a]	1765.45 ± 52.60 [a]
Benzoic acid	Protocatechuic acid	280	15.75 ± 0.09 [a]	15.85 ± 0.79 [a]	16.52 ± 1.18 [a]	17.37 ± 2.25 [a]	18.03 ± 2.33 [a]
	4-hydroxybenzoic acid		56.77 ± 0.93	84.81 ± 57.03	90.72 ± 47.72	35.72 ± 47.18	12.45 ± 6.57
Flavanone	Naringenin	280	9.74 ± 0.40 [a]	0 [#b]	0 [#b]	0 [#b]	0 [#b]
Flavone	Luteolin	320	235.44 ± 8.64 [a]	362.26 ± 22.19 [a]	301.59 ± 71.29 [a]	265.14 ± 101.79 [a]	260.38 ± 96.03 [a]
	Apigenin		53.49 ± 2.32 [a]	41.92 ± 3.33 [ab]	34.18 ± 4.73 [b]	31.05 ± 7.51 [b]	28.94 ± 6.60 [b]
Cinnamic acid	Chlorogenic acid	320	43.71 ± 0.37 [a]	13.71 ± 2.69 [b]	0 [#c]	0 [#c]	0 [#c]
	p-Coumaric acid		138.67 ± 4.19 [ab]	173.01 ± 13.66 [a]	90.74 ± 48.62 [bc]	33.79 ± 17.46 [cd]	21.65 ± 14.56 [d]

(**A**) Average ± standard deviation of three biological replicates. Different letters indicate statistical differences ($p \leq 0.05$) within each row. [#] For values below detection threshold, a value of 0 (zero) was assigned for statistical analysis. (**B**) Average ± standard deviation of three biological replicates. Different letters indicate statistical differences ($p \leq 0.05$) within each row. [#] For values below detection threshold, a value of 0 (zero) was assigned for statistical analysis. (**C**) Average ± standard deviation of three biological replicates. Different letters indicate statistical differences ($p \leq 0.05$) within each row. [#] For values below detection threshold, a value of 0 (zero) was assigned for statistical analysis. NQ: Apigenidin was not quantifiable when 5-methoxyluteolinidin was present due to peak overlap.

Flavonoids: 3-Deoxyanthocyanidins

In the non-acidified extract, luteolinidin eluted at 18.25 min, apigenidin at 21.42 min, and 7-methoxyapigenidin at 23.8–24 min. Prominent, unknown peaks at 21.95, 22.16, 23.37, and 27.10 min were observed at the 510 nm wavelength (Figure S1.9). The peak at 21.945 min was identified as 5-methoxyluteolinidin by LC-MS. As shown in Figure 6, this unknown compound (RT: 6.75 min) had the molecular ion at m/z 283.0608 ((M − H) − calculated 283.0605) under high-resolution negative ESI mode, which showed one more methyl unit than luteolinidin (m/z 269.0450). The typical fragment ions at m/z 268.0375, 240.0425, and 196.0526 of this unknown compound are similar to those of 5-methoxyluteolinidin reported from sorghum (Peak 75 in Xiong and coworkers [24]). Therefore, this compound was tentatively characterized as 5-methoxyluteolinidin.

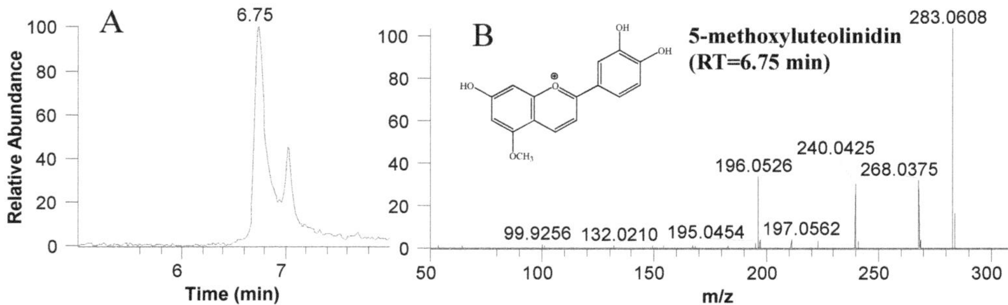

Figure 6. Ion chromatogram (**A**) under selected ion mode and MS^2 spectrum (**B**) of 5-methoxyluteolinidin in sorghum samples at negative ESI mode.

In Sumac bran, the luteolinidin increased from 60.63 µg/g to 132.30 µg/g (Table 1A) after acidification at 4% *v/v* HCl, which was significant compared to the control. The luteolinidin in acidified PI570366 bran ranged from 1067.4 to 1197.85 µg/g, which was significantly higher than in the non-acidified extracts (613 µg/g). There were no statistical differences among the HCl treatments (Table 1B). The luteolinidin concentration in non-acidified and acidified PI570366 extracts (Table 1B) was higher than the 319.9 µg/g obtained after the microwave-assisted extraction of non-tannin black sorghum genotypes, as reported by others [9].

In the SC991 leaf extracts, the luteolinidin increased after acidification, but the difference was not significant. According to Mizuno et al. [36], the RedforGreen (RG) mutant genotype contains over 1000-fold more 3-DA than the wild sorghum genotypes. However, the luteolinidin found by Petti et al. [37] in the RG genotype (1768 µg/g) is comparable to the SC991 leaf extracts (1486.01–1595.43 µg/g) reported in Table 1C. This contradicts Mizuno et al.'s [36] findings and support the findings of Petti et al. [37] that some natural genotypes of sorghum may also serve as low-cost sources of 3-DAs, i.e., pigments. 3-DAs provide superior properties such as improved stability after long storage, resistance to thermal processing, acidification, and color bleaching when compared to other naturally derived pigments such as carotenoids [31,34,38].

After acidification, the apigenidin peak at 22.16 min was dominated by 5-methoxyluteolinidin (identified by LC-MS) in SC991 leaf and leaf sheaths. This made apigenidin not quantifiable using the current methods after the appearance of the 5-methoxyluteolinidin peak. Figure S1.10, available in the Supplementary Materials, shows the peak separation after spiking the acidified SC991 extract with apigenidin.

The apigenidin increased twofold after acidification in the Sumac bran extracts after 0.5% *v/v* HCl treatment with no additional significant increase at higher HCl concentrations. In the PI570366 bran extracts, the apigenidin significantly increased at 1% *v/v* HCl with no additional increase at higher HCl concentrations. In the SC991 extracts, apigenidin was not quantified after acid treatment due to the reasons stated previously.

In the Sumac bran extract, 7-methoxyapigenidin was not detected. In the PI570366 bran and SC991 leaf extracts, 7-methoxyapigenidin significantly increased 5-fold and 10-fold with HCl treatments. There was no additional significant increase in 7-methoxyapigenidin at HCl concentrations higher than 0.5% v/v. 5-methoxyluteolinidin was only detected in SC991 leaf tissue after the acid treatments; therefore, the amount of 5-methoxyluteolinidin increased from non-detectable to a range of 1733.41–1881.75 µg/g raw material after acid treatment. There were no statistical differences between the acid treatment concentrations.

Overall, the acid treatments greatly improved the quantifiable levels of 3-DA. The increase in the 3-DA concentrations was achievable using 0.5% v/v HCl.

Flavonoids: Other Classes

In the Sumac and PI570366 extracts, cyanidin could only be detected after acidification (Figures S1.3 and S1.6), and the highest significance was found at 4% HCl (v/v) for both genotypes (Table 1A,B).

A non-identified compound that eluted at 6.55 min (Figure S1.1) and taxifolin (261.32 µg/g) were the most prominent compounds in the non-acidified extracts of Sumac at 280 nm. In Sumac, the acidification significantly decreased taxifolin from 261.32 µg/g to 54.61–88.14 µg/g and catechin from 67.48 µg/g to 3.88–12.16 µg/g.

In the PI570366 extracts, a significant catechin increase was observed at 4% HCl (Table 1B). Condensed tannin monomers such as catechin could not be found in SC991, which was different from the findings of Oboh and coworkers, who detected 16.09 mg/g catechin in Nigerian sorghum stems [39].

In the non-acidified extracts of Sumac and PI570366, a prominent peak at approximately 34.17 min with different spectra (Figure S1.1) was observed, and after acidification, the peak was replaced by naringenin at 33.99 min. In Sumac, no statistical differences in naringenin were found across the HCl treatments. In PI570366, naringenin could only be detected in the acidified extracts (Table 1B). In PI570366, 0.5%, 2%, and 4% HCl significantly increased the naringenin.

Glycosylation, or the hydrolysis of glycosides, converts the theoretical phenolic glycosides present in the extracts into aglycones such as naringenin, which is an aglycone form of naringin. According to Mizuno et al. [36], naringenin is considered the branching point of the metabolic pathway of 3-DA. The acidification favored the release of both naringenin and 3-DA in PI570366. Interestingly, no naringenin was found in the SC991 leaf extracts after acidification (Table 1C), even with the substantial increase in 3-DA.

In the Sumac and PI570366 extracts, no statistical difference in apigenin content was observed within the HCl treatments (Table 1A,B), whereas for the SC991 extracts, apigenin decreased significantly after acidification.

In the grain extracts, luteolin significantly increased in Sumac from 27.87 µg/g in the non-acidified extract to 46.83 µg/g at 4% HCl (Table 1A), and in PI570366 from 32.22 µg/g in the non-acidified extract to 56.70 µg/g at 2% HCl (Table 1B). In the SC991 extracts, no additional significant luteolin increase was observed with increasing HCl percentage, from which the content ranged from 260.38 µg/g to 362.26 µg/g (Table 1C).

After acidification, there were no significant differences in eriodictyol concentrations in Sumac (Table 1A), whereas in PI570366, 0.5%, 2%, and 4% eriodictyol significantly increased from 12.11 µg/g in the non-acidified extract to 75.40 µg/g in 0.5% HCl (Table 1B). No eriodictyol was detected in the SC991 extracts.

Non-Flavonoids

In Sumac, the effect of acidification on the protocatechuic acid was most significant at 0.5% HCl (Table 1A). In PI570366, acidification increased protocatechuic acid from 7.16 µg/g in the non-acidified extract to 16.39–18.84 µg/g in the acidified extracts, representing an increase of over 50% (Table 1B). In SC991, no significant changes in protocatechuic acid were detected throughout the acidification (Table 1C).

In the Sumac extracts, no significant increase was found in p-coumaric after acidification. For PI570366, a 33% increase in the p-coumaric acid content was observed from 2.19 µg/g in the non-acidified extract to 2.93 µg/g detected in the extract acidified with 0.5% HCl, and a significant decrease to zero in the p-coumaric acid concentration was observed with increases toward a higher HCl content (Table 1B). In the SC991 extracts, p-coumaric increased from 138.67 µg/g in the non-acidified extract to 173.01 µg/g in the acidified extract at 0.5% HCl, followed by decrease with the HCl concentration (Table 1C).

Caffeic acid was only detected in the Sumac bran, from which a significant decrease was observed from 9.25 µg/g in the non-acidified extract to 0 µg/g at 4% HCl (Table 1A).

In the Sumac and PI570366 extracts, no significant increase in ferulic acid was observed after HCl treatment (Table 1A,B).

Acidification reduced chlorogenic acid in all products studied. Chlorogenic acid reductions to 0 µg/g were observed in Sumac and SC991 with HCl concentrations. In PI570366, acidification decreased the chlorogenic acid significantly; however, no additional reductions were observed at higher HCl concentrations (Table 1B).

3.2.5. Thin-Layer Chromatography (TLC)

Multiple bands of compounds were detected in non-acidified extracts of SC991 and PI570366 at 366 nm. However, in Sumac, pale purple spots were observed at visible light. The separation of the bands in the stationary phase decreased according to the HCl concentration. The silica gel used as the stationary phase is a polar, and slightly acidic adsorbent. Therefore, the HCl present in some extracts negatively affected the binding ability of the analytes to the stationary phase. There are no analytical standards available for 5-methoxyluteolinidin, so we were not able to establish qualitative TLC comparisons with HPLC results for this compound.

Pure luteolinidin, after elution, resulted in a long purple spot at $R_F = 0.27$ (Figure 7a). Acidification increased the TLC intensity of luteolinidin in SC991 followed by the PI570366 extracts. However, in Sumac, the luteolinidin streaks were almost imperceptible (Figure 7a). Like the HPLC analysis, the TLC showed that luteolinidin is the dominant 3-DA in PI570366 and SC991 extracts. The low concentration of luteolinidin in Sumac detected by HPLC justifies its lack of intensity in the TLC plates.

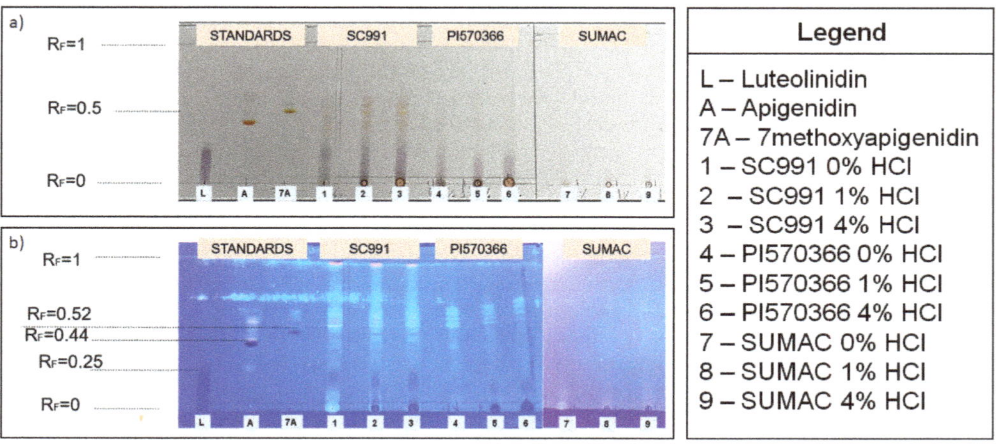

Figure 7. Thin-layer chromatography of 3-deoxyanthocyanidins and the selected extracts at (**a**) visible light and (**b**) 366 nm.

Apigenidin standard elution resulted in a red ($R_F = 0.45$) and pale purple ($R_F = 0.60$ and 0.63) spot under visible light that emitted dark blue and orange fluorescent colors at 366 nm. In the non-acidified SC991 extracts, the orange fluorescent bands matched with apigenidin at R_F = 0.63. In the acidified extracts, we found that the prominent red spot's intensity decreased (Figure 7b). At R_F = 0.44, red spots of low intensity under visible light and at R_F = 0.52, orange fluorescent spots at 366 nm were observed in the SC991 extracts, which were qualitatively comparable with the apigenidin standard (Figure 7b). In the acidified PI570366 extract at 0.5%, it was possible to observe a pale red spot at R_F = 0.44, comparable to the apigenidin standard. Although detected in Sumac via HPLC, apigenidin could not be qualitatively detected in the TLC plates.

The pure 7-methoxyapigenidin eluted as a dark blue spot at R_F = 0.52, which could be visualized in the acidified SC991 extracts at low intensity, but could not be visualized in the PI570366 and Sumac extracts.

4. Conclusions

In this work, a two-step strategy was developed with aim of (1) extracting sorghum phenolics in a high-throughput process, (2) improving sorghum phenolic detectability with post extraction acid treatments, and (3) evaluating the concentration of acids used for post extraction treatment to improve green chemistry practices.

The color analysis indicated that acidification decreased the luminosity and increased the redness of the extracts. Even with almost no statistical increase found in TPC and antioxidant capacity with acidification, the HPLC analysis showed us diversity in the phenolic species present in each genotype and how the release of the same compounds differed after acidification across the genotypes studied. For instance, in the SC991 extracts, apigenidin was not detected in the acidified extracts by HPLC because 5-methoxyluteolinidin became dominant. Increasing the HCl concentration above 0.5% did not significantly increase the anthocyanidin or the 3-DA. For this reason, our results show that the addition of HCl at concentrations of 0.5% *v/v* would be enough for the detection of the purple colorant cyanidin and 5-methoxyluteolinidin and to increase the concentration of luteolinidin and 7-methoxyapigenidin twofold.

Interestingly, cyanidin was not detected in the Sumac and PI570366 extracts until the acid treatment. Overall, the chlorogenic acid, caffeic acid, taxifolin, and quercetin significantly decreased after the acidification of the extracts. In addition, there were no statistical differences after acidification for protocatechuic and ferulic acids.

The qualitative (TLC) and quantitative (HPLC) analysis of 3-deoxyanthocyanidins showed that luteolinidin is the most abundant compound and its quantified amount increased with HCl addition, highlighting the byproducts of sorghum as a low-cost source of red colorants with desirable properties for industry.

Supplementary Materials: The following supporting information can be downloaded at: https://www.mdpi.com/article/10.3390/pr11072079/s1. References [23–25,40] are cited in the supplementary materials. Figure S1.1. The chromatograms at 280 nm registered for Sumac extracts. Figure S1.2. The chromatograms at 320 nm registered for Sumac extracts. Figure S1.3. The chromatograms at 510 nm registered for Sumac extracts. Figure S1.4 The chromatograms at 280 nm registered for PI570366 extracts. Figure S1.5. The chromatograms at 320 nm registered for PI570366 extracts. Figure S1.6. The chromatograms at 510 nm registered for PI570366 extracts. Figure S1.7. The chromatograms at 280 nm registered for SC991 extracts. Figure S1.8. The chromatograms at 320 nm registered for SC991 extracts. Figure S1.9. The chromatograms at 510 nm registered for SC991 extracts. Figure S1.10. The acidified extract of SC991 (1%HCl) before and after apigenidin (0.045mg/mL) spiking (a) and the comparisons within the spectra (b). Figure S2.1 The representative chromatograms of blanks, the spiked extracts and the external standards recorded at 280 nm (A, A.1 and A.2), 320 nm (B, B.1 and B.2) and 510 nm (C, C.1 and C.2). Group 1 (or G1) of standards consisted of: protocatechuic acid, 4-hydroxybenzoic acid, caffeic acid, ferulic acid, luteolinidin, 7-methoxyapigenidin, luteolin and naringenin. Group 2 (or G2) of standards consisted of: catechin, chlorogenic acid, p-coumaric acid, taxifolin, cyanidin, apigenidin, quercetin, and apigenin. The concentration of standards was 0.009 mg/mL. Figure S2.2 The representative chromatograms of spiked extracts and the external standards recorded at 280 nm for the Group 2 of standards (A) the spectrum of 4-hydroxybenzoic acid ((B) detected in the spiked sample and the spectrum of catechin standard (C). Group 2 (or G2) of standards consisted of: catechin, chlorogenic acid, p-coumaric acid, taxifolin, cyanidin, apigenidin, quercetin, and apigenin. The concentration of standards was 2.25 µg/g. Figure S2.3 The representative chromatograms of non-spiked extract mixture used for method validation at 280 nm (A), 320 nm (B) and 510 nm (C). Table S2.1. The calibration curve, correlation factor (R^2), limit of detection (LOD) and limit of quantification (LOQ). Table S2.2. Within day and between day precision evaluated at (2.25, 12.5, 22.5 µg/g) in sorghum.

Author Contributions: Conceptualization: Á.L.S., J.P., K.S. and D.S. Methodology: Á.L.S., J.P., K.S., R.P., C.H., S.S. and D.S. Validation: Á.L.S., J.P., R.P., C.H., S.S., K.S. and D.S. Formal analysis: Á.L.S., J.P., R.P., C.H., S.S., K.S. and D.S. Investigation: Á.L.S., J.P., K.S., R.P. and D.S. Resources: K.S. and D.S. Data curation: Á.L.S., J.P., R.P., C.H., S.S., K.S. and D.S. Writing—original draft: Á.L.S., J.P., K.S., R.P., C.H., S.S. and D.S. Writing—review and editing: Á.L.S., J.P., K.S., R.P., C.H., S.S. and D.S. Visualization: Á.L.S., J.P., K.S. and D.S. Supervision: K.S. and D.S. Project administration: K.S. and D.S. Funding acquisition: D.S. All authors have read and agreed to the published version of the manuscript.

Funding: The funding for this work was provided by the USDA-ARS under non-assistance cooperative agreement numbers 3020-43440-002-018S and 3020-43440-002-020S.

Data Availability Statement: Data will be uploaded into https://data.nal.usda.gov/ no later than 12 months after the publication of the manuscript.

Conflicts of Interest: The authors declare no conflict of interest.

Disclaimer: Mention of trade names or commercial products in this publication is solely for the purpose of providing specific information and does not imply recommendation or endorsement by the US Department of Agriculture. USDA is an equal opportunity provider and employer.

Abbreviations

ASE	Accelerated solvent extraction
HCl	Hydrochloric acid
HPLC	High-performance liquid chromatography
LC-MS	Liquid chromatography–mass spectrometry
PLE	Pressurized liquid extraction
TLC	Thin-layer chromatography
TPC	Total phenolic content
3-DA	3-deoxyanthocyanidins

References

1. Gleave, G.L.; Roethe, N.J.; Kemp, D.W.; Richter, B.E.; Ezzel, J.L. Automated Accelerated Solvent Extraction Apparatus and Method. U.S. Patent 5,785,856, 28 July 1998.
2. Cardenas-Toro, F.P.; Forster-Carneiro, T.; Rostagno, M.A.; Petenate, A.J.; Maugeri Filho, F.; Meireles, M.A.A. Integrated supercritical fluid extraction and subcritical water hydrolysis for the recovery of bioactive compounds from pressed palm fiber. *J. Supercrit. Fluids* **2014**, *93*, 42–48. [CrossRef]
3. Lachoz-Perez, D.; Brown, A.B.; Mudhoo, A.; Martinez, J.; Timko, M.T.; Rostagno, M.A.; Forster-Carneiro, T. Applications of subcritical and supercritical water conditions for extraction, hydrolysis, gasification, and carbonization of biomass: A critical review. *Biofuel Res. J.* **2017**, *14*, 611–626. [CrossRef]
4. Luo, X.; Cui, J.; Zhang, H.; Duan, Y. Subcritical water extraction of polyphenolic compounds from sorghum (*Sorghum bicolor* L.) bran and their biological activities. *Food Chem.* **2018**, *262*, 14–20. [CrossRef]
5. Dykes, L.; Seitz, L.M.; Rooney, W.L.; Rooney, L.W. Flavonoid composition of red sorghum genotypes. *Food Chem.* **2009**, *116*, 313–317. [CrossRef]
6. Cox, S.; Noronha, L.; Herald, T.; Bean, S.; Lee, S.-H.; Perumal, R.; Wang, W.; Smolensky, D. Evaluation of ethanol-based extraction conditions of sorghum bran bioactive compounds with downstream anti-proliferative properties in human cancer cells. *Heliyon* **2019**, *5*, e01589. [CrossRef]
7. Hou, F.; Su, D.; Xu, J.; Gong, Y.; Zhang, R.; Wei, Z.; Chi, J.; Zhang, M. Enhanced Extraction of Phenolics and Antioxidant Capacity from Sorghum (*Sorghum bicolor* L. Moench) Shell Using Ultrasonic-Assisted Ethanol–Water Binary Solvent. *J. Food Process. Preserv.* **2016**, *40*, 1171–1179. [CrossRef]
8. Wu, Y.; Wang, Y.; Liu, Z.; Wang, J. Extraction, Identification and Antioxidant Activity of 3-Deoxyanthocyanidins from Sorghum bicolor L. Moench Cultivated in China. *Antioxidants* **2023**, *12*, 468. [CrossRef] [PubMed]

9. Herrman, D.A.; Brantsen, J.F.; Ravisankar, S.; Lee, K.-M.; Awika, J.M. Stability of 3-deoxyanthocyanin pigment structure relative to anthocyanins from grains under microwave assisted extraction. *Food Chem.* **2020**, *333*, 127494. [CrossRef]

10. Barros, F.; Dykes, L.; Awika, J.M.; Rooney, L.W. Accelerated solvent extraction of phenolic compounds from sorghum brans. *J. Cereal Sci.* **2013**, *58*, 305–312. [CrossRef]

11. Hefni, M.E.; Amann, L.S.; Witthöft, C.M. A HPLC-UV Method for the Quantification of Phenolic Acids in Cereals. *Food Anal. Methods* **2019**, *12*, 2802–2812. [CrossRef]

12. İlbay, Z.; Şahin, S.; Büyükkabasakal, K. A novel approach for olive leaf extraction through ultrasound technology: Response surface methodology versus artificial neural networks. *Korean J. Chem. Eng.* **2014**, *31*, 1661–1667. [CrossRef]

13. Chaves, J.O.; Souza, M.C.; Silva, L.C.; Lachoz-Perez, D.; Torres-Mayanga, P.C.; Machado, A.P.F.; Forster-Carneiro, T.; Vazquez-Espinoza, M.; Gonzalez-de-Peredo, A.V.; Barbero, G.F.; et al. Extraction of Flavonoids From Natural Sources Using Modern Techniques. *Front. Chem.* **2020**, *8*, 507887. [CrossRef] [PubMed]

14. Wizi, J.; Wang, L.; Hou, X.; Tao, Y.; Ma, B.; Yang, Y. Ultrasound-microwave assisted extraction of natural colorants from sorghum husk with different solvents. *Ind. Crops Prod.* **2018**, *120*, 203–213. [CrossRef]

15. Paunović, D.Đ.; Mitić, S.S.; Kostić, D.A.; Mitić, M.N.; Stojanović, B.T.; Pavlović, J.L. Kinetics and Thermodynamics of the Solid-Liquid Extraction Process of Total Polyphenols From Barley. *Adv. Technol.* **2015**, *3*, 58–63. [CrossRef]

16. Ju, Z.Y.; Howard, L.R. Effects of Solvent and Temperature on Pressurized Liquid Extraction of Anthocyanins and Total Phenolics from Dried Red Grape Skin. *J. Agric. Food Chem.* **2003**, *51*, 5207–5213. [CrossRef]

17. Thermo Fisher Scientific. *Dionex ASE 350*; Thermo Fisher Scientific: Waltham, MA, USA, 2011.

18. Herald, T.J.; Gadgil, P.; Tilley, M. High-throughput micro plate assays for screening flavonoid content and DPPH-scavenging activity in sorghum bran and flour. *J. Sci. Food Agric.* **2012**, *92*, 2326–2331. [CrossRef]

19. Plank, D.W.; Szpylka, J.; Sapirstein, H.; Woolard, D.; Zapf, C.M.; Lee, V.; Chen, C.-Y.O.; Liu, R.H.; Tsao, R.; Dusterloch, A.; et al. Determination of Antioxidant Activity in Foods and Beverages by Reaction with 2,2′-Diphenyl-1-Picrylhydrazyl (DPPH): Collaborative Study First Action 2012.04. *J. AOAC Int.* **2012**, *95*, 1562–1569. [CrossRef] [PubMed]

20. Irakli, M.N.; Samanidou, V.F.; Biliaderis, C.G.; Papadoyannis, I.N. Simultaneous determination of phenolic acids and flavonoids in rice using solid-phase extraction and RP-HPLC with photodiode array detection. *J. Sep. Sci.* **2012**, *35*, 1603–1611. [CrossRef]

21. Lee, H.-S.; Santana, Á.L.; Peterson, J.; Yucel, U.; Perumal, R.; De Leon, J.; Lee, S.-H.; Smolensky, D. Anti-Adipogenic Activity of High-Phenolic Sorghum Brans in Pre-Adipocytes. *Nutrients* **2022**, *14*, 1493. [CrossRef]

22. Speranza, S.; Knechtl, R.; Witlaczil, R.; Schönlechner, R. Reversed-Phase HPLC Characterization and Quantification and Antioxidant Capacity of the Phenolic Acids and Flavonoids Extracted From Eight Varieties of Sorghum Grown in Austria. *Front. Plant Sci.* **2021**, *5*, 769151. [CrossRef] [PubMed]

23. Irakli, M.N.; Samanidou, V.F.; Biliaderis, C.G.; Papadoyannis, I.N. Development and validation of an HPLC-method for determination of free and bound phenolic acids in cereals after solid-phase extraction. *Food Chem.* **2012**, *134*, 1624–1632. [CrossRef]

24. Xiong, Y.; Zhang, P.; Warner, R.D.; Shen, S.; Johnson, S. HPLC-DAD-ESI-QTOF-MS/MS qualitative analysis data and HPLC-DAD quantification data of phenolic compounds of grains from five Australian sorghum genotypes. *Data Br.* **2020**, *33*, 106584. [CrossRef] [PubMed]

25. Bae, H.; Jayaprakasha, G.K.; Jifon, J.; Patil, B.S. Extraction efficiency and validation of an HPLC method for flavonoid analysis in peppers. *Food Chem.* **2012**, *130*, 751–758. [CrossRef]

26. Shapla, U.M.; Solayman, M.; Alam, N.; Khalil, M.; Gan, S.H. 5-Hydroxymethylfurfural (HMF) levels in honey and other food products: Effects on bees and human health. *Chem. Cent. J.* **2018**, *12*, 35. [CrossRef] [PubMed]

27. Brantsen, J.F.; Hermann, D.; Ravisankar, S.; Awika, J.M. Effect of tannins on microwave-assisted extractability and color properties of sorghum 3-deoxyanthocyanins. *Food Res. Int.* **2021**, *148*, 110612. [CrossRef] [PubMed]

28. Kayodé, A.P.P.; Bara, C.A.; Dalodé-Vieira, G.; Linnemann, A.R.; Mout, M.J.R. Extraction of antioxidant pigments from dye sorghum leaf sheaths. *LWT Food Sci. Technol.* **2012**, *46*, 49–55. [CrossRef]

29. Przybylska-Balcerek, A.; Frankowski, J.; Stuper-Szablewska, K. The influence of weather conditions on bioactive compound content in sorghum grain. *Eur. Food Res. Technol.* **2020**, *246*, 13–22. [CrossRef]

30. Di Meo, F.; Lemaur, V.; Cornil, J.; Lazzaroni, R.; Duroux, J.-L.; Olivier, Y.; Trouillas, P. Free Radical Scavenging by Natural Polyphenols: Atom versus Electron Transfer. *J. Phys. Chem. A* **2013**, *117*, 2082–2092. [CrossRef]

31. Zhao, J.; Ou, S.; Ding, S.; Wang, Y.; Wang, Y. Effect of activated charcoal treatment of alkaline hydrolysates from sugarcane bagasse on purification of p-coumaric acid. *Chem. Eng. Res. Des.* **2011**, *89*, 2176–2181. [CrossRef]

32. Nuutila, A.M.; Kammiovirta, K.; Oksman-Caldentey, K.-M. Comparison of methods for the hydrolysis of flavonoids and phenolic acids from onion and spinach for HPLC analysis. *Food Chem.* **2002**, *76*, 519–525. [CrossRef]

33. Wu, G.; Johnson, S.K.; Fang, Z. Sorghum non-extractable polyphenols: Chemistry, extraction and bioactivity. In *Non-Extractable Polyphenols and Carotenoids: Importance in Human Nutrition and Health*; Saura-Calixto, F., Pérez-Jiménez, J., Eds.; Royal Society of Chemistry: London, UK, 2018; pp. 326–344.

34. Ofosu, F.K.; Elahi, F.; Daliri, E.B.M.; Yeon, S.-J.; Ham, H.J.; Kim, J.-H.; Han, S.-I.; Oh, D.-W. Flavonoids in Decorticated Sorghum Grains Exert Antioxidant, Antidiabetic and Antiobesity Activities. *Molecules* **2020**, *25*, 2854. [CrossRef] [PubMed]

35. Pontieri, P.; Pepe, G.; Campiglia, P.; Mercial, F.; Basilicata, M.G.; Smolensky, D.; Calcagnile, M.; Troisi, J.; Romano, R.; Del Giudice, F.; et al. Comparison of Content in Phenolic Compounds and Antioxidant Capacity in Grains of White, Red, and Black Sorghum Varieties Grown in the Mediterranean Area. *ACS Food Sci. Technol.* **2021**, *1*, 1109–1119. [CrossRef]

36. Mizuno, H.; Yazawa, T.; Kasuga, S.; Sawada, Y.; Ogata, J.; Ando, Y.; Kanamori, H.; Yonemaru, J.-I.; Wu, J.; Hirai, M.Y.; et al. Expression level of a flavonoid 3′-hydroxylase gene determines pathogen-induced color variation in sorghum. *BMC Res. Notes* **2014**, *7*, 761. [CrossRef]

37. Petti, C.; Kushwaha, R.; Tateno, K.; Harman-Ware, A.E.; Crocker, M.; Awika, J.; DeBolt, S. Mutagenesis Breeding for Increased 3-Deoxyanthocyanidin Accumulation in Leaves of *Sorghum bicolor* (L.) Moench: A Source of Natural Food Pigment. *J. Agric. Food Chem.* **2014**, *62*, 1227–1232. [CrossRef] [PubMed]

38. Xiong, Y.; Zhang, P.; Warner, R.D.; Fang, Z. In vitro and cellular antioxidant activities of 3-deoxyanthocyanidin colourants. *Food Biosci.* **2021**, *42*, 101171. [CrossRef]

39. Oboh, G.; Adewuni, T.M.; Ademosun, A.O.; Olasehinde, T.A. Sorghum stem extract modulates Na^+/K^+-ATPase, ecto-5′-nucleotidase, and acetylcholinesterase activities. *Comp. Clin. Path.* **2016**, *25*, 749–756. [CrossRef]

40. Parkes, R.; McGee, D.; McDonnell, A.; Gillespie, E.; Touzet, N. Rapid screening of phenolic compounds in extracts of photosynthetic organisms separated using a C18 monolithic column based HPLC-UV method. *J. Chromatogr. B* **2022**, *1213*, 123521. [CrossRef] [PubMed]

Article

Optimization of Supercritical Carbon Dioxide Fluid Extraction of Medicinal Cannabis from Quebec

Hinane Boumghar [1,2], Mathieu Sarrazin [2], Xavier Banquy [3,4], Daria C. Boffito [1], Gregory S. Patience [1] and Yacine Boumghar [2,*]

[1] Chemical Engineering Department, Polytechnique Montreal, 2500 Chemin de Polytechnique, Montreal, QC H3T 1J4, Canada; hboumghar@cmaisonneuve.qc.ca (H.B.); daria-camilla.boffito@polymtl.ca (D.C.B.); gregory-s.patience@polymtl.ca (G.S.P.)

[2] Centre d'Études des Procédés Chimiques du Québec, College Maisonneuve, 6220 Sherbrooke Est Street, Montreal, QC H1N 1C1, Canada; msarrazin@cmaisonneuve.qc.ca

[3] Faculty of Pharmacy, Axe Formulation et Analyse du Médicament (AFAM), Université de Montréal, 2900 Edouard Montpetit Blvd, Montreal, QC H3T 1J4, Canada; xavier.banquy@umontreal.ca

[4] Institut en Génie Biomédical, Faculté de Médecine, Université de Montréal, 2900 Edouard Montpetit Blvd, Montreal, QC H3T 1J4, Canada

* Correspondence: yboumghar@cmaisonneuve.qc.ca; Tel.: +1-514-255-4444 (ext. 6259)

Abstract: Research on cannabis oil has evolved to encompass the pharmaceutical industry for the therapeutic potential of the active compounds for pathologies such as Alzheimer, auto-immune disorders, and cancer. These debilitating diseases are best treated with cannabinoids such as tetrahydrocannabinol (Δ^9-THC), cannabigerol (CBG), and cannabinol (CBN), which relieve neuropathic pain and stimulate the immune system. We extracted cannabinoids from plants with supercritical CO_2 and produced an extract with a total yield close to 26%. The three-level Box–Behnken experimental design considered four factors: Temperature, pressure, CO_2 flow rate, and processing time, with predetermined parameters at low, medium, and high levels. The mathematical model was evaluated by regression analysis. The yield of Δ^9-THC and CBG reached a maximum after 2 h and 15 g/min of CO_2, 235 bar, 55 °C (64.3 g THC/100 g of raw material and 4.6 g CBG/100 g of raw material). After another 2 h of extraction time, the yield of CBN reached 2.4 g/100 g. The regression analysis identified pressure and time as the only significant factors for total yield while pressure was the only significant factor for Δ^9-THC and CBG. Time, temperature, pressure, and flow rate were all significant factors for CBN.

Keywords: cannabinoids; Box–Behnken; optimization; supercritical carbon dioxide

Citation: Boumghar, H.; Sarrazin, M.; Banquy, X.; Boffito, D.C.; Patience, G.S.; Boumghar, Y. Optimization of Supercritical Carbon Dioxide Fluid Extraction of Medicinal Cannabis from Quebec. *Processes* **2023**, *11*, 1953. https://doi.org/10.3390/pr11071953

Academic Editors: Maria Angela A. Meireles, Ádina L. Santana and Grazielle Nathia Neves

Received: 30 May 2023
Revised: 23 June 2023
Accepted: 25 June 2023
Published: 28 June 2023

1. Introduction

The cannabis plant was originally used in central Asia during the Neolithic period. Western medicine adopted medicinal cannabis in the 1800s, when the Irish physician William O'Shaughnessy and French psychiatrist Jacques-Joseph Moreau attempted to treat tetanus, rabies, and mental disorders with it [1]. For example, dronabinol, a synthetic tetrahydro-cannabinoid (Δ^9-THC), is effective for anorexia, nausea, and vomiting, and was initially approved by the FDA on 31 May 1985 [2]. The interest in cannabis interest grew, especially for the psychoactive ingredients of the plant in other medical contexts, such as weight gain in HIV-positive patients [3].

Chronic pain affects 30% of the adult population, with prognostic factors of age, baseline pain, mental health complaints, and genetic factors. In fact, patients experience the same pain at multiple sites in the body [4]. Ointments and tinctures were the first forms of medical cannabis applied to relieve soldier's pain during the American civil war [1] but they also came as oils and resins [5]. In addition to these medicinal forms, people smoke cannabis for recreational purposes or apply it as a wax [6].

In 2019, cannabis plant seeds and female flowers were recognized as pharmaceuticals and nutraceuticals by Health Canada [7]. Results from clinical trials published in 2022 confirmed the safety, tolerance, and pain reduction properties of medical cannabis for older adults [8].

Pain costs society billions of dollars in lower productivity and affects millions of people [9]. However, the scale and standardization of cannabis-based product is still a challenge. There is an urgent need to establish the efficacy of medical cannabis and standardize its properties for adjunctive therapy [10]. The European Pain Federation advises patients to use medical cannabis as an oil extract [11].

For the cannabis industry, choosing a technology depends on the composition of the natural plant, but also on the end use (recreational vs. pharmaceutical, for example). Supercritical fluid extraction (SFE) is a technology already commercialized for cannabis oil extraction for recreational purposes. It is considered a green technique because of its minimal environmental impact and the absence of non-hazardous solvents. However, the lack of data concerning the interactions between the operational parameters and biomass source remains an active area of research.

Recent studies focus on pressure and temperature, but the experimental design must also consider objectives beyond extraction yield end use, cost, and time. Several studies use supercritical CO_2 to produce pharmaceutical-grade extracts with varying pressure and temperature while maintaining the same CO_2 densities [12]: For 100 g of CBD-dominant biomass, the total extraction yield was 8.8%. Qamar and co-workers [13] reported among the highest total yields (29.7%), using 500 g of the same type of biomass, which was double their earlier studies with 1 g of biomass [14]. On a similar Indica plant, Pattnaik et al. [15] reported a 7.3% total yield for 100 g feed after 1.7 h processing time, while Rochfort et al. [5] added a decarboxylation step and processed 1 kg of the same biomass type for 10 h and obtained a similar total yield of 7.1%.

Adding ethanol as a co-solvent increases the polarity of CO_2 and yield. Kargili and Aytaç [16] increased the total yield to 9.7%, with 2% ethanol on 100 g of CBD biomass after 2 h, but decarboxylated the biomass before the extraction. Grijo [17] and Fernandez [18] also added ethanol as a co-solvent, decarboxylated, and reported 31% and 18% total yield, respectively.

Ethanol increases total yield even without the decarboxylation step [19–21] and reaches up to 22%. However, decarboxylation activates the pharmacological properties of cannabinoids [22], which is missing in many studies.

From the above-mentioned studies, only four (Grijo, Rovetto, Lewis, and Fernandez) tested Δ^9-THC-dominant biomass, reporting 37%, 17%, 26%, and 21% total yields, respectively, with correspondingly large differences in the mass treated (6 g, 500 g, 0.25–3.75 g, and 3.7–5.1 g). Ethanol was the co-solvent in all cases. Although the extraction conditions were similar, two factors that account for the differences in total yield are the CO_2-to-feed ratio and the CO_2 density. The yield from 600 g of THC dominant biomass was 7.2% with 17% ethanol and 1 h processing time [23], while for 8 g of biomass, it reached 26% with 5% ethanol and 4 h processing time [24], both without decarboxylation. The yield from decarboxylated THC-CBD balanced biomass was 18% with 5% ethanol and 16% in the absence of ethanol [25]. Ethanol increases the yield and concentration of cannabinoids but considering the International Council for Harmonisation threshold for solvent residue in the pharmaceutical formulations, which is 0.005 v/v, the quality of the product should be prioritized on the quantity and total yield.

Here, we optimize the supercritical CO_2 conditions to extract Δ^9-THC, CBG, and CBN from 15 g of LSD-balanced hybrid, THC-rich cannabis biomass from Quebec, targeted by a Canadian company to initiate standardization of a nutraceutical grade natural extract, with the perspective of providing products to meet the increasing demand of the Canadian market. This contribution to standardization is an important step in the Goods Manufacturing Practices (GMP) compliance process.

2. Materials and Methods

2.1. Plant Material

We purchased 1 kg of dried and crushed Quebec LSD-type Cannabis flowers (18–28% THC) from QCGold TECH, (Saint-André-Avellin, QC, Canada). The cannabis was stored in the dark, at room temperature.

2.2. Chemicals

CO_2 (99% purity) used for the supercritical extraction was from Oxymed (Montréal, QC, Canada). HPLC solvents, methanol (MeOH), and formic acid (HCOOH) were of analytical grade from Techni Science (Oisterwijk, The Netherlands). The standards used for the HPLC quantification are THCA, Δ^9-THC, CBD, CBG, and CBN, of 1.0 mg/mL in methanol, bought from Sigma Aldrich (Oakville, ON, Canada).

2.3. Experimental Method

2.3.1. Decarboxylation

The cannabis plants were decarboxylated at 120 °C for 90 min in a Thermo Electron Corp. (Waltham, MA, USA) oven (model 6520 series) that employs gravity convection as a method of heat transfer.

2.3.2. Supercritical Fluid Extraction (SFE) Protocol

A half-liter Thar Technologies extraction vessel was equipped with a removable basket that facilitated the transfer of material to and from the system. The basket was 230 mL. The reactor was also equipped with a controlled heating element.

In each run, 15 g of dried decarboxylated cannabis was placed in the extraction vessel (Figure 1). The CO_2 was filled from a 50 L cylinder and compressed to the desired pressure by a model P50 CO_2 pump from Thar Technologies (Pittsburgh, PA, USA). A manual backpressure regulator maintained the pressure at the prescribed settings (BPRV, model BP66-1A11QEQ151, 0–10,000 PSIG, GO Regulator, Wajax, Montreal, QC, Canada). The extract was separated from the solvent when the pressure dropped to 50 bar in the receiver vessel.

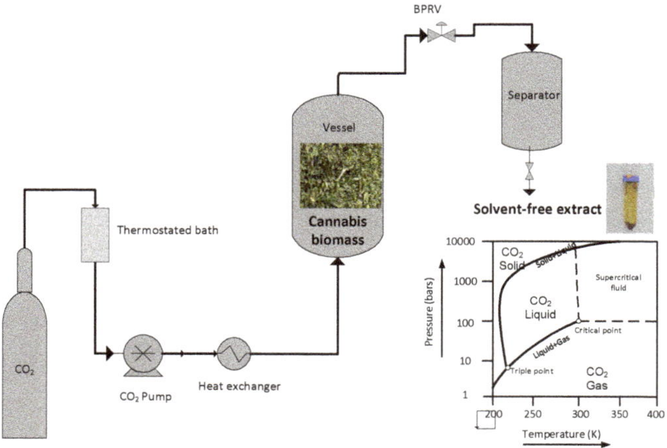

Figure 1. Supercritical carbon dioxide extraction unit.

After extraction, an analytical balance weighed the estimated yield, which was expressed as a mass percentage on a dry basis (% d.b.).

The total yield (Y) is the ratio of the mass of extract, M_e, to the mass of raw material loaded to the reaction, M_{rm},

$$Y = \frac{M_e}{M_{rm}} * 100 \tag{1}$$

The yield for each cannabinoid in g/100 g raw material, Y_i, is the ratio of the mass percentage of the cannabinoid, M_i ($X_i M_e/100$), to M_{rm}

$$Y_{Cannabinoid,i} = \frac{M_i}{M_{rm}} * 100 \tag{2}$$

$$X_i = \frac{C_i}{C_s} \cdot 100 \tag{3}$$

$$C_s = \frac{M_a}{V_s} \tag{4}$$

where:

X_i: Mass percentage of the cannabinoid i, in g/100 g of extract.
C_i: Concentration of cannabinoid i, measured by HPLC.
C_s: Concentration of the analyzed solution of cannabinoid i.
M_a: Mass of analyzed cannabinoid i.
V_s: Volume of analyzed solution of cannabinoid i.

2.3.3. Box–Behnken Experimental Design

We applied a Box–Behnken experimental design with 4 factors at 3 levels (Table 1). Ranges for the 4 factors were chosen based on literature data [26]. A randomization factor was added to the experimental plan to restrict the variability of the response due to external factors, for a total of 27 experimental runs. The central point was repeated three times. All responses are expressed in a second-order polynomial equation, as a function of independent variables, according to Equation (5):

$$Z = a_0 + a_1X_1 + a_2X_2 + a_3X_3 + a_4X_4 + a_{11}X_1{}^2 + a_{22}X_2{}^2 + a_{33}X_3{}^2 + a_{44}X_4{}^2 + a_{12}X_1X_2 + a_{13}X_1X_3 + a_{14}X_1X_4 + \\ a_{23}X_2X_3 + a_{24}X_2X_4 + a_{34}X_3X_4 \tag{5}$$

where Z is the response, a_0, a_1, a_2, a_3, and a_4 are the linear coefficients, a_{11}, a_{22}, a_{33}, and a_{44} are the quadratic coefficients, and a_{12}, a_{13}, a_{14}, a_{23}, a_{24}, and a_{34} are the interaction coefficients for the independent variables X_1 (temperature), X_2 (pressure), X_3 (flowrate), and X_4 (time). Statistical analysis was performed using Statistica® software (14.0.0.15). Analysis of an experiment with three level factors for Box–Behnken designs ($\alpha < 0.05$) and a model of 2-way interactions (linear.quadratic.) were used to evaluate the model's fit (the block effect was excluded).

Table 1. Range and variables for the experimental design.

Independent Variables	Levels		
	Low (−1)	Average (0)	High (+1)
T (°C)	40	55	70
P (bar)	150	235	320
$Q\,CO_2$ (g/min)	5	10	15
t (h)	2	3	4

The correlation between the predicted and observed data was established using the coefficient of determination (R^2). The accuracy of the model was evaluated using the values of R^2 and R^2 adjusted. The model equation can be used for interpolation in the experimental domain because it defines the true behavior of the system and has acceptable values of R^2 [27].

2.3.4. HPLC Analysis

All chromatographic analyses were performed using an Agilent 1260 Infinity Quaternary HPLC (Agilent Technologies Canada, Montreal, QC, Canada), including a quaternary pump, a solvent degasser, an autosampler, and a column temperature regulation module. A 1260 Agilent photodiode-array detector (DAD) with a Phenomenex Kinetex® (Torrance, CA, USA) C18 100 Å column (50 × 2.1 mm ID and 2.6 µm particle size) measured the concentration of the extract at a wavelength of 220 nm. Data acquisition and integration were performed with MassHunter Quantitative Analysis Software (6.0.388.0).

The mobile phase A was a mixture of 5% MeOH, 94.9% H_2O, and 0.1% HCOOH. Mobile phase B was 99.9% MeOH and 0.1% HCOOH. The column temperature was maintained at 40 °C with a mobile phase flow rate of 0.4 mL/min.

We injected 7 µL and the total run time was 26 min. A variable gradient was used, starting with 48% B, gradually increasing to 88% B over 18 min, then to 100% B after 1 min, and decreasing to 48% B after 7 min.

Standards at 100, 50, 25, 12.5, 6.25, and 3.125 mg/L were prepared for the calibration curve, for each cannabinoid (THCA, Δ^9-THC, CBDA, CBD, CBN, and CBG).

The 7 µL sample was drawn from a solution of 25 mg of extract charged to a 25 mL flask with 20 mL of methanol. All samples were filtered and loaded into the sample vial, then injected into the HPLC. For Δ^9-THC analysis, we diluted the sample by a factor of 10 with methanol, since LSD cannabis is a THC-rich plant.

3. Results

3.1. Effects of Decarboxylation

Decarboxylation of the cannabis biomass was performed to convert the acidic cannabinoids into their neutral forms, making them more extractable because of their lower polarity [20] and biopharmacologically active [22]. According to a study that compared SC-CO_2 extraction with and without decarboxylation, the first extract contained 5- to 10-fold higher CBD and Δ^9-THC content [25].

Decarboxylation increased Δ^9-THC yield and reduced total THCA (Table 2). Moreover, the quantity of CBN increased due to the oxidation of CBNA and Δ^9-THC. CBG increased as a product of CBGA oxidation.

Table 2. Cannabinoids concentration before and after decarboxylation.

Cannabinoids	CBN	THCA	Δ^9-THC	CBG	CBDA	CBD
	mg/100 g Raw Material (RM)					
Before	0.31	22.4	1.11	/	0.2	0
After	0.67	0	15.7	0.48	0	0.07

In the three natural synthetic pathways for the most studied cannabinoids [12,22,28], (Figure 2), (1) CBGA decarboxylates to CBG and biosynthesizes CBDA, CBCA, and THCA; (2) CBCA decarboxylates to CBC, CBDA to CBD, and THCA to Δ^9-THC, but also oxidizes to CBNA; and (3) CBNA potentially decarboxylates to CBN, and finally, Δ^9-THC also oxidizes to CBN.

3.2. Supercritical Fluid Extraction Yields

In this study, the effects of four parameters (pressure, temperature, time, and CO_2 flow rate) on cannabinoid extraction are investigated. Table 3 summarizes all extraction results. The yield is the total extract weight recovery, and cannabinoids (THC, CBG, and CBN) content is expressed as a weight percentage of each one in the dry extract.

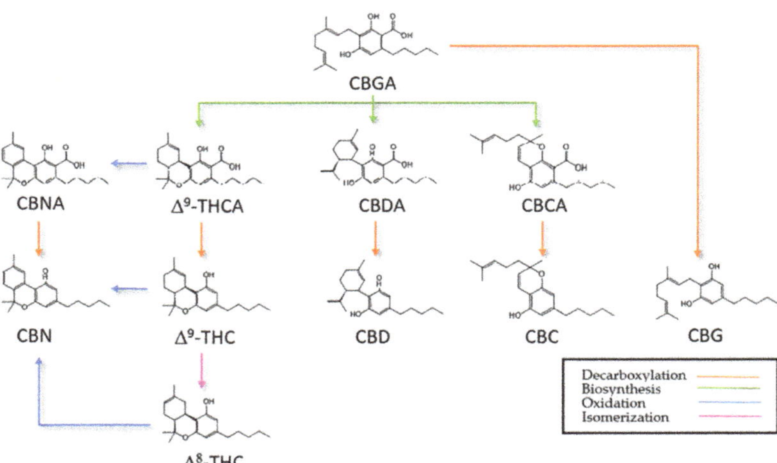

Figure 2. Simplified cannabinoid synthetic pathway: Decarboxylation, biosynthesis [29].

Table 3. Extraction yield and cannabinoid content.

Run	T °C	P Bar	Q g/min	t h	Yield %	Δ^9-THC	CBN	CBG
						g/100 g Extract		
1	40	150	10	3	11.1	62.0	2.1	3.7
2	70	150	10	3	18.2	51.1	1.7	4.1
3	40	320	10	3	11.6	48.2	1.8	2.6
4	70	320	10	3	22.9	50.3	1.8	4.6
5	55	235	5	2	7.4	62.6	2.2	3.9
6	55	235	15	2	22.7	64.3	2.1	4.6
7	55	235	5	4	16.0	59.1	2.2	4.0
8	55	235	15	4	25.9	52.3	2.4	4.1
9	55	235	10	3	11.3	45.0	1.8	3.1
10	40	235	10	2	15.2	55.7	1.5	5.8
11	70	320	10	2	18.0	51.9	1.9	3.8
12	40	235	10	4	21.1	48.9	1.8	4.1
13	70	235	10	4	21.7	55.0	1.9	4.4
14	55	150	5	3	12.0	62.0	2.0	3.8
15	55	320	5	3	11.0	60.7	2.2	4.6
16	55	150	15	3	6.7	62.6	2.1	4.0
17	55	320	15	3	19.0	49.8	1.5	3.9
18	55	235	10	3	9.3	53.9	2.0	6.0
19	40	235	5	3	12.1	52.8	1.9	4.1
20	70	235	5	3	8.5	57.8	2.2	5.1
21	40	320	15	3	19.5	61.1	1.8	3.7
22	70	235	15	3	12.7	49.7	1.8	4.5
23	55	150	10	2	2.4	50.2	1.6	4.0
24	55	320	10	2	16.2	50.2	1.8	3.8
25	55	150	10	4	11.0	54.9	2.1	4.8
26	55	320	10	4	24.9	51.6	1.8	3.2
27	55	235	10	3	12.2	59.7	2.1	3.6

The conditions of run 8 (235 bar, 55 °C, and a CO_2 flow rate of 15 g/min for 4 h) had the highest yield at 25.9%. This yield is comparable to the maximum yield obtained by Gallo-Molina and co-workers (26.4%) [24], with ethanol as a co-solvent. These conditions demonstrate the effectiveness of the supercritical solvent, attributed to the elevated extraction pressure. Lower temperatures can enhance the extraction yield [30] due to the increase in SC-CO_2 solubility by increasing its density [16], but this impact becomes less significant at high pressure. This phenomenon, called the cross-over region, also affects the selectivity of the SC-CO_2 extractions [25], thus increasing the cannabinoids' volatility at higher temperatures [16], which leads us to question if variations of the extraction conditions affect the recovery of cannabinoids of interest in the range of optimal CO_2 density since it affects its polarity [25].

The yield of run 26 was also high (24.9%) at 320 bar, 55 °C, and 10 g/min of CO_2 flow rate for 2 h. Similarly, run 6 achieved a high yield of 22.7% using the same pressure, temperature, and CO_2 flow rate as run 8, but with a shorter extraction time of 2 h instead of 4 h.

Previous studies [18,21] achieved comparable yields under similar pressure conditions, but with different biomass compositions (CBD-rich and THC-rich, respectively), and employed ethanol as a co-solvent in the latter case. Gallo-Molina previously explained how the raw material composition and extraction conditions impact cannabinoid content [24]. These similar recoveries, despite the use of co-solvent and shorter processing time, could be due to the variation of the CO_2-to-feed ratio, considering the CO_2 density is the same (same pressure and temperature).

The highest extraction yield of 25.9% was achieved by employing a combination of the longest extraction time (4 h), the highest flow rate (15 g/min), and a medium pressure of 235 bar. These conditions also resulted in high Δ^9-THC recoveries. However, due to the limited number of studies focusing on cannabinoid content using neat CO_2, it is premature to draw definitive conclusions about the optimal Δ^9-THC recovery conditions for different biomass. Fernandez et al. quantified Δ^9-THC at 63.6 g/100 g extract, using similar biomass [18], but the inclusion of ethanol as a co-solvent enhances CO_2 solubility, compromising a direct comparison. In contrast, the lowest extraction yields (2.4%, 6.7%, and 7.4%) were obtained using lower to medium flow rates, shorter extraction times of 2–3 h, and pressures in the low to medium range.

Interestingly, the three center point runs (9-18-27) resulted in low extraction yields but high Δ^9-THC content, suggesting potential medicinal value under these specific process conditions. Thus, we confirm that CO_2 density may have a limited impact on the cannabinoid's solubility, even at the optimal CO_2 density of 818 kg/m^3 as reported by Qamar [25].

ANOVA analyses and effect estimates confirmed that pressure and time have a significant impact ($p < 0.05$) on extraction yield (Table 4). The fitted equation for Δ^9-THC content in the cannabis extract shows that only pressure positively affected the content, while temperature, flow rate, and time had no effect. However, a median temperature (55 °C) and pressure (235 bar) with high flow rate and low time were associated with high Δ^9-THC and CBG yields. When time increases, Δ^9-THC and CBG yields decrease, whereas CBN yield increases.

Table 4. Effect estimates for the total yield.

Factors	Effect	*p*-Value	Coefficient
Mean interaction	16	0.0000	16
T	0.4	0.89	0.2
P	8	0.014	4
Q	5.5	0.081	2.7
t	−4.2	0.049	−2.1

Moreover, according to the three best extraction conditions for CBN yield, high temperature increases CBN yield, while, conversely, the Δ^9-THC and CBG yield decreases (Table 3). This leads us to choose a point where the extract contains the maximum of the three cannabinoids simultaneously.

The optimal SC-CO_2 runs were selected using the desirability value after varying the constraints. Run 6 (235 bar, 55 °C, 15 g/min, 2 h) was optimal and was used for verification.

Conditions in run 8 (235 bar, 55 °C, 15 g/min, 4 h) are the same as run 6, but with an extended processing time.

The first-order term of temperature ($X1$), flow rate ($X3$), and time ($X4$) had a significant effect ($p < 0.05$) on CBN in raw extract (Table 5). The first-order term of pressure ($X2$) had a significant effect ($p < 0.05$) on Δ^9-THC, CBG, and CBN extraction yields. The second-order interactions between time and flowrate ($X3\ X4$) and pressure and flowrate ($X2\ X3$) also had a significant effect ($p < 0.05$) on CBN content in raw extract.

Table 5. Analysis of variance.

Factor	Δ^9-THC			CBG			CBN		
	SS	MS	*p*-Value	SS	MS	*p*-Value	SS	MS	*p*-Value
T	0.01	0.01	0.97	0.001	0.001	0.77	0.01	0.01	0.04
P	39.8	39.8	0.04	0.06	0.06	0.03	0.04	0.04	0.02
Q	14.9	14.9	0.19	0.01	0.01	0.24	0.05	0.05	0.02
t	11.5	11.5	0.25	0.02	0.02	0.15	0.15	0.15	0.01
t(Quad) × Q(Lin)	/	/	/	/	/	/	0.16	0.16	0.01
P(Quad) × Q(Lin)	/	/	/	/	/	/	0.21	0.21	0.01

The impact of pressure was linear on the Δ^9-THC content, while temperature had no significant effect based on the statistical model (Table 6). Δ^9-THC recovery does change with temperature, which is likely due to differences in temperature-induced polarity. This observation leads us to investigate the impact of CO_2 density and the CO_2-to-biomass feed ratio to identify the optimal CO_2 density, maximizing cannabinoid extraction [25], and determine if it is independent of the solvent-to-feed ratio.

Table 6. Regression model.

Response	Model	Equation	*p*-Value
Extract yield	Linear	$Y = 16 + 4.0P - 2.1t$	0.05
Δ^9-THC	Linear	$Y = 8.6 + 2.0P$	0.05
CBG	Linear	$Y = 0.296 + 0.075P$	0.05
CBN	Quadratic	$Y = 0.657 + 0.148t + 0.183T^2Q + 0.199P^2t$	0.009

4. Discussion

Pressure is a significant factor in both the total extraction yield and the quantity of bioactive compounds extracted. Increasing the pressure from 150 to 320 bar increases the yield of cannabis extract. However, it increases total yield at the expense of Δ^9-THC, CBN, and CBG content. Higher pressures increase solvent strength and decrease extraction selectivity [24]. Recent studies [5,16,24,31,32] indicate that pressure impacts the extraction process, regardless of whether or not ethanol is a co-solvent. This impact is significant when the objective is the extraction of cannabinoids, particularly Δ^9-THC. Temperature affects extraction yield but not cannabinoid content [32].

The cannabinoid yield is influenced by the operating conditions, which depend on their respective solubilities. Perrotin et al. determined the solubility of four cannabinoids in supercritical carbon dioxide [33,34]. Molar solubility for Δ^9-THC lies between 0.20 and 2.95×10^{-4}, 42 and 72 °C, and 130 and 250 bar. The molar solubility of CBN (1.26 to 4.16×10^{-4}) and CBG (1.17 to 1.91×10^{-4}) was determined in the range of 41 to 61 °C and 113 to 206 bars.

The optimal extraction parameters (235 bar, 55 °C, and 15 g/min) produce the most Δ^9-THC and CBG at 2 h processing, but at 4 h, the CBN yield increases at the expense of Δ^9-THC and CBG. This trend agrees with kinetic models suggesting that both CBG and Δ^9-THC degrade to CBN [22]. Therefore, Δ^9-THC and CBG formulations should not exceed 2 h, while if CBN is the target molecule, longer times are better.

In our attempt to establish a correlation among the studied parameters (pressure, temperature, time, and flowrate), we hypothesize that the combination of temperature and pressure can be effectively represented by the density, while the combination of time and flowrate can be represented by the CO_2-to-feed ratio. Further investigation is needed to validate this hypothesis and explore the precise relationships between these parameters.

Our research specifically targets the extraction of THC, CBG, and CBN for medical purposes, considering their complementary therapeutic effects [28] and similar solubilities, which facilitate their simultaneous extraction [20,34]. However, during this study, our focus is primarily on THC, as a preliminary test, when examining the CO_2 density and solvent-to-feed ratio.

Figure 3 illustrates that, regardless of the CO_2 density (796.94 kg/m^3), the solvent's ability to extract THC varies, resulting in both maximum and minimum THC yield. This observation holds true for all other tested densities. This can be attributed to the increase in vapor pressure, where the temperature has a greater influence on solubility compared to density, particularly at pressures above 200 bar [25]. Consequently, we conclude that CO_2 density only plays a minor role to optimize cannabinoid extraction.

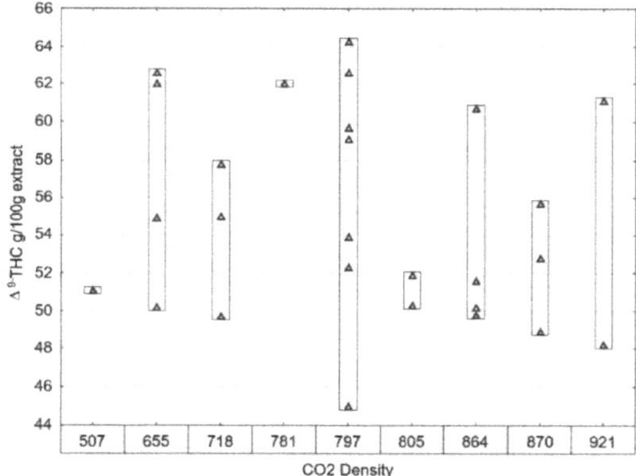

Figure 3. Variability plot of Δ^9-THC as a function of CO_2 density.

Similar to density, the CO_2-to-feed ratio as a single factor is incapable of accounting for the variance in the data (Figure 4). Specifically, for a fixed ratio of 120, the CO_2 extraction yields both the maximum and minimum Δ^9-THC concentrations. Grijo et al. [17] reported a similar crossover behavior in their investigation of solvent-to-feed ratios for cannabinoid-rich plants, indicating that higher densities may result in lower Δ^9-THC concentrations. Rovetto et al. [20] explained that while consuming the same amount of CO_2, higher pressures are associated with an expected increase in yield. Therefore, in cannabinoid extraction, the interplay between the solvent-to-feed ratio, density, and pressure should be considered. Including these factors, the extraction conditions for our Quebec LSD-type cannabis flowers of maximum Δ^9-THC, CBG, and CBN are 235 bar, 55 °C, 15 g/min, 4 h, 797 kg/m^3 of CO_2 density, and a 120 solvent-to-feed ratio.

Rochfort et al. [5] conducted a medicinal cannabis extraction study and emphasized the complexity of Δ^9-THC recoveries compared to CBD, indicating that the interaction between time and pressure plays a role. Additionally, longer extraction times may not always lead to increased total yield under the same pressure and temperature conditions. These findings support our approach of prioritizing the balance between CO_2 density and the solvent-to-feed ratio over the initial parameters studied.

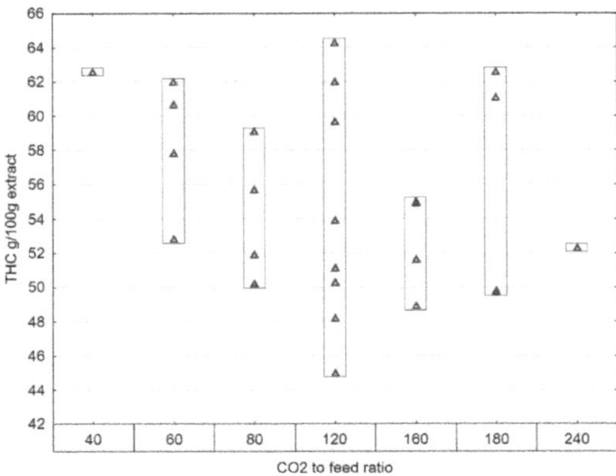

Figure 4. Δ^9-THC as a function of CO_2-to-feed ratio.

Prior to undertaking any process development, it is crucial to thoroughly test and optimize the extraction conditions specific to the composition of different cannabis plants, including cannabinoids and other compounds [33]. Consequently, we present this experimental research focusing on the LSD cannabis indica type from Quebec, aiming to support the industry in gathering data and establishing Good Manufacturing Practices (GMPs) for their facilities. This study not only sheds light on the impact of varying extraction conditions on Δ^9-THC, CBG, and CBN cannabinoids but also examines their interrelation. While the optimization primarily emphasizes Δ^9-THC yield, it can serve as a valuable starting point for investigations targeting the extraction of CBG or CBN.

5. Conclusions

Supercritical carbon dioxide is a selective extraction method for cannabinoids. Literature supports the observation that the concentration of cannabinol (CBN) increases over time due to the degradation of delta-9-tetrahydrocannabinol (Δ^9-THC) and cannabigerol (CBG), demonstrating that reaction kinetics are involved. The optimal conditions for extracting Δ^9-THC and CBG from the Quebec LSD cannabis plant, were 55 °C, 235 bar, 15 g/min CO_2, and 2 h. The optimal conditions for CBN extraction were the same but with a longer duration of 4 h. When targeting cannabinoid extraction, it is important to consider the CO_2 density (818 kg/m^3) and the CO_2-to-feed ratio (120). The pressure was identified as the primary factor influencing Δ^9-THC yield, while time was found to be a limiting factor due to Δ^9-THC degradation into CBN. Additionally, time is a significant factor in increasing total yield. The results of our experimental work, based on the Box–Behnken design of experiments, align with the existing literature: SC-CO_2 can achieve maximum extract yields of 24% (w/w) and a Δ^9-THC content of 64 g/100 g of extract. The CO_2-to-feed ratio and CO_2 density are also two key factors to balance in order to extract the highest concentrations of cannabinoids.

Author Contributions: H.B., Y.B., M.S., G.S.P., D.C.B. and X.B., conceptualization; H.B., Y.B. and M.S., methodology; Y.B. and M.S., validation; H.B., formal analysis; H.B., investigation; Y.B., resources; H.B. and Y.B., data curation; H.B., writing—original draft preparation; Y.B., G.S.P. and D.C.B., writing—review and editing; H.B., visualization; Y.B., supervision; H.B. and Y.B., project administration; Y.B. and D.C.B., funding acquisition. All authors have read and agreed to the published version of the manuscript.

Funding: NSERC College and Community Innovation program—Innovation Enhancement grants (CÉPROCQ, 544301-2019) the hosting laboratory, and Zollaris Laboratories Corporation the industrial partner, are both gratefully acknowledged for the financial support for this project. This project was also supported by Mitacs through the Mitacs Accelerate program (IT25903).

Data Availability Statement: Not applicable.

Acknowledgments: This work was supported by the Natural Sciences and Engineering Research Council of Canada (stipend allocated to Hinane Boumghar via the NSERC-CREATE PrEEmiuM program). This research was undertaken, in part, thanks to funding from the Canada Research Chair program.

Conflicts of Interest: The authors declare no conflict of interest. The funders had no role in the design of the study; in the collection, analyses, or interpretation of data; in the writing of the manuscript; or in the decision to publish the results.

References

1. Song, Y.X.; Furtose, A.; Fuoco, D.; Boumghar, Y.; Patience, G.S. Meta-analysis and review of cannabinoids extraction and purification techniques. *Can. J. Chem. Eng.* **2023**, *101*, 3108–3131. [CrossRef]
2. *Schedules of Controlled Substances: Rescheduling of the Food and Drug Administration Approved Product Containing Synthetic Dronabinol [(-)—[DELTA] Less than 9 Greater than—(Trans)-Tetrahydrocannabinol] in Sesame Oil and Encapsulated in Soft Gelatin Capsules from Schedule II to Schedule III*; Final rule; Federal Register; Department of Justice (DOJ): Washington, DC, USA; Drug Enforcement Administration (DEA): Springfield, VA, USA, 1999; Volume 64, pp. 35928–35930.
3. Tibbo, P.; Crocker, C.E.; Lam, R.W.; Meyer, J.; Sareen, J.; Aitchison, K.J. Implications of cannabis legalization on youth and young adults. *Can. J. Psychiatry* **2018**, *63*, 65–71. [CrossRef] [PubMed]
4. Landmark, T.; Dale, O.; Romundstad, P.; Woodhouse, A.; Kaasa, S.; Borchgrevink, P.C. Development and course of chronic pain over 4 years in the general population: The HUNT pain study. *Eur. J. Pain* **2018**, *22*, 1606–1616. [CrossRef] [PubMed]
5. Rochfort, S.; Isbel, A.; Ezernieks, V.; Elkins, A.; Vincent, D.; Deseo, M.A.; Spangenberg, G.C. Utilisation of design of experiments approach to optimise supercritical fluid extraction of medicinal cannabis. *Sci. Rep.* **2020**, *10*, 9124. [CrossRef]
6. Goodman, S.; Wadsworth, E.; Leos-Toro, C.; Hammond, D. Prevalence and forms of cannabis use in legal vs. illegal recreational cannabis markets. *Int. J. Drug Policy* **2020**, *76*, 102658. [CrossRef]
7. Classifying Cannabis in the Canadian Statistical System. 2019. Available online: https://www150.statcan.gc.ca/n1/pub/11-621-m/11-621-m2018105-eng.htm (accessed on 13 October 2022).
8. MacNair, L.; Kalaba, M.; Peters, E.N.; Feldner, M.T.; Eglit, G.M.L.; Rapin, L.; El Hage, C.; Prosk, E.; Ware, M.A. Medical cannabis authorization patterns, safety, and associated effects in older adults. *J. Cannabis Res.* **2022**, *4*, 50. [CrossRef]
9. Mackey, S. Future directions for pain management: Lessons from the institute of medicine pain report and the national pain strategy. *Hand Clin.* **2016**, *32*, 91–98. [CrossRef]
10. Blake, A.; Wan, B.A.; Malek, L.; De Angelis, C.; Diaz, P.; Lao, N.; Chow, E.; O'Hearn, S. A selective review of medical cannabis in cancer pain management. *Ann. Palliat. Med.* **2017**, *6*, S215–S222. [CrossRef]
11. Häuser, W.; Finn, D.P.; Kalso, E.; Krcevski-Skvarc, N.; Kress, H.G.; Morlion, B.; Perrot, S.; Schäfer, M.; Wells, C.; Brill, S. European pain federation (EFIC) position paper on appropriate use of cannabis-based medicines and medical cannabis for chronic pain management. *Eur. J. Pain* **2018**, *22*, 1547–1564. [CrossRef]
12. Jokic, S.; Jerkovic, I.; Pavic, V.; Aladic, K.; Molnar, M.; Kovac, M.J.; Vladimir-Knezevic, S. Terpenes and cannabinoids in supercritical CO_2 extracts of industrial hemp inflorescences: Optimization of extraction, antiradical and antibacterial activity. *Pharmaceuticals* **2022**, *15*, 1117. [CrossRef]
13. Qamar, S.; Torres, Y.J.M.; Parekh, H.S.; Falconer, J.R. Fractional factorial design study for the extraction of cannabinoids from CBD-dominant cannabis flowers by supercritical carbon dioxide. *Processes* **2022**, *10*, 93. [CrossRef]
14. Qamar, S.; Manrique, Y.J.; Parekh, H.S.; Falconer, J.R. Development and optimization of supercritical fluid extraction setup leading to quantification of 11 cannabinoids derived from medicinal cannabis. *Biology* **2021**, *10*, 481. [CrossRef] [PubMed]
15. Pattnaik, F.; Hans, N.; Patra, B.R.; Nanda, S.; Kumar, V.; Naik, S.N.; Dalai, A.K. Valorization of wild-type *Cannabis indica* by supercritical CO_2 extraction and insights into the utilization of raffinate biomass. *Molecules* **2023**, *28*, 207. [CrossRef]
16. Kargılı, U.; Aytaç, E. Supercritical fluid extraction of cannabinoids (Δ^9-THC and CBD) from four different strains of cannabis grown in different regions. *J. Supercrit. Fluids* **2022**, *179*, 105410. [CrossRef]
17. Ribeiro Grijó, D.; Vieitez Osorio, I.A.; Cardozo-Filho, L. Supercritical extraction strategies using CO_2 and ethanol to obtain cannabinoid compounds from Cannabis hybrid flowers. *J. CO2 Util.* **2018**, *28*, 174–180. [CrossRef]
18. Fernández, S.; Carreras, T.; Castro, R.; Perelmuter, K.; Giorgi, V.; Vila, A.; Rosales, A.; Pazos, M.; Moyna, G.; Carrera, I.; et al. A comparative study of supercritical fluid and ethanol extracts of cannabis inflorescences: Chemical profile and biological activity. *J. Supercrit. Fluids* **2022**, *179*, 105385. [CrossRef]
19. Rovetto, L.J.; Aieta, N.V. Supercritical carbon dioxide extraction of cannabinoids from *Cannabis sativa* L. *J. Supercrit. Fluids* **2017**, *129*, 16–27. [CrossRef]
20. Lewis-Bakker, M.M.; Yang, Y.; Vyawahare, R.; Kotra, L.P. Extraction of medical cannabis cultivars and the role of decarboxylation in optimal receptor responses. *Cannabis Cannabinoid Res.* **2019**, *4*, 183–194. [CrossRef]
21. Da Porto, C.; Voinovich, D.; Decorti, D.; Natolino, A. Response surface optimization of hemp seed (*Cannabis sativa* L.) oil yield and oxidation stability by supercritical carbon dioxide extraction. *J. Supercrit. Fluids* **2012**, *68*, 45–51. [CrossRef]

22. Moreno, T.; Montanes, F.; Tallon, S.J.; Fenton, T.; King, J.W. Extraction of cannabinoids from hemp (*Cannabis sativa* L.) using high pressure solvents: An overview of different processing options. *J. Supercrit. Fluids* **2020**, *161*, 104850. [CrossRef]

23. Monton, C.; Chankana, N.; Leelawat, S.; Suksaeree, J.; Songsak, T. Optimization of supercritical carbon dioxide fluid extraction of seized cannabis and self-emulsifying drug delivery system for enhancing the dissolution of cannabis extract. *J. Supercrit. Fluids* **2022**, *179*, 105423. [CrossRef]

24. Gallo-Molina, A.C.; Castro-Vargas, H.I.; Garzón-Méndez, W.F.; Martínez Ramírez, J.A.; Rivera Monroy, Z.J.; King, J.W.; Parada-Alfonso, F. Extraction, isolation and purification of tetrahydrocannabinol from the *Cannabis sativa* L. plant using supercritical fluid extraction and solid phase extraction. *J. Supercrit. Fluids* **2019**, *146*, 208–216. [CrossRef]

25. Qamar, S.; Torres, Y.J.M.; Parekh, H.S.; Falconer, J.R. Effects of ethanol on the supercritical carbon dioxide extraction of cannabinoids from near equimolar (Δ^9-THC and CBD balanced) cannabis flowers. *Separations* **2021**, *8*, 154. [CrossRef]

26. Baldino, L.; Scognamiglio, M.; Reverchon, E. Supercritical fluid technologies applied to the extraction of compounds of industrial interest from *Cannabis sativa* L. and to their pharmaceutical formulations: A review. *J. Supercrit. Fluids* **2020**, *165*, 104960. [CrossRef]

27. Quinino, R.C.; Reis, E.A.; Bessegato, L.F. Using the coefficient of determination R2 to test the significance of multiple linear regression. *Teach. Stat.* **2013**, *35*, 84–88. [CrossRef]

28. Al Ubeed, H.M.S.; Bhuyan, D.J.; Alsherbiny, M.A.; Basu, A.; Vuong, Q.V. A comprehensive review on the techniques for extraction of bioactive compounds from medicinal cannabis. *Molecules* **2022**, *27*, 604. [CrossRef]

29. Moreno, T.; Dyer, P.; Tallon, S. Cannabinoid decarboxylation: A comparative kinetic study. *Ind. Eng. Chem. Res.* **2020**, *59*, 20307–20315. [CrossRef]

30. Gaspar, F. Extraction of essential oils and cuticular waxes with compressed CO_2: Effect of extraction pressure and temperature. *Ind. Eng. Chem. Res.* **2002**, *41*, 2497–2503. [CrossRef]

31. Qamar, S.; Torres, Y.J.M.; Parekh, H.S.; Falconer, R.J. Extraction of medicinal cannabinoids through supercritical carbon dioxide technologies. *J. Chromatogr. B* **2021**, *1167*, 122581. [CrossRef]

32. Reverchon, E.; De Marco, I. Supercritical fluid extraction and fractionation of natural matter. *J. Supercrit. Fluids* **2006**, *38*, 146–166. [CrossRef]

33. Perrotin-Brunel, H.; Perez, P.C.; Van Roosmalen, M.J.E.; Van Spronsen, J.; Witkamp, G.J.; Peters, C.J. Solubility of Δ^9-tetrahydrocannabinol in supercritical carbon dioxide: Experiments and modeling. *J. Supercrit. Fluids* **2010**, *52*, 6–10. [CrossRef]

34. Perrotin-Brunel, H. Sustainable Production of Cannabinoids with Supercritical Carbon Dioxide Technologies. Ph.D. Thesis, TU Delft, Delft, The Netherlands, 2011.

Article

Modelling and Scaling-Up of a Supercritical Fluid Extraction of Emulsions Process

Diego F. Tirado [1], Albertina Cabañas [2] and Lourdes Calvo [3,*]

[1] Dirección Académica, Universidad Nacional de Colombia, Sede de La Paz, La Paz 202017, Colombia
[2] Department of Physical Chemistry, Universidad Complutense de Madrid, Av. Complutense s/n, 28040 Madrid, Spain
[3] Department of Chemical and Materials Engineering, Universidad Complutense de Madrid, Av. Complutense s/n, 28040 Madrid, Spain
* Correspondence: lcalvo@ucm.es; Tel.: +34-913-944-185

Abstract: Supercritical CO_2 ($scCO_2$) is utilized in the supercritical fluid extraction of emulsions (SFEE) to swiftly extract the organic phase (O) from an O/W emulsion. The dissolved substances in the organic phase precipitate into small particles and remain suspended in the water (W) with the aid of a surfactant. The process can be continuously conducted using a packed column in a counter-current flow of the emulsion and $scCO_2$, at moderate pressure (8–10 MPa) and temperature (37–40 °C). To ensure the commercial viability of this technique, the organic solvent must be separated from the CO_2 to facilitate the recirculation of both streams within the process while minimizing environmental impact. Thus, the aim of this work was to design a plant to produce submicron materials using SFEE, integrating the recovery of both solvents. First, experimental equilibrium data of the ternary system involved (CO_2/ethyl acetate/water) were fitted with a proper thermodynamic model. Then, simulations of the whole integrated process at different scales were carried out using Aspen Plus®, along with economical evaluations. This work proposes the organic solvent separation with a distillation column. Thus, the two solvents can be recovered and recycled to the process in almost their entirety. Furthermore, the particles in the aqueous raffinate are produced free of solvents and sterilized for further safe use. The costs showed an important economy scale-up. This work could ease the transfer of the SFEE technology to the industry.

Keywords: counter-current packed column; scale-up; continuous process; supercritical carbon dioxide; supercritical fractionation

Citation: Tirado, D.F.; Cabañas, A.; Calvo, L. Modelling and Scaling-Up of a Supercritical Fluid Extraction of Emulsions Process. *Processes* **2023**, *11*, 1063. https://doi.org/10.3390/pr11041063

Academic Editors: Maria Angela A. Meireles, Ádina L. Santana and Grazielle Nathia Neves

Received: 23 February 2023
Revised: 27 March 2023
Accepted: 29 March 2023
Published: 1 April 2023

1. Introduction

Supercritical fluids are substances that are heated and pressurized to a temperature and pressure beyond their critical point, where they exist in a state that is neither purely liquid nor purely gas [1]. The higher diffusivity and lower viscosity of these fluids compared to liquid solvents ensure a better mass transfer, which is often a limiting factor in extraction processes [2]. Among all supercritical fluids, supercritical carbon dioxide ($scCO_2$) is the most utilized in the formation of micro- and nanocarriers, owing to their distinct thermodynamic and fluid-dynamic properties [3,4]. Furthermore, $scCO_2$ has proven to be versatile in such technologies thanks to its non-flammability, non-toxicity, low cost, and availability [2,5]. These beneficial properties make supercritical fluids, and particularly $scCO_2$, popular in a wide range of particle formation and design processes [4].

Supercritical fluid extraction of emulsions (SFEE) technology has been proposed and successfully used to micronize particles and encapsulate compounds among all the available supercritical techniques [6,7]. This method capitalizes on the benefits associated with conventional emulsions and the unique properties of supercritical fluids to manufacture personalized micro- and nanoparticles and embed compounds in such carriers. By modifying the initial emulsion composition, SFEE can encapsulate either hydrophilic or lipophilic compounds. Lipophilic compounds can be encapsulated with an oil-in-water (O/W) emulsion, whereas a water-in-oil-in-water (W/O/W) emulsion can be used to enclose hydrophilic materials [8]. SFEE uses CO_2 as a solvent mainly because it has a near-ambient critical temperature (T_C, 31.1 °C) and a relatively low critical pressure (P_C, 7.4 MPa) compared to other fluids; it is innocuous, completely removable from the extract and raffinate via simple decompression, and it has also been given a "Generally Recognized As Safe" (GRAS) status, encouraging its use in the pharmaceutical and food industries [9].

SFEE is founded on the principle of utilizing $scCO_2$ to rapidly extract the organic phase (O) from an emulsion, typically an O/W emulsion, in which a bioactive compound and a coating polymer have already been dissolved. When the solvent is removed, both materials precipitate, generating a suspension of particles in the water (W) [10,11]. Aggregates are usually produced with this technology [10]; however, carriers with a core–shell structure can be also created when the compound is liquid [11]. The particles produced have a controlled size and morphology due to the use of the emulsion and the rapid kinetics of the $scCO_2$ extraction. In addition, particle agglomeration in the water phase is avoided as the particles are stabilized by the surfactant [6].

SFEE has proven to be an effective technique for encapsulating sensitive bioactive compounds without compromising their functionality [10]. This makes it a suitable method for micronizing and encapsulating compounds, such as pharmaceuticals (for protection and/or controlled release), nutraceuticals, bioactive and probiotics (for their incorporation into functional foods), cosmetics, pigments, explosives, and semiconductor precursors, among others [8]. Additionally, SFEE is highly versatile [10,12]: (*i*) the fast kinetics of the process allow for a controlled particle size and morphology; (*ii*) non-aggregated spheres are produced, and the particle size can be precisely adjusted by modifying the initial emulsion formulation; (*iii*) the produced particles remain stable over extended periods of storage; and (*iv*) the technology can be easily scaled up as a counter-current process of liquid mixtures with supercritical or near-critical fluids.

In a previous work reported by our research group [10], a stainless-steel packed column was used to encapsulate astaxanthin in ethyl cellulose, in which CO_2 was fed from the bottom and the emulsion was delivered from the top of the column. By fixing the oil–water ratio at 20/80 (ethyl acetate/water), the process was conducted at 8.0 MPa and 38 °C with a liquid to supercritical gas (L/G) ratio of 0.1 (CO_2 flow rate of 1.4 kg h^{-1}). Spherical particles with a mean diameter of 360 nm and a narrow particle size distribution (polydispersity index of 0.16) were obtained. The yielded particles had a high encapsulation efficiency of roughly 85% with an average polymer recovery of around 90%. Accordingly, the loading ratio astaxanthin/polymer was 2.1 mg astaxanthin g powder^{-1}. The aforementioned work claimed that the use of a counter-current packed column during the SFEE process offered the possibility of a high production capacity of micro- and nanoparticles with a small volume plant in only a few minutes, along with greater product homogeneity and recovery.

Although the operation has shown to be of promise, no widely implemented industrial application has been found even though industrial fractionation units do exist, which could be related to the complexity and the significant economic cost of performing basic and engineering studies on an industrial scale [10,11]. Moreover, SFEE is strongly protected by the patents of Ferro Corporation [13,14] and by the group of Professor Ernesto Reverchon from the University of Salerno [15], which could jeopardize its commercialization. Hence, performing a rigorous economic evaluation would be crucial in assessing the viability of utilizing this technology for commercial production.

The promising results of SFEE make its study and extension as a counter-current supercritical fluid fractionation technology interesting while keeping in mind that for nearly all industrial processes, the solvent, i.e., CO_2, must be recycled. Additionally, as all other technologies involving high-pressure CO_2, the industrial SFEE process does not contribute to the environmental CO_2 problem [16].

Complex problems in the industry are often not solved by hand for two reasons: associated costs and equipment availability. Additionally, in response to this, there are several simulation programs used in the chemical engineering field depending on the application and desired simulation products. In this regard, when used to its full potential, Aspen Plus® (Aspen Technology, Bedford, MA, USA) can be a very powerful tool for a chemical engineer in a variety of fields, including high-pressure processes [17]. Moreover, process integration has become increasingly important [18].

The phase equilibrium relevant to the SFEE column involves at least a ternary mixture, not including the solute and carrier, which are typically present in low concentrations [11]. Advanced thermodynamic models, utilizing equations of state with activity coefficients, are necessary to describe these complex and non-ideal mixtures at high pressure [19]. Therefore, the primary objective of this study was to accurately model the experimental equilibrium data of the ternary system involved in particle formation by SFEE. Specifically, the pseudo-ternary mixture of CO_2/ethyl acetate/water was investigated. Ultimately, this research aimed to establish a reliable method for the development of an industrial-scale SFEE process capable of producing submicron structures for the inclusion of bioactive compounds in the formulation of cosmetics, para-pharmaceuticals, and hydrophilic foods. In order to carry this out, previous laboratory-scale experimental data obtained with a counter-current packed column [10] were simulated in Aspen Plus®. Afterward, pilot- and semi-industrial scale units were simulated in the same process simulation tool, along with economical evaluations.

2. Materials and Methods

2.1. Block Diagram

Figure 1 shows the block diagram of the SFEE process of this study. The solvent (scCO$_2$) and emulsion (ethyl acetate + water) streams entered the first stage of the process (extraction). A stream exhausted in ethyl acetate, along with the newly formed nanoparticles, exited this extraction stage. In the second stage (solvent recovery), the scCO$_2$ and ethyl acetate were separated and then recycled to extraction otherwise, the process would not be profitable nor environmentally sustainable. If the particles obtained are to be used as solids, they would have to be dried. However, this last step was not considered in this simulation.

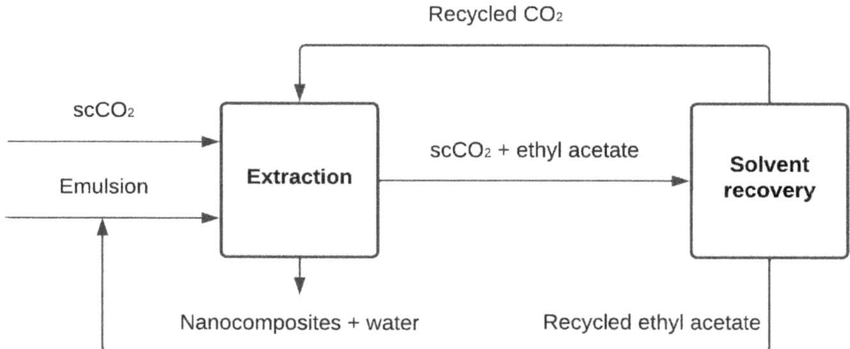

Figure 1. Block diagram of the supercritical fluid extraction of emulsion (SFEE) process.

2.2. Definition of Purities, Recoveries, and Key Components

The emulsion fed to the extraction had a mass fraction composition of 80% water and 20% ethyl acetate, as in the laboratory-scale experimental work [10]. In order to produce particles suitable for food and pharma applications, the maximum residual ethyl acetate concentration in the water suspension had to be less than 5000 ppm [20]. Additionally, the purity of the recovered organic solvent had to be at least the ethyl acetate concentration in the emulsion (mass fraction $\geq$ 80%). Finally, the CO_2 stream had to be recirculated pure to the process ($\geq$99.9%) to maintain a high solvent capacity.

2.3. Thermodynamic Modelling

To design an SFEE process, a reliable phase equilibrium model had to be established for the main column. The complex system involved was handled as a pseudo-ternary system consisting of the three major components of the system: CO_2, ethyl acetate, and water. The bioactive compound and the coating polymer were not considered since their concentration in the emulsion was very low (<1%) [10,11]. The property method selection assistant in Aspen Plus® V11 was used to select the most appropriate one. We also built on our research group's previous experience simulating high-pressure processes involving CO_2, water, and organic solvents [19,21]. Thus, the Schwartzentruber–Renon (SR-Polar), Predictive Soave–Redlich–Kwong (PSRK), and electrolyte non-random two-liquid (eNRTL) thermodynamics model were tested to describe the ternary equilibrium. The available experimental equilibrium data [22] were compared with the simulation software results, and the most accurate model was chosen to represent the experimental phase equilibrium data for simulating the SFEE column. Additionally, the best thermodynamic model predicting equilibrium data of the CO_2-ethyl acetate mixture under typical separator conditions were investigated.

2.4. High-Pressure Column Simulation

SFEE can be managed as a multistage counter-current separation. In such a case, the components distribute between the solvent (extract phase) and the liquid (raffinate phase), which counter-currently flow through the separation column. Thus, the SFEE unit was simulated in the Aspen Plus® environment using the Extract block available in the equipment model library. Extract is a rigorous model to simulate liquid–liquid extractors [23]. It was used in counter-current flow without reflux (since no high-purity extract was sought, but an exhausted raffinate) where the upper inlet stream was the emulsion, and the solvent stream was the continuous phase entering at the bottom of the column. The liquid phase exhausted in ethyl acetate was obtained as the bottom product. Meanwhile, the phase containing the CO_2 and the ethyl acetate left the column as the top product stream. To simulate, it was mandatory to specify before components, mass flow and mass fraction composition of the emulsion and CO_2, the thermodynamic model, inlet streams, and the extract column itself.

3. Results and Discussion

This study aimed to design a plant to produce submicron materials using SFEE, integrating the downstream processing with the same solvent (CO_2). Since the continuous particle production by SFEE was previously studied by our research group [10], this work was focused on the design and simulation of the process using a counter-current packed column. First, a prior selection of the thermodynamic model that better described the CO_2–ethyl acetate–water equilibrium data was carried out. Then, a proposal for the flowsheet of the plant was provided. Additionally, simulation results from different scales (lab, pilot, and semi-industrial) were performed. Finally, an economical evaluation of the process was presented.

3.1. Selection of the Thermodynamic Model

To the best of our knowledge, only the study performed by Luther et al. [22] has reported enough CO_2–ethyl acetate–water phase equilibrium data at the operating conditions used in this study. A comparison was performed between equilibrium data from those authors [22] and data retrieved from simulations employing the thermodynamic models mentioned in Section 2.3.

As a result of the comparison, PSRK was the model that best predicted the CO_2–ethyl acetate–water behavior at the condition explored. Figure 2 shows a comparison between the experimental high-pressure vapor–liquid equilibrium data at 8.5 MPa and 37 °C of the ternary system CO_2–ethyl acetate–water [22] and the fit generated by Aspen Plus® with the PSRK thermodynamic model. As can be seen in Figure 2, PSRK correctly predicted the type II behavior [24] of the mixture with less than 5% error between predicted and experimental compositions of the light (CO_2-rich) phase, over the entire ethyl acetate concentration range. Additionally, at the lowest molar fractions of ethyl acetate, the fitting in the water-rich phase was found to be even better, which further improves the predictions, as low fractions of ethyl acetate were found when the column was operated.

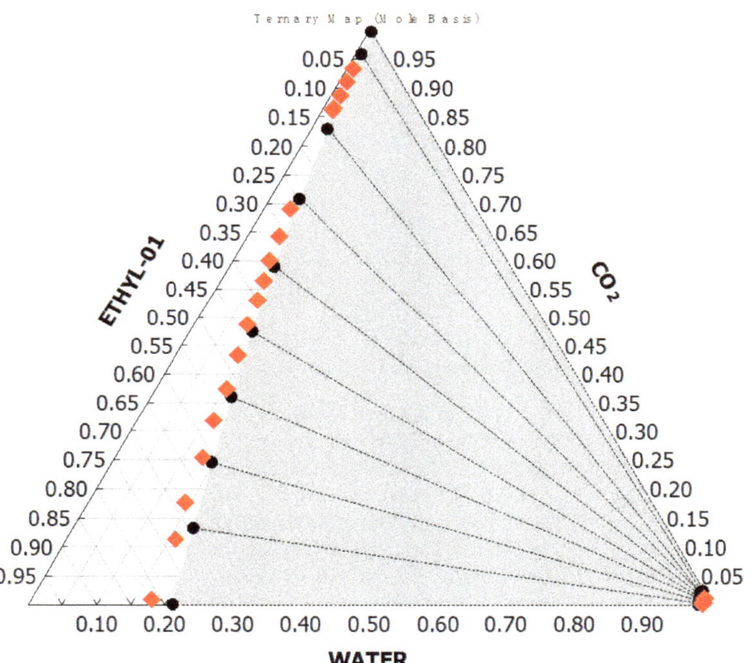

Figure 2. Ternary phase diagram (mol basis) of the system CO_2–ethyl acetate–water at 8.5 MPa and 37 °C provided by Aspen Plus® using the PSRK model (•) and compared with experimental phase equilibrium data (♦) [22].

Equilibrium data of the CO_2-ethyl acetate mixture were also needed at the typical conditions of flash separators (~22 °C and 6 MPa) as these are the room conditions of the CO_2 make-up from the storage (*i.e.*, at its vapor pressure). Experimental data under these conditions were found in the literature [25] and in the Aspen Plus® V11 database as BVLE010.

As can be seen in Figure 3, the Peng–Robinson equation of state with Wong–Sandler mixing rules correctly predicted the CO_2–ethyl acetate behavior at the separator. The load of gaseous CO_2 in ethyl acetate was relatively high in all ranges of pressures. Unfortunately, large amounts of CO_2 will dissolve in the ethyl acetate-rich phase at the same time. Therefore, the separation of ethyl acetate and CO_2 is difficult.

3.2. Operating Conditions of the Fractionation Column

Optimal operating conditions should be employed to facilitate the organic solvent extraction from the emulsion while preventing the dissolution of the bioactive compound and the polymer in sc-CO_2, as well as avoiding the loss of the emulsion through washout in the scCO$_2$ stream.

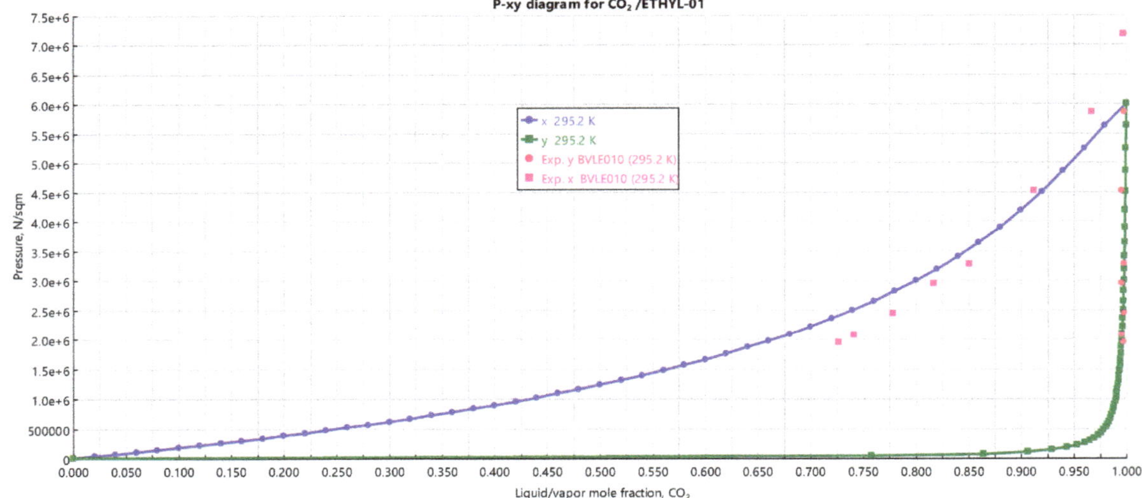

Figure 3. Experimental [25] and correlated data with Peng–Robinson equation of state with Wong–Sandler mixing rules for the system CO$_2$/ethyl acetate, at 22 °C (295 K).

The simulation proposed in this work was conducted under the same conditions as the previous experimental work [10], *i.e.*, at 8.0 MPa and 38 °C. At these conditions, the mixture of ethyl acetate/scCO$_2$ is supercritical [26] and water is only slightly soluble in scCO$_2$ [27], thus enhancing the ethyl acetate extraction from the emulsion. Additionally, 38 °C was below the glass transition temperature (T_g) of ethyl cellulose, which is roughly 133 °C [28]. Using a higher pressure was not considered since microspheres coalescence was observed in previous work [29]. Low pressure also helps reduce energy costs, which improves the viability of the process on a commercial scale.

The decision to operate at the lowest possible pressure was also influenced by the larger density difference between the scCO$_2$ (304 kg m^{-3}) and the emulsion (965 kg m^{-3}) at the column inlet [30], which promoted the formation of two separate phases in the column, which was critical for the efficient operation of the system as it prevented flooding.

3.3. Aspen Plus® Process Simulation

The proposed flowsheet for the continuous SFEE integrated process with the organic solvent recovery and the recirculation of scCO$_2$ to the process is shown in Figure 4. This same flowsheet was used to simulate the SFEE process at pilot and industrial scales.

In Figure 4, the emulsion (stream 1; mass fraction composition of 80% water and 20% ethyl acetate) entered the extract column (C-101) at the first stage (top of the column) and the scCO$_2$ (stream 2) came recirculated from the distillation column and introduced at the bottom of the column. The extract phase containing scCO$_2$ and ethyl acetate left the column as the top product stream (stream 4) and the liquid phase exhausted in the organic solvent was obtained as the bottom product (stream 3).

3.3.1. Number of Theoretical Stages of Extract Unit

It was necessary to employ only three equilibrium stages adiabatic columns using a solvent-to-feed ratio of 10 (kg kg^{-1}), to reduce the amount of ethyl acetate in the water-rich raffinate (stream 6) below 5000 ppm, as recommended [20], operating at 38 °C and 8.0 MPa.

In fact, from the simulation in Aspen Plus, the residual concentration of ethyl acetate in the obtained particles (stream 6) was roughly 37 ppm. In this regard, particles with a residual ethyl acetate concentration of 40 ppm were obtained with the laboratory-scale experiments [10], which meant that the experimental conditions were reproduced by the simulation. In the aforementioned investigation [10], the column was formed by three AISI 316 stainless steel cylindrical sections of 30 cm height connected by four cross-unions and was packed with 0.16-inch Pro-Pak® [31]. This means that the height equivalent to a theoretical stage (HETS value) would be 30 cm.

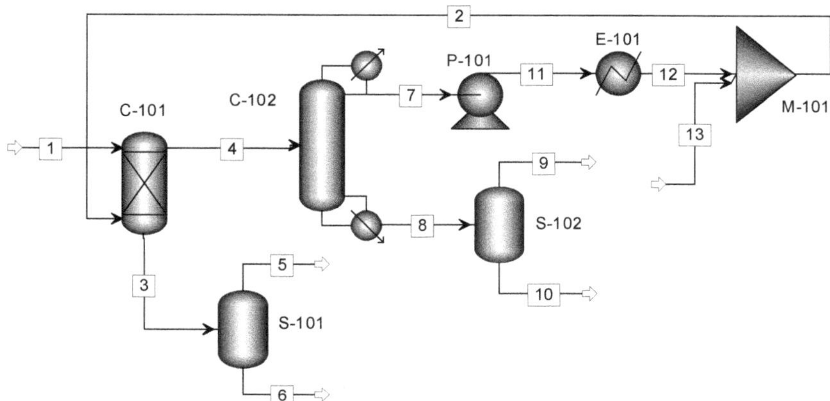

Figure 4. Proposed flowsheet of the SFEE plant.

3.3.2. Design of the Integrated Process

The economic and environmental viability of this technology on a commercial scale is heavily reliant on successfully separating the organic solvent from the $scCO_2$ for both streams to be recirculated to the process. However, ethyl acetate's high solubility in CO_2, even at low pressures [22], makes complete separation challenging. Traditional methods, such as reducing pressure in a flash tank to 5–6 MPa, which are effective in the recovery of "heavy" extracts in other supercritical extraction processes [16,32], are insufficient for complete separation. In fact, the binary analysis of the ethyl acetate-$scCO_2$ mixture showed that the concentration of ethyl acetate in the gas phase was 0.2–0.4% for all pressures below, as shown in Figure 3. Similar results were reported by Budich and Brunner [32] for the ethanol–$scCO_2$ system.

To obtain ethyl acetate (stream 8) and $scCO_2$ with a purity of over 99.9% (stream 7), an Aspen Plus® simulated distillation column as a RadFrac block (C-102 in Figure 4) with five stages, a reflux ratio of 0.2, and a distillate to feed ratio of 0.85 were used. The extract stream (stream 4) was fed to the distillation column. The resulting $scCO_2$ was compressed (stream 11) using a pump (P-101) up to 10 MPa and heated up to 38 °C (stream 12) in a heat exchanger (E-101). A pure CO_2 stream (stream 13) was required to compensate for the lost CO_2 from the raffinate and extract. Stream 13 (0.19 kg h^{-1} at lab scale) was combined with stream 12 in a mixer (M101) to recirculate the resultant stream (stream 2) as feed to the extraction column C-101. Stream 8 obtained from C-102 was split in a flash separator (S-102) at atmospheric pressure and 5 °C. The resulting stream 10 had a purity of approximately 91% ethyl acetate, with the remaining composition being 6.2% water and 2.8% CO_2. After the CO_2 vaporized, stream 10 was used directly in the formulation of the starting emulsion, resulting in a stream with 94% ethyl acetate and the rest water. Finally, the raffinate (stream 6) obtained from C-101 bottoms contained mainly particles suspended in water and had a remaining ethyl acetate concentration of around 37 ppm, as previously mentioned.

3.4. Scale-Up

The mass flow rate of lab-scale inlet streams 1, 2 and 13 in Figure 4 was increased according to the scaling factors for the pilot and industrial plants of Atelier Fluides Super-critiques from France. The characteristics of such equipment are described in [33]. The pilot plant has two pumps to supply up to 5 L and 20 L of liquid, respectively; and 80 kg of

CO_2 up to 30 MPa. The industrial plant allows a maximum liquid capacity of 50 L h^{-1} and a top CO_2 capacity of 600 kg h^{-1}. Thus, the production capacity (stream 6 in Figure 4) was increased to 2.2 kg h^{-1} for the pilot plant (a scale factor of 20), and to 33.1 kg h^{-1} (a scale factor of 300) for the industrial plant. The dimensions, packings, and capacity of the three installations are compared in Table 1.

Table 1. Dimensions, packings, capacities, and costs of the scaled SFEE installations.

	Lab	**Pilot** [a]	**Industrial** [a]
Diameter (m)	0.013	0.058	0.126
Height (m)	1	4	8
Volume (L)	0.13	11	100
Maximum liquid capacity [a] (L h^{-1})		5–20	50
Top CO_2 capacity [a] (kg h^{-1})		80	600
Packing	0.16-inch Pro-Pak	VFF Interpack (10–15 mm)	Sulzer CY
Production capacity (stream 6)	0.11	2.2	33
Equipment cost (EUR)	150,000 [b]	976,000 [c]	3,205,000 [c]
Utilities cost (EUR /year) [d]	37,284	74,568	123,736

[a] Equipment specifications of the Atelier Fluids Supercritique. [b] Given by supercritical equipment providers. [c] Estimated using the Perrut equation. [d] Provided by Aspen Plus® V11.

3.5. Economic Evaluation

We have asked manufacturers of CO_2 plants for bids for the lab-scale installation, ranging from EUR 125,000 (with only one separator) to EUR 180,000. Using an intermediate value as a basis, we used Michel Perrut's equation to estimate the price of the pilot and industrial scale facilities. Based on his experience in SEPAREX, building supercritical fluid extraction, fractionation, impregnation, and particle formation installations at all scales, Professor Perrut concluded that all prices approximate a straight line with a slope of 0.24 versus the logarithm of the product of the total net volume (V_T) of the column by the design CO_2 flow rate (Q) [34]. This means that there is a huge economy of scale in capital expenditure (CAPEX) in supercritical CO_2-based processes. With the resulting column volumes from the simulations of the pilot and industrial scale installations and considering that the CO_2 flow rates were ten times higher than the liquid feed in the respective scales, we arrived at the costs shown in Table 1.

Utility costs estimated using the APEA tool of Aspen Plus would drop from EUR 47 kg^{-1} at lab scale to EUR 0.52 kg^{-1} product for the largest scale, showing also very important savings with production capacity. If the particles had to be dried/lyophilized, this additional cost would have to be considered apart. Installation costs, assuming a payback period of 10 years, would increase the product price by 30 cents per kg for the laboratory scale, but by less than 5 cents for the industrial scale (operating 300 days per year, 24 h per day).

4. Conclusions

The SFEE column and the solvent recovery separator were designed to achieve a specified amount of residual organic in the leaving streams, according to the tight requirements of the food and pharmaceutical industries [20]. The most difficult part is the organic solvent separation from the CO_2 as they are very soluble. However, it is possible with a distillation column. Thus, the two solvents can be recovered and recycled to the process in almost their entirety. Furthermore, the particles in the aqueous raffinate are produced free of solvents. Because of the biocidal and virucidal power of the $scCO_2$, the particle suspension is sterile [21], thus easing the direct further packaging. Based on the utilities and installation costs for increasing production capacities, the importance of economy of scale in supercritical CO_2 installations is verified. On the other hand, CO_2 fractionation in packed columns is already commercial, *e.g.*, Atelier Fluids Supercritique in France, https://www.atelier-fsc.com/ (accessed on 19 January 2023) or Solutex in Spain, https://solutexcorp.com/technology/platforms/#flutex (accessed on 19 January 2023). The SFEE process could be carried out in similar installations, facilitating the technology transfer of the SFEE procedure for particle production to the industry.

Author Contributions: Conceptualization, D.F.T. and L.C.; Data curation, D.F.T. and L.C.; Formal analysis, D.F.T. and L.C.; Funding acquisition, D.F.T., A.C. and L.C.; Investigation, D.F.T.; Methodology, D.F.T. and L.C.; Project administration, A.C. and L.C.; Resources, A.C. and L.C.; Software, D.F.T. and L.C.; Supervision, L.C.; Validation, D.F.T.; Visualization, D.F.T.; Writing—original draft, D.F.T. and L.C.; Writing—review and editing, D.F.T. and L.C. All authors have read and agreed to the published version of the manuscript.

Funding: This work was financed by the Ministry of Science, Innovation and Universities of Spain (project ref. RTI2018-097230-B-I00); the Universidad Nacional de Colombia (project ref. 56099) and by a REACT-EU grant from the Comunidad de Madrid to the ANTICIPA project of Complutense University of Madrid. The funders had no role in study design, data collection and analysis, preparation of the manuscript or decision to publish.

Data Availability Statement: The data that support the findings of this study are available from the corresponding author, upon reasonable request.

Conflicts of Interest: The authors declare no conflict of interest. The funders had no role in the design of the study; in the collection, analyses, or interpretation of data; in the writing of the manuscript, or in the decision to publish the results.

References

1. Lukic, I.; Pajnik, J.; Nisavic, J.; Tadic, V.; Vági, E.; Szekely, E.; Zizovic, I. Application of the integrated supercritical fluid extraction–impregnation process (SFE-SSI) for development of materials with antiviral properties. *Processes* **2022**, *10*, 680. [CrossRef]
2. Kyzas, G.Z.; Matis, K.A. Green separation and extraction processes: Part I. *Processes* **2020**, *8*, 374. [CrossRef]
3. Drago, E.; Campardelli, R.; De Marco, I.; Perego, P. Optimization of PCL polymeric films as potential matrices for the loading of alpha-tocopherol by a combination of innovative green processes. *Processes* **2021**, *9*, 2244. [CrossRef]
4. Bagheri, H.; Hashemipour, H.; Rahimpour, E.; Rahimpour, M.R. Particle size design of acetaminophen using supercritical carbon dioxide to improve drug delivery: Experimental and modeling. *J. Environ. Chem. Eng.* **2021**, *9*, 106384. [CrossRef]
5. Marcus, Y. Some advances in supercritical fluid extraction for fuels, bio-materials and purification. *Processes* **2019**, *7*, 156. [CrossRef]
6. Prieto, C.; Calvo, L. The encapsulation of low viscosity omega-3 rich fish oil in polycaprolactone by supercritical fluid extraction of emulsions. *J. Supercrit. Fluids* **2017**, *128*, 227–234. [CrossRef]
7. Prieto, C.; Calvo, L. Supercritical fluid extraction of emulsions to nanoencapsulate vitamin E in polycaprolactone. *J. Supercrit. Fluids* **2017**, *119*, 274–282. [CrossRef]
8. Chattopadhyay, P.; Huff, R.; Shekunov, B.Y. Drug encapsulation using supercritical fluid extraction of emulsions. *J. Pharm. Sci.* **2006**, *95*, 667–679. [CrossRef] [PubMed]
9. Wang, W.; Rao, L.; Wu, X.; Wang, Y.; Zhao, L.; Liao, X. Supercritical carbon dioxide applications in food processing. *Food Eng. Rev.* **2021**, *13*, 570–591. [CrossRef]
10. Tirado, D.F.; Palazzo, I.; Scognamiglio, M.; Calvo, L.; Della Porta, G.; Reverchon, E. Astaxanthin encapsulation in ethyl cellulose carriers by continuous supercritical emulsions extraction: A study on particle size, encapsulation efficiency, release profile and antioxidant activity. *J. Supercrit. Fluids* **2019**, *150*, 128–136. [CrossRef]
11. Tirado, D.F.; Latini, A.; Calvo, L. The encapsulation of hydroxytyrosol-rich olive oil in Eudraguard® protect via supercritical fluid extraction of emulsions. *J. Food Eng.* **2021**, *290*, 110215. [CrossRef]
12. Prieto, C.; Duarte, C.M.M.; Calvo, L. Performance comparison of different supercritical fluid extraction equipments for the production of vitamin E in polycaprolactone nanocapsules by supercritical fluid extraction of emulsionsc. *J. Supercrit. Fluids* **2017**, *122*, 70–78. [CrossRef]
13. Chattopadhyay, P.; Huff, R.W.; Seitzinger, J.S.; Shekunov, B.Y. Composite Particles and Method for Preparing. US6966990B2, 22 November 2005.
14. Chattopadhyay, P.; Huff, R.W.; Seitzinger, J.S.; Shekunov, B.Y. Particles from Supercritical Fluid Extraction of Emulsion. US6998051B2, 14 February 2006.
15. Reverchon, E.; Della Porta, G. Continuous Process for Microspheres Production by Using Expanded Fluids. US20100203145A1, 14 January 2014.
16. Brunner, G. Counter-current separations. *J. Supercrit. Fluids* **2009**, *47*, 574–582. [CrossRef]
17. Calvo, L.; Prieto, C. The teaching of enhanced distillation processes using a commercial simulator and a project-based learning approach. *Educ. Chem. Eng.* **2016**, *17*, 65–74. [CrossRef]

18. Li, Z.; Tian, S.; Zhang, D.; Chang, C.; Zhang, Q.; Zhang, P. Optimization study on improving energy efficiency of power cycle system of staged coal gasification coupled with supercritical carbon dioxide. *Energy* **2022**, *239*, 122168. [CrossRef]
19. Prieto, C.; Calvo, L.; Duarte, C.M.M. Continuous supercritical fluid extraction of emulsions to produce nanocapsules of vitamin E in polycaprolactone. *J. Supercrit. Fluids* **2017**, *124*, 72–79. [CrossRef]
20. *ICH Impurities: Guideline for Residual Solvents Q3C(R6)*; ICH: London, UK, 2016; Available online: https://database.ich.org/sites/default/files/Q3C-R6_Guideline_ErrorCorrection_2019_0410_0.pdf (accessed on 22 February 2023).
21. Martín-Muñoz, D.; Tirado, D.F.; Calvo, L. Inactivation of *Legionella* in aqueous media by high-pressure carbon dioxide. *J. Supercrit. Fluids* **2022**, *180*, 105431.
22. Luther, S.K.; Schuster, J.J.; Leipertz, A.; Braeuer, A. Non-invasive quantification of phase equilibria of ternary mixtures composed of carbon dioxide, organic solvent and water. *J. Supercrit. Fluids* **2013**, *84*, 146–154. [CrossRef]
23. Aspen-Technology Aspen Plus V 11.0 2020. Available online: https://www.aspentech.com/en/products/engineering/aspen-plus (accessed on 22 February 2023).
24. Brunner, G. *Gas Extraction*; Topics in Physical Chemistry; Steinkopff: Heidelberg, Germany, 1994; Volume 4, ISBN 978-3-662-07382-7.
25. da Silva, M.V.; Barbosa, D.; Ferreira, P.O.; Mendonça, J. High pressure phase equilibrium data for the systems carbon dioxide/ethyl acetate and carbon dioxide/isoamyl acetate at 295.2, 303.2 and 313.2 K. *Fluid Phase Equilib.* **2000**, *175*, 19–33. [CrossRef]
26. Smith, R.L.; Yamaguchi, T.; Sato, T.; Suzuki, H.; Arai, K. Volumetric behavior of ethyl acetate, ethyl octanoate, ethyl laurate, ethyl linoleate, and fish oil ethyl esters in the presence of supercritical CO_2. *J. Supercrit. Fluids* **1998**, *13*, 29–36. [CrossRef]
27. King, M.B.; Mubarak, A.; Kim, J.D.; Bott, T.R. The mutual solubilities of water with supercritical and liquid carbon dioxides. *J. Supercrit. Fluids* **1992**, *5*, 296–302. [CrossRef]
28. Sakellariou, P.; Rowe, R.C.; White, E.F.T. The thermomechanical properties and glass transition temperatures of some cellulose derivatives used in film coating. *Int. J. Pharm.* **1985**, *27*, 267–277. [CrossRef]
29. Della Porta, G.; Falco, N.; Reverchon, E. Continuous supercritical emulsions extraction: A new technology for biopolymer microparticles production. *Biotechnol. Bioeng.* **2011**, *108*, 676–686. [CrossRef]
30. National Institute of Standards and Technology NIST Chemistry webBook. Available online: https://webbook.nist.gov/chemistry/fluid/ (accessed on 6 July 2022).
31. Falco, N.; Reverchon, E.; Della Porta, G. Continuous supercritical emulsions extraction: Packed tower characterization and application to poly(lactic- co -glycolic acid) + insulin microspheres production. *Ind. Eng. Chem. Res.* **2012**, *51*, 8616–8623. [CrossRef]
32. Budich, M.; Brunner, G. Supercritical fluid extraction of ethanol from aqueous solutions. *J. Supercrit. Fluids* **2003**, *25*, 45–55. [CrossRef]
33. Atelier Fluides Supercritiques Equipment. Available online: https://www.atelier-fsc.com/Equipment_a23.html (accessed on 6 July 2022).
34. Perrut, M. Supercritical fluid applications: Industrial developments and economic issues. *Ind. Eng. Chem. Res.* **2000**, *39*, 4531–4535. [CrossRef]

Article

Fatty Acid Alkyl Ester Production by One-Step Supercritical Transesterification of Beef Tallow by Using Ethanol, Iso-Butanol, and 1-Butanol

Ricardo García-Morales [1], **Francisco J. Verónico-Sánchez** [2], **Abel Zúñiga-Moreno** [3], **Oscar A. González-Vargas** [4], **Edgar Ramírez-Jiménez** [1] and **Octavio Elizalde-Solis** [1,*]

[1] Instituto Politécnico Nacional, Escuela Superior de Ingeniería Química e Industrias Extractivas, Departamento de Ingeniería Química Petrolera and Sección de Estudios de Posgrado e Investigación, UPALM, Ed. 8, Lindavista 07738, Mexico

[2] Tecnológico de Monterrey, Escuela de Ingeniería y Ciencias, Carretera Lago de Guadalupe km. 3.5 col. Margarita Maza de Juárez, Atizapán de Zaragoza 52926, Mexico

[3] Instituto Politécnico Nacional, Escuela Superior de Ingeniería Química e Industrias Extractivas, Departamento de Ingeniería Química Industrial, Laboratorio de Investigación en Fisicoquímica y Materiales, Edif. Z-5, 2° piso, UPALM, Lindavista 07738, Mexico

[4] Departamento de Ingeniería en Control y Automatización, ESIME-Zacatenco, Instituto Politécnico Nacional, UPALM, Lindavista 07738, Mexico

[*] Correspondence: oelizalde@ipn.mx; Tel.: +52-55-5729-6000 (ext. 55120 or 55124)

Citation: García-Morales, R.; Verónico-Sánchez, F.J.; Zúñiga-Moreno, A.; González-Vargas, O.A.; Ramírez-Jiménez, E.; Elizalde-Solis, O. Fatty Acid Alkyl Ester Production by One-Step Supercritical Transesterification of Beef Tallow by Using Ethanol, Iso-Butanol, and 1-Butanol. *Processes* **2023**, *11*, 742. https://doi.org/10.3390/pr11030742

Academic Editors: Maria Angela A. Meireles, Ádina L. Santana and Grazielle Nathia Neves

Received: 3 February 2023
Revised: 21 February 2023
Accepted: 24 February 2023
Published: 2 March 2023

Abstract: The effect of temperature was studied on the synthesis of fatty acid alkyl esters by means of transesterification of waste beef tallow using ethanol and, iso-butanol and 1-butanol at supercritical conditions. These alcohols are proposed for the synthesis of biodiesel in order to improve the cold flow properties of alkyl esters. Alcohol–beef tallow mixtures were fed to a high-pressure high-temperature autoclave at a constant molar ratio of 45:1. Reactions were carried out in the ranges of 310–390 °C and 310–420 °C for ethanol and iso-butanol, respectively; meanwhile, synthesis using 1-butanol was assessed only at 360 °C. After separation of fatty acid alkyl esters, these samples were characterized by nuclear magnetic resonance (NMR) and gas chromatography coupled to mass spectrometry (GC-MS) to quantify yields, chemical composition, and molecular weight. Results indicated that yields enhanced as temperature increased; the maximum yields for fatty acid ethyl esters (FAEEs) were attained at 360 °C, and for fatty acid butyl esters (FABEs) were achieved at 375 °C; beyond these conditions, the alkyl ester yields reached equilibrium. Concerning the physicochemical properties of biodiesel, the predicted cetane number and cloud point were enhanced compared to those of fatty acid methyl esters.

Keywords: fatty acid ethyl ester; fatty acid butyl ester; 1-butanol; ethanol; iso-butanol; supercritical; waste beef tallow; transesterification

1. Introduction

Energy consumption changes according to the world population density and technological advances, as well as human needs and comforts. According to a projection for the year 2050 published in the latest report of the U.S. Energy Information Administration, energy utilization is going to increase in sectors such as electric power, transportation and industry in the next years. Moreover, residential and commercial segments are expected to keep the same consumption energy rate. In the early years, the three most produced energy sources are expected to be natural gas, oil, and renewable materials; furthermore, the maximum fossil oil demand will be achieved, followed by variable consumption in benefit of other energy sources. Natural gas and renewable materials will have a constant growing demand between the years 2022 and 2050, with a production of about 35 to 44 and 10 to 22 quadrillion of BTU, respectively [1].

Biodiesel is a liquid biofuel derived from biomass and is constituted by fatty acid alkyl esters mainly obtained by transesterification of fatty acids with methanol (FAME) in which the final product (100% biodiesel, B100) has similar properties to fossil diesel [2]. Biodiesel development has been focused on FAMEs, however exploration and development to use higher alcohols, such as ethanol and butanol, to obtain fatty acid ethyl and fatty acid butyl esters are also relevant [3,4]. Ethanol and butanol are gaining attention recently since these are potential additives for fuels and especially due to the fact that these could be produced by fermentation; therefore, a combination of biodiesel and green alcohol technologies would strengthen the development of sustainable fuels [5–7].

Transportation engines commonly use blends of biodiesel (B20: 20%) + diesel (80%) [8]. Fatty acids are supplied from feedstocks classified as renewable materials such as vegetable oils, animal fats, waste oil or fat, algae, among others [9,10]. Vegetable oils are the major source used around the world but are also destined for human consumption; then, the other materials are taking growing participation, i.e., beef tallow [11]. Non-edible tallow feedstock had a remarkable role in Brazil (13%) in 2018 and has been gaining presence with about 7% in the United States since 2021 where soybean oil was the main feedstock, at 50% [12–14].

Biodiesel production from non-edible tallow, under certain pathways, carries some benefits, such as low-cost feedstocks, and overwhelms obstacles associated with the high content of free fatty acids (FFA) and moisture that reduce yields and affects its properties. Low yields on fat synthesis attract the interest of improving transesterification to attain new insights in basic, acid and enzymatic catalysis; each pathway has its own characteristics, and the conditions include moderated temperatures, low-cost catalysts, and novel catalysts [15–24].

Transesterification with alcohols at supercritical conditions is a viable option carried out at higher temperatures and pressure among the abovementioned methods [24]. Production of fatty acid alkyl esters from beef tallow with high FFA content can be enhanced since the mass transfer and the homogeneous medium are promoted at supercritical conditions. Among the alcohols, methanol is the most used in transesterification for several feedstocks, for instance fresh or used vegetable oils [25–46]. Ethanol and butanol are other alternative solvents for fatty acid alkyl esters production, and both can be transformed from chemical and biochemical processes [47,48]; bioethanol and biobutanol are generated as main or secondary by-products using biological methods from first, second and third generation feedstocks. These solvents have attracted much attention; hence, new developments have been reported recently [49–53].

Insights about the transesterification of animal fats with supercritical long carbon-chain alcohols are limited [29,42–45,54], and to the best of our knowledge, two contributions are specialized on beef tallow [34,55]. Marulanda-Buitrago and Marulanda-Cardona [55] evaluated the effect of temperature (320–400 °C), alcohol excess (9–15) in the feed molar ratio and time (8–40 min) in the transesterification of beef tallow with supercritical ethanol; the highest conditions were claimed to promote the ester production. Bolonio et al. [34] evaluated the fatty acid ethyl esters manufacture from beef tallow by two methods: one-step transesterification under supercritical ethanol, and two-step reactions consisting of hydrolysis and esterification. The one-step method was explored in ranges of 300–350 °C and 40–120 min, and 20 or 40 mol of ethanol per mol of animal fat. For the two-step method, reactions lasted 60 min and temperature ranged from 200 to 350 °C; hydrolysis needed a high water/tallow volume ratio (2/1) and esterification operated at 7/1 of ethanol/free fatty acids molar ratio. Both methods had their own advantages: the first was simple, the esters had low content of polyunsaturated compounds and reported few degradations, while the second generated FAEEs rich in unsaturated chemicals.

Due to the fact that the raw material composition affects the properties of the final product, our contribution is aligned with the valorization of beef tallow waste for the fatty acid alkyl esters production. The aim of this work was to study the effect of alcohols and temperature on the fatty acid alkyl ester yields using three different alcohols (ethanol, iso-butanol, 1-butanol) at a fixed molar ratio of alcohol/tallow (45/1) during 60 min based on our previous findings using supercritical methanol [56] and the literature compilation [34,55]. This research also presents the quantitative characterization of fatty acid alkyl esters.

2. Materials and Methods

2.1. Materials

ACS grade iso-butanol and n-butanol were provided by Reactivos Química Meyer, having 0.1% of water content. Commercial ethanol had a water content of 3.6%. A sample lot of beef tallow was picked up from different local markets in the State of Mexico, this waste did not have any further valorization. Then, meat residues were separated from beef tallow by a simple filtration following a previous melting stage. The external standard (49454-U) for quantification of fatty acid ethyl ester was provided by Sigma-Aldrich. Fatty acid content and physicochemical properties of waste beef tallow have been previously reported and are listed in Table 1. This raw material was analyzed by Fourier-transform infrared spectroscopy (FTIR) and nuclear magnetic resonance (^{1}H-NMR and ^{13}C-NMR). Detailed analytical conditions can be found in an earlier contribution [56].

Table 1. Physicochemical properties of waste beef tallow (WBT) [56].

Property	Value
ρ at 308.15 K/g cm^{-3}	0.890
MW_{WBT}/g mol^{-1}	737.7 ± 2.0 [a]
MW_{WBT}/g mol^{-1}	742.5 ± 1.8 [b]
HHV/J g^{-1}	39,143.7
water content/ppm	200
SI	189.1
% $FFA_{C16:0}$	3.55 ± 0.04
% $FFA_{C18:0}$	3.94 ± 0.03
% $FFA_{C18:1\ n-9}$	3.91 ± 0.04

[a] HPLC-MS; [b] Cryoscopy.

2.2. Transesterification

A batch autoclave made of Inconel 625 alloy with a volume capacity of 170 cm^3 was operated to implement the synthesis at supercritical conditions. The procedure briefly consisted of pouring 150 cm^3 of alcohol–waste beef tallow blend to the autoclave with a molar ratio of 45/1, sealing, degassing, and wrapping the autoclave in an electrical heating resistance, turning on the magnetic shaft stirrer (1000 rpm) coupled to a three-blade propeller. Then, autoclave temperature was set above the critical point of the alcohol (ethanol: T_c = 240.77 °C, p_c = 61.48 bar; iso-butanol: T_c = 274.63 °C, p_c = 43 bar; 1-butanol: T_c = 289.9 °C, p_c = 44.23 bar) [57]. Furthermore, the temperature was monitored by a calibrated K-type thermocouple vertically immersed near the internal bottom of the autoclave. A gradual heating rate was established with the aim of minimizing the thermal degradation, particularly at elevated temperatures. Once desire temperature was reached, the synthesis lasted 60 min in order to guarantee the maximum yield [34,55,56]. Pressure changes in the constant volume autoclave were dependent on temperature and volume feedstock variations; moreover, the addition of a pressurizing agent (hydrogen, nitrogen, or carbon dioxide) was not considered since this component modifies the phase diagram. Thereafter, the heating was immediately stopped to inhibit further reactions by circulating cold water throughout an internal coil and removing the autoclave from the electrical resistance. Finally, the collected liquid sample was distilled for alcohol recovery, and the residual section was poured in a separation funnel. After stabilization of the residual, the glycerol-rich phase located on the bottom phase and the fatty acid alkyl esters in the upper phase were separated. Esters were stored into amber bottles for subsequent analyses. Reactions were performed thrice for each temperature by separating the total amount of waste beef tallow, thus, to avoid different composition on raw material for each set of experiments.

2.3. Quantification

The recovered fatty acid alkyl esters were characterized by nuclear magnetic resonance spectroscopy (^{1}H-NMR) in a Bruker Ascend 750 MHz to determine yields, as well as gas chromatography coupled to mass spectrometry (GC-MS) to generate FAEEs and FABEs profiles and their molecular weight. Physicochemical properties, cloud point (CP), calculated cetane number (CN), color Gardner (CG) and free fatty acid content (FFA_i) were also reported with basis on standard methods and correlations detailed elsewhere [56].

3. Results and Discussion

Fatty acid alkyl esters were obtained at different temperatures and pressures above the critical point of the alcohol. Each reaction was performed thrice and the corresponding first run is summarized in Table 2. The transesterification reversibility caused the loading molar ratio, between the alcohol and waste beef tallow, to be established in excess (45/1); thus, the synthesis direction was displaced to promote esters production while a homogeneous phase was probably coexisting at the required conditions [54,58,59].

Table 2. Properties for each transesterification at 45:1 of molar ratio for A:BT.

Reaction	$T/^\circ$C	P/bar	I_{FAEEs} Area	I_{FAEEs} Interval	I_{TAG} Area	I_{TAG} Interval	Y_{FAEEs}/%	$CP/^\circ$C	CN	CG	FFA_{SA} /%
					Ethanol						
E1S7	310	152.0	95,455.9	4.18–4.09	19,980.7	4.35–4.17	55.74	20	79.7	3.6	8.1
E2S7	320	171.6	101,315.3	4.17–4.10	18,443.4	4.35–4.17	59.96	18	79.2	4.0	7.2
E3S7	330	217.3	94,003.9	4.17–4.10	15,789.8	4.34–4.17	62.28	18	86.8	4.1	7.6
E4S7	340	254.9	86,045.8	4.17–4.09	7509.2	4.26–4.17	77.71	19	79.7	3.5	7.6
E5S7	350	280.4	77,754.0	4.17–3.97	3585.4	4.22–4.17	87.33	17	79.5	3.8	8.3
E6S7	360	344.6	141,698.7	4.17–4.09	2565.2	4.34–4.17	94.76	16	80.1	4.1	9.2
E7S7	360	344.2	83,930.6	4.17–4.07	1410.7	4.36–4.17	95.12	16	80.8	4.1	9.2
E8S7	370	397.1	81,931.6	4.17–4.09	4318.1	4.34–4.17	85.70	14	80.8	5.4	10.9
E9S7	380	397.9	50,956.1	4.17–4.10	2745.7	4.35–4.17	85.41	14	81.1	5.7	15.0
E10S7	390	397.1	70,722.1	4.18–4.10	9636.4	4.36–4.17	67.88	14	81.2	7.4	18.7
					Iso-butanol						
E1S9	310	63.7	70,200.5	3.87–3.77	25,722.0	4.35–3.89	76.47	18		4.6	6.0
E2S9	335	147.1	73,763.8	3.90–3.79	5817.2	4.34–3.90	94.73	17		2.7	5.3
E3S9	350	205.9	68,245.6	3.90–3.79	7571.5	4.33–3.88	93.06	14		2.4	4.7
E4S9	375	269.6	65,417	3.89–3.77	1266.7	4.34–3.88	96.65	10		4.8	6.2
E6S9	390	279.4	71,736.0	3.89–3.80	1946.8	4.35–3.88	94.80	9		5.7	8.2
E5S9	420	343.2	34,260.4	3.90–3.78	12,063.0	4.37–3.89	35.55	11		7.3	8.2
					1-butanol						
E1S10	360	210.7	237,318.6	3.01–4.10	20,289.96	4.31–4.11	87.70	8		5.0	6.4

3.1. Fatty Acid Ethyl Esters (FAEEs)

The isolated FAAEs were dissolved in deuterated chloroform and subject to ^{1}H-NMR analyses, and the corresponding spectra are depicted in Figure 1. The triplet (A) in the interval of 2.2–2.4 ppm denoted hydrogens of α–CH$_2$. Three multiplets ranging from 1.55 to 1.70 ppm (B), 1.1 to 1.46 ppm (C) and 5.2 to 5.4 ppm (D) corresponded to the hydrogens of β–CH$_2$, –(CH$_2$)$_n$– end and –CH$_2$–CH=, in the same order; this last hydrogen with a double carbon chain was also identified by the multiplet within 1.92–2.07 ppm (F). The CH$_3$ hydrogens were detected in two zones, one in a triplet ranging from 0.83 to 0.93 ppm (E) and the second peak in a multiplet at 0.3 ppm (K). Subsequently, a quartet in the interval from 3.97 to 4.18 ppm (EE) was related to CH$_3$-CH$_2$-OOC- hydrogens.

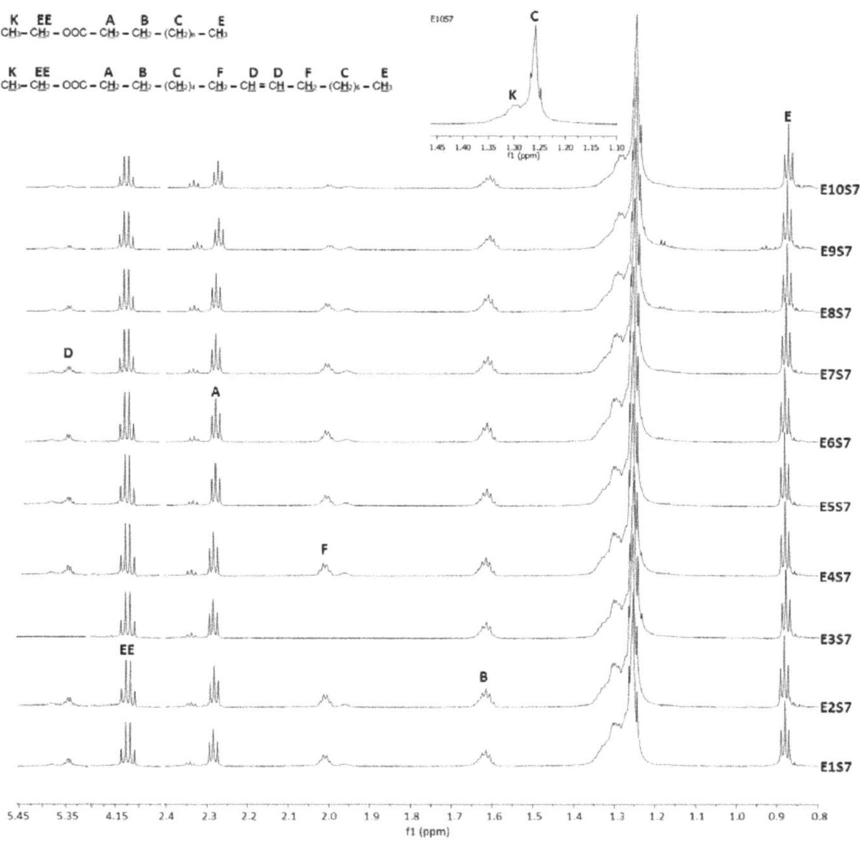

Figure 1. ^{1}H-NMR spectra for FAEE obtained from beef tallow by using supercritical ethanol.

Yield for FAEEs (Y_{FAEEs}) listed in Table 2 were estimated by integrating peak areas from the ^{1}H-NMR analyses and formulating the expression proposed by Ghesti et al. [60]:

$$Y_{FABEs}/\% = \left(\frac{4 \left(I_{FABEs} - I_{TAG} \right)}{4 \left(I_{FABEs} - I_{TAG} \right) + 6 \left(2 \, I_{TAG} \right)} \right) \times 100 \tag{1}$$

where the number 4 in Equation (1) denoted the four glycerol methylene hydrogens in triacylglycerides or triacylglycerol (TAG), while the number 6 expressed the six hydrogens in the three FAEEs. From Figure 1, I_{FAEEs} corresponded to the quartet area observed in the range of 3.97–4.18 ppm and indicated as EE; it was assumed to belong to the two hydrogens of the CH_3-CH_2-OOC- group presented both in FAEEs and TAG. Finally, I_{TAG} was evaluated from the double doublet area in the interval of 4.17–4.36 ppm, and was related to the hydrogens of -CH_2 hydrogens from glycerol in the TAG samples.

The temperature induced a promising effect on the Y_{FAEEs}, whose values enhanced in the range of 310–360 °C, the yield was 55.74% at 310 °C, then a high temperature led to a strong upgrading Y_{FAEEs}= 77.71% (at 340 °C), and the maximum yield (95.12%) was achieved at 360 °C, assigned as the proper temperature. Conversely, the performance of fatty acid ethyl esters was interfered beyond 370 °C since the yield decreased to 67.88%.

Quantification of FAEE constituents was investigated by GC-MS analyses and a compositional standard. The identified components were ethyl palmitate, ethyl stearate and ethyl oleate. Composition of different samples studied are reported in Table 3. The synthesis named "E3S7" carried out at 330 °C did not generate the unsaturated ester (ethyl oleate) as can be seen in Table 3. It also presented in the ^{1}H-NMR spectrum shown in Figure 1, where the multiplet (F) was not detectable. The absence of the unsaturated ester could be related to the initial feedstock: oleic acid was probably not contained at the beginning regardless of the pretreatment of the feedstock for achieving a homogeneous sample lot. Out of this, ethyl stearate had the highest composition for all the reactions. The chemicals contribution below 360 °C had the following order: ethyl stearate > ethyl oleate > ethyl palmitate, while the order of chemicals of ethyl stearate > ethyl palmitate > ethyl oleate was noted at temperatures over 370 °C. The average molecular weight for FAEEs oscillated in the range of 302.84–304.22 g/mol, but the FAEEs for the "E3S7" experiment had 306.41 g/mol due to the absence of the unsaturated ester, as reported in Table 3.

Table 3. FAEE composition from beef tallow transesterification.

Experiment	Chemical	Formula	MW	wt%	mol%	MW_{FAEEs}
E1S7	Ethyl palmitate	$C_{18}H_{36}O_2$	284.4772	26.492	28.306	303.960
	Ethyl stearate	$C_{20}H_{40}O_2$	312.5304	41.598	40.458	
	Ethyl oleate	$C_{20}H_{38}O_2$	310.5145	31.910	31.236	
E2S7	Ethyl palmitate	$C_{18}H_{36}O_2$	284.4772	25.856	27.641	304.108
	Ethyl stearate	$C_{20}H_{40}O_2$	312.5304	40.281	39.196	
	Ethyl oleate	$C_{20}H_{38}O_2$	310.5145	33.862	33.164	
E3S7	Ethyl palmitate	$C_{18}H_{36}O_2$	284.4772	20.239	21.800	306.415
	Ethyl stearate	$C_{20}H_{40}O_2$	312.5304	79.761	78.200	
E4S7	Ethyl palmitate	$C_{18}H_{36}O_2$	284.4772	26.650	28.472	303.921
	Ethyl stearate	$C_{20}H_{40}O_2$	312.5304	41.835	40.683	
	Ethyl oleate	$C_{20}H_{38}O_2$	310.5145	31.515	30.846	
E5S7	Ethyl palmitate	$C_{18}H_{36}O_2$	284.4772	26.338	28.145	303.990
	Ethyl stearate	$C_{20}H_{40}O_2$	312.5304	41.008	39.887	
	Ethyl oleate	$C_{20}H_{38}O_2$	310.5145	32.654	31.968	
E6S7	Ethyl palmitate	$C_{18}H_{36}O_2$	284.4772	25.985	27.782	304.143
	Ethyl stearate	$C_{20}H_{40}O_2$	312.5304	43.961	42.782	
	Ethyl oleate	$C_{20}H_{38}O_2$	310.5145	30.054	29.437	
E7S7	Ethyl palmitate	$C_{18}H_{36}O_2$	284.4772	25.927	27.727	304.226
	Ethyl stearate	$C_{20}H_{40}O_2$	312.5304	47.416	46.156	
	Ethyl oleate	$C_{20}H_{38}O_2$	310.5145	26.658	26.118	
E8S7	Ethyl palmitate	$C_{18}H_{36}O_2$	284.4772	29.913	31.868	303.069
	Ethyl stearate	$C_{20}H_{40}O_2$	312.5304	43.568	42.249	
	Ethyl oleate	$C_{20}H_{38}O_2$	310.5145	26.519	25.883	
E9S7	Ethyl palmitate	$C_{18}H_{36}O_2$	284.4772	30.763	32.749	302.844
	Ethyl stearate	$C_{20}H_{40}O_2$	312.5304	43.861	42.502	
	Ethyl oleate	$C_{20}H_{38}O_2$	310.5145	25.376	24.749	
E10S7	Ethyl palmitate	$C_{18}H_{36}O_2$	284.4772	27.426	29.291	303.817
	Ethyl stearate	$C_{20}H_{40}O_2$	312.5304	47.416	46.094	
	Ethyl oleate	$C_{20}H_{38}O_2$	310.5145	25.158	24.615	

3.2. Fatty Acid Butyl Esters (FABEs)

Regarding transesterification in the presence of iso-butanol, the spectra of fatty acid isobutyl esters analyzed by ^{1}H-NMR are illustrated in Figure 2. Peaks for the FABEs samples were similar to those formed on FAEEs identified with the symbols from (A) to (F): the triplet in 2.2–2.4 ppm (A) ascribed to α–CH$_2$ hydrogens, the respective three multiplets in 1.55–170 ppm (B), 1.1–1.46 ppm (C) and 5.2–5.4 ppm (D) associated with hydrogens of β–CH$_2$, –(CH$_2$)$_n$– structure end and double carbon chain, the triplet within 0.83–0.93 ppm (E) assigned to the CH$_3$ hydrogens, and the multiplet 1.95–2.07 ppm (F) represented the CH$_2$ within the double carbon chain. The distinctive peaks for FABE were identified with the symbols (J), (BE), (L), and were classified as hydrogens using the same order: the doublet ranging from 0.9–0.96 ppm denoted CH$_3$ hydrogens, the doublet within 3.8–3.88 ppm belonged to CH$_2$ in the >CH-CH$_2$-OOC- structure, and the multiplet from 1.89 to 1.95 ppm indicated the -CH< hydrogens.

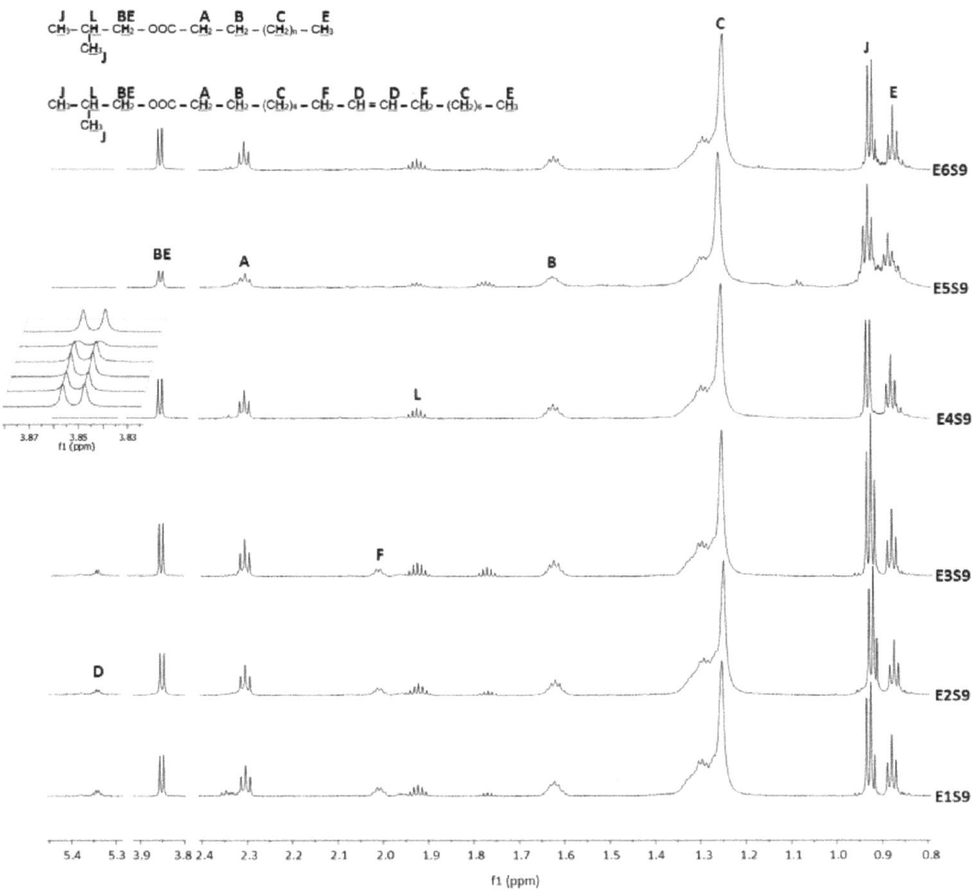

Figure 2. ^{1}H-NMR spectra for FABE obtained from beef tallow by using supercritical iso-butanol.

The production of fatty acid isobutyl esters was evaluated in terms of the yield (Y_{FABEs}). The calculated values listed in Table 2 were approached by using the Equation (2) and the integration of peak areas from ^{1}H-NMR analyses. The proposed equation was developed under the same considerations and notation given for yields of fatty acid ethyl esters:

$$Y_{\text{FABEs}}/\% = \left(\frac{4\,(I_{\text{FABEs}} - I_{\text{TAG}})}{4\,(I_{\text{FABEs}} - I_{\text{TAG}}) + 6\,(2\,I_{\text{TAG}})} \right) \times 100 \tag{2}$$

where the number 4 expresses the glycerol methylene hydrogens in triacylglycerides or triacylglycerol (TAG), the number 6 represents the hydrogens in the three FABEs, the area of the hydrogens in the >CH–CH$_2$–OOC– structure from produced FABEs, and unreacted TAG is indicated with the I_{FABEs} notation that appears within 3.77–3.90 ppm. The hydrogens of –CH$_2$ hydrogens from glycerol contained in TAG are symbolized with I_{TAG} and are in the range of 3.88–4.37 ppm. The highest yield ($Y_{\text{FABEs}} = 96.65\%$) was attained at 375 °C, which was superior to the maximum value accomplished by using ethanol as solvent at 360 °C. Critical temperature for each alcohol played an important role in FAAEs synthesis; fatty acid isobutyl esters production was not satisfactory at 310 °C ($Y_{\text{FABEs}} = 76.47\%$) but it was superior to that reported for the FAEEs ($Y_{\text{FAEEs}} = 55.74\%$), and this temperature was close to the critical temperature of iso-butanol.

Correspondingly, the FABEs content via 1-butanol was assessed under the same conditions as the reported for the other alcohols at 375 °C. The yield calculated with Equation (2) was reported as 87.7%, which was lower than the produced FAAEs using iso-butanol (93.06% at 350 °C and 96.65% at 365 °C) or ethanol (94.76% at 360 °C). A higher effect of the thermal degradation was taking part in the involved system as a consequence of the additional competing reactions. The identified peaks from Figure 3 for the transesterification with 1-butanol were the same as those identified in the reaction using iso-butanol for the (A)–(F) and L notation. The exceptions were the multiplets (G) and (H), the triplet (J) and the doublet (BE) that denoted the hydrogens of CH_2, CH_2, CH_3, CH_2 assigned on the CH_3–CH_2–CH_2–CH_2–OOC– structure.

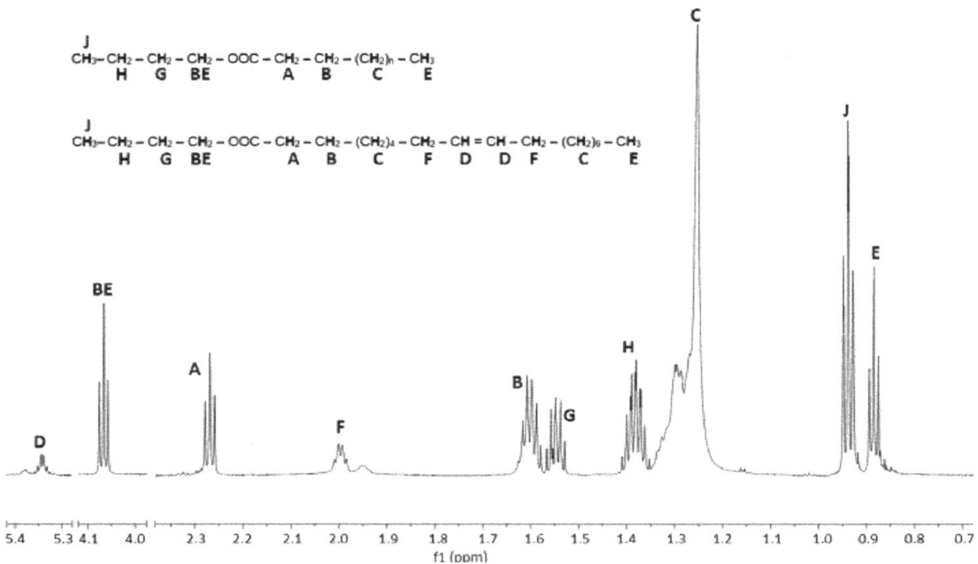

Figure 3. ^{1}H-NMR spectra for FABE obtained from beef tallow by supercritical 1-butanol.

The components identified in the fatty acid isobutyl ester samples are summarized in Table 4. The existence of esters was confirmed with a carbon chain length from eight to twenty-two; with reference to the relative area, the major esters were those of high molecular weight: isobutyl stearate, isobutyl oleate, isobutyl palmitate, isobutyl margarate and isobutyl myristate. Other esters were produced above 390 °C where the yield did was not enhanced. At the maximum temperature (420 °C) used, the lowest yield 35.55% was obtained, and the formation of 4-methyl-2ethyl-pentanol and some alkanes confirmed the appearance of additional reactions; these chemicals could be obtained from simultaneous thermal cracking reactions apart from transesterification. Concerning the FABEs sample using 1-butanol, the identified species by GC-MS were 2-methyl propyl octadecenoate, butyl hexadecanoate, pentyl octadecenoate and butyl myristate. Their composition followed the above sequence in terms of the relative area from high to low content as listed in Table 5.

Table 4. FABE profile from beef tallow transesterification using iso-butanol.

Experiment:			E1S9	E2S9	E3S9	E4S9	E6S9	E5S9
Chemical	Formula	*MW*	Area %					
isobutyl butyrate	$C_8H_{16}O_2$	144.2114					1.293	1.698
4-methyl-2ethyl-pentanol	$C_8H_{18}O$	130.1800						3.237
isobutyl caproate	$C_{10}H_{20}O_2$	172.2646				1.469	4.537	4.132
butyl heptanoate	$C_{11}H_{22}O_2$	186.2912						1.567
isobutyl enanthate	$C_{11}H_{22}O_2$	186.2912				1.133	1.765	
isobutyl caprylate	$C_{12}H_{24}O_2$	200.3178	0.067	0.119		0.735	1.472	1.523
3-methyl-undecane	$C_{12}H_{26}$	170.3300						2.958
isobutyl pelargonate	$C_{13}H_{26}O_2$	214.3443				0.687	1.226	1.424
tridecane	$C_{13}H_{28}$	184.3700						1.867
isobutyl caprate	$C_{14}H_{28}O_2$	228.3709	0.095	0.168		0.972	1.418	1.914
tetradecane	$C_{14}H_{30}$	198.3900						1.225
isobutyl undecylenate	$C_{15}H_{30}O_2$	242.4030					1.798	1.555
n-butyl laurate	$C_{16}H_{32}O_2$	256.4241						1.314
isobutyl laurate	$C_{16}H_{32}O_2$	256.4241	0.158	0.247			1.953	
hexadecane	$C_{16}H_{34}$	226.4100						4.342
isobutyl tridecanoate	$C_{17}H_{34}O_2$	270.4570				1.559	4.458	4.528
isobutyl myristate	$C_{18}H_{36}O_2$	284.4772	6.065	5.477	5.019	5.016	4.010	4.211
isobutyl pentadecanoate	$C_{19}H_{38}O_2$	298.5038	1.291	3.106	2.918	1.131	5.148	4.165
nonadecane	$C_{19}H_{40}$	268.5100						4.981
2-methyl-octadecane	$C_{19}H_{40}$	268.5100						0.874
isobutyl palmitate	$C_{20}H_{40}O_2$	312.5304	27.119	27.507	27.652	25.951	28.426	18.358
isobutyl margarate	$C_{21}H_{42}O_2$	326.5570	4.738	6.105	5.938	5.182		
2-methyl-eicosane	$C_{21}H_{44}$	296.6000						0.797
isobutyl oleate	$C_{22}H_{42}O_2$	338.5677	24.679	22.784	23.499			
isobutyl stearate	$C_{22}H_{44}O_2$	340.5836	33.536	31.609	32.185	47.921	41.353	32.125

Table 5. FABE profile from beef tallow transesterification using 1-butanol.

Experiment	Chemical	Formula	*MW*	Area %
E1S10	Butyl myristate	$C_{18}H_{36}O_2$	284.4772	1.904
	Butyl hexadecanoate	$C_{20}H_{40}O_2$	312.5304	25.549
	2-Methyl propyl octadecanoate	$C_{22}H_{44}O_2$	340.592	50.282
	Pentyl octadecenoate	$C_{23}H_{44}O_2$	352.5940	22.264

Our findings could be explained based on the phase diagram for the mixture where non-catalytic transesterification takes advantage of the FAAEs synthesis made at supercritical conditions. Aside from the alcohol excess with respect to the oil or fat, the possible location of the complex mixture out of a homogenous phase impeded mass transfer; then, the mixture was feasible to coexist in the vicinity of its phase transition at the lowest temperature and pressure studied in the reactions [24,59,61].

Concerning the yields falling beyond optimal temperatures, cracking of chemicals was competing with transesterification along with other kinds of reactions, such as those associated with a possible polymerization and isomerization. Under the same experimental conditions of alcohol/waste beef tallow in molar ratio, the generated fatty acid alkyl esters with different alcohols as a function of temperature are compared in Figure 4. The maximum yields were accomplished at 365 °C for methanol (98.3%) [56], 360 °C for ethanol (95.12%) and 375 °C for iso-butanol (96.65%); meanwhile the additional reaction made under 1-butanol seemed to provide a low ester content, but an acceptable yield (87.7%). The high ester portion at supercritical conditions is suggested to occur within the homogenous phase. This was verified by the reduction of the hydrogen bonds of the alcohol occurring as the temperature was increased, where its dielectric constant diminished at supercritical conditions; then, a higher solvation of the non-polar chemicals in the alcohol generated a unique fluid phase where the particle interaction was enhanced [54,58,62]. The longer carbon chain for ethanol and butanol could contribute to a weakening of hydrogen bonds and obstructing interactions between functional groups. Furthermore, the higher yield reached via methanol could be ascribed to its easier reactivity with lipids due to the shorter carbon chain compared to higher alcohols; thus, the interaction is simpler between the oxygen atom from methanol with the carbon atom of the carbonyl functional group from successive glycerides (tri-, di- and mono-) [63]. This mechanism allowed the generation of the respective chemicals: diglycerides, monoglycerides and glycerol. Comparing FAAEs values, those obtained via butanol and methanol were similar but the reaction via ethanol seemed to be less effective, probably caused by its high-water content; the valuable advantage via ethanol and butanol over methanol was the lower glycerol formation observed at the end of the reaction.

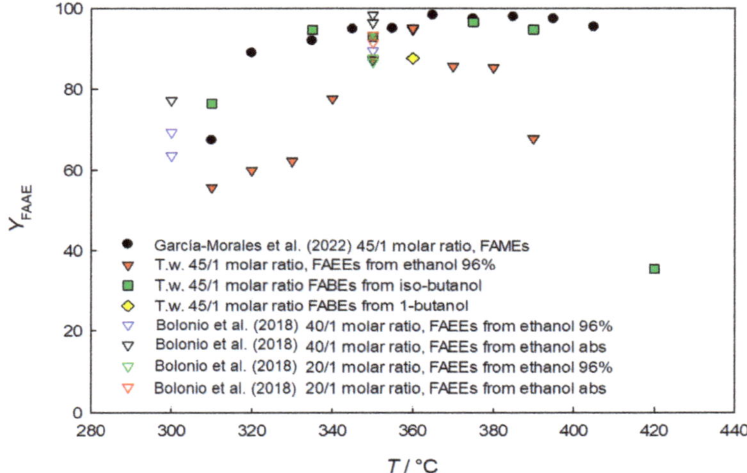

Figure 4. Production of fatty acid alkyl esters from waste beef tallow with different alcohols and feedstock ratios [34,56].

The yield results from this work are in agreement with those insights reported by Bolonio et al. [34] for a one-step reaction performed at 300 and 350 °C, as depicted in Figure 4; water as the main impurity in ethanol (4%) and low ethanol/beef tallow molar ratio (20/1) reduced the FAEEs yield. Moreover, the successful ester production via supercritical alcohols in this work was also contrasted with the catalytic methods, which have pros and cons; for instance, the basic catalysis that commonly used KOH and 6/1 (methanol/beef tallow) molar ratio yielded >91% beyond 35 min [20,64], the acid catalysis lasted 90 min with a 6/1 (methanol/beef tallow) molar ratio, and reached a yield of 96.3% [65], and the enzymatic catalysis performed during 48 h and 50 °C that yielded 89.7% [66].

3.3. Fatty Acid Alkyl Ester Properties

The cloud point (*CP*), cetane number, color Gardner and free fatty acid content are summarized in Table 2. Cloud points for esters tended to diminish as the temperature of the reaction was increased, even for the samples with the lowest FAEEs content. The exception was detected for the highest temperature (420 °C) for FABEs via iso-butanol, whose cloud point (*CP* = 11 °C) was superior to that reported at reactions undertaken at 390 °C (*CP* = 9 °C); the generation of other functional groups on the reaction performed at 420 °C could contribute on the opalescence. Cloud point values for FAEEs started at 20 °C and were lowered at 16 °C for the highest yield; the cloud point lowered to 14 °C for the reactions that were prone to thermal degradation. FABEs via iso-butanol attained a *CP* ranging from 18 to 9 °C. Cloud points from FAEEs and FABEs indicated good transport properties under cold ambient temperatures since these values were lower than those reported for fatty acid methyl esters (FAMEs) that ranged from 18 to 20 °C using the same raw material [56].

The cetane number (CN) for FAEEs was estimated by adding the contribution of the cetane number of pure fatty acid ethyl ester as reported in the literature [56] and the composition $(CN = \sum w_i\, CN_i)$ as reported elsewhere [67,68]. The esters from the optimal reaction temperature (360 °C) achieved a $CN = 80.8$, where the cetane number varied within 79.2 to 86.8. Under the same feedstock, the produced fatty acid ethyl esters seemed to have a CN superior to that reported for FAMEs in our preceding contribution (65.5 to 74.9) [56]. Concerning fatty acid butyl esters, this parameter was not able to be estimated under the same method because of the absence of mass composition and the insufficient cetane number data for pure esters (FABEs); based on our literature review, the existing cetane number values for butyl esters were: 73 for n-butyl laurate, 84.8 for isobutyl palmitate, 59.6 for isobutyl oleate, 99.3 for isobutyl stearate, 69.4 for butyl myristate, 82.6 for butyl hexadecanoate, and 86.3 for 2-methyl propyl octadecenoate [63,69].

The color Gardner with values of 7.3 and 7.4 demonstrated thermal decomposition at high temperature, where the production of esters diminished. The color Gardner below 5.7 seemed to have an acceptable appearance before it reached darker above 380 °C. Equal to the first property, the color Gardner had lower values for FAEEs and FABEs in contrast with FAMEs.

Finally, a certain content of free fatty acids, appraised as stearic acid (FFASA) [70], was still contained in the esters, as indicated in Table 2. An increase in the free fatty acids was noticeable with rising temperature, and the ethyl esters exhibited superior values in comparison with the butyl esters. Hydrolysis as a possible additional reaction at supercritical conditions took place among the lipids and fatty acid alkyl esters: a simultaneous hydrolysis of glycerides generated free fatty acids, which was competing with transesterification, while the produced fatty acid alkyl esters were also hydrolyzed to generate again free fatty acids [71,72]. This was ascribed to the elapsed thermal exposure and the water content of materials, mainly for ethanol, which contained 3.6 wt% of water. The formation of FFA is beneficial to supercritical transesterification since these could be converted to FAAEs, but this was not possible in our experiments; the free fatty acid formation from FAMEs during biodiesel obtainment was also proposed by catalytic thermal decomposition throughout a different mechanism of successive reactions [73]. Therefore, the analysis of short residence times should be considered to avoid secondary reactions which are related to thermal exposure.

4. Conclusions

The effect was analyzed of three supercritical alcohols, ethanol, iso-butanol and 1-butanol, accompanied by temperature variations, on the conversion of waste beef tallow via transesterification for the fatty acid alkyl esters production. Ester content decreased dramatically at the highest temperature in each case, and this drop in yields might be caused by the appearance of additional reactions attributed to long thermal exposure. The maximum yields were attained at 360 and 375 °C for ethanol and iso-butanol (temperatures were lower than those via methanol) without any apparent thermal degradation and the preservation of the yellowish coloring on the samples; physicochemical properties for FABEs via 1-butanol were kept but with a lower yield against with the reported for iso-butanol. Our findings suggested a better solvation to the detriment of the restriction of the mass transfer associated with the existence of a homogenous phase. Transesterification of waste beef tallow via supercritical ethanol and butanol seemed to be better than using supercritical methanol under the same conditions in terms of the physicochemical properties. The longer carbon chain alcohols reduced the glycerol formation and allowed the FAAEs synthesis with low cloud points (for instance 14 °C for FAEEs, 10 °C and 8 for FABEs) and a high predicted cetane number (80.8 for FAEEs); nevertheless, the produced esters via methanol were slightly higher than the other alcohols, emphasizing that the dissolved water in ethanol did not restrain yields dramatically.

Author Contributions: Conceptualization, R.G.-M. and O.A.G.-V.; methodology, R.G.-M. and F.J.V.-S.; formal analysis, E.R.-J.; investigation, O.A.G.-V. and E.R.-J.; resources, F.J.V.-S.; data curation, R.G.-M. and E.R.-J.; writing—original draft preparation, O.E.-S. and A.Z.-M.; writing—review and editing, O.E.-S., A.Z.-M. and F.J.V.-S.; visualization, E.R.-J.; supervision, O.A.G.-V.; project administration, A.Z.-M.; funding acquisition, O.E.-S. and A.Z.-M. All authors have read and agreed to the published version of the manuscript.

Funding: This research was funded by the Instituto Politécnico Nacional through project number 20221456.

Institutional Review Board Statement: Not applicable.

Data Availability Statement: Not applicable.

Acknowledgments: All authors acknowledge financial support from the Instituto Politécnico Nacional of México. R. García-Morales thanks CONACyT-SENER (Project number 185183) for the doctoral scholarship awarded.

Conflicts of Interest: The authors declare no conflict of interest.

References

1. U.S. Energy Information Administration. Annual Energy Outlook 2022. Washington, DC, USA . March 2022. Available online: https://www.eia.gov/outlooks/aeo/pdf/AEO2022_ChartLibrary_full.pdf (accessed on 1 June 2022).
2. Gharehghani, A.; Fakhari, A.H. Biodiesel as a clean fuel for mobility. In *Clean Fuels for Mobility. Energy, Environment, and Sustainability*; Di Blasio, G., Agarwal, A.K., Belgiorno, G., Shukla, P.C., Eds.; Springer: Singapore, 2022; pp. 141–168. [CrossRef]
3. Gotovuša, M.; Pucko, I.; Racar, M.; Faraguna, M. Biodiesel produced from propanol and longer chain alcohols—Synthesis and properties. *Energies* **2022**, *15*, 4996. [CrossRef]
4. Kurczyński, D.; Wcisło, G.; Leśniak, A.; Kozak, M.; Łagowski, P. Production and testing of butyl and methyl esters as new generation biodiesels from fatty wastes of the leather industry. *Energies* **2022**, *15*, 8744. [CrossRef]
5. Canabarro, N.I.; Silva-Ortiz, P.; Nogueira, L.A.H.; Cantarella, H.; Maciel-Filho, R.; Souza, G.M. Sustainability assessment of ethanol and biodiesel production in Argentina, Brazil, Colombia, and Guatemala. *Renew. Sustain. Energy Rev.* **2023**, *171*, 113019. [CrossRef]
6. Dutta, N.; Usman, M.; Ashraf, M.A.; Luo, G.; El-Din, M.G.; Zhang, S. Methods to convert lignocellulosic waste into biohydrogen, biogas, bioethanol, biodiesel and value-added chemicals: A review. *Environ. Chem. Lett.* **2022**. [CrossRef]
7. Culaba, A.B.; Mayol, A.P.; San Juan, J.L.G.; Ubando, A.T.; Bandala, A.A.; Concepcion, R.S., II; Alipio, M.; Chen, W.H.; Show, P.L.; Chang, J.S. Design of biorefineries towards carbon neutrality: A critical review. *Biores. Technol.* **2023**, *369*, 128256. [CrossRef] [PubMed]
8. Dahiya, A. *Bioenergy. Biomass to Biofuels and Waste to Energy*, 2nd ed.; Academic Press: London, UK, 2020. [CrossRef]
9. Carlucci, C. An Overview on the production of biodiesel enabled by continuous flow methodologies. *Catalysts* **2022**, *12*, 717. [CrossRef]
10. Vernier, L.J.; Barrachini-Nunes, A.L.; Albarello, M.; de Castilhos, F. Continuous production of fatty acid methyl esters from soybean oil deodorized distillate and methyl acetate at supercritical conditions. *J. Supercrit. Fluids* **2022**, *186*, 105603. [CrossRef]
11. da Silva, P.R.; Alhadeff, E.M. Biodiesel from beef tallow: A technological patent mapping. *Braz. J. Dev.* **2022**, *8*, 38061–38075. [CrossRef]
12. dos Reis Carraro, A.; da Silva César, A.; Conejero, M.A.; Ribeiro, E.C.B.; Mozer, T.S. The potential use of beef tallow for biodiesel production in Brazil. *Revista Valore* **2021**, *6*, e-6006. [CrossRef]
13. U.S. Energy Information Administration. Monthly Biofuels Capacity and Feedstocks Update. Washington, DC. May 2022. Available online: https://www.eia.gov/biofuels/update/table2.pdf (accessed on 1 June 2022).
14. Vignesh, P.; Kumar, A.R.P.; Ganesh, N.S.; Jayaseelan, V.; Sudhakar, K. Biodiesel and green diesel generation: An overview. *Oil Gas Sci. Technol. Rev. IFP Energies Nouvelles* **2021**, *76*, 6. [CrossRef]
15. Toldrá-Reig, F.; Mora, L.; Toldrá, F. Developments in the use of lipase transesterification for biodiesel production from animal fat waste. *Appl. Sci.* **2020**, *10*, 5085. [CrossRef]
16. Pinotti, L.M.; Salomão, G.S.B.; Benevides, L.C.; Antunes, P.W.P.; Cassini, S.T.A.; de Oliveira, J.P. Lipase-catalyzed biodiesel production from grease trap. *Arab. J. Sci. Eng.* **2022**, *47*, 6125–6133. [CrossRef]
17. Toldrá-Reig, F.; Mora, L.; Toldrá, F. Trends in biodiesel production from animal fat waste. *Appl. Sci.* **2020**, *10*, 3644. [CrossRef]
18. Samanta, S.; Sahoo, R.R. Waste cooking (palm) oil as an economical source of biodiesel production for alternative green fuel and efficient lubricant. *BioEnergy Res.* **2021**, *14*, 163–174. [CrossRef]

19. Farrokheh, A.; Tahvildari, K.; Nozari, M. Comparison of biodiesel production using the oil of chlorella vulgaris micro-algae by electrolysis and reflux methods using CaO/KOH-Fe$_3$O$_4$ and KF/KOH-Fe$_3$O$_4$ as magnetic nano catalysts. *Waste Biomass Valorization* **2021**, *12*, 3315–3329. [CrossRef]
20. Jambulingam, R.; Srinivasan, G.R.; Palani, S.; Munir, M.; Saeed, M.; Mohanam, A. Process optimization of biodiesel production from waste beef tallow using ethanol as co-solvent. *SN Appl. Sci.* **2020**, *2*, 1454. [CrossRef]
21. Rosson, E.; Sgarbossa, P.; Pedrielli, F.; Mozzon, M.; Bertani, R. Bioliquids from raw waste animal fats: An alternative renewable energy source. *Biomass Convers. Biorefin.* **2021**, *11*, 1475–1490. [CrossRef]
22. Erchamo, Y.S.; Mamo, T.T.; Workneh, G.A.; Mekonnen, Y.S. Improved biodiesel production from waste cooking oil with mixed methanol–ethanol using enhanced eggshell-derived CaO nano-catalyst. *Sci. Rep.* **2021**, *11*, 6708. [CrossRef]
23. Demir, V.; Akgün, M. New catalysts for biodiesel production under supercritical conditions of alcohols: A comprehensive review. *ChemistrySelect* **2022**, *7*, e202104459. [CrossRef]
24. Ghosh, N.; Halder, G. Current progress and perspective of heterogeneous nanocatalytic transesterification towards biodiesel production from edible and inedible feedstock: A review. *Energy Convers. Manag.* **2022**, *270*, 116292. [CrossRef]
25. Mahlia, T.M.I.; Syazmi, Z.A.H.S.; Mofijur, M.; Abas, A.E.P.; Bilad, M.R.; Ong, H.C.; Silitonga, A.S. Patent landscape review on biodiesel production: Technology updates. *Renew. Sustain. Energy Rev.* **2020**, *118*, 109526. [CrossRef]
26. Rathnam, V.M.; Madras, G. Conversion of Shizochitrium limacinum microalgae to biodiesel by non-catalytic transesterification using various supercritical fluids. *Bioresour. Technol.* **2019**, *288*, 121538. [CrossRef]
27. Vega-Guerrero, D.B.; Gómez-Castro, F.I.; López-Molina, A. Production of biodiesel with supercritical ethanol: Compromise between safety and costs. *Chem. Eng. Res. Des.* **2022**, *184*, 79–89. [CrossRef]
28. Sakdasri, W.; Sawangkeaw, R.; Ngamprasertsith, S. An entirely renewable biofuel production from used palm oil with supercritical ethanol at low molar ratio. *Braz. J. Chem. Eng.* **2017**, *34*, 1023–1034. [CrossRef]
29. Yuliana, M.; Santoso, S.P.; Soetaredjo, F.E.; Ismadji, S.; Ayucitra, A.; Angkawijaya, A.E.; Ju, Y.H.; Nguyen, P.L.T. A one-pot synthesis of biodiesel from leather tanning waste using supercritical ethanol: Process optimization. *Biomass Bioenergy* **2020**, *142*, 105761. [CrossRef]
30. Hegel, P.E.; Martín, L.A.; Popovich, C.A.; Damiani, C.; Leonardi, P.I. Biodiesel production from Halamphora coffeaeformis microalga oil by supercritical ethanol transesterification. *Chem. Eng. Process. Process Intensif.* **2019**, *145*, 107670. [CrossRef]
31. Poudel, J.; Karki, S.; Sanjel, N.; Shah, M.; Oh, S.C. Comparison of biodiesel obtained from virgin cooking oil and waste cooking oil using supercritical and catalytic transesterification. *Energies* **2017**, *10*, 546. [CrossRef]
32. Sanjel, N.; Gu, J.H.; Oh, S.C. Transesterification kinetics of waste vegetable oil in supercritical alcohols. *Energies* **2014**, *7*, 2095–2106. [CrossRef]
33. Akkarawatkhoosith, N.; Kaewchada, A.; Jaree, A. Production of biodiesel from palm oil under supercritical ethanol in the presence of ethyl acetate. *Energy Fuels* **2019**, *33*, 5322–5331. [CrossRef]
34. Bolonio, D.; Marco Neu, P.; Schober, S.; García-Martínez, M.J.; Mittelbach, M.; Canoira, L. Fatty acid ethyl esters from animal fat using supercritical ethanol process. *Energy Fuels* **2018**, *32*, 490–496. [CrossRef]
35. Farobie, O.; Leow, Z.Y.M.; Samanmulya, T.; Matsumura, Y. In-depth study of continuous production of biodiesel using supercritical 1-butanol. *Energy Convers. Manag.* **2017**, *132*, 410–417. [CrossRef]
36. dos Santos, K.C.; Hamerski, F.; Voll, F.A.P.; Corazza, M.L. Experimental and kinetic modeling of acid oil (trans)esterification in supercritical ethanol. *Fuel* **2018**, *224*, 489–498. [CrossRef]
37. Sun, Y.; Reddy, H.K.; Muppaneni, T.; Ponnusamy, S.; Patil, P.D.; Li, C.; Jiang, L.; Deng, S. A comparative study of direct transesterification of camelina oil under supercritical methanol, ethanol and 1-butanol conditions. *Fuel* **2014**, *135*, 530–536. [CrossRef]
38. Geuens, J.; Kremsner, J.M.; Nebel, B.A.; Schober, S.; Dommisse, R.A.; Mittelbach, M.; Tavernier, S.; Kappe, C.O.; Maes, B.U.W. Microwave-assisted catalyst-free transesterification of triglycerides with 1-butanol under supercritical conditions. *Energy Fuels* **2008**, *22*, 643–645. [CrossRef]
39. Singh, C.S.; Kumar, N.; Gautam, R. Supercritical transesterification route for biodiesel production: Effect of parameters on yield and future perspectives. *Environ. Prog. Sustain. Energy* **2021**, *40*, e13685. [CrossRef]
40. Tobar, M.; Núñez, G.A. Supercritical transesterification of microalgae triglycerides for biodiesel production: Effect of alcohol type and co-solvent. *J. Supercrit. Fluids* **2018**, *137*, 50–56. [CrossRef]
41. Hoang, D.; Bensaid, S.; Saracco, G.; Pirone, R.; Fino, D. Investigation on the conversion of rapeseed oil via supercritical ethanol condition in the presence of a heterogeneous catalyst. *Green Process. Synth.* **2017**, *6*, 91–101. [CrossRef]
42. Trentini, C.P.; Postaue, N.; Cardozo-Filho, L.; Reisa, R.R.; Sampaio, S.C.; da Silva, C. Production of esters from grease trap waste lipids under supercritical conditions: Effect of water addition on ethanol. *J. Supercrit. Fluids* **2019**, *147*, 9–16. [CrossRef]
43. Poudel, J.; Shah, M.; Karki, S.; Oh, S.C. Qualitative analysis of transesterification of waste pig fat in supercritical alcohols. *Energies* **2017**, *10*, 265. [CrossRef]
44. Trentini, C.P.; Fonseca, J.M.; Cardozo-Filho, L.; Reis, R.R.; Sampaio, S.C.; da Silva, C. Assessment of continuous catalyst-free production of ethyl esters from grease trap waste. *J. Supercrit. Fluids* **2018**, *136*, 157–163. [CrossRef]
45. Shah, M.; Poudel, J.; Kwak, H.; Oh, S.C. Kinetic analysis of transesterification of waste pig fat in supercritical alcohols. *Process Saf. Environ. Prot.* **2015**, *98*, 239–244. [CrossRef]

46. Nematian, T.; Fatehi, M.; Hosseinpour, M.; Barati, M. One-pot conversion of sesame cake to low N-content biodiesel via nano-catalytic supercritical methanol. *Renew. Energy* **2021**, *170*, 964–973. [CrossRef]

47. Sherpa, K.C.; Kundu, D.; Banerjee, S.; Ghangrekar, M.M.; Banerjee, R. An integrated biorefinery approach for bioethanol production from sugarcane tops. *J. Clean. Prod.* **2022**, *352*, 131451. [CrossRef]

48. Su, G.; Chan, C.; He, J. Enhanced biobutanol production from starch waste via orange peel doping. *Renew. Energy* **2022**, *193*, 576–583. [CrossRef]

49. Maity, S.; Mallick, N. Trends and advances in sustainable bioethanol production by marine microalgae: A critical review. *J. Clean. Prod.* **2022**, *345*, 131153. [CrossRef]

50. Devi, A.; Bajar, S.; Kour, H.; Kothari, R.; Pant, D.; Singh, A. Lignocellulosic biomass valorization for bioethanol production: A circular bioeconomy approach. *BioEnergy Res.* **2022**, *15*, 1820–1841. [CrossRef]

51. Iyyappan, J.; Bharathiraja, B.; Varjani, S.; Praveenkumar, R.; Muthu Kumar, S. Anaerobic biobutanol production from black strap molasses using Clostridium acetobutylicum MTCC11274: Media engineering and kinetic analysis. *Bioresour. Technol.* **2022**, *346*, 126405. [CrossRef]

52. Das, M.; Maiti, S.K. Current knowledge on cyanobacterial biobutanol production: Advances, challenges, and prospects. *Rev. Environ. Sci. Biotechnol.* **2022**, *21*, 483–516. [CrossRef]

53. Tse, T.J.; Wiens, D.J.; Reaney, M.J.T. Production of bioethanol—A review of factors affecting ethanol yield. *Fermentation* **2021**, *7*, 268. [CrossRef]

54. Andreo-Martínez, P.; Ortiz-Martínez, V.M.; Salar-García, M.J.; Veiga-del-Baño, J.M.; Chica, A.; Quesada-Medina, J. Waste animal fats as feedstock for biodiesel production using non-catalytic supercritical alcohol transesterification: A perspective by the PRISMA methodology. *Energy Sustain. Dev.* **2022**, *69*, 150–163. [CrossRef]

55. Marulanda-Buitrago, P.A.; Marulanda-Cardona, V.F. Supercritical transesterification of beef tallow for biodiesel production in a batch reactor. *Cienc. Tecnol. Futuro* **2014**, *5*, 55–73. [CrossRef]

56. García-Morales, R.; Zúñiga-Moreno, A.; Verónico-Sánchez, F.J.; Domenzain-González, J.; Pérez-López, H.I.; Bouchot, C.; Elizalde-Solis, O. Fatty acid methyl esters from waste beef tallow using supercritical methanol transesterification. *Fuel* **2022**, *313*, 122706. [CrossRef]

57. Reid, R.C.; Prausnitz, J.M.; Poling, B.E. *The Properties of Gases and Liquids*, 4th ed.; McGraw-Hill: New York, NY, USA, 1987.

58. Postaue, N.; Borba, C.E.; da Silva, C. Transesterification under high pressure as a sequential step from pressurized liquid extraction: Effect of operational parameters and characterization. *J. Supercrit. Fluids* **2023**, *193*, 105814. [CrossRef]

59. Corazza, M.L.; Fouad, W.A.; Chapman, W.G. Application of molecular modeling to the vapor–liquid equilibrium of alkyl esters (biodiesel) and alcohols systems. *Fuel* **2015**, *161*, 34–42. [CrossRef]

60. Ghesti, G.F.; de Macedo, J.L.; Resck, J.A.; Dias, I.S.; Dias, S.C.L. FT-Raman spectroscopy quantification of biodiesel in a progressive soybean oil transesterification reaction and its correlation with 1H NMR spectroscopy methods. *Energy Fuels* **2007**, *21*, 2475–2480. [CrossRef]

61. Osmieri, L.; Alipour Moghadam Esfahani, R.; Recasens, F. Continuous biodiesel production in supercritical two-step process: Phase equilibrium and process design. *J. Supercrit. Fluids* **2017**, *124*, 57–71. [CrossRef]

62. Roze, F.; Pignat, P.; Ferreira, O.; Pinho, S.P.; Jaubert, J.N.; Coniglio, L. Phase equilibria of mixtures involving fatty acid ethyl esters and fat alcohols between 4 and 27 kPa for bioproduct production. *Fuel* **2021**, *306*, 121304. [CrossRef]

63. Farobie, O.; Matsumura, Y. A comparative study of biodiesel production using methanol, ethanol, and tert-butyl methyl ether (MTBE) under supercritical conditions. *Bioresour. Technol.* **2015**, *191*, 306–311. [CrossRef]

64. Teixeira, L.S.G.; Assis, J.C.R.; Mendonça, D.R.; Santos, I.T.V.; Guimarães, P.R.B.; Pontes, L.A.M.; Teixeira, J.S.R. Comparison between conventional and ultrasonic preparation of beef tallow biodiesel. *Fuel Process. Technol.* **2009**, *90*, 1164–1166. [CrossRef]

65. Ehiri, R.C.; Ikelle, I.I.; Ozoaku, O.F. Acid-catalyzed transesterification reaction of beef tallow for biodiesel production by factor variation. *Am. J. Eng. Res.* **2014**, *3*, 174–177.

66. Da Rós, P.C.M.; Silva, G.A.M.; Mendes, A.A.; Santos, J.C.; de Castro, H.F. Evaluation of the catalytic properties of Burkholderia cepacia lipase immobilized on non-commercial matrices to be used in biodiesel synthesis from different feedstocks. *Bioresour. Technol.* **2010**, *101*, 5508–5516. [CrossRef]

67. Knothe, G. A comprehensive evaluation of the cetane numbers of fatty acid methyl esters. *Fuel* **2014**, *119*, 6–13. [CrossRef]

68. Ramírez-Verduzco, L.F.; Rodríguez-Rodríguez, J.E.; Jaramillo-Jacob, A.R. Predicting cetane number, kinematic viscosity, density and higher heating value of biodiesel from its fatty acid methyl ester composition. *Fuel* **2012**, *91*, 102–111. [CrossRef]

69. Yanowitz, J.; Ratcliff, M.A.; McCormick, R.L.; Taylor, J.D.; Murphy, M.J. Compendium of Experimental Cetane Numbers. In *National Renewable Energy Lab Technical Report NREL/TP-5400-67585*; National Renewable Energy Laboratory: Golden, CO, USA, February 2017. [CrossRef]

70. Official Mexican Standard. NMX-F-101-SCFI-2012. Foods—Vegetable or Animal Fats and Oils—Determination of Free Fatty Acids—Test Method. Official Gazette of the Federation. Mexico. 2012. Available online: http://www.economia-nmx.gob.mx/normas/nmx/2010/nmx-f-101-scfi-2012.pdf (accessed on 13 January 2023).

71. Kusdiana, D.; Saka, S. Effects of water on biodiesel fuel production by supercritical methanol treatment. *Bioresour. Technol.* **2004**, *91*, 289–295. [CrossRef]

72. Xu, J.; Jiang, Z.; Li, L.; Fang, T. A review of multi-phase equilibrium studies on biodiesel production with supercritical methanol. *RSC Adv.* **2014**, *4*, 23447–43455. [CrossRef]

73. Seames, W.; Luo, Y.; Ahmed, I.; Aulich, T.; Kubátová, A.; Štávová, J.; Kozliak, E. The thermal cracking of canola and soybean methyl esters: Improvement of cold flow properties. *Biomass Bioenergy* **2010**, *34*, 939–946. [CrossRef]

Article

Pilot-Plant-Scale Extraction of Antioxidant Compounds from Lavender: Experimental Data and Methodology for an Economic Assessment

Encarnación Cruz Sánchez, Jesús Manuel García-Vargas, Ignacio Gracia, Juan Francisco Rodríguez and María Teresa García *

Department of Chemical Engineering, Facultad de Ciencias y Tecnologías Químicas, University of Castilla-La Mancha, Avda. Camilo José Cela 12, 13071 Ciudad Real, Spain
* Correspondence: teresa.garcia@uclm.es

Abstract: The techno-economic feasibility of lavender essential oil supercritical CO_2 extraction was studied. The process was scaled up to a pilot plant, and the extraction yield, composition, and antioxidant potential of the extracts were evaluated at 60 °C and 180 bar or 250 bar, achieving a maximum yield of 6.9% and a percentage inhibition of the extracts of more than 80%. These results drove the development of a business plan for three scenarios corresponding to different extraction volumes (20, 50, and 100 L) and annual production. The SWOT matrix showed that this is a promising business idea. The COM was calculated and an investment analysis was performed. The profitability of this process was demonstrated by means of a financial analysis for 8 years, considering a selling price of 1.38 EUR/g for the extract from the 20 L plant and 0.9 EUR/g for industrial-scale plants, supported by the price curve. The sensitivity analysis showed that the price of the equipment was the factor that could most influence the robustness of the project and the business strategy, and the financial ratios evaluation resulted in a ROE value above 57% in all cases, indicating the economic attractiveness of the process.

Keywords: lavender essential oil; economic evaluation; supercritical fluid extraction (SFE); price curve; financial analysis; antioxidant potential

Citation: Cruz Sánchez, E.; García-Vargas, J.M.; Gracia, I.; Rodríguez, J.F.; García, M.T. Pilot-Plant-Scale Extraction of Antioxidant Compounds from Lavender: Experimental Data and Methodology for an Economic Assessment. *Processes* **2022**, *10*, 2708. https://doi.org/10.3390/pr10122708

Academic Editors: Maria Angela A. Meireles, Ádina L. Santana and Grazielle Nathia Neves

Received: 17 November 2022
Accepted: 12 December 2022
Published: 15 December 2022

Publisher's Note: MDPI stays neutral with regard to jurisdictional claims in published maps and institutional affiliations.

1. Introduction

The continuously growing demand for natural products in sectors such as food and pharmaceuticals has led to the search for natural sources rich in bioactive substances with beneficial properties for human health to synthesise drugs and develop nutraceuticals [1–3].

One of the main commercial products derived from nature is essential oils [4,5]. Traditionally, one of the most commonly used essential oils is lavender. It stands out for having compounds with a high antioxidant and anti-inflammatory capacity such as linalool [6,7]. Conventionally, this substance predominates in cosmetic applications such as soaps and perfumes. However, in recent years, the therapeutic potential of this molecule has been tested in different ways with successful results, such as an alternative to traditional topical drugs [8]. The above properties would make it an effective substance for the treatment of skin diseases such as atopic dermatitis.

In addition, these bioactive, antioxidant, and anti-inflammatory substances play an important role in the progression of reactive oxygen species and antioxidative function, which helps to reduce the development of some types of cancer and mitochondrial dysfunction [9–11].

Regarding the method of obtaining bioactive substances, one of the most used techniques for the extraction of essential oils is distillation, specifically the type based on steam extraction. It is a simple and cost-effective technique. Nevertheless, this method has lower yields than others. Another traditional alternative is extraction with organic solvents, which is carried out at ambient pressure but has several drawbacks: the need for solvent recovery, the energy-intensive operation, and the thermal degradation and loss of valuable compounds in the purification step [12]. In recent years, new technologies have been implemented to reduce the environmental impact, such as microwave and supercritical fluid extraction (SFE), the latter being a sustainable alternative that provides extracts with high purity. Microwave-assisted extraction is a technique that combines microwave extraction and traditional solvent extraction. Although it reduces the use of solvents and offers high extraction yields, it does not avoid the use of solvents and involves significant thermal degradation of the extracts [13].

SFE overcomes the limitations of traditional methods, as it is more efficient and reduces the environmental impact by avoiding the use of organic solvents. CO_2 is usually employed as the solvent in supercritical conditions due to its non-flammable, non-toxic, and non-corrosive nature [14]. Moreover, the use of SFE has proven to be effective for obtaining bioactive compounds and mixtures such as essential oils from natural matrixes [15] and, in particular, lavender essential oil and derived compounds [16–18].

By contrast, the main disadvantage of employing this technique is that, due to the high-pressure technology required, the operating costs and the investment needed for the equipment is higher than that of other conventional processes [19], although at present the development of industrial-scale units enables us to lower equipment costs associated with SFE processes [20–22]. However, its relatively energy-intensive needs make it an economically demanding process, and usually it is restricted to the use of supercritical CO_2 extraction for high-value applications [23,24].

Some studies have proposed the extraction of lavender essential oil with CO_2 under supercritical conditions. Reverchon et al. compared this process with hydrodistillation [16]. Different modelling and working conditions have also been analysed [17] as well as water extraction techniques [25]. Moreover, some research has focused on the antioxidant property and composition of lavender extracts obtained by supercritical fluids [18].

To properly analyse the economic feasibility of an SFE process, its special characteristics should be considered. An easy approach traditionally used to predict the economic viability of this type of process is to determine the cost of manufacturing (COM) of extracts [26]. Many authors have used this method to develop techno-economic studies of supercritical extraction processes for different raw materials [21,27,28].

However, this estimation is not sufficient to make a comprehensive economic assessment. Supercritical extracts are characterised by their exclusivity, higher quality, and customised design, thus not simple substitutes of conventional extracts. Hence, the higher-quality market niche in which these products could be placed has to be strategically determined. Therefore, other economic characteristics such as profitability over the time horizon, their flexibility to price changes, and their convenience for particular strategies have to be assessed before their introduction on the market [24]. Consequently, a complete study of the economic viability of this type of process must be based on a business plan.

The aim of this study was to demonstrate the technical and economic feasibility of extracting lavender essential oil with supercritical CO_2. Based on a preliminary study at laboratory scale, the extractions offering the highest yield were carried out in a pilot plant. The extraction yield, composition, and antioxidant potential of the supercritical extracts were evaluated, data on which the economic study was based. This study involved evaluating three different annual productions, 20 L, 50 L, and 100 L. After carrying out a strengths, weaknesses, opportunities, and threats (SWOT) matrix to capture the ideas in which the interest of the project lies, the business plan was started. First, the necessary investment was calculated, formed by the cost of the equipment and facilities as well as the necessary capital and the COM. After that, the financial analysis was developed, consisting of an analysis of the income statement, the sales curve price, and financial ratios such as the return on equity (ROE), which are crucial in considering the economic value of this project. Economic profitability is not enough to ensure economic success, and for that reason a sensitivity test was performed in order to validate the robustness of the idea by considering aspects that could affect the return on investment such as global or local events, changes in the interest rate, and underestimation of items [29]. This last piece of information is highly valuable to determine the market strategy of the company.

2. Materials and Methods

2.1. Materials

Dried lavender flowers (Peñarrubia del Alto Guadiana S. L., Albacete, Spain) and CO_2 of 99.8% purity (Carburos Metálicos, Barcelona, Spain) were required for the extraction. For the chromatographic analysis, lavender essential oil and Sigma-Aldrich Spain sponsors were needed, as well as diethyl ether for analysis (Scharlab, Barcelona, Spain). For the antioxidant potential assay, ethanol absolute HPLC grade (Scharlab, Barcelona, Spain), 1,1-diphenyl-2-picrylhydrazyl (DPPH) (Alfa-Aesar, Thermo Fisher, Ward Hill, MA, USA), and L-(+) ascorbic acid (Scharlab, Barcelona, Spain) were used.

2.2. SFE Process

The supercritical CO_2 extraction tests were carried out in semi-continuous mode, according to the diagram in Figure 1. The main part is the extraction vessel, whose nominal volume was 20 L. In the process, the raw material was fed through the upper part by means of a basket, and CO_2, after subcooling, was pumped and thermally conditioned to reach the working conditions of pressure and temperature. Once the pressure in the extractor was reached, the opening of the outlet was initiated and the fluid passed to the two separators, where the process conditions were modified to cause the precipitation of those substances that may have been recovered together with the CO_2 and to allow the solvent to recirculate.

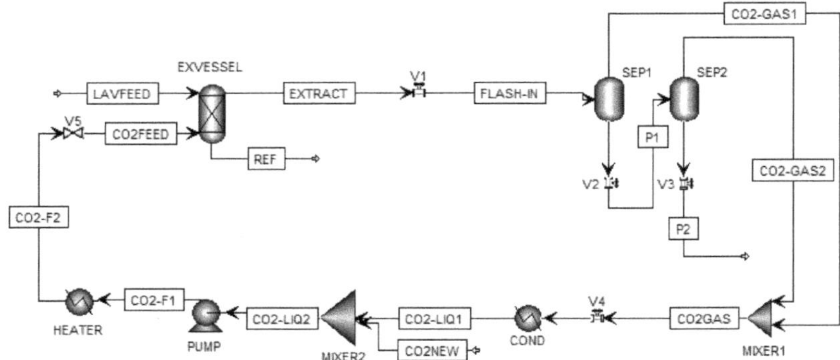

Figure 1. Supercritical extraction process flow diagram and equipment.

To develop the economic feasibility study, a solvent cycle was proposed in order to make the process more efficient. The pressure of the supercritical CO_2 was decreased after extraction until reaching 50 bar for separation of the extract, and then, in a condenser, it had to be cooled until obtaining liquid phase and conditioned in order to feed it back to the extraction vessel. The solvent cycle was fed by the two gaseous streams of the separators, and consisted of a mixer, a condenser, another mixer in which a fresh stream of CO_2 entered (assuming losses of 10% of solvent in the process [30,31]), a pump, and a heater that conditioned the CO_2 for entering the extraction vessel.

The working pressures and temperature were 180 and 250 bar and 60 °C, respectively. These conditions were selected based on previous experiments at lab scale that showed they offered the highest extraction yields. The extractions lasted 90 min with a CO_2 flow rate of 60 kg/h. Equation (1) was used to quantify the extraction yield.

$$(\%)\ Extraction\ Yield = \frac{weight\ of\ collected\ extract}{weight\ of\ fed\ lavender} \times 100 \tag{1}$$

2.3. Extracts Characterisation

Gas chromatography coupled to mass spectrometry (GC-MS) was used to evaluate the composition of the lavender extracts. The equipment used was a Varian Saturn 2000 GC-MS equipped with a HP-5 capillary column (30 m, 0.25 mm i.d., 0.25 μm film thickness). The method of analysis applied is summarised below [32]. The column temperature was set to 60 °C for 5 min. Then, it was increased to 160 °C at a rate of 4 °C/min. Finally, it was raised to 240 °C at 15 °C/min. The injector temperature was 250 °C and the detector temperature was 280 °C. An electron ionisation system with an energy of 70 eV was used. The diluted samples were analysed with 1:100 diethylether and 1 μL of the diluted samples was injected in splitless mode. The carrier gas was helium at a flow rate of 1 mL/min.

The 1,1-diphenyl-2-picrylhydrazyl (DPPH) assay was used to estimate the antioxidant capacity of the extracts [33]. DPPH• is a stable nitrogen-centred free radical which is conventionally used to determine the free radical scavenging activities of antioxidants present in plant extracts or synthetic compounds. The reduction capability of DPPH• radical is determined by the decrease in absorbance at 517 nm induced by the antioxidant.

The procedure described in L. T. Danh et al. [18] was used. Briefly, 40 μL of extracted essential oil was mixed with 0.4 mL of 0.5 mM DPPH• solution in ethanol and the final volume was adjusted to 1.5 mL with ethanol. The control solution was prepared with 0.4 mL of the ethanolic DPPH solution with 1.1 mL ethanol. In addition, a sample of the standard antioxidant, ascorbic acid, was prepared to compare the antioxidant potential of the supercritical extracts. The absorbance of the solutions was measured at 517 nm, with three replicates, after 30 min of stirring in darkness to allow a complete reaction.

To quantify the antioxidant potential of the supercritical extracts, the inhibition percentage (I%), which corresponds to the radical scavenging activity of the extracts, was calculated with Equation (2), where A_0 corresponds to the absorbance of the control sample and A_s to that of the extracted oil samples.

$$(\%) \; Inhibition = \frac{A_0 - A_s}{A_0} \times 100 \tag{2}$$

2.4. Economic Evaluation

For the economic evaluation of the supercritical CO_2 extraction of lavender essential oil, the annual production was used as a design basis. The study was conducted for three process scales concerning the volume of the extractor: 20 L (scenario 1), which corresponds to pilot plant scale, and 50 L (scenario 2) and 100 L (scenario 3), which are more like industrial production. In addition, the production data were calculated taking into account the maximum yield obtained in a laboratory-scale extraction, which is reported in another study [34].

Economic evaluations based on the business plan strategy follow a methodology in which the process is first evaluated by a SWOT analysis (Figure 2). This method takes the information from an environmental analysis and separates it into internal (strengths and weaknesses) and external issues (opportunities and threats) to obtain the SWOT matrix.

Through the analysis of the matrix, it is possible to know without very strict calculations whether a project is interesting or whether any changes leading to improvements must be made before an economic investment is considered. If a positive conclusion is drawn from the matrix analysis, the business plan should be continued with the analysis of the total capital investment of the plant [35].

Once the necessary investment has been determined, the annual costs and revenues are calculated and then the profit and loss account. In addition, before moving on to the assessment of financial ratios, the COM will be calculated. As mentioned previously, this is a simple method for calculating the cost per kg of extract that provides useful information for classifying production costs. This information is required in the search for investors (Figure 2).

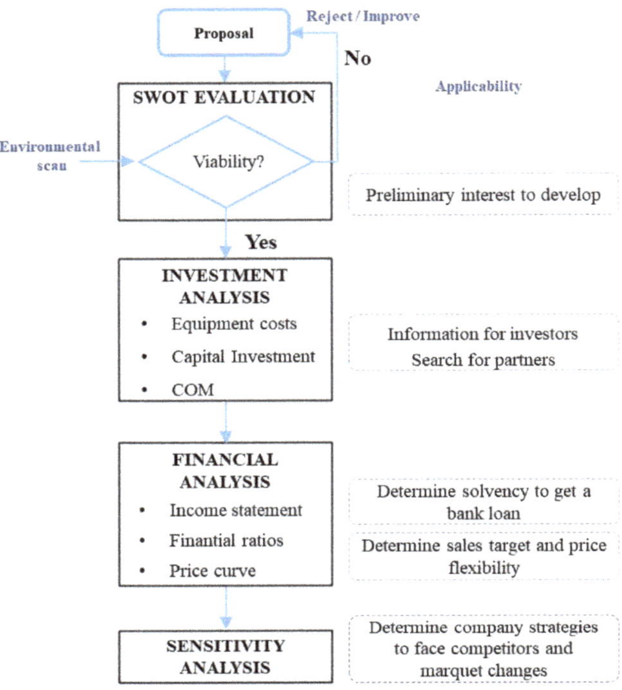

Figure 2. Flow diagram of the economic analysis based on the business plan strategy.

The financial analysis defines, for a given income statement associated with sales, a set of financial ratios used for the determination of solvency in applying for a bank loan, such as the ROE, return on investment (ROI), and earnings before interest, taxes, depreciation, and amortisation (EBITDA). Financial analysis also provides the price curve, required for the sales department to define the price flexibility and sales needs in the context of competitors. Finally, a sensitivity analysis is performed to check the influence of possible variations in factors such as equipment costs, salaries, raw materials, electricity, and interest rates on the profitability of the project to better define the robustness of the business idea in market situations not considered to be affected by global or local changes. This information would be used by the chief executive office to define the company strategies.

2.5. Data Reproducibility Analysis

Extraction experiments were carried out under two different conditions, 180 bar and 60 °C and 250 bar and 60 °C, both for a duration of 90 min and a solvent flow rate of 60 kg/h. The extraction experiments were performed in triplicate and the extraction yield value obtained was averaged. The difference in the value obtained was less than 1% in all cases.

As for the characterisation of the extracts, GC-MS analysis and a DPPH assay was performed for two different extract samples from each experiment. Concentration and absorbance were measured in triplicate to check reproducibility.

3. Results and Discussion

The technical viability of the SFE of lavender essential oil at pilot plant scale will first be discussed based on the lavender essential oil extraction yield obtained for the two different working pressures. In addition, the composition and antioxidant potential of the obtained extracts were analysed, which are considered key factors that encourage the idea of designing a commercially attractive product. From these results, the business plan can be started, and this is discussed in the corresponding section.

3.1. Lavender Essential Oil Supercritical CO_2 Extraction

For an operating pressure of 180 bar, an extraction yield of 6.2% was obtained, while for 250 bar it was 6.9%. Although this is not a very significant increase, it follows the usual trend in supercritical CO_2 extraction processes: the higher the pressure, the higher the CO_2 density and therefore the higher the solubility of the interesting compounds in the supercritical solvent [36].

Table 1 shows the mass% data for the main compounds identified in the extracts. As studies on the composition of lavender (L. Angustifolia) essential oil indicate [37], the two major components are linalool and linalyl acetate. The extracts obtained in this work concur with these results, and it can be seen that the mass percentage of linalool is approximately 32% and linalyl acetate 43%. These are interesting results from the point of view of the commercial application of these supercritical extracts, as the mass percentage of these two main compounds in the extracts is higher than in the commercial essential oil (Sigma-Aldrich) that was also analysed and which showed a mass percentage of linalool and linalyl acetate of 28 and 31%, respectively. As can be seen, the pressure increase had no clear influence on the composition. Eucalyptol, camphor, endoborneol, terpinen-4-ol, α-terpineol, nerol acetate, caryophyllene, and β-Famesene were also identified as minor compounds present in the supercritical extracts.

Table 1. Composition of supercritical extracts.

% wt	60 °C	
	180 Bar	**250 Bar**
Eucalyptol	3.48	4.11
Linalool	32.07	32.20
Camphor	4.24	4.58
Endoborneol	4.07	3.91
Terpinen-4-ol	3.60	3.58
α-terpineol	1.49	1.43
Linalyl acetate	43.03	43.01
Nerol acetate	2.44	2.34
Caryophyllene	2.95	2.58
β-Famesene	2.64	2.25

Regarding the study of the antioxidant capacity, by means of the DPPH method, it was observed that the two extracts showed an inhibition percentage higher than 80%, values very close to the inhibition percentage of ascorbic acid, a substance considered as an antioxidant standard. It can be stated, therefore, that the lavender essential oil obtained by supercritical technology has a high antioxidant potential. Figure 3 shows the results of the DPPH assay for each of the samples analysed.

Regarding the influence of pressure on the percentage of inhibition obtained, a small increase in its value can be observed (Figure 3) with increasing working pressure. This can be attributed to the increase in the concentration of the compounds of the essential oil to which the antioxidant characteristics are attributed.

The technological feasibility of lavender essential oil extraction with CO_2 under supercritical conditions can now be considered. High-purity extracts can be obtained without the need to apply high temperatures and by avoiding the use of organic solvents. Moreover, the antioxidant potential allows considering these natural extracts for the development of pharmaceuticals and cosmetics, which requires an economic feasibility study.

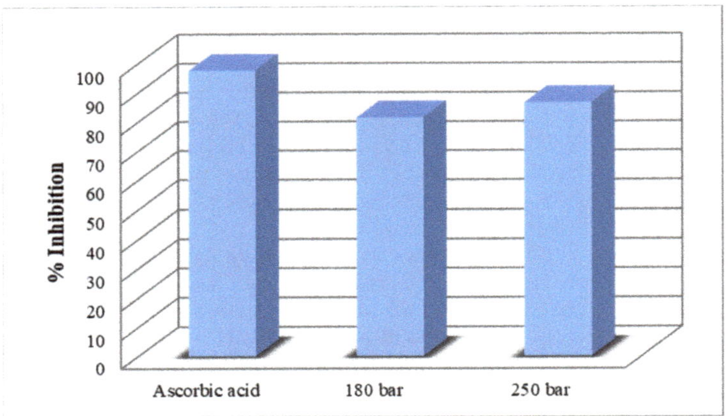

Figure 3. % Inhibition of ascorbic acid and lavender essential oil supercritical extracts.

3.2. SWOT Analysis

Table 2 shows the SWOT matrix obtained for the process of extracting lavender essential oil with supercritical CO_2. It should be noted that the SWOT matrix is a tool that depends on the current situation and future market trends [38].

Table 2. SWOT matrix for the supercritical extraction of lavender essential oil.

Strengths	Weaknesses
S1. Great availability of high-quality lavender in the area. S2. Large knowledge of the extraction process using supercritical technology and characterisation of the extracts obtained. S3. Cost advantages of patented know-how. S4. Obtention of a product with high added value and beneficial properties for health.	W1. High-cost structures. W2. Need for safety mechanisms due to high working pressures. W3. Limited financial resources. W4. Lack of access to distribution channels.
Opportunities	**Threats**
O1. Economic and social promotion of the surrounding area. O2. Use of the extracts in new and growing pharmaceutical and nutraceutical products. O3. Use of green solvents that promote compliance with health and food standards. O4. Unsatisfied customer needs.	T1. Minimum control in the final distribution. T2. High dependence on market developments. T3. Customers' ability to differentiate the new product from the traditional one. T4. Strong competitors. T5. Emergence of substitute products.

To design a SWOT matrix, internal factors of strengths and weaknesses and external factors of opportunities and threats must be taken into account.

The strengths of this business plan are the competitive advantages of the project, such as the wide availability of lavender at a good price in the area of Castilla-La Mancha, where the research was carried out, and the extensive experience of the research group with supercritical extraction. The most important advantage is the possibility of developing a novel product of high purity. Weaknesses that could compromise the viability of the project include the high investment costs of the facilities, working at high pressures, and the limited economic resources, as well as the challenge of finding a distribution channel.

On the other hand, the study of external factors has shown that this project is highly dependent on market developments and that there are many products whose properties can be considered similar to those obtained by this high-pressure process. Therefore, the main threat lies in competition. However, the use of green solvents that comply with the imposed standards, the economic and social promotion of the area, and the unmet need of customers are opportunities that make this process suitable to enter the growing market of natural products dedicated to human health care. Furthermore, it could provide an economic and social boost to the region. In summary, after the analysis of all the factors in the SWOT matrix, it is considered that the extraction of lavender essential oil with supercritical CO_2 can be a good business idea and should not be ruled out a priori.

3.3. Investment Analysis

To calculate the investment in supercritical equipment required at various sizes, information is usually available for one of the sizes and the power law (Equation (3)) is used to scale the costs of equipment with different capacities [39,40], where C_1 is the cost of equipment with capacity Q_1, C_2 is the cost of equipment with capacity Q_2, and M is a constant that refers to the type of equipment and is available in the literature [41].

$$C_1 = C_2 \times \left(\frac{Q_1}{Q_2} \right)^M \tag{3}$$

The costs associated with the three production plants proposed in this study have been estimated on the basis of the costs relating to the volume of a 1 L extractor as in the research of other authors [42]. Table 3 lists the data used to estimate the cost of the three proposed scenarios as well as the total cost for each scenario.

Table 3. Base cost for the equipment of the supercritical unit.

Item	M [a]	Unit Base Cost (EUR) [b,c]	Quantity (un.)	Total Cost Scenario 1 (EUR)	Total Cost Scenario 2 (EUR)	Total Cost Scenario 3 (EUR)
Jacketed extraction vessel	0.82	4361.00	1	50,866.36	107,830.47	190,364.55
CO_2 electrical pump	0.55	10,399.65	1	54,023.81	89,423.62	130,923.84
Cooler	0.59	1210.40	1	7088.21	12,170.86	18,320.15
Heater	0.59	382.70	1	2241.13	3848.14	5792.40
Separation vessel	0.49	979.00	2	8498.01	13,313.98	18,698.75
Manometer	0.00	62.30	3	186.90	186.90	186.90
Blocking valve	0.60	53.40	4	1288.90	2233.49	3385.33
Backpressure valve	0.60	1157.00	1	6981.54	12,098.05	18,337.21
Safety valve	0.60	80.10	1	483.34	837.56	1269.50
Flowmeter	0.60	249.20	4	6014.87	10,422.94	15,798.22
Temperature controller	0.60	160.20	5	4833.38	8375.57	12,694.99
Piping, connectors, mixers	0.40	587.40		1946.91	2808.81	3706.24
Structural material	0.40	267.00		884.96	1276.73	1684.66
Total cost of SFE plant				145,338.31	264,827.10	421,162.75

[a] M constant [41], [b] based on an operating plant with extraction and separation vessels of 1 L, [c] direct quotation for reference year of 2018.

In order to be able to determine the investment required for the establishment of a supercritical CO_2 extraction plant for lavender essential oil, the scale of the plant has to be taken into account, in particular the volume of the extraction vessel, which is closely linked to the production. For scenario 1, the price of the electric pump exceeds that of the extraction vessel. In contrast, in scenarios 2 and 3, it is the cost of the container that is the largest, accounting for approximately 40% and 45% of the total plant cost, respectively. These results are consistent with studies on extraction plants published by other authors [42,43].

3.4. Cost of Manufacturing (COM)

To determine the COM in processes using supercritical fluids, the methodology described by Turton et al. [44] (Equation (4)) is commonly used, where the *COM* is a function of five main costs: fixed capital of investment (*FCI*), cost of operational labour (*COL*), cost of utilities (*CUT*), cost of waste treatment (*CWT*), and cost of raw material (*CRM*). Table 4 indicates the parameters used for this estimation. Updated prices to 2021 were used to calculate the costs of different items.

$$COM = 0.304 \times FCI + 2.73 \times COL + 1.23 \times (CUT + CWT + CRM) \qquad (4)$$

Table 4. Parameters for estimating the COMs of extracts.

Economic Parameter	20 L Scale	50 L Scale	100 L Scale	Units
Fixed Capital Investment (FCI)				
Tangible fixed assets	384,419.84	700,467.69	1,113,975.46	EUR
Depreciation rate	10	10	10	%
Maintenance and others rate	6	6	6	%
Cost of Raw Material (CRM)				
Lavender	11.50	11.50	11.50	EUR/kg
Commercial CO_2	5.33	5.33	5.33	EUR/kg
Cost of Operational Labour (COL)				
Worker	3	3	3	Worker/year
Wages	90,000.00	90,000.00	90,000.00	EUR/year
Cost of Utilities (CUT)				
Electricity	0.27	0.27	0.27	EUR/kW.h
Water	0.33	0.33	0.33	EUR/m^3
Cooling fluid	1.83	1.83	1.83	EUR/L

The COM has been estimated for the lavender essential oil extraction process in the three proposed dimensions. The cost of raw material (CRM) includes the amount of lavender flowers and CO_2 needed for one year of operation, and the cost of operational labour (COL) refers to the salary of the workers. The number of workers will vary depending on the production. In the utilities part (CUT), the costs for electricity, water, and coolant are included. In addition, the cost of waste treatment (CWT) must be excluded. This is because the refined and exhausted lavender from the extraction, which is the only solid waste, can be used for various purposes, such as biomass or sold to companies for fertiliser or animal feed.

The fixed capital investment cost (FCI) consists not only of the cost of the equipment but both the installation costs of the extraction plant and the investment necessary for the operation of the plant, and it was calculated using the "Percentage Method" [45]. Moreover, an annual rate of 10% for depreciation and 6% for maintenance and other costs has to be taken into account [21].

In order to analyse the COM values obtained for each of the proposed scenarios, it is also useful to evaluate the productivity for each case. This value corresponds to the annual production of lavender essential oil in each scenario and was calculated on the basis of the maximum extraction yield obtained in the experimental work [34]. The results for the COM estimation are shown in Table 5.

Table 5. Specific costs, productivity, and COM.

	20 L Scale	50 L Scale	100 L Scale
Productivity (kg extract/year)	322.96	807.39	1614.78
FCI (EUR/kg extract)	1380.77	1089.56	866.01
CRM (EUR/kg extract)	168.30	168.30	168.30
COL (EUR/kg extract)	241.52	96.61	48.30
CUT (EUR/kg extract)	210.94	211.51	210.49
COM (EUR/kg extract)	1512.43	1035.98	840.26
COM (EUR/g extract)	1.51	1.04	0.84

As shown in Table 5, an increase in the volume of the extraction vessel leads to a decrease in COM, thus justifying the economic viability of large-scale projects, particularly concerning supercritical processes [46]. The specific costs related to raw materials, utilities, CRM, and CUT are virtually independent of scale. Conversely, with increasing production capacity, fixed equipment costs and operating costs with operators, FCI, and COL decrease.

The distribution of the specific costs in the value of the COM is represented in Figure 4, which shows the distribution of each of the factors involved in the total cost for each volume of the 20 L, 50 L, and 100 L extraction plants. It has to be taken into account that the 20 L volume could be considered as pilot plant size, so it cannot be compared in terms of investment and production to the 50 L and 100 L plants.

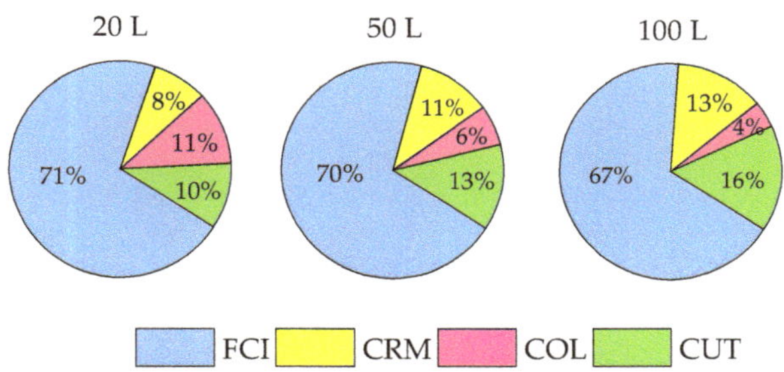

Figure 4. Distribution of specific costs according to the different plant sizes.

The distribution of relative costs (Figure 4) shows that it is the relative cost of capital (FCI) that represents the highest percentage, between 71 and 67%, being higher in the plant pilot scale.

In the case of operators (COL), the percentage in the case of the 20 L extractor is 11% while in the larger scales it decreases. In addition, it should be noted that for the 20 L plant the COL has the highest economic importance after the investment, which can be attributed to the fact that in this case the number of workers is the same as in the other two scenarios.

The decreasing impact of COL and FCI allows the percentage of CUT and CRM to increase with the increase in scale, although their specific cost is constant (Table 5). The percentage of CRM input increases from 8% to 13% from 20 L to 100 L, and the value of CUT varies from 10% to 16%. This fact confirms the theories of other authors about the trend of COM in scale-up analyses [47,48].

3.5. Financial Analysis

3.5.1. Selling Price Assessment

The sale prices have been set on the basis of current market research as well as promoting the idea that these are exclusive products which have been obtained through environmentally sustainable technology and in a selective way [24]. Furthermore, the business strategy is in line with the idea of entering an exclusive niche, with high added value, which represents an alternative for health care [49].

In this study, to establish the selling price to be used for the financial study, the prices of the main sellers of lavender essential oil on the market were evaluated. Table 6 shows the four references considered. These are lavender essential oils with similar characteristics to the essential oil studied in this project and are intended for cosmetic and nutraceutical applications.

Table 6. Sales prices of the most popular essential oils on the market.

	Price (EUR/g)
Lavender Essential Oil 1 [a]	0.50
Lavender Essential Oil 2 [b]	0.80
Lavender Essential Oil 3 [c]	1.99
Lavender Essential Oil 4 [d]	4.98

[a] Mundo Dos Oleos [50], [b] Pranarom [51], [c] Apivita [52], [d] Alqvimia [53].

As shown in Table 6, lavender essential oil prices vary greatly depending on the brand. This is because of the different markets targeted and the scale of production. The highest prices correspond to brands that promote their exclusivity and for functions more related to cosmetics. Conversely, lower prices are attributed to brands that produce on a larger scale and prefer to expand their market on demand for any application.

In the case of the essential oil in this study, despite being a product with exclusive properties and an application related to the pharmaceutical or nutraceutical market, an average value will be taken to make initial economic estimates.

In addition to the prices of the most recognised essential oils on the market, in order to establish the selling price of lavender essential oil to be obtained in supercritical extraction plants, the investment and operating costs associated with the production of lavender essential oil have to be considered.

Considering the above information, the selling price for the following calculations will be set at 1.38 EUR/g for the extract from the pilot plant (20 L) and 0.9 EUR/g for the 50 and 100 L industrial-scale plants. Reasonably, production at pilot-plant scale results in higher costs, so the selling price would have to be higher to ensure the viability of the business. Moreover, an attempt has been made to establish a selling price that is in line with the prices of other manufacturers.

3.5.2. Income Statement

The income statement is an essential part of an economic evaluation; it shows the income for a given period and also the related costs and expenses such as depreciation, amortisation of assets, and taxes. In this sense, the income statement shows the development of a company's assets and liabilities, i.e., it indicates whether investors are earning or losing money over the set time horizon.

For the preparation of the profit and loss account for each of the scenarios to produce lavender essential oil, a linear amortisation of 8 years (duration of the project) was established, and a leverage of 70% was considered, with an 8-year loan at 5% interest, and the remaining 30% obtained from own resources. VAT of 21% [54] and an average tax rate of 21.1% [55] were taken into account. Inflation of 6.5% [56] was also considered to analyse the evolution over the time horizon of the project [57]. The profit and loss account report is shown in Table 7.

In year 0, no cash flow or net profit is calculated, as no income is generated. This year is considered the start-up year in which the investment is fully realised. For the rest of the years, the gross margin is obtained by subtracting the variable costs, referring to raw materials and utilities, from the total sales, if all the essential oil obtained is sold.

EBITDA is determined by also considering fixed costs such as salaries, supplies, advertising, repairs and maintenance, and transport services or insurance. In addition, the earnings before interest and taxes (EBIT) and the earnings before taxes (EBT) were calculated, showing that both increase for each exercise.

Table 7. Income statement over the time horizon of the project for 3 scenarios.

Scenario	Financial Year	1	2	3	4	5	6	7	8
1	Gross Profit	323,200	344,208	366,581	390,409	415,786	442,812	471,594	502,248
	EBITDA [a]	214,200	235,208	257,581	281,409	306,786	333,812	362,594	393,248
	EBIT [b]	165,784	186,792	209,165	232,993	258,370	285,396	314,179	344,832
	EBT [c]	148,871	171,650	195,883	221,664	249,091	278,270	309,313	342,340
	Net Income	117,459	135,432	154,552	174,893	196,533	219,555	244,048	270,106
	NPV [d]	161,044	183,682	202,962	223,308	244,948	267,971	292,464	318,522
2	Gross Profit	419,997	447,297	476,371	507,336	540,312	575,433	612,836	652,670
	EBITDA [a]	310,997	338,297	367,371	398,336	431,312	466,433	503,836	543,670
	EBIT [b]	222,777	250,077	279,151	310,115	343,092	378,212	415,615	455,450
	EBT [c]	191,987	222,512	254,971	289,490	326,200	365,239	406,758	191,987
	Net Income	151,478	175,562	201,172	228,408	257,372	288,174	320,932	355,770
	NPV [d]	232,717	263,545	289,385	316,628	345,592	376,394	409,152	443,991
3	Gross Profit	841,643	896,349	954,612	1,016,662	1,082,745	1,153,123	1,228,076	1,307,901
	EBITDA [a]	732,643	787,349	845,612	907,662	973,745	1,044,123	1,119,076	1,198,901
	EBIT [b]	592,343	647,050	705,312	767,362	833,445	903,823	978,776	1,058,601
	EBT [c]	543,398	603,230	666,875	734,576	806,592	883,201	964,696	1,051,389
	Net Income	428,741	475,949	526,164	579,580	636,401	696,845	761,145	829,546
	NPV [d]	552,467	615,695	666,446	719,880	776,701	837,145	901,445	969,846

[a] EBITDA: earnings before interest, taxes, depreciation, and amortisation, [b] EBIT: earnings before interest and taxes, [c] EBT: earnings before taxes, [d] NPV: net present value at 3% interest.

To conclude the income statement, the net income was determined by deducting the value of the income tax provision from the EBT. Positive results were obtained for all scenarios, increasing this value every year and with an increase in scale, which can be taken as an indicator that this is a viable process that can be scaled up and a promising business plan.

On the other hand, we also applied other methods to judge the viability of the investment, such as the calculation of net present value (NPV) [58]. This method takes into account the timing of cash flows, and therefore uses the discounting or discounting procedure in order to homogenise the amounts of money received at different points in time [35]. These results again indicate that this is a healthy investment. The maximum value of NPV reached was EUR 969,846 in year 8 of scenario 3, which indicates that it is the most profitable scenario and the most attractive investment to scale the extraction process of lavender essential oil with supercritical CO_2.

3.5.3. Financial Ratios

The financial ratios are used to further investigate the profitability of the project in addition to the realisation of the income statement. This information is mandatory to obtain a bank loan. Table 8 shows the main financial ratios studied together with their generally considered healthy values. The investment will be attractive to investors if the result of these ratios for each financial year is above their healthy value [29]. These ratios are calculated for the three proposed scenarios and their results are shown in Table 9.

These results are grouped into categories. First, those related to profitability were calculated: ROE, ROI, and EBITDA to sales. These values indicate how assets are used to cover costs and achieve maximum profitability. It is shown that all these parameters increase as the time horizon progresses and become higher the larger the scale. Of note, the ROE value obtained reaches a high value, indicating the advantages of investing capital in this business instead of a bank deposit. The maximum value for this ratio is acquired in year 8 of each of the scenarios and has a value of 76.74, 71.65, and 77.65 for scenario 1, scenario 2, and scenario 3, respectively.

Table 8. Summary of the main financial ratios of economic analyses.

Financial Ratio	Equation	Definition	Healthy Value
Return on equity (ROE)	$\text{ROE} = \frac{\text{Net Income}}{\text{Shareholder's Equity}}$	Profit generated with the money that shareholders have invested	>13%
Return on investment (ROI)	$\text{ROI} = \frac{\text{Net Income}}{\text{Total capital}}$	To compare the efficiency of a number of different investments	>15%
EBITDA to sales	$\text{EBITDA to Sales} = \frac{\text{EBITDA}}{\text{Revenue}}$	Company's operational profitability by comparing its revenue with earnings	>10%
Solvency	$\text{Solvency} = \frac{\text{After Tax Net Profit} + \text{Depreciation}}{\text{Long Term} + \text{Short Term Liabilities}}$	Company's ability to meet long-term obligations	>0.2
Acid test	$\text{Acid test} = \frac{\text{Cash Accounts Receivable} + \text{Short Term Invests.}}{\text{Current Liabilities}}$	Enough short-term assets to cover its immediate liabilities	>1
Debt to capital	$\text{Debt to capital} = \frac{\text{Total Debts}}{\text{Shareholder's Equity} + \text{Debt}}$	Ability to absorb asset reductions without jeopardising the interest of creditors	Low
Manoeuvre fund	$\text{Manoeuvre fund} = \text{Current assets} - \text{Current Liabilities}$	Company's efficiency and its short-term financial health	High
Break-even point Margin of safety	$\text{Margin of safety} = \frac{\text{Sales}}{\text{Break even Point}}$	The point at which a business begins to make profits	>1.1

Table 9. Main financial ratios for 3 scenarios.

Scenario	Financial Year	1	2	3	4	5	6	7	8
1	ROI (%)	41.26	43.45	45.50	47.40	49.19	50.86	52.42	53.89
	ROE (%)	63.21	65.57	67.76	69.81	71.73	73.51	75.18	76.74
	EBITDA to sales (%)	48.06	49.55	50.96	52.27	53.51	54.67	55.76	56.78
	Solvency	3.57	4.11	4.70	5.34	6.01	6.71	7.43	8.17
	Acid test	1.04	2.07	3.10	4.12	5.14	6.14	7.13	8.10
	Debt-to-capital ratio (%)	34.68	34.77	34.85	34.92	34.99	35.04	35.09	35.13
	Manoeuvre fund	5384	160,010	336,384	535,920	760,127	1,010,609	1,289,077	1,597,351
	Break-even point	174,328	172,557	170,697	168,745	166,694	164,541	162,281	159,907
	Margin of safety	2.56	2.75	2.96	3.19	3.44	3.71	4.01	4.33
2	ROI (%)	35.95	38.02	39.98	41.84	43.61	45.28	46.86	48.37
	ROE (%)	57.70	60.02	62.22	64.31	66.30	68.18	69.96	71.65
	EBITDA to sales (%)	42.80	43.71	44.57	45.38	46.14	46.85	47.52	48.15
	Solvency	4.05	4.35	4.72	5.15	5.64	6.17	6.74	7.33
	Acid test	0.90	1.80	2.72	3.63	4.54	5.45	6.36	7.25
	Debt-to-capital ratio (%)	32.96	33.18	33.38	33.57	33.75	33.91	34.06	34.19
	Manoeuvre fund	−20,521.54	175,523.59	399,792.51	654,122.70	940,471.22	1,260,922.50	1,617,696.61	2,013,158.10
	Break-even point	228,009.48	224,785.16	221,399.62	217,844.81	214,112.26	210,193.08	206,077.94	201,757.05
	Margin of safety	3.19	3.44	3.72	4.03	4.37	4.74	5.15	5.60

Table 9. *Cont.*

Scenario	Financial Year	1	2	3	4	5	6	7	8
3	ROI (%)	47.36	48.85	50.28	51.64	52.94	54.18	55.35	56.48
	ROE (%)	67.08	68.82	70.47	72.05	73.56	74.99	76.35	77.65
	EBITDA to sales (%)	50.41	50.87	51.30	51.70	52.08	52.44	52.77	53.09
	Solvency	3.86	4.66	5.47	6.29	7.12	7.96	8.79	9.63
	Acid test	1.32	2.59	3.84	5.05	6.23	7.37	8.48	9.56
	Debt-to-capital ratio (%)	32.18	32.39	32.58	32.76	32.93	33.08	33.23	33.36
	Manoeuvre fund	127,159	684,196	1,297,872	1,971,874	2,710,126	3,516,811	4,396,383	5,353,584
	Break-even point	298,244	293,118	287,737	282,086	276,152	269,922	263,380	256,512
	Margin of safety	4.87	5.28	5.73	6.22	6.77	7.38	8.05	8.80

Moreover, the ratios for cash flow and solvency were then calculated as well as those related to debts such as the debt-to-capital ratio. As for the manoeuvre fund, as shown in Table 9, it has a negative value in year 1 of scenario 2. The change in scale may imply a reduction in the amount of liquid cash held by the company in the first year. Fortunately, this value becomes positive from the second year onwards. In general, this value acquires a positive value in all cases and is greater as the scale increases, which means that in scenario 3 the company could face short-term debts in a much more efficient way and without losing liquid assets.

Other ratios, such as the break-even point, were also calculated. Its value decreases over the years for all cases, indicating that the company's situation improves over the years and it has to sell less to cover costs. In addition, it is worth noting that the value of the safety margin is always above the healthy value, which affirms the robustness of the investment.

3.5.4. Price Curve

Additionally, to assess how the sales strategy could evolve in function to obtain the same profit, a price curve was made for scenario 3 (Figure 5), which corresponds to industrial-scale production.

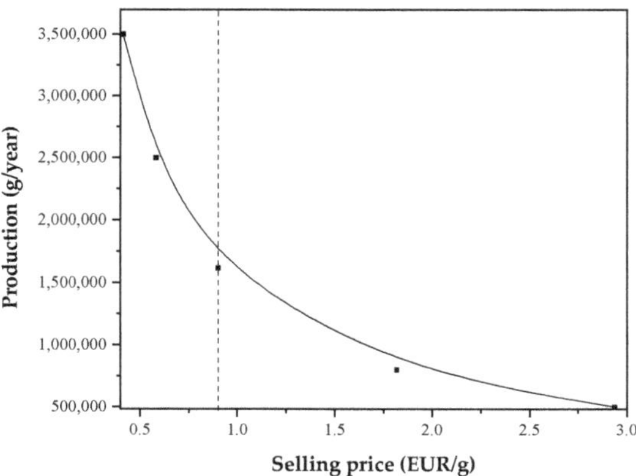

Figure 5. Price curve (scenario 3).

The price curve gives information about sales department objectives (increase production and hence sales) when prices must be decreased in order to fight competitors. From Figure 5, it can be observed that reducing prices lower that 0.9 EUR/g (selling price set for scenario 3) means that sales must be increased exponentially to reach a similar profitability, so this situation should be avoided as its dangerous for the company.

3.6. Sensitivity Analysis

In order to analyse the robustness of the business idea, a sensitivity study was carried out. This study analyses changes in factors not considered previously due to external or internal circumstances, providing information about commercial strategies and alternatives to face them. This study consisted of assessing the ROE profitability based on a 10% overestimation of equipment prices, wages, raw materials, electricity, and interest on money for scenario 3, assuming an average value of ROE. Table 10 shows the average ROE results for the 8-year time horizon. In addition, these results are plotted in Figure 6.

Table 10. Sensitivity analysis for scenario 3.

	ROE (%)	Variation (%)
Initial situation	72.62	-
+10% Raw material	71.89	−1
+10% Wages	72.48	−0.19
+10% Interest	72.62	0
+10% Electricity	72.62	0
+10% Equipment	71.71	−1.25

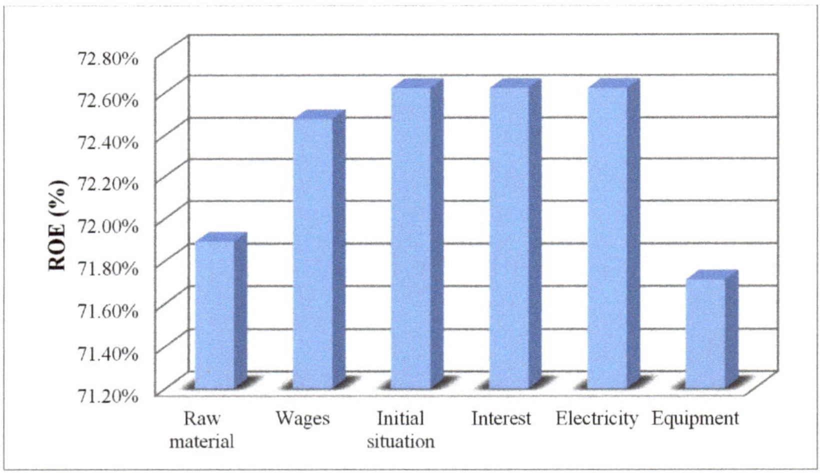

Figure 6. Sensitivity analysis for scenario 3 with the 10% overestimation in the price of raw materials, wages, interest, and equipment.

From the results shown in Table 10 and Figure 6, it can be assumed that the interest rate and wages has less influence on the profitability of the project. Furthermore, the increase in the price of electricity would hardly imply any change in the profitability of the process [59]. On the other hand, the increase in the price of equipment and raw materials, particularly the latter with a decrease of more than 1% compared to the initial situation, do have a significant influence on the ROE value.

Therefore, it can be concluded that the two factors that will determine the commercial strategy of the lavender essential oil extraction business with supercritical CO_2 over the time horizon are the price of raw materials and equipment. These results are in line with those reported by the COM determination. In this sense, it should be considered that these will be the parameters to which the profitability of this project is most sensitive to any possible variation in them.

4. Conclusions

The technical feasibility of supercritical CO_2 extraction of lavender essential oil was tested at pilot plant scale, with an extraction temperature of 60 °C and working pressures of 180 and 250 bar. The extracts obtained had linalool and linalyl acetate as their main compounds and showed an inhibition percentage higher than 80%, which shows that this extract could be promising in the current market. After these experiments, the economic viability of this supercritical process was studied by means of an 8-year business plan for three scenarios that included production from pilot plant scale to industrial scale. The SWOT matrix showed that it is a promising business idea, and by means of the price curve and the study of the market niche it was possible to establish the ideal prices for the success of the project: 1.38 EUR/g for the extract from the pilot plant (20 L) and 0.9 EUR/g for the 50 and 100 L industrial-scale plants. The profit and loss account was drawn up, and by calculating the financial ratios a detailed study of the profitability could be performed, paying special attention to the ROE value, which always acquired a value above 57%, indicating that it is an interesting project from an economic point of view. In addition, to determine the robustness of the idea and find out the cost items that most influence the commercial strategy to be followed, a sensitivity analysis was carried out, from which it could be seen that the cost of the equipment is the item that most influences the ROE value. The business plan provided results that postulate the production of lavender essential oil and other natural extracts by means of supercritical technology as an attractive technology for investment and one capable of satisfying the needs of a growing market niche.

Author Contributions: Conceptualisation, I.G. and M.T.G.; data curation, E.C.S., J.M.G.-V., I.G. and M.T.G.; formal analysis E.C.S., J.M.G.-V. and M.T.G.; funding acquisition I.G., J.F.R. and M.T.G.; investigation, E.C.S.; methodology, E.C.S., I.G. and M.T.G.; project administration, M.T.G.; resources, M.T.G.; software, E.C.S. and I.G.; supervision, J.M.G.-V., I.G., J.F.R. and M.T.G.; validation, E.C.S., J.M.G.-V. and M.T.G.; visualisation, E.C.S., J.M.G.-V. and M.T.G.; writing—original draft, E.C.S.; writing—review and editing, E.C.S., J.M.G.-V., I.G. and M.T.G. All authors have read and agreed to the published version of the manuscript.

Funding: This research was funded by the Project Ref. SBPLY/17/180501/000311 from Consejería de Educación, Cultura y Deportes of Junta de Comunidades de Castilla La Mancha.

Data Availability Statement: Not applicable.

Acknowledgments: We gratefully acknowledge the Consejería de Educación, Cultura y Deportes of Junta de Comunidades de Castilla La Mancha for funding the project.

Conflicts of Interest: The authors declare no conflict of interest; they have no known competing financial interests or personal relationships that could have appeared to influence the work reported in this paper.

References

1. Serra, A.T.; Seabra, I.J.; Braga, M.E.M.; Bronze, M.R.; de Sousa, H.C.; Duarte, C.M.M. Processing cherries (*Prunus avium*) using supercritical fluid technology. Part 1: Recovery of extract fractions rich in bioactive compounds. *J. Supercrit. Fluids* **2010**, *55*, 184–191. [CrossRef]
2. Moore, J.; Yousef, M.; Tsiani, E. Anticancer Effects of Rosemary (*Rosmarinus officinalis* L.) Extract and Rosemary Extract Polyphenols. *Nutrients* **2016**, *8*, 731. [CrossRef] [PubMed]
3. Chauhan, B.; Kumar, G.; Kalam, N.; Ansari, S.H. Current concepts and prospects of herbal nutraceutical: A review. *J. Adv. Pharm. Technol. Res.* **2013**, *4*, 4–8. [PubMed]
4. Edris, A. Pharmaceutical and Therapeutic Potentials of Essential Oils and Their Individual Volatile Constituents: A Review. *Phytother. Res.* **2007**, *21*, 308–323. [CrossRef]
5. Miguel, M.G. Antioxidant and anti-inflammatory activities of essential oils: A short review. *Molecules* **2010**, *15*, 9252–9287. [CrossRef]
6. Peana, A.; Moretti, L. Linalool in Essential Plant Oils: Pharmacological Effects. *Bot. Med. Clin. Pract.* **2008**, *10*, 716–724.
7. Peana, A.T.; D'Aquila, P.S.; Panin, F.; Serra, G.; Pippia, P.; Moretti, M.D.L. Anti-inflammatory activity of linalool and linalyl acetate constituents of essential oils. *Phytomedicine* **2002**, *9*, 721–726. [CrossRef]

8. Aboutaleb, N.; Jamali, H.; Abolhasani, M.; Pazoki Toroudi, H. Lavender oil (*Lavandula angustifolia*) attenuates renal ischemia/reperfusion injury in rats through suppression of inflammation, oxidative stress and apoptosis. *Biomed. Pharmacother.* **2019**, *110*, 9–19. [CrossRef]

9. Liu, Y.; Chen, C.; Wang, X.; Sun, Y.; Zhang, J.; Chen, J.; Shi, Y. An Epigenetic Role of Mitochondria in Cancer. *Cells* **2022**, *11*, 2518. [CrossRef]

10. Chen, K.; Zhang, J.; Beeraka, N.M.; Tang, C.; Babayeva, Y.V.; Sinelnikov, M.Y.; Zhang, X.; Zhang, J.; Liu, J.; Reshetov, I.V.; et al. Advances in the Prevention and Treatment of Obesity-Driven Effects in Breast Cancers. *Front. Oncol.* **2022**, *12*, 820968. [CrossRef]

11. Chen, K.; Lu, P.; Beeraka, N.M.; Sukocheva, O.A.; Madhunapantula, S.V.; Liu, J.; Sinelnikov, M.Y.; Nikolenko, V.N.; Bulygin, K.V.; Mikhaleva, L.M.; et al. Mitochondrial mutations and mitoepigenetics: Focus on regulation of oxidative stress-induced responses in breast cancers. *Semin. Cancer Biol.* **2022**, *83*, 556–569. [CrossRef] [PubMed]

12. Azwanida, N.N. A review on the extraction methods use in medicinal plants, principle, strength and limitation. *Med. Aromat. Plants* **2015**, *4*, 1–6.

13. Cassol, L.; Rodrigues, E.; Zapata Noreña, C.P. Extracting phenolic compounds from Hibiscus sabdariffa L. calyx using microwave assisted extraction. *Ind. Crop. Prod.* **2019**, *133*, 168–177. [CrossRef]

14. Uwineza, P.A.; Waśkiewicz, A. Recent Advances in Supercritical Fluid Extraction of Natural Bioactive Compounds from Natural Plant Materials. *Molecules* **2020**, *25*, 3847. [CrossRef]

15. Yousefi, M.; Rahimi-Nasrabadi, M.; Pourmortazavi, S.M.; Wysokowski, M.; Jesionowski, T.; Ehrlich, H.; Mirsadeghi, S. Supercritical Fluid Extraction of Essential Oils. *TrAC Trends Anal. Chem.* **2019**, *118*, 182–193. [CrossRef]

16. Reverchon, E.; Della Porta, G.; Senatore, F. Supercritical CO_2 Extraction and Fractionation of Lavender Essential Oil and Waxes. *J. Agric. Food Chem.* **1995**, *43*, 1654–1658. [CrossRef]

17. Varona, S.; Martin, A.; Cocero, M.J.; Gamse, T. Supercritical carbon dioxide fractionation of Lavandin essential oil: Experiments and modeling. *J. Supercrit. Fluids* **2008**, *45*, 181–188. [CrossRef]

18. Danh, L.T.; Triet, N.D.A.; Han, L.T.N.; Zhao, J.; Mammucari, R.; Foster, N. Antioxidant activity, yield and chemical composition of lavender essential oil extracted by supercritical CO_2. *J. Supercrit. Fluids* **2012**, *70*, 27–34. [CrossRef]

19. Perrut, M. Supercritical Fluid Applications: Industrial Developments and Economic Issues. *Ind. Eng. Chem. Res.* **2000**, *39*, 4531–4535. [CrossRef]

20. Fernández-Ponce, M.T.; Parjikolaei, B.R.; Lari, H.N.; Casas, L.; Mantell, C.; Martínez de la Ossa, E.J. Pilot-plant scale extraction of phenolic compounds from mango leaves using different green techniques: Kinetic and scale up study. *Chem. Eng. J.* **2016**, *299*, 420–430. [CrossRef]

21. Klein, E.J.; Carvalho, P.I.N.; Náthia-Neves, G.; Vardanega, R.; Meireles, M.A.A.; da Silva, E.A.; Vieira, M.G.A. Techno-economical optimization of uvaia (*Eugenia pyriformis*) extraction using supercritical fluid technology. *J. Supercrit. Fluids* **2021**, *174*, 105239. [CrossRef]

22. Cerón-Martínez, L.J.; Hurtado-Benavides, A.M.; Ayala-Aponte, A.; Serna-Cock, L.; Tirado, D.F. A Pilot-Scale Supercritical Carbon Dioxide Extraction to Valorize Colombian Mango Seed Kernel. *Molecules* **2021**, *26*, 2279. [CrossRef] [PubMed]

23. Teja, A.S.; Eckert, C.A. Commentary on Supercritical Fluids: Research and Applications. *Ind. Eng. Chem. Res.* **2000**, *39*, 4442–4444. [CrossRef]

24. Gracia, I. Prospective and Opportunities of High Pressure Processing in the Food, Nutraceutical and Pharmacy Market. *Food Eng. Ser.* **2015**, 479–508.

25. Filly, A.; Fabiano-Tixier, A.S.; Louis, C.; Fernandez, X.; Chemat, F. Water as a green solvent combined with different techniques for extraction of essential oil from lavender flowers. *Comptes Rendus Chim.* **2016**, *19*, 707–717. [CrossRef]

26. Rosa, P.T.V.; Meireles, M.A.A. Rapid estimation of the manufacturing cost of extracts obtained by supercritical fluid extraction. *J. Food Eng.* **2005**, *67*, 235–240. [CrossRef]

27. Zabot, G.L.; Moraes, M.N.; Carvalho, P.I.N.; Meireles, M.A.A. New proposal for extracting rosemary compounds: Process intensification and economic evaluation. *Ind. Crop. Prod.* **2015**, *77*, 758–771. [CrossRef]

28. Chañi-Paucar, L.O.; Johner, J.C.F.; Zabot, G.L.; Meireles, M.A.A. Technical and economic evaluation of supercritical CO_2 extraction of oil from sucupira branca seeds. *J. Supercrit. Fluids* **2022**, *181*, 105494. [CrossRef]

29. Fernández-Ronco, M.P.; de Lucas, A.; Rodríguez, J.F.; García, M.T.; Gracia, I. New considerations in the economic evaluation of supercritical processes: Separation of bioactive compounds from multicomponent mixtures. *J. Supercrit. Fluids* **2013**, *79*, 345–355. [CrossRef]

30. Soares, M.C.; Machado, P.R.; Guinosa, R.E. Supercritical Extraction of Essential Oils from Dry Clove: A Technical and Economic Viability Study of a Simulated Industrial Plant. *Environ. Sci. Proc.* **2022**, *13*, 11.

31. De Melo, M.M.R.; Barbosa, H.M.A.; Passos, C.P.; Silva, C.M. Supercritical fluid extraction of spent coffee grounds: Measurement of extraction curves, oil characterization and economic analysis. *J. Supercrit. Fluids* **2014**, *86*, 150–159. [CrossRef]

32. Daferera, D.J.; Ziogas, B.N.; Polissiou, M.G. GC-MS analysis of essential oils from some Greek aromatic plants and their fungitoxicity on Penicillium digitatum. *J. Agric. Food Chem.* **2000**, *48*, 2576–2581. [CrossRef] [PubMed]

33. Brand-Williams, W.; Cuvelier, M.E.; Berset, C. Use of a free radical method to evaluate antioxidant activity. *LWT-Food Sci. Technol.* **1995**, *28*, 25–30. [CrossRef]

34. Cruz, E.; Jesús, M.; García-Vargas; Ignacio Gracia, J.; Francisco Rodríguez, M.T.G. Optimization, modelling and scaling-up of linalool supercritical extraction from lavender essential oil. In Proceedings of the 18th European Meeting on Supercritical Fluids, Online, 4–6 May 2021.
35. De Lucas Martínez, A. *Bases de Economía Para la Función Directiva del Ingeniero Químico*; Universidad de Castilla La Mancha: Ciudad Real, Spain, 2016.
36. Çelik, H.T.; Gürü, M. Extraction of oil and silybin compounds from milk thistle seeds using supercritical carbon dioxide. *J. Supercrit. Fluids* **2015**, *100*, 105–109. [CrossRef]
37. Erland, L.A.E.; Mahmoud, S.S. Lavender (*Lavandula angustifolia*) Oils. *Essent. Oils Food Preserv. Flavor Saf.* **2016**, 501–508.
38. Terrados, J.; Almonacid, G.; Hontoria, L. Regional energy planning through SWOT analysis and strategic planning tools: Impact on renewables development. *Renew. Sustain. Energy Rev.* **2007**, *11*, 1275–1287. [CrossRef]
39. Green, D.W.; Perry, R.H. *Perry's Chemical Engineers' Handbook*, 8th ed.; McGraw-Hill Education: New York, NY, USA, 2008; ISBN 9780071422949.
40. King, C.F. Analysis, Synthesis, and Design of Chemical Processes. Richard Turton, Richard Bailie, Wallace Whiting, Joseph Shaeiwitz Prentice Hall, 1998. *Chemie Ing. Technol.* **1999**, *71*, 1319–1320. [CrossRef]
41. Peters, M.S.; Timmerhaus, K.D.; West, R.E. *Plant Design and Economics for Chemical Engineers*; McGraw-Hill: New York, NY, USA, 2003; Volume 4.
42. Johner, J.; Hatami, T.; Zabot, G.; Meireles, M.A. Kinetic behavior and economic evaluation of supercritical fluid extraction of oil from pequi (*Caryocar brasiliense*) for various grinding times and solvent flow rates. *J. Supercrit. Fluids* **2018**, *140*, 188–195. [CrossRef]
43. Veggi, P.C.; Cavalcanti, R.N.; Meireles, M.A.A. Production of phenolic-rich extracts from Brazilian plants using supercritical and subcritical fluid extraction: Experimental data and economic evaluation. *J. Food Eng.* **2014**, *131*, 96–109. [CrossRef]
44. Turton, R.; Bailie, R.C.; Whiting, W.B.; Shaeiwitz, J.A. *Analysis, Synthesis and Design of Chemical Processes*; Pearson Education: London, UK, 2008; ISBN 0132459183.
45. Kobe, K.A. Plant Design and Economics for Chemical Engineers (Peters, Max S.). *J. Chem. Educ.* **1958**, *35*, A506. [CrossRef]
46. Carvalho, P.I.N.; Osorio-Tobón, J.F.; Rostagno, M.A.; Petenate, A.J.; Meireles, M.A.A. Techno-economic evaluation of the extraction of turmeric (*Curcuma longa* L.) oil and ar-turmerone using supercritical carbon dioxide. *J. Supercrit. Fluids* **2015**, *105*, 44–54. [CrossRef]
47. Viganó, J.; Zabot, G.L.; Martínez, J. Supercritical fluid and pressurized liquid extractions of phytonutrients from passion fruit by-products: Economic evaluation of sequential multi-stage and single-stage processes. *J. Supercrit. Fluids* **2017**, *122*, 88–98. [CrossRef]
48. Olivera-Montenegro, L.; Best, I.; Bugarin, A.; Berastein, C.; Romero-Bonilla, H.; Romani, N.; Zabot, G.; Marzano, A. Techno-Economic Evaluation of the Production of Protein Hydrolysed from Quinoa (*Chenopodium quinoa* Willd.) Using Supercritical Fluids and Conventional Solvent Extraction. *Biol. Life Sci. Forum* **2021**, *6*, 55.
49. Mark-Herbert, C. Innovation of a new product category—Functional foods. *Technovation* **2004**, *24*, 713–719. [CrossRef]
50. Mundo Dos Óleos. Available online: https://www.mundodosoleos.com (accessed on 13 January 2022).
51. PRANAROM. Available online: https://www.pranarom.es (accessed on 13 January 2022).
52. Apivita. Available online: https://www.apivita.com (accessed on 13 January 2022).
53. Alqvimia. Available online: https://www.alqvimia.com (accessed on 13 January 2022).
54. Agencia Tributaria. Available online: https://www.agenciatributaria.es (accessed on 13 January 2022).
55. Infoautónomos. Available online: https://www.infoautonomos.com (accessed on 13 January 2022).
56. Instituto Nacional de Estadística. Available online: https://www.ine.es (accessed on 13 January 2022).
57. Sequeira, R.S.; Miguel, S.P.; Cabral, C.S.D.; Moreira, A.F.; Ferreira, P.; Correia, I.J. Development of a poly(vinyl alcohol)/lysine electrospun membrane-based drug delivery system for improved skin regeneration. *Int. J. Pharm.* **2019**, *570*, 118640. [CrossRef]
58. Osorio-Tobón, J.F.; Meireles, M.A.A.; Rostagno, M.A.; Carvalho, P.I.N. Process integration for turmeric products extraction using supercritical fluids and pressurized liquids: Economic evaluation. *Food Bioprod. Process.* **2016**, *98*, 227–235. [CrossRef]
59. Dimopoulou, M.; Offiah, V.; Falade, K.; Smith, A.M.; Kontogiorgos, V.; Angelis-Dimakis, A. Techno-Economic Assessment of Polysaccharide Extraction from Baobab: A Scale Up Analysis. *Sustainability* **2021**, *13*, 9915. [CrossRef]

Article

Carotenoids Recovery Enhancement by Supercritical CO$_2$ Extraction from Tomato Using Seed Oils as Modifiers

Mihaela Popescu, Petrica Iancu *, Valentin Plesu and Costin Sorin Bildea

Department of Chemical and Biochemical Engineering, University POLITEHNICA of Bucharest, 1, Gh. Polizu Street, Building A, Room A056, RO-011061 Bucharest, Romania
* Correspondence: p_iancu@chim.upb.ro; Tel.: +40-21-4023-916

Abstract: The food, cosmetic and pharmaceutical industries have strong demands for lycopene, the carotenoid with the highest antioxidant activity. Usually, this carotenoid is extracted from tomatoes using various extraction methods. This work aims to improve the quantity and quality of extracts from tomato slices by enhancing the recovery of the carotenoids from the solid matrix to the solvent using 20 *w/w*% seeds as modifiers and supercritical CO$_2$ extraction with optimal parameters as the method. *Tomato* (TSM), *camelina* (CSM) and *hemp* (HSM) seeds were used as modifiers due to their quality (polyunsaturated fatty acids content of 53–72%). A solubility of ~10 mg carotenoids/100 g of oil was obtained for CSM and HSM, while, for TSM, the solubility was 28% higher (due to different compositions of long carbon chains). An increase in the extraction yield from 66.00 to 108.65 g extract/kg dried sample was obtained in the following order: TSM < HSM < CSM. Two products, an oil rich in carotenoids (203.59 mg/100 g extract) and ω3-linolenic acid and a solid oleoresin rich in lycopene (1172.32 mg/100 g extract), were obtained using SFE under optimal conditions (450 bar, 70 °C, 13 kg/h and CSM modifier), as assessed by response surface methodology. A recommendation is proposed for the use of these products in the food industry based on their quality.

Keywords: supercritical extraction technology; extraction enhancement; seed oils as modifiers; carotenoids; polyunsaturated fatty acids

Citation: Popescu, M.; Iancu, P.; Plesu, V.; Bildea, C.S. Carotenoids Recovery Enhancement by Supercritical CO$_2$ Extraction from Tomato Using Seed Oils as Modifiers. *Processes* **2022**, *10*, 2656. https://doi.org/10.3390/pr10122656

Academic Editors: Maria Angela A. Meireles, Ádina L. Santana and Grazielle Nathia Neves

Received: 14 November 2022
Accepted: 7 December 2022
Published: 9 December 2022

Publisher's Note: MDPI stays neutral with regard to jurisdictional claims in published maps and institutional affiliations.

1. Introduction

Tomatoes are a rich source of natural bioactive compounds, such as carotenoids (in peels, pulp and the whole fruit) and polyunsaturated fatty acids (PUFA) (in seeds), that can be used in the food, cosmetic and pharmaceutical industries due to their natural antioxidant, preservative and colourant abilities [1].

Carotenoids are non-polar compounds, soluble in non-polar organic solvents and oils. Incorporating carotenoids into vegetable oils enhances their bioavailability and favours their absorption in the body [1]. Carotenoids dissolved in vegetable oils can be used as such in functional food products and cosmetic formulations [2]. The quality of the oils into which the carotenoids are incorporated is also important. The PUFA content and the ω6/ω3 ratio between ω6-linoleic acid (C18:2 ω6) and ω3-linolenic acid (C18:3ω3) are important characteristics of vegetable oils owing to their ability to treat and prevent several health conditions [1,3]. The only essential PUFA that cannot be synthesized by the human body are ω6-linoleic acid (C18:2ω6) and ω3-linolenic acid (C18:3ω3). However, these are found in vegetable oils and can be introduced into the human body through the diet [3]. Moreover, it has been observed that oils protect carotenoids from degradation, isomerization and oxidation reactions [4].

The food, cosmetic and pharmaceutical industries all have a strong demand for lycopene, the carotenoid with the highest antioxidant activity. Lycopene is one of the ten most-commercialized carotenoids because of its capacity to quench singlet molecular oxygen [5], with a global economic value estimated at 1.5 USD billion in 2017 and a compound annual growth rate between 2017 and 2022 of 2.3% [1]. Lycopene and β-carotene are used in food and supplements and are produced on an industrial scale. Due to rising demand for natural carotenoids as food colourants and advancements in carotenoid extraction, the carotenoid market is anticipated to increase from USD 1.5 billion in 2019 to USD 2.0 billion in 2026 [6].

For the food, cosmetic and pharmaceutical industries, it is desirable that the extraction process can ensure environmentally friendly carotenoid extraction with minimal loss of bioactivity [7]. Supercritical fluid extraction (SFE) is a green extraction method with many advantages over conventional methods that use organic solvents [6]. Supercritical carbon dioxide ($scCO_2$) is an ideal solvent in the health industry due to its green nature and because produces extracts with a high degree of purity, free of traces of toxic solvents [8]. Concerning the operating parameters, the extraction yield increases with pressure, temperature and CO_2 flow rate [9,10]. For the extraction temperature, it has been found that an increase in temperature is associated with a decrease in the $scCO_2$ density and in the quality of the extract. However, high temperatures facilitate the release of target compounds from the plant matrix due to the increase in the carotenoids' volatility and diffusivity and tomato structure breakage releasing the solute in the extraction solvent [7,9,11,12].

The extraction pressure influences the $scCO_2$ properties. An increase in pressure is associated with an increase in $scCO_2$ density, which favours the solubility and the extraction of solutes due to the improved solvating power of the solvent. Additionally, by using greater pressures, it may be possible to achieve the same extraction yields with less CO_2 [9,10]. Additionally, high extraction pressure and temperature increase the solubility of triglycerides and carotenoids in $scCO_2$ [13]. However, very high conditions could cause the degradation and isomerization of carotenoids. The optimal conditions for carotenoid extraction using the SFE process are represented by extraction pressures of 200–450 bar and extraction temperatures of 40–70 °C [6,14,15].

The CO_2 flow rate ensures the solvent velocity that influences the residence time between the solvent and the pant matrix and the mass transfer. Higher extraction yields can be obtained at high flow rates because the external resistance to mass transfer decreases and the internal diffusion increases. However, very high flow rates do not improve the extraction process, but only increase the operating costs [9].

The extraction yield in carotenoid recovery from tomatoes can be improved when co-solvents or modifiers are added to the vegetable matrix. Tomatoes contain seeds that are an endogenous source of oil that participates as a co-solvent during the extraction process [5]. Tomato seeds ensure a certain amount of oil in the extraction process, improving the extraction efficiency [16]. Modifiers are added directly to the plant sample and their action decreases with extraction time, their concentration being high at the beginning of the process [11]. Their purpose is to enhance the solubility of target compounds in $scCO_2$ and consequently improve the extraction efficiency. Additionally, edible oil modifiers do not require separation from the extract and can be used as a product due to their quality [9,13]. Several studies have been concerned with the extraction of carotenoids from plant matrices using different types of vegetable oils as solvents. Vegetable oils are capable of extracting non-polar compounds, such as carotenoids, and represent an ecologic, economic and green alternative to traditional organic solvents [2].

Studies concerning carotenoid extraction with SFE and seeds or vegetable oils from different types of tomato samples (pulp, peels and pomace) used tomato oil [5,12], hazelnut oil [8,17,18], avocado oil [19], canola oil [7], soybean oil [13,18] and linseed, corn, sesame, sunflower, rapeseed and olive oils [18]. The quality of these oils (triglyceride and carotenoid contents) influence SFE extraction. Thus, the evaluation of different seeds was the first step in this study. Three types of seeds were evaluated as possible sources of oil to be used in the SFE process to improve the extraction efficiency.

Tomato seeds (*Lycopersicum esculentum*) contain 17–23% oil with around 55% ω6-linoleic acid (C18:2ω6) and 24% ω9-oleic acid (C18:1ω9) as major compounds [3,20,21].

The carotenoid content of this oil varies with the amount of peel and pulp residues found in the seed sample subjected to extraction.

Hemp seeds (*Cannabis sativa* L.) are free of psychoactive substances and can be used in the food, pharmaceutic and cosmetic industries [22]. Hemp seeds contain 28–37.3% edible oil [23,24] that contains ω6-linoleic acid (C18:2ω6) and ω3-linolenic acid (C18:3ω3) PUFA as majority compounds. Hemp seed oil is unique and rare among the common plant oils due to its low ω6:ω3 ratio (3:1) and due to the presence of ω6-linolenic acid (C18:3ω6) [25]. The hemp seed oil ω6:ω3 ratio falls within the ideal range of 1:1–5:1, which has been claimed to be optimal for human consumption [22,26], while the consumption of ω6-linolenic acid is associated with the maintenance of hormonal balance [24]. Regarding the carotenoids, hemp seed oil presents β-carotene, its content ranging between 6.22–12.65 mg/kg oil [22].

Camelina seeds (*Camelina sativa* L.) contain 28–40% oil [26], with 50–60% PUFA, from which 35–40% consists of ω3-linolenic acid (C18:3ω3) and 15–20% ω6-linoleic acid (C18:2ω6) [27]. Moreover, it has a high content of ω9-gondoic acid (C20:1ω9), which plants very rarely possess, and a low content, below 5%, of ω9-erucic acid (C22:1ω9), which is undesirable in oils [28]. The ω6:ω3 ratio value is below 1, which shows that camelina oil is beneficial in the human diet [26].

The aim of this work was to investigate the impact on the extraction yield and carotenoid content of oils extracted in the dynamic mode by supercritical CO_2 extraction from tomatoes. Powders of tomato, camelina and hemp seeds were added to the initial tomato samples to be used as sources of oils, increasing the carotenoids' solubility in supercritical fluid. The solubility of carotenoids from tomato samples was checked before supercritical extraction to determine the oil contribution. The extracts obtained by extraction with supercritical CO_2 and oil extracted from the tomato sample with seed powders were analysed and their effects were presented. The optimal effect of modifiers combined with the effects of the pressure and CO_2 flow rate was assessed by the design of the experiments. Four quadratic models (yield of oil and solid extracts and their compositions in carotenoids and lycopene) were formulated based on 15 experimental runs and the optimal values of the extraction parameters were identified. The qualities of the extracts obtained from tomatoes by supercritical extraction under optimal conditions were compared with extracts from tomato pomace and their potential applications were highlighted.

2. Materials and Methods

2.1. Chemicals and Standards

The Supelco 37 Component FAME Mix, 2,2-diphenyl-1-picrylhydrazyl (DPPH), 6-hydroxy-2,5,7,8-tetra methylbroman-2-carboxylic acid (Trolox), sodium hydroxide, boron trifluoride–methanol (BF_3-MeOH) (10–14%) complex solution, anhydrous magnesium sulphate, acetone, methanol, n-heptane and hexane used in this study were of analytical grade and obtained from Sigma-Aldrich (Munich, Germany). CO_2 with 99.9% purity was acquired from Linde Gaz (Bucharest, Romania).

2.2. Plant Material

Rila tomatoes, harvested in June 2022 and farmed in Colibași, Giurgiu County, camelina and hemp seeds were purchased from a local market. For the preparation of tomato slices (TS), ripe tomatoes were washed of soil traces, dried with paper tissues and manually cut into around 5 mm-thickness slices. For the preparation of tomato seeds (TSM), clean tomatoes were manually pressed with a squeezer (Ertone, model MN503) to isolate the pomace from the juice. The seeds from the pomace were separated with a sieve. Tomato slices (TS), tomato seeds (TSM), camelina seeds (CSM) and hemp seeds (HSM) were dried in a food dehydrator (Hendi Profi Line, model 229026) at 50 °C for 48 h and ground before extraction using a grinder (Tarrington House, model KM150S). The initial seed content of the TS sample in this study varied from 6–8%.

2.3. Soxhlet Extraction (SE)

The Soxhlet extraction method was used to recover the oils from TSM, CSM and HSM seeds. Tomato seed oil (TSO), camelina seed oil (CSO) and hemp seed oil (HSO) were extracted using acetone/hexane (*v/v*, 1:1) (AH) for 6 h. For each extraction experiment, 25 g of dried and ground seeds and 250 mL of solvent were necessary. At the end of each extraction experiment, the solvent was separated from the extract using a rotary evaporator (Hahnvapor, model HS-2000NS) to isolate the oil and to recover the solvent. The collected vegetable oils were weighed and stored in the freezer at $-20\,^{\circ}C$ until analysis. The results were expressed in g oil/100 g dried seeds.

2.4. Maceration (M)

Maceration was used to extract carotenoids from tomato slices using three different solvents, which were vegetable oils (TSO, HSO and CSO). The carotenoids' solubilities in these oils were determined and compared. For each extraction experiment, 0.5 g of dried and ground TS and 3 mL of solvent (TS:oil ratio of 1:6) were introduced to a 15 mL centrifuge tube and vortexed for 15 min at 3000 rpm (Velp Scientifica, model ZX4). Next, the obtained mixtures were centrifuged for 30 min at 8000 rpm (Hettich centrifuge, model EBA 200S) to isolate the extract (the supernatant phase) from the TS sample. The extracts were analysed by the UV–VIS method to determine the composition of carotenoids. The collected extracts were stored in the freezer at $-20\,^{\circ}C$ until analysis. The results are presented in mg carotenoids/100 g extract.

2.5. Supercritical CO$_2$ Extraction (SFE)

The supercritical CO_2 extraction method was used to perform the extraction experiments to evaluate the influence of modifiers on mass transfer and for the determination of the optimal parameters. The extraction experiments were carried out in a laboratory-scale High-Pressure Extraction Unit (HPE-CC 500, Eurotechnica GmbH). The values used for the operating parameters were a pressure of 350–450 bar, CO_2 flow rate of 9–13 kg/h, temperature of 70 $^{\circ}C$ and an extraction time of 600 min to obtain the full extraction curve. For each experiment, 180 g of sample (150 g of dried and ground TS and 30 g of dried and ground TSM, HSM or CSM) was loaded into the extractor and the air from the system was purged with CO_2. Next, the heating was switched on, the extraction temperature was set and the extractor was pressurized with CO_2 using the high-pressure CO_2 pump at the desired value. The pressure and CO_2 flow rate were monitored and kept constant using the pressure control valve and by adjusting the pump stroke. The CO_2 flow rate was measured using a mass flow meter. During the experiment, the extract was collected every 30 min. The extracts were centrifuged for 30 min at 8000 rpm to isolate the SFE-A and SFE-B fractions according to a previous study [21]. The collected extracts were stored in a freezer at $-20\,^{\circ}C$ until analysis. The results were expressed in g extract/100 g dried tomato sample.

2.6. GC–MS Analysis of Vegetable Oils

The vegetable oils (TSO, HSO and CSO) were transesterified to obtain fatty acid methyl esters (FAME) using the BF_3–MeOH complex as a catalyst in an acid-catalysed procedure [29]. The gas chromatography coupled with mass spectrometry (GC–MS) method, using a gas chromatograph (Agilent Technologies 7890A) equipped with a mass spectrometer detector (Agilent Technologies 5975C), was used to determine the FAME components in the analysed oils according to the procedure described in a previous study [30]. The results were expressed as the percentage of FAMEs.

2.7. UV–VIS Analysis of the Extracts

The UV–VIS spectrometry method was used to analyse the carotenoid contents of the extracts with a Helios UV–Visible spectrophotometer (Helios beta, Thermo Spectronic). The extracts' spectra were recorded in the 325–575 nm wavelength range. The quantification of carotenoids was performed using the IPM-II-WG6 method [31] based on the sample absorbances and specific extinction coefficients of lycopene and β-carotene in acetone/hexane (AH) at the isosbestic point (461 nm) a = 0.194 and lycopene maximum wavelength (504 nm) a = 0.261. The results were expressed in mg carotenoid/100 g extract.

2.8. Antioxidant Activity of the Extracts

The total antioxidant activity of tomato extract samples was assessed by the DPPH (2,2-diphenyl-1-picrylhydrazyl) free radical-scavenging assay using Trolox (6-hydroxy-2,5,7,8-tetramethylbroman-2- carboxylic acid) as a reference antioxidant compound according to the procedure described in a previous study [21]. The results were expressed as the percentage inhibition of free radicals, inhibition % DPPH.

2.9. Statistical Analysis

Extraction (SE, M and SFE), centrifugation and transesterification experiments were performed in duplicate. The UV-VIS and GC-MS analyses of extracts for the quali-quantitative determination of carotenoids and FAME were carried out in triplicate. The results were presented as the mean values ± standard deviation (SD). All experimental datasets were subjected to ANOVA analysis to determine the variability between the samples and within the samples. The extraction yields, carotenoid concentrations and FAME compositions of the extracts were validated using the homogeneity of variances for individual sets, applying Hartley's Fmax test [32].

2.10. Design of Experiments

The response surface methodology (RSM) was used as a tool to determine the effects of different operating parameters used for supercritical CO_2 extraction and their interactions on some productivity parameters (extraction yield and compounds' concentrations) based on a number of experiments. The number of experiments was influenced by the design type, the number of factors and the number of replicates in the central point according to Equation (1) [33,34].

$$N = 2\,k\,(k - 1) + C_0, \tag{1}$$

where N is the total number of experiments, k is the factor number and C_0 is the number of replicates at the central point. The selected factors were related to the operating parameters and to the composition of the vegetal matrix, while the evaluated responses were related to the quali-quantitative profiles of the extracts. The Box–Behnken surface experimental design p^k (BBD) was used as the design type in this study to evaluate the effects of three selected factors (k = 3) with three levels (p = 3), including the extraction pressure (350, 400 and 450 bar), seed type (TSM, CSM and HSM) and CO_2 flow rate (9, 11 and 13 kg/h). Four responses, such as the SFE-A (oily extract) yield, SFE-B (solid extract) yield, total carotenoid content in SFE-A and lycopene content in SFE-B, were proposed to evaluate the effects of these three factors. Considering three replicates in the central point and BBD design, the number of experiments was 15, with each experiment being performed in duplicate. The selected factors and responses with their coded and uncoded levels are presented in Tables 1 and 2.

Table 1. Independent variables (factors) for the BBD experimental design.

Independent Variables (Factors)	Symbol	Range of Coded Levels of Variables		
		Low	Medium	High
		−1	0	+1
Extraction pressure/(bar)	X_1	350	400	450
Seed type	X_2	TSM	CSM	HSM
CO_2 flow rate/(kg/h)	X_3	9	11	13

Table 2. Dependent variables (responses) for the BBD experimental design.

Dependent Variables (Responses)	Symbol
SFE-A yield/(g/100 g dried sample)	Y_1
SFE-B yield/(g/100 g dried sample)	Y_2
SFE-A carotenoids/(mg/100 g extract)	Y_3
SFE-B lycopene/(mg/100 g extract)	Y_4

To predict the effect of each factor that affects the chosen responses of the SFE process using tomato slices, four second-order polynomial models (Equation (2)) for k independent variables (X_1, X_2 and X_3) were developed:

$$Y_i = \beta_{0,i} + \beta_{1,i} \cdot X_1 + \beta_{2,i} \cdot X_2 + \beta_{3,i} \cdot X_3 + \beta_{4,i} \cdot X_1 \cdot X_2 + \beta_{5,i} \cdot X_2 \cdot X_3 + \beta_{6,i} \cdot X_1 \cdot X_3 \\ + \beta_{7,i} \cdot X_1^2 + \beta_{8,i} \cdot X_2^2 + \beta_{9,i} \cdot X_3^2 \tag{2}$$

where X_1, X_2 and X_3 are the selected factors and β_{ji} are the intercept ($\beta_{0,i}$), linear ($\beta_{1,i}$, $\beta_{2,i}$ and $\beta_{3,i}$), quadratic ($\beta_{4,i}$, $\beta_{5,i}$ and $\beta_{6,i}$) and interaction ($\beta7_{,i}$, $\beta_{8,i}$ and= $\beta_{9,i}$) coefficients that described the factors' effects on the chosen responses (Y_i). Statistical validation of the model was performed using analysis of variance (ANOVA) to verify that the model correctly described the relationship between factors and responses. The accuracy of the model was checked through the coefficient of determination (R^2) and lack-of-fit test. Finally, the determination of the optimal conditions was accomplished using the desirability function and response surface plots. Response surface plots were generated using the function of two factors and keeping the third factor constant [34].

3. Results and Discussion

3.1. Extraction of Carotenoids with Vegetable Oils

Vegetable oils used for the extraction of carotenoids from TS were obtained by the SE method with organic solvents (AH) using three types of seeds: TSM, CSM and HSM. The characteristics of the seeds before and after grinding (A and B), the extracts mixed with the solvent (C) and the isolated oils (D) are illustrated in Figure 1.

Figure 1. Vegetable oils (TSO, CSO and HSO) obtained using Soxhlet extraction with acetone:hexane (1:1, *v/v*): A—seeds sample, B—ground seeds and C—extract with solvent, D—oil.

These oils were chosen due to their particular compositions of ω6-linoleic acid (C18:2ω6) and ω3-linolenic acid (C18:3ω3). Additionally, they contain compounds that provide different colours. Due to their different compositions of carotenoids and chlorophyll, TSO is orange, CSO is yellow and HSO is green.

The oil contents of the seeds are presented in Table 3. The extraction yield for TSO was 19.17 ± 0.21 g TSO/100 g dried seeds, being in the range of 17–23%, as reported in the literature [20]. From HSM seeds, 30.85 ± 0.29 g HSO/100 g dried seeds was extracted, higher than that from TSO, with 61%. The highest amount of oil, 41.85 ± 0.16 g CSO/100 g dried seeds, was obtained from CSM seeds, twice the yield from TSM seeds. Similar yield values of 28–35% for HSO [23] and 28–40% for CSO [26] were presented in other studies.

Table 3. Extraction yields of vegetable oils (g oil/100 g dried seeds $\pm$ SD).

Extract ID	Extraction Method *	Extraction Solvent *	Extraction Parameters	Extraction Yield **
TSO				19.17 ± 0.21
CSO	SE	AH	$m_{sample} = 25$ $V_{solvent} = 250$	41.85 ± 0.16
HSO				30.85 ± 0.29

* SE—Soxhlet extraction, AH—acetone:hexane (1:1, v/v), m_{sample}/(g)—mass of the sample, $V_{solvent}$/(mL)—volume of the solvent. ** there were no statistically significant differences between extraction yields for sets variances according to Hartley's Fmax test ($p < 0.05$) at the $\alpha = 0.05$ significance level.

The composition of fatty acids of the vegetable oils extracted with AH is presented in Table 4. The fatty acids profile of these oils was similar to that in other studies [3,25,28]. For the selected oils, a difference between SFA (saturated fatty acids), MUFA (monounsaturated fatty acids) and PUFA compositions was observed. HSO contained a higher concentration of PUFA (71.54%), while TSO and CSO had similar PUFA concentrations (53.19% and 55.06%). MUFA were present in higher concentrations in CSO (38.08%), while their content in HSO was twice lower (17.59%). The major fatty acids found in these oils were ω6-linoleic acid (C18:2ω6) in TSO (55.06%) and HSO (55.93%), and ω3-linolenic acid (C18:3ω3) in CSO (33.90%), which was not present in TSO. Due to their composition, these oils can be used in the food industry.

Table 4. Composition of fatty acids of the vegetable oils (% $\pm$ SD).

Fatty Acid Profile	Composition *		
	TSO	CSO	HSO
Palmitic acid (C16:0)	15.06 ± 0.03 [a]	6.42 ± 0.05 [b]	6.43 ± 0.02 [c]
Stearic acid (C18:0)	6.36 ± 0.09 [a]	2.32 ± 0.04 [b]	3.40 ± 0.03 [c]
Oleic acid (C18:1ω9)	23.52 ± 0.22 [a]	18.70 ± 0.09 [b]	16.59 ± 0.06 [c]
Linoleic acid (C18:2ω6)	55.06 ± 0.17 [a]	19.29 ± 0.10 [b]	55.93 ± 0.10 [c]
Linolenic acid (C18:3ω6)	-	-	2.37 ± 0.03 [c]
Linolenic acid (C18:3ω3)	-	33.90 ± 0.07 [b]	13.25 ± 0.03 [c]
Arachidic acid (C20:0)	-	-	1.04 ± 0.01 [c]
Gondoic acid (C20:1ω9)	-	17.22 ± 0.06 [b]	1.00 ± 0.02 [c]
Erucic acid (C22:1ω9)	-	2.17 ± 0.04 [b]	-
SFA	21.42	8.74	10.87
MUFA	23.52	38.08	17.59
PUFA	55.06	53.19	71.54

* means $\pm$ SD followed by a letter (a–c) indicate that there were no statistically significant differences between the fatty acid compositions for sets variances with the same superscript letter according to Hartley's Fmax test ($p < 0.05$) at the $\alpha = 0.05$ level of significance.

The presence of carotenoids in the extracted oils was determined by the UV-VIS spectrophotometric method. The lycopene and β-carotene contents of these oils are presented in Table 5. CSO contained traces of both carotenoids (less than 1 mg/100 g oil), while TSO had the highest content (5.89 mg carotenoids/100 g oil) due to the presence of peels and pulp residues from the tomato seed sample subjected to the SE method. β-carotene was present in a high concentration of 5.95 mg/100 g oil in HSO, being within the range of 3.15–12.54 mg/100 g oil reported by another author [35].

Table 5. Carotenoid contents of vegetable oils and TS extracts (mg/100 g extract ± SD).

Extract ID	Extraction Method *	Extraction Solvent *	Lycopene Content **	β-Carotene Content **
TSO	SE	AH	3.26 ± 0.05	2.63 ± 0.07
CSO	SE	AH	0.08 ± 0.01	0.50 ± 0.03
HSO	SE	AH	0.11 ± 0.01	5.95 ± 0.08
TSO-TS	M	TSO	11.35 ± 0.06	1.42 ± 0.03
CSO-TS	M	CSO	8.94 ± 0.07	1.15 ± 0.05
HSO-TS	M	HSO	8.42 ± 0.25	1.48 ± 0.03

* SE—Soxhlet extraction, AH—acetone:hexane (1:1, *v/v*), M—maceration. ** there were no statistically significant differences between carotenoid contents for sets variances according to Hartley's F_{max} test ($p < 0.05$) at the α = 0.05 level of significance.

To assess the solubility of carotenoids in the three selected oils, the maceration method was used. Ground TS samples (A and B) were mixed with the oils (C), vortexed (D) and centrifuged in fractions (E), as shown in Figure 2. After the same extraction times, three oily extracts (TSO-TS, CSO-TS and HSO-TS) were obtained and changes in the oils' colours were observed: TSO-TS changed from orange to red, CSO-TS changed from yellow to orange and HSO-TS changed from green to brown (Figure 3a,b). These changes in colours were associated with the carotenoid content extracted from TS, even if the vegetable oils were also pigmented.

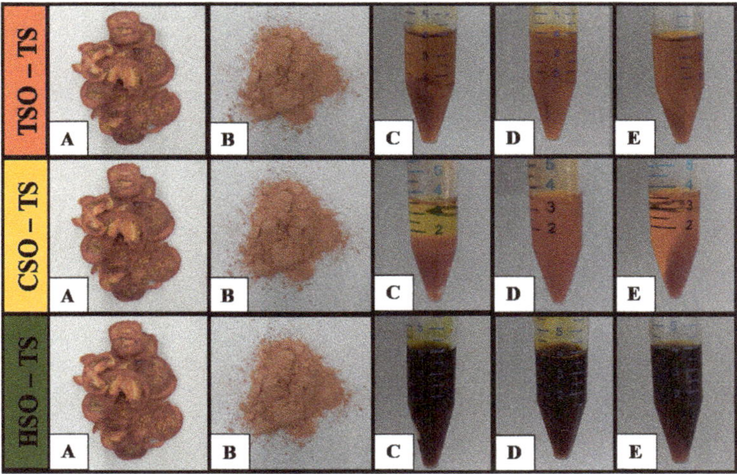

Figure 2. Samples and extracts from tomato slices (TS) for determining carotenoid solubility in vegetable oils (TSO, CSO and HSO): A—TS sample, B—ground TS, C—TS with oil, D—vortexed sample, E—centrifuged extract.

The extracts' contents of carotenoids (lycopene and β-carotene), without the oil contribution, are presented in Table 5. The solubility of both carotenoids in CSO and HSO was similar (~10 mg carotenoids/100 g oil), while, in TSO, the solubility was higher, at 28%. Lycopene's solubility at 25 °C in TSO was 11.35 mg/100 g oil, close to the values of 7–8 mg/100 g oil reported in another study [36]. An amount of 7–10 mg lycopene/100 g oil was also reported in a recent study as its solubility in five vegetable oils, including sunflower, palm, soybean, olive and coconut oils [37]. The maximum solubility of pure lycopene in tomato oil is 46 mg/100 g oil, as reported by Squillace et al. [5]. These results show that the carotenoids' solubility in oils is a very important factor for improving the extraction efficiency.

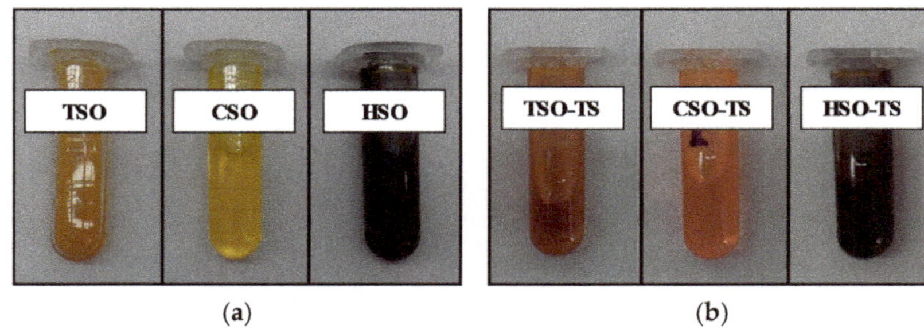

(a)   (b)

Figure 3. Carotenoids' solubility in vegetable oils (TSO, CSO and HSO): (**a**) vegetable oils; (**b**) extracts from TS.

The lycopene recovery with TSO was higher, with 21% in CSO and with 26% in HSO. This behaviour may be caused by the chain length of the triacylglyceride fatty acids present in the selected oils. As also reported by Borel et al. [38] the solubility of carotenoids in oils increases with a decrease in the chain length. As can be observed from Table 4, TSO contains triacylglycerides with C16 and C18 fatty acid groups, while CSO and HSO also contain C20 and C22 fatty acid groups.

3.2. Extraction of Carotenoids with Supercritical CO_2 and Seed Oils as Modifiers

The supercritical CO_2 extraction method was used to recover carotenoids from the TS samples enriched with 20% extra seeds (TSM, CSM and HSM) as modifiers and to investigate their influence on the extraction efficiency. The final seed contents of the samples were around 26–28% (as the tomato pomace seed content). The extraction parameters of 450 bar extraction pressure, 70 °C extraction temperature, 11 kg/h CO_2 flow rate and extraction time of 10 h were chosen according to the results of a previous study [21] that established them as being favourable for carotenoid recovery. The seed content of the tomato samples, the extraction pressure and the CO_2 flow rate or extraction time influenced the extraction efficiency and the extracts' qualities expressed by the yield, carotenoid recovery and antioxidant activity for supercritical CO_2 extraction of carotenoids from tomatoes.

Figure 4 presents the characteristics of the initial tomato sample and seeds (A), the ground sample (B) and the obtained centrifuged extracts TSM-TS, CSM-TS and HSM-TS (C) by SFE. Two extract phases were separated: red to orange oil (SFE-A) and dark-red solid oleoresin (SFE-B).

3.2.1. Modifiers' Effects on Extraction Yield

Supercritical fluid extraction is a contact equilibrium separation process between a supercritical fluid and solid matrix, where mass transfer kinetics and solubility play important roles. The variations in the amount of the extract (g) with the amount of solvent (kg) per kg of sample (SFE extraction curves) for three samples (TSM, CSM and HSM) are presented in Figure 5a. From a mass transfer point of view, these curves resemble a typical extraction curve with three regions: a solubility-controlled phase (0–200 kg CO_2/kg sample), transition phase (200–350 kg CO_2/kg sample) and diffusion-controlled phase (350–600 kg CO_2/kg sample) for the same operating conditions (pressure, temperature and flow rate) [39].

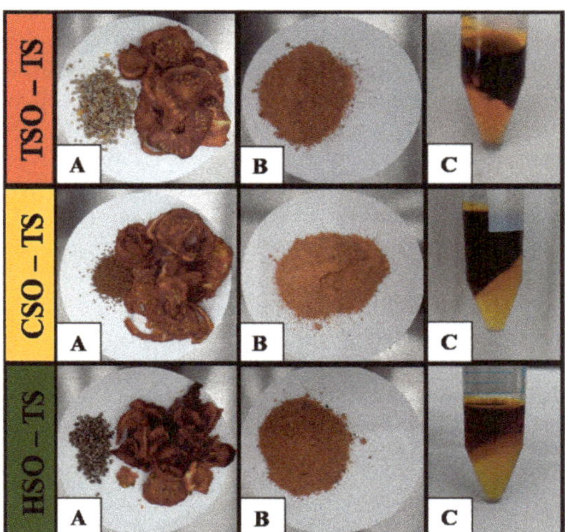

Figure 4. SFE samples and extracts: A—initial sample, B—ground sample, C—centrifuged extracts.

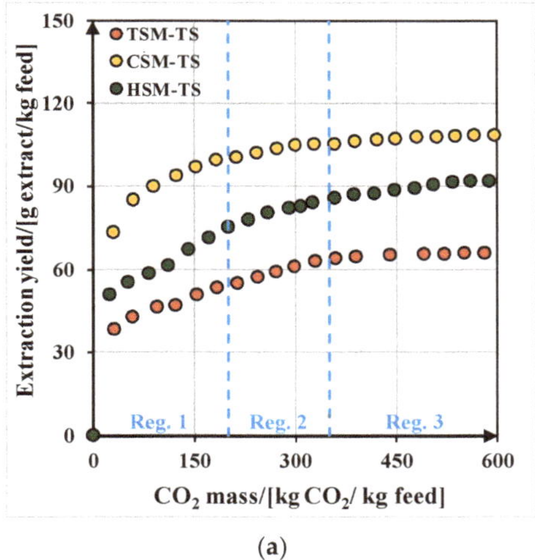

(a)

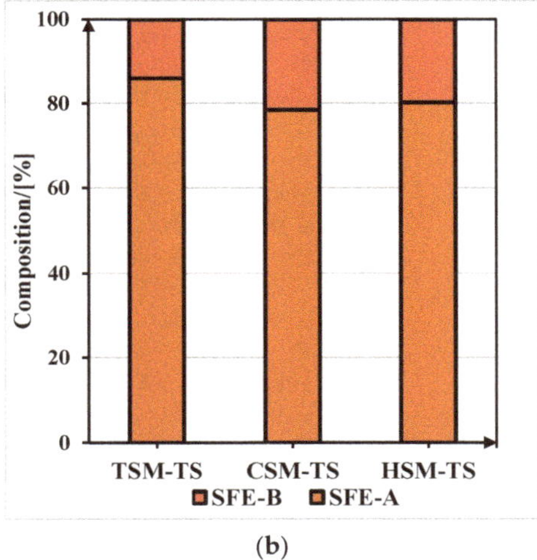

(b)

Figure 5. Supercritical CO_2 extraction from tomato slices (TS) using modifiers (TSM, CSM and HSM): (**a**) extraction curves (Reg. 1—solubility-controlled phase, Reg. 2—transition phase, Reg. 3—diffusion-controlled phase); (**b**) extracts' compositions in oil (SFE-A) and solid oleoresin (SFE-B).

In the first region (Reg. 1), the slope of the curve was linear and the amount of the extract was influenced by the solubility in supercritical CO_2. The compounds from the surface of the solid were solubilised in solvent. Lipids and carotenoids are compounds with high molecular weight, slight polarity and low volatility, and need higher-density conditions to be solubilised in CO_2. At higher pressures (450 bar) and moderate temperatures (70 °C), the CO_2 density is similar to that of a liquid (850–900 kg/m^3). In this region, it was observed that the extraction yield decreased in the order of CSM > HSM > TSM, from 101 to 55 g extract/kg of dried sample. In this region, between 82% and 93% of the total extract is obtained. The transition between solubility-controlled to diffusion-controlled phases occurred in the second region (Reg. 2). The extract recovery was between 5% and 13% for TSM, HSM and CSM. In the third region (Reg. 3), the supercritical CO_2-extracted compounds from the inside of the solid diffused to the medium at a slow rate [9,40]. Even if more CO_2 was added, the extraction yield did not increase by more than 3–9%. Therefore, no more than 450 kg of CO_2/kg feed is needed for a significant extraction yield. This CO_2 consumption corresponds to an extraction time of about 450–480 min. During the performed SFE experiments, the CO_2 was recirculated and reused in the plant; therefore, for each SFE experiment, only around 3–4 kg of CO_2 was consumed when the plant was depressurised.

Adding ground TSM, CSM and HSM seeds to the initial samples of tomato slices, the extraction efficiency increased by 39% for HSM (91.8 g extract/kg dried sample) and by 65% for CSM seeds (108.7 g extract/kg dried sample) from that of TSM seeds (66 g extract/kg dried sample). This increase was caused by the oil content of the seed modifiers. These results are better than those reported by other authors who used other types of modifiers for different tomato samples (pulp, peels and pomace) to improve the supercritical CO_2 extraction efficiency. For solid modifiers, such as tomato seeds [5,12], hazelnut powder [17] or avocado pulp added directly to tomato samples, the same effect was observed, with the extraction yield increasing from 8.04% to 24.4% [19]. The addition of liquid modifiers and co-solvents, such as different types of edible canola oil [7], soybean oil [13,18], hazelnut oil [8,18], linseed, corn, sesame, sunflower, rapeseed and olive oils [18], had the same effect.

3.2.2. Modifiers' Effect on Extracts' Compositions

In the presence of oil and supercritical CO_2, the carotenoid composition of the extracts improved. Sovova and Stateva [13] studied the effect of oil as a co-solvent on the carotenoid composition of extracts and reported that β-carotene solubility in supercritical CO_2 increased by 1.3–1.5% when soybean oil was used as a co-solvent (mixed with supercritical CO_2 before entering the extractor), while, when the oil was used as a liquid modifier (being added directly to the sample), the solubility was even higher.

The presence of modifiers in solids subjected to supercritical extraction led to the extraction of oils together with carotenoids. The obtained SFE extracts were centrifuged to separate two fractions, such as the SFE-A fraction with the consistency of a pigmented oil rich in carotenoids and SFE-B fraction with the consistency of a pigmented solid oleoresin rich in lycopene. In Figure 5b, the extracts' compositions of both fractions, SFE-A (79–86% oil) and SFE-B (14–21% solid oleoresin), are presented. It can be observed that the highest amounts of SFE-A and SFE-B were found in the TSM-TS extract (6.1 g SFE-A:1 g SFE-B) and CSM-TS extract (3.8 g SFE-A:1 g SFE-B). The carotenoid contents of the SFE-A and SFE-B fractions, determined by the UV-VIS method, are presented in Table 6. The SFE-A fractions contained mostly β-carotene, with values between 101.35 and 159.73 mg/100 g oil, while SFE-B fractions were enriched in lycopene, with values in the range of 947.92–1212.68 mg/100 g solid oleoresin.

Table 6. Carotenoid contents of TSM-TS, CSM-TS and HSM-TS SFE extract fractions (mg/100 g extract ± SD).

Extract ID	Extraction Method	Extraction Solvent	Lycopene Content *	β-Carotene Content *
TSM-TS-SFE-A	SFE	$scCO_2$	43.81 ± 0.40 [a]	159.73 ± 1.31 [a]
CSM-TS-SFE-A	SFE	$scCO_2$	71.35 ± 0.24 [a]	139.70 ± 0.79 [a]
HSM-TS-SFE-A	SFE	$scCO_2$	61.79 ± 0.14 [a]	101.35 ± 0.23 [a]
TSM-TS-SFE-B	SFE	$scCO_2$	1212.68 ± 0.69 [b]	133.78 ± 0.76 [a]
CSM-TS-SFE-B	SFE	$scCO_2$	1073.94 ± 3.45 [a]	93.52 ± 8.49 [c]
HSM-TS-SFE-B	SFE	$scCO_2$	947.92 ± 4.36 [a]	123.41 ± 2.11 [a]

* means ± SD followed by a letter (a–c) indicate that there were no statistically significant differences between the lycopene and β-carotene contents for sets variances with the same superscript letter according to Hartley's F_{max} test ($p < 0.05$) at the $\alpha = 0.05$ level of significance.

This behaviour was also observed in other studies (the presence of both fractions), but with lower carotenoid contents than those reported in the present work. By the extraction of carotenoids from tomato pulp with supercritical CO_2, 80 mg β-carotene/100 g oil fraction (SFE-A) and 581.6 mg lycopene/100 g solid fraction (SFE-B) were reported by Longo et al. [17]. Valecilla-Yepez and Ciftici [41] extracted lycopene from a mixture of tomato peels and seeds (70:30), and the extract fractions contained 220–300 mg lycopene/100 g oil fraction (SFE-A) and 330–900 mg lycopene/100 g insoluble fraction (SFE-B). When hazelnut oil was used as a co-solvent in SFE, lycopene recovery from tomato pulp increased from 1.2% to 21.6% [18], while the addition of 37% tomato seeds as a modifier to tomato peels in SFE increased lycopene and β-carotene recovery from 17.5% and 37% to 46% and 68%, respectively [12].

Additionally, the modifiers directly influenced the composition of the SFE-A fraction in fatty acids. As shown in Table 4, CSO and HSO had superior quality to TSO due to the presence of ω3-linolenic acid (C18:3ω3).

Based on the presented results, samples of CSM or HSM mixed with TS could be used as raw materials to extract carotenoids with the same quality (900–1100 mg lycopene/100 g extract and 90–120 mg β-carotene/100 g extract) as those obtained from tomato pomace. The oily fraction separated from the extract was rich in carotenoids (40–70 mg lycopene/100 g extract and 100–160 mg β-carotene/100 g extract) and in essential ω3-linolenic acid (C18:3ω3), which is very rare in common oils.

3.3. Optimal Parameters for SFE Extraction with Seed Oils as Modifiers

Twelve experiments were performed to extract carotenoids by the supercritical CO_2 extraction method using tomato slices mixed with three types of seeds (TSM, CSM and HSM). Three other experiments were also performed in the central point (pressure of 400 bar, 20% CSM and 11 kg/h CO_2 flow rate). The values of the dependent variables are presented in Table 7. The extraction yields at different experimental runs varied between 2.97 and 9.14 g/100 g dried sample for the SFE-A fraction (oily extract) and between 0.22 and 2.81 g/100 g dried sample for the SFE-B fraction (solid extract). For the extracts' compositions, the total carotenoid contents in the oily fractions (SFE-A) varied between 86.17 and 206.15 mg/100 g extract, while the lycopene content in the solid fractions was between 659.17 and 1212.68 mg/100 g extract.

Table 7. Box–Behnken experimental design matrix for carotenoid SFE extraction with modifiers.

	Independent Variables			Dependent Variables *			
Run	X_1 Extraction Pressure	X_2 Seeds Type	X_3 CO_2 Flow Rate	Y_1 SFE-A Yield	Y_2 SFE-B Yield	Y_3 SFE-A Carotenoids	Y_4 SFE-B Lycopene
1	−1	−1	0	2.97	0.22	147.82	814.97
2	+1	−1	0	5.67	0.93	203.54	1212.68
3	−1	+1	0	4.76	0.75	86.17	659.17
4	+1	+1	0	7.36	1.82	163.14	947.92
5	−1	0	−1	4.32	0.73	98.57	756.22
6	+1	0	−1	8.08	1.90	198.58	1055.26
7	−1	0	+1	5.95	1.39	115.66	819.64
8	+1	0	+1	9.14	2.81	206.15	1167.62
9	0	−1	−1	3.43	0.30	172.34	956.10
10	0	+1	−1	5.27	1.09	138.35	739.99
11	0	−1	+1	4.56	0.53	189.80	1059.35
12	0	+1	+1	6.46	1.30	142.11	831.25
13	0	0	0	7.10	1.55	160.59	920.59
14	0	0	0	6.62	1.74	163.01	918.46
15	0	0	0	7.04	1.77	158.63	916.21

* SFE-A yield/(g/100 g dried sample), SFE-B yield/(g/100 g dried sample), SFE-A carotenoids/(mg/100 g extract), SFE-B lycopene/(mg/100 g extract).

In Table 8, the regression coefficients of four quadratic models that adequately described the effects of independent variables X_1–X_3 through the chosen Y_1–Y_4 responses of the SFE process are presented. The significant terms of the models from a statistical point of view ($p > 0.05$ at a significance level of $\alpha = 0.05$) for the extraction yields of SFE-A (Y_1) and SFE-B (Y_2) were the linear terms of the pressure (X_1), seed type (X_2) and CO_2 flow rate (X_3) and the quadratic term of the seed type (X_2^2). For the first response (Y_1), it can be seen that the quadratic effect of the seed type (X_2^2) and the linear effects of the extraction pressure (X_1) and CO_2 flow rate (X_3) were positively correlated with the extraction yield of the SFE-A fraction (run 8 Table 7). This means that using CSM as a modifier, 450 bar as the extraction pressure and a 13 kg/h CO_2 flow rate resulted in high extraction yields. The same behaviour was found for the second response (Y_2), the extraction yield of the SFE-B fraction. Regarding the Y_3 and Y_4 responses, it can be seen that the linear terms of the extraction pressure (X_1) and seed type (X_2) significantly affected the recovery of carotenoids from the SFE-A and SFE-B fractions, as shown by the regression coefficient values. The effect of the seed type was negative, indicating that the use of the TSM modifier had the greatest effect, while the effect of the pressure was positive, indicating the use of high pressure values (450 bar). For all of the responses, it was observed that the extraction pressure (X_1) and CO_2 flow rate (X_3) had positive effects, while the seed type (X_2) had a complex effect, being positive for the Y_1 and Y_2 responses and negative for the Y_3 and Y_4 responses. The use of CSM led to high extraction yields and lower carotenoid concentrations, while the use of TSM led to higher recovery of carotenoids and lower yields. Thus, RSM analysis confirmed the importance of the type of seeds used as modifiers in carotenoid recovery from tomato slices through the SFE process.

Table 8. Models' regression coefficients.

Statistical Data	Y_1 = SFE-A Yield	Y_2 = SFE-B Yield	Y_3 = SFE-A Carotenoids	Y_4 = SFE-B Lycopene
β_0 (Intercept)	6.893	1.688	160.690	918.419
β_1 (X_1)	1.531	0.547	40.398	166.686
β_2 (X_2)	0.902	0.373	−22.978	−108.097
β_3 (X_3)	0.626	0.252	5.735	46.287
β_4 (X_1X_2)	*	*	5.312	−27.241
β_5 (X_2X_3)	*	*	*	*
β_6 (X_1X_3)	*	*	*	12.235
β_7 (X_1^2)	*	*	−8.237	21.640
β_8 (X_2^2)	−1.832	−0.830	*	−31.374
β_9 (X_3^2)	*	*	*	9.627
df Lack of fit	8	3	7	4
p-value Lack of fit	0.565	0.154	0.056	0.053
df Pure error	2	2	2	2
Pure Error	0.068	0.014	4.830	4.793
R^2	0.983	0.961	0.968	0.999
Adjusted R^2	0.976	0.889	0.950	0.998
Predicted R^2	0.942	0.861	0.922	0.985

* Statistically insignificant ($p > 0.05$) at the $\alpha = 0.05$ level of significance.

For the statistical validation of the proposed models, the lack-of-fit values were evaluated for a confidence level of $\alpha = 0.05$ (5% risk is considered significant). Coefficients with $p > 0.05$ were considered insignificant from a statistical point of view and were removed from the model. Additionally, the precision of the predictive models was also verified by the coefficient of determination (R^2). The R^2 values were higher than 0.96, indicating a good fit between the experimental and predicted data, as can also be seen in Figure 6a–d for the Y_1–Y_4 responses. The predicted R^2 values show how well regression models make predictions. Response surface graphs between two factors keeping the third factor constant were generated to analyse the effect of them on the four responses (Figure 7). The CO_2 flow rate was kept constant at the maximum value of 13 kg/h because, as shown above, it had the lowest effect among the analysed factors.

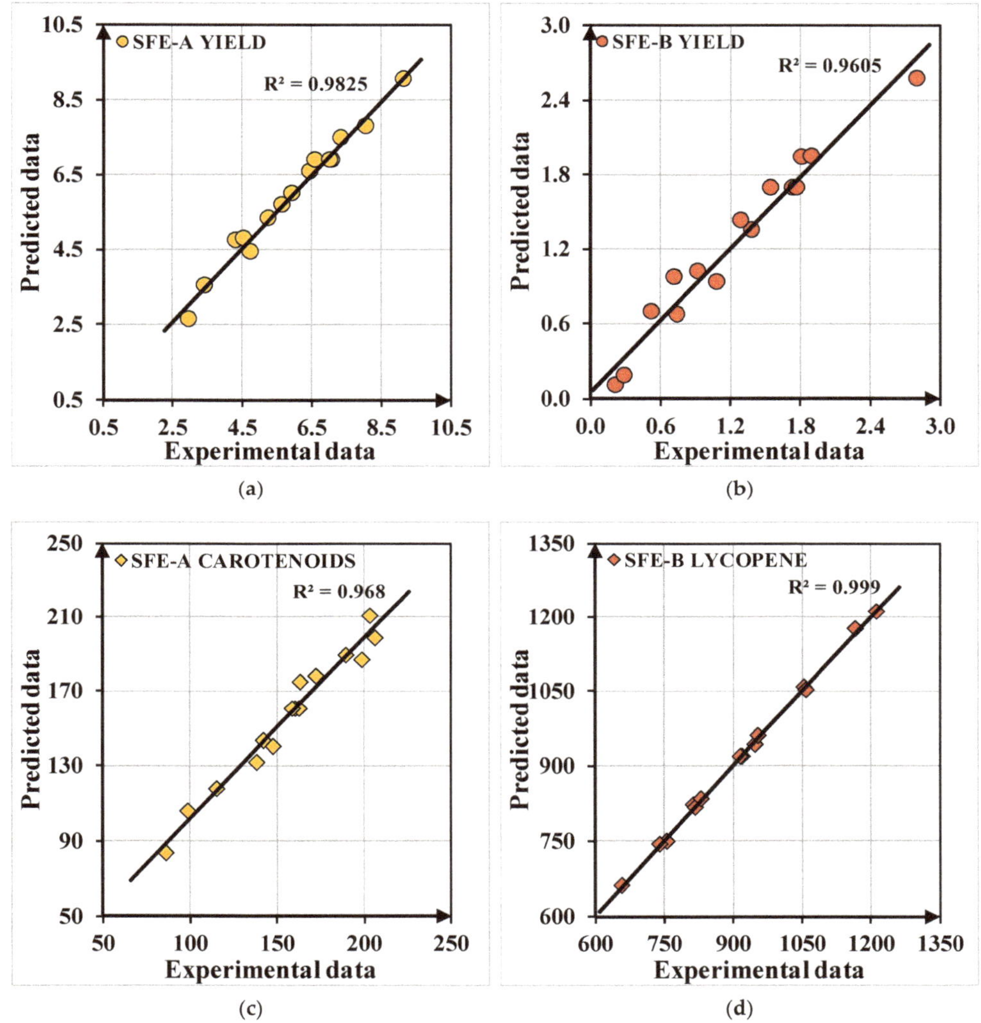

Figure 6. Predicted versus experimental data for responses: (**a**) Y_1—SFE-A extraction yield/(g/100 g dried sample); (**b**) Y_2—SFE-B extraction yield/(g/100 g dried sample); (**c**) Y_3—SFE-A total carotenoid content/(mg/100 g extract); (**d**) Y_4—SFE-B lycopene content/(mg/100 g extract).

For the extraction yields of the SFE-A (Figure 7a) and SFE-B (Figure 7b) fractions, it can be seen that an increase in pressure and the use of CSM seeds led to higher yields. Yield values of 7–8 g/100 g dried sample for SFE-A and 2–2,5 g/100 g dried sample for SFE-B could be obtained using pressures between 400 and 450 bar, $X_1 = (0, 1)$ and 20% CSM seeds ($X_2 = 0$), a mixture of 10% CSM + 10%TSM seeds ($X_2 = -0.5, 0$) or 10% CSM + 10% HSM seeds ($X_2 = 0, 0.5$).

For the total carotenoid content of SFE-A (Figure 7c) and lycopene content of SFE-B (Figure 7d), an increase in pressure and the use of TSM seeds led to higher amounts of carotenoids. Total carotenoid values of 180–200 mg/100 g extract for SFE-A and 1100–1300 mg/100 g extract for SFE-B were obtained using pressures between 375 and 450 bar, $X_1 = (-0.5, 1)$ and 20% TSM seeds ($X_2 = -1$), a mixture of 10% TSM + 10%CSM seeds ($X_2 = -1, 0$) or a mixture of 10% TSM + 5% CSM +5% HSM seeds ($X_2 = -1, 0.5$).

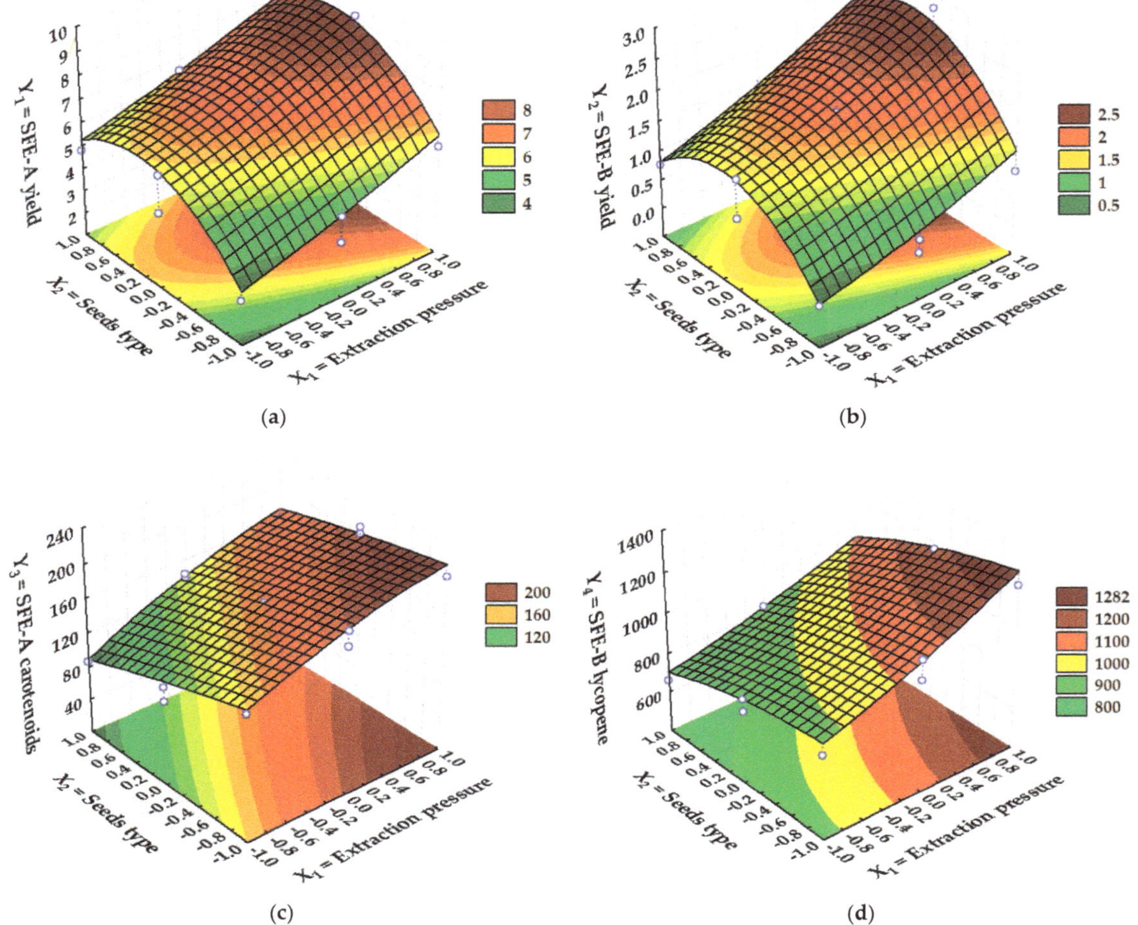

Figure 7. Response surface plots for Y_1–Y_4 responses: (**a**) X_1X_2 effects, $X_3 = 1$ (ct) for Y_1; (**b**) X_1X_2 effects, $X_3 = 1$ (ct) for Y_2; (**c**) X_1X_2 effects, $X_3 = 1$ (ct) for Y_3; (**d**) X_1X_2 effects, $X_3 = 1$ (ct) for Y_4.

It is desirable to obtain both high extraction yields and extracts rich in carotenoids; thus a trade-off is needed. For this purpose, the desirability function was used to obtain the optimum extraction conditions for all responses optimized simultaneously. In Figure 8a–c, the optimal desirability plots for the proposed factors (X_1, X_2 and X_3) are presented. A desirability of 93.31% was obtained when 450 bar, CSM and 13 kg/h were used as the independent variables. Under these conditions, the obtained optimum response values were 8.89 g SFE-A/100 g dried sample, 2.57 g SFE-B/100 g dried sample, 198.49 mg total carotenoids/100 g SFE-A and 1174.90 mg lycopene/100 g SFE-B.

3.4. Quality of Products

By using SFE as an environmentally friendly extraction technique and vegetal samples with large productivity (tomato and camelina seeds), two natural products were obtained: a solid oleoresin (SFE-B) enriched in lycopene, which can be used as a natural colourant or additive in the food industry, and tomato and camelina oil enriched in carotenoids and ω3-linolenic acid (C18:3ω3), which can be used for consumption.

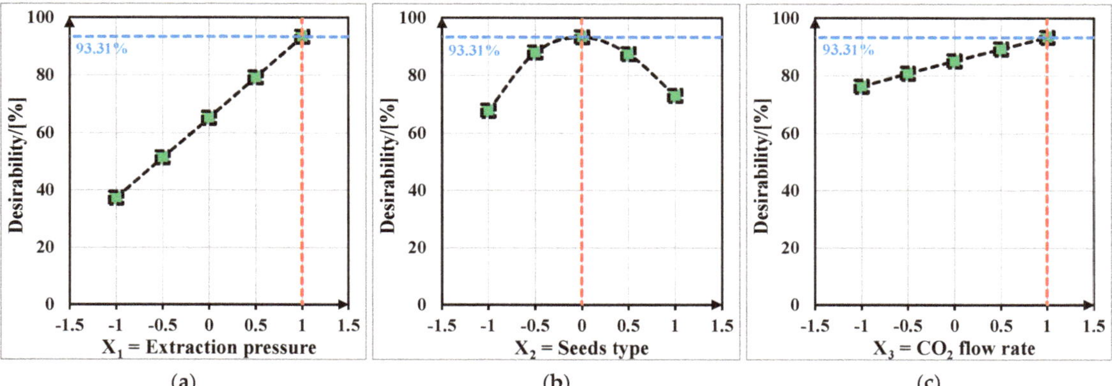

Figure 8. Optimal desirability for (**a**) X_1—extraction pressure; (**b**) X_2—seed type; (**c**) X_3—CO$_2$ flow rate.

In Table 9, a comparison between the quality of the extracts obtained by SFE from tomato slices with camelina seeds as a modifier and other extracts obtained from tomato slices and pomace is presented. The extract conditions were as follows: pressure of 450 bar, temperature of 70 °C, CO$_2$ flowrate of 13 kg/h and extraction time of 10 h. A quantitative analysis of the extracts showed that, when using camelina seeds, the extraction yield of the SFE-A product was higher by 107% than that when tomato slices and in the range of that when using tomato pomace. The extraction yield of the SFE-B product was 8.7 times higher than that obtained using tomato slices and 1.41 times higher than that obtained using tomato pomace. Regarding the products' qualities, the total carotenoid content in the SFE-A product was almost half the content in the SFE-A product obtained from tomato pomace. However, the presence of ω3-linolenic acid (C18:3ω3) increased the value of this product's antioxidant activity from 38.41 to 50.16%. For the SFE-B product, the lycopene content and the antioxidant activity were the highest from the analysed samples. In the study by Szabo et al. [1], HSO and flaxseed oil, with ω6/ω3 ratios of 3:1 and 0.3:1, respectively, were enriched with carotenoids recovered from tomato residue to increase their antioxidant activities, and their results showed that these products were suitable to be used in various food matrices.

Table 9. Quality of the products obtained by SFE from TS enriched with a seed modifier.

Extract ID *	Extraction Yield **		Carotenoid Content **		SFE-A PUFA Content **		Antioxidant Activity **		Ref.
	SFE-A	SFE-B	SFE-A	SFE-B	C18:2ω6	C18:3ω3	SFE-A	SFE-B	
CSM-TS	9.05 ± 1.02	2.61 ±0.16	203.59 ± 13.12	1172.32 ± 15.98	19.29 ± 0.10	33.90 ± 0.07	50.16 ± 4.19	71.23 ± 5.03	this study
TS	4.38 ± 0.86	0.30 ± 0.03	198.36 ± 15.45	916.50 ± 16.47	55.59 ± 0.12	-	49.65 ± 4.21	58.85 ± 4.09	[21]
TP	8.95 ± 1.07	1.85 ± 0.11	404.08 ± 26.04	1016.94 ± 12.03	55.59 ± 0.12	-	38.41 ± 3.04	67.02 ± 5.11	

* CSM-TS extract from tomato slices with 6–8% tomato seed and 20% camelina seed, TS extract from tomato slices with 12–17% tomato seed, TP extract from tomato pomace with 22–28% tomato seed, ** extraction yield/(g/100 g dried sample), carotenoid content/(mg/100 g extract), PUFA content/(g/100 g oil), antioxidant activity/(% inhibition DPPH).

Due to the existence of a wide range of bioactive compounds and the synergistic effects of these molecules, using camelina seeds as a modifier increases the added value of supercritical CO$_2$ extracts, allowing them to keep up with current trends in the field of health-related products.

4. Conclusions

The major compounds with antioxidant activities from tomato slice samples were carotenoids (lycopene and β-carotene) and ω-PUFA (ω6-linoleic acid). Three types of seeds, including tomato, camelina and hemp seeds, were chosen to be used as modifiers in the supercritical CO_2 extraction of carotenoids from tomato slices.

A quali-quantitative analysis of the seeds was performed to determine the amount of oil from the seeds and the oil quality expressed by the FAME composition and carotenoid solubility. The oil quantity of selected seeds, determined by Soxhlet extraction with AH solvent, increased from 19.17 to 41.85 g oil/100 g dried seed in the order of TSM < HSM < CSM. FAME analysis showed that camelina oil contained ω3-linolenic, gondoic and erucic acids, while hemp oil contained ω3-linolenic, ω6-linolenic, arachidic and gondoic acids and a low ω6/ω3 ratio. To determine the carotenoids' solubility, the maceration of tomato slices in selected oils was performed and, as a result, ~10 mg carotenoids/100 g of oil was extracted from camelina and hemp oils, while ~13 mg carotenoids/100 g oil was extracted from tomato oil.

For the supercritical CO_2 extraction process at 450 bar, 70 °C and 11 kg/h CO_2 flow rate for 10 h applied to tomato slices enriched with 20 *w/w*% seed modifier, the extraction yield increased from 66.00 to 108.65 g extract/kg dried sample in the order of TSM < HSM < CSM. The extracts were centrifuged to separate two products, an oily fraction (79–86%) and a solid fraction (14–21%). Products analysis showed similar total carotenoid concentrations (163.14–211.05 mg/100 g extract) in the oil fraction and similar values of lycopene (947.92–1212.68 mg/100 g extract) in the solid fraction. Further, a Box–Behnken experimental design was applied to analyse the effects of three independent variables (extraction pressure, seed type and CO_2 flow rate) on four dependent variables (oil and solid yields and carotenoid composition) to identify the optimal extraction conditions for the high recovery of the two products and their quality. With the optimal SFE conditions of 450 bar, 70 °C and 13 kg/h, the CSM modifier resulted in the following yields of the two products: 90.52 g oil/kg dried sample with 203.59 mg carotenoid/100 g oil, 19.29 g ω3-linoleic acid and 33.90 ω3-linolenic, and 26.13 g solid/kg dried sample with 1172.32 mg lycopene/100 g solid. The quality of these products means that they can be used as valuable products in the pharmaceutical, cosmetic and food industries.

Author Contributions: Conceptualization, P.I., M.P. and V.P.; methodology, P.I. and M.P.; validation, P.I., V.P. and C.S.B.; formal analysis, P.I. and V.P.; investigation, M.P. and P.I.; resources, V.P.; data curation, V.P.; writing—original draft preparation, P.I. and M.P.; writing—review and editing, P.I. and M.P.; visualization, P.I., V.P. and C.S.B.; supervision, C.S.B.; project administration, P.I. and C.S.B.; funding acquisition, M.P. and P.I. All authors have read and agreed to the published version of the manuscript.

Funding: This research received no external funding.

Institutional Review Board Statement: Not applicable.

Informed Consent Statement: Not applicable.

Data Availability Statement: Not applicable.

Acknowledgments: This work has been funded by the European Social Fund from the Sectorial Operational Programme Human Capital 2014–2020 through the Financial Agreement with the title "Training of PhD students and postdoctoral researchers in order to acquire applied research skills—SMART", Contract no. 13530/16.06.2022—SMIS code: 153734.

Conflicts of Interest: The authors declare no conflict of interest.

References

1. Szabo, K.; Teleky, B.-E.; Ranga, F.; Roman, I.; Khaoula, H.; Boudaya, E.; Ltaief, A.B.; Aouani, W.; Thiamrat, M.; Vodnar, D.C. Carotenoid Recovery from Tomato Processing By-Products through Green Chemistry. *Molecules* **2022**, *27*, 3771. [CrossRef] [PubMed]
2. Potillo-Lopez, R.; Morales-Contreras, B.; Lozano-Guzman, E.; Basilio-Heredia, J.; Muy-Rangel, M.D.; Ochoa-Martinez, L.A.; Rosas-Flores, W.; Moraes-Castro, J. Vegetable oils as green solvents for carotenoid extraction from pumpkin (Cucurbita argyrosperma Huber) byproducts: Optimization of extraction parameters. *J. Food Sci.* **2021**, *86*, 3122–3136. [CrossRef] [PubMed]
3. Szabo, K.; Dulf, F.V.; Teleky, B.-E.; Eleni, P.; Boukouvalas, C.; Krokida, M.; Kapsalis, N.; Rusu, A.V.; Socol, C.T.; Vodnar, D.C. Evaluation of the Bioactive Compounds Found in Tomato Seed Oil and Tomato Peels Influenced by Industrial Heat Treatments. *Foods* **2021**, *10*, 110. [CrossRef] [PubMed]
4. Kehili, M.; Sayadi, S.; Frikha, F.; Zammel, A.; Allouche, N. Optimization of lycopene extraction from tomato peels industrial by-product using maceration in refined olive oil. *Food Bioprod. Process.* **2019**, *117*, 321–328. [CrossRef]
5. Squillace, P.; Adani, F.; Scaglia, B. Supercritical CO_2 extraction of tomato pomace: Evaluation of the solubility of lycopene in tomato oil as limiting factor of the process performance. *Food Chem.* **2020**, *315*, 126–224. [CrossRef]
6. Kultys, E.; Kurek, M.A. Green Extraction of Carotenoids from Fruit and Vegetable Byproducts: A Review. *Molecules* **2022**, *27*, 518. [CrossRef]
7. Saldana, M.A.D.; Temelli, F.; Guigard, S.E.; Tomberli, B.; Gray, C.G. Apparent solubility of lycopene and b-carotene in supercritical CO_2, CO_2 + ethanol and CO_2 + canola oil using dynamic extraction of tomatoes. *J. Food Eng.* **2010**, *99*, 1–8. [CrossRef]
8. Vasapollo, G.; Longo, L.; Rescio, L.; Ciurlia, L. Innovative supercritical CO_2 extraction of lycopene from tomato in the presence of vegetable oil as co-solvent. *J. Supercrit. Fluids* **2004**, *29*, 87–96. [CrossRef]
9. Aniceto, J.P.S.; Rodrigues, V.H.; Portugal, I.; Silva, C.M. Valorization of Tomato Residues by Supercritical Fluid Extraction. *Processes* **2022**, *10*, 28. [CrossRef]
10. Pellicano, T.M.; Sicari, V.; Loizzo, M.R.; Leporini, M.; Falco, T.; Poiana, M. Optimizing the supercritical fluid extraction process of bioactive compounds from processed tomato skin by-products. *Food Sci. Technol.* **2020**, *40*, 692–697. [CrossRef]
11. Dominguez, R.; Gullon, P.; Pateiro, M.; Munekata, P.E.S.; Zhang, W.; Lorenzo, J.M. Tomato as Potential Source of Natural Additives for Meat Industry. A Review. *Antioxidants* **2020**, *9*, 73. [CrossRef] [PubMed]
12. Machmudah, S.; Zakaria; Winardi, S.; Sasaki, M.; Goto, M.; Kusumoto, N.; Hayakawa, K. Lycopene extraction from tomato peel by-product containing tomato seed using supercritical carbon dioxide. *J. Food Eng.* **2012**, *108*, 290–296. [CrossRef]
13. Sovova, H.; Stateva, R.P. New developments in the modelling of carotenoids extraction from microalgae with supercritical CO_2. *J. Supercrit. Fluids* **2019**, *148*, 93–103. [CrossRef]
14. Saini, R.K.; Moon, S.H.; Keum, Y.S. An updated review on use of tomato pomace and crustacean processing waste to recover commercially vital carotenoids. *Food Res. Int.* **2018**, *108*, 516–529. [CrossRef] [PubMed]
15. Rizk, E.M.; El-Kady, A.T.; El-Bialy, A.R. Charactrization of carotenoids (lyco-red) extracted from tomato peels and its uses as natural colorants and antioxidants of ice cream. *Ann. Agric. Sci.* **2014**, *59*, 53–61. [CrossRef]
16. Hatami, T.; Meireles, M.A.A.; Ciftici, O.N. Supercritical carbon dioxide extraction of lycopene from tomato processing by-products: Mathematical modelling and optimization. *J. Food Eng.* **2018**, *241*, 18–25. [CrossRef]
17. Longo, C.; Leo, L.; Leone, A. Carotenoids, Fatty Acid Composition and Heat Stability of Supercritical Carbon Dioxide-Extracted-Oleoresins. *Int. J. Mol. Sci.* **2012**, *13*, 4233–4254. [CrossRef]
18. Watanabe, Y.; Honda, M.; Higashiura, T.; Fukaya, T.; Machmudah, S.; Wahyudiono; Kanda, H.; Goto, M. Rapid and Selective Concentration of Lycopene Z-isomers from Tomato Pulp by Supercritical CO_2 with Co-solvents. *Solvent Extr. Res. Dev. Jpn.* **2018**, *25*, 47–57. [CrossRef]
19. Barros, H.D.F.Q.; Grimaldi, R.; Cabral, F.A. Lycopene-rich avocado oil obtained by simultaneous supercritical extraction from avocado pulp and tomato pomace. *J. Supercrit. Fluids* **2017**, *120*, 1–6. [CrossRef]
20. Porretta, S. *Tomato Chemistry, Industrial Processing and Product Development*; Royal Society of Chemistry: Cambridge, UK, 2019.
21. Popescu, M.; Iancu, P.; Plesu, V.; Todasca, M.C.; Isopencu, G.O.; Bildea, C.S. Valuable Natural Antioxidant Products Recovered from Tomatoes by Green Extraction. *Molecules* **2022**, *27*, 4191. [CrossRef]
22. Grijo, D.R.; Piva, G.K.; Osorio, I.V.; Cardozo-Filho, L. Hemp (*Cannabis sativa* L.) seed oil extraction with pressurized n-propane and supercritical carbon dioxide. *J. Supercrit. Fluids* **2019**, *143*, 268–274. [CrossRef]
23. Kostic, M.D.; Jokovic, N.M.; Stamenkovic, O.S.; Rajkovic, K.M.; Milic, P.S.; Veljkovic, V.B. Optimization of hempseed oil extraction by n-hexane. *Ind. Crops Prod.* **2013**, *48*, 133–143. [CrossRef]
24. Devi, V.; Khanam, S. Comparative study of different extraction processes for hemp (Cannabis sativa) seed oil considering physical, chemical and industrial-scale economic aspects. *J. Clean. Prod.* **2019**, *207*, 645–657. [CrossRef]
25. Da Porto, C.; Voinobich, D.; Decorti, D.; Natolino, A. Response surface optimization of hemp seed (*Cannabis sativa* L.) oil yield and oxidation stability by supercritical carbon dioxide extraction. *J. Supercrit. Fluids* **2012**, *68*, 45–51. [CrossRef]
26. Belayneh, H.D.; Wehling, R.L.; Cahoon, E.; Ciftici, O.N. Extraction of omega-3-rich oil from Camelina sativa seed using supercritical carbon dioxide. *J. Supercrit. Fluids* **2015**, *104*, 153–159. [CrossRef]
27. Moslavac, T.; Jovic, S.; Subaric, D.; Aladic, K.; Vukoja, J.; Prce, N. Pressing and supercritical CO_2 extraction of Camelina sativa oil. *Ind. Crops Prod.* **2014**, *54*, 122–129. [CrossRef]
28. Popa, A.-L.; Jurcoane, S.; Dumitriu, B. Camelina sativa oil—A review. *Sci. Bulletin. Ser. F Biotechnol.* **2017**, *11*, 233–238.

29. Li, Y.; Watkins, B.A. Analysis of Fatty Acids in Food Lipids. In *Current Protocols in Food Analytical Chemistry*; Whitaker, J., Ed.; Wiley: New York, NY, USA, 2001.

30. Iancu, P.; Stefan, N.G.; Plesu, V.; Toma, A.; Stepan, E. Advanced high vacuum techniques for ω-3 polyunsaturated fatty acids esters concentration. *Rev. Chim.* **2015**, *66*, 911–917.

31. Popescu, M.; Iancu, P.; Plesu, V.; Bildea, C.S.; Todasca, M.C. Different spectrophotometric methods for simultaneous quantification of lycopene and β-carotene from a binary mixture. *LWT-Food Sci. Technol.* **2022**, *160*, 113238. [CrossRef]

32. Konieczka, P.; Namiesnik, J. *Quality Assurance and Quality Control in the Analytical Chemical Laboratory: A Practical Approach*; CRC Press: Boca Raton, FL, USA, 2009.

33. Lawson, J.; Erjavec, J. *Basic Experimental Strategies and Data Analysis for Science and Engineering*; CRC Press: Boca Raton, FL, USA, 2017.

34. Tirado-Kulieva, V.A.; Sanchez-Chero, M.; Villegasyarleque, M.; Aguilar, G.F.V.; Carrion-Barco, G.; Santa Cruz, A.G.Y.; Sanchez-Chero, J. An Overview on the Use of Response Surface Methodology to Model and Optimize Extraction Processes in the Food Industry. *Curr. Res. Nutr. Food Sci.* **2021**, *9*, 745–754. [CrossRef]

35. Aladic, K.; Jovic, S.; Moslavac, T.; Tomas, S.; Vidovic, S.; Vladic, J.; Subaric, D. Cold Pressing and Supercritical CO_2 Extraction of Hemp (Cannabis sativa) Seed Oil. *Chem. Biochem. Eng. Q.* **2014**, *28*, 481–490. [CrossRef]

36. Mieliauskaite, D.; Viskelis, P.; Noreika, R.-K.; Viskelis, J. Lycopene Solubility and Its Extraction from Tomatoes and Their By-Products. In Proceedings of the CIGR Section VI International Symposium on Food Processing, Monitoring Technology in Bioprocesses and Food Quality Management, Potsdam, Germany, 31 August–2 September 2009.

37. Kunthakudee, N.; Sunsandee, N.; Chutvirasakul, B.; Ramakul, P. Extraction of lycopene from tomato with environmentally benign solvents: Box-Behnken design and optimization. *Chem. Eng. Commun.* **2019**, *207*, 574–583. [CrossRef]

38. Borel, P.; Grolier, P.; Armand, M.; Partier, A.; Lafont, H.; Lairon, D.; Azais-Braesco, V. Carotenoids in biological emulsions: Solubility, surface-to-core distribution, and release from lipid droplets. *J. Lipid Res.* **1996**, *37*, 250–261. [CrossRef]

39. Pawliszyn, J. *Comprehensive Sampling and Sample Preparation: Analytical Techniques for Scientists*, 1st ed.; Academic Press: Waterloo, ON, Canada, 2012.

40. Kehili, M.; Kammlott, M.; Choura, S.; Zammel, A.; Zetzl, C.; Smirnova, I.; Allouche, N.; Sayadi, S. Supercritical CO_2 extraction and antioxidant activity of lycopene and -carotene-enriched oleoresin from tomato (*Lycopersicum esculentum* L.) peels by-product of a Tunisian industry. *Food Bioprod. Process.* **2017**, *102*, 340–349. [CrossRef]

41. Vallecilla-Yepez, L.; Ciftici, O.N. Increasing cis-lycopene content of the oleoresin from tomato processing byproducts using supercritical carbon dioxide. *LWT-Food Sci. Technol.* **2018**, *95*, 354–360. [CrossRef]

processes

Article

Efficacy of Supercritical Fluid Decellularized Porcine Acellular Dermal Matrix in the Post-Repair of Full-Thickness Abdominal Wall Defects in the Rabbit Hernia Model

Yen-Lung Chiu [1,†], Yun-Nan Lin [2,†], Yun-Ju Chen [3], Srinivasan Periasamy [3], Ko-Chung Yen [3] and Dar-Jen Hsieh [3,*]

[1] Department of Life Sciences, National Cheng Kung University, Tainan 70101, Taiwan
[2] Division of Plastic Surgery, Department of Surgery, Kaohsiung Medical University Hospital, Kaohsiung Medical University, Kaohsiung City 80708, Taiwan
[3] Center of Research and Development, ACRO Biomedical Co., Ltd., Kaohsiung City 82151, Taiwan
* Correspondence: dj@acrobiomedical.com; Tel.: +886-7-695-5569
† These authors contributed equally to this work.

Abstract: Damage to abdominal wall integrity occurs in accidents, infection and herniation. Repairing the hernia remains to be one of the most recurrent common surgical techniques. Supercritical carbon dioxide ($SCCO_2$) was used to decellularize porcine skin to manufacture acellular dermal matrix (ADM) for the reparation of full-thickness abdominal wall defects and hernia. The ADM produced by $SCCO_2$ is chemically equivalent and biocompatible with human skin. The ADM was characterized by hematoxylin and eosin (H&E) staining, 4,6-Diamidino-2-phenylindole, dihydrochloride (DAPI) staining, residual deoxyribonucleic acid (DNA) contents and alpha-galactosidase (α-gal staining), to ensure the complete decellularization of ADM. The ADM mechanical strength was tested following the repair of full-thickness abdominal wall defects (4 × 4 cm) created on the left and right sides in the anterior abdominal wall of New Zealand White rabbits. The ADM produced by $SCCO_2$ technology revealed complete decellularization, as characterized by H&E, DAPI staining, DNA contents (average of 26.92 ng/mg) and α-gal staining. In addition, ADM exhibited excellent performance in the repair of full-thickness abdominal wall defects. Furthermore, the mechanical strength of the reconstructed abdominal wall after using ADM was significantly ($p < 0.05$) increased in suture retention strength (30.42 ± 1.23 N), tear strength (63.45 ± 7.64 N and 37.34 ± 11.72 N) and burst strength (153.92 ± 20.39 N) as compared to the suture retention (13.33 ± 5.05 N), tear strength (6.83 ± 0.40 N and 15.27 ± 3.46 N) and burst strength (71.77 ± 18.09 N) when the predicate device materials were concomitantly tested. However, the efficacy in hernia reconstruction of ADM is substantially equivalent to that of predicate material in both macroscopic and microscopic observations. To conclude, ADM manufactured by $SCCO_2$ technology revealed good biocompatibility and excellent mechanical strength in post-repair of full-thickness abdominal wall defects in the rabbit hernia model.

Keywords: supercritical carbon dioxide; $scCO_2$; acellular dermal matrix; ADM; full-thickness abdominal wall defects; hernia; mechanical strength

Citation: Chiu, Y.-L.; Lin, Y.-N.; Chen, Y.-J.; Periasamy, S.; Yen, K.-C.; Hsieh, D.-J. Efficacy of Supercritical Fluid Decellularized Porcine Acellular Dermal Matrix in the Post-Repair of Full-Thickness Abdominal Wall Defects in the Rabbit Hernia Model. *Processes* 2022, 10, 2588. https://doi.org/10.3390/pr10122588

Academic Editors: Maria Angela A. Meireles and Alessandro Trentini

Received: 26 October 2022
Accepted: 1 December 2022
Published: 4 December 2022

Publisher's Note: MDPI stays neutral with regard to jurisdictional claims in published maps and institutional affiliations.

1. Introduction

Abdominal wall surgery includes primarily techniques and surgical procedures focused to repair hernia defects. Every year, more than 20 million hernia surgeries are carried out worldwide [1]. In the US alone 700,000 inguinal hernia surgeries are performed in one year. The incisional hernias no less often may ascend after a laparotomy, in which between 11 and 20% of laparotomies may progress to an incisional hernia [2]. The tension-free hernia repair technique using biomaterial in a rabbit model is the keystone in the field of surgical procedures intended to repair abdominal wall defects [3,4].

Acellular dermal matrix (ADM) is a biomaterial derived from decellularized animal and human skin and tissues [5]. The ADM's properties such as structural, mechanical, and biochemical functions are credited to the extracellular matrix (ECM), the main component of ADM [6]. The major element of ECM is collagen I, accountable for hemostasis on the wound bed and for modulating the wound-healing process [7]. In wounds, the role of type I collagen in the wound bed is to avoid the worsening of wounds by attaching to free radicals, proteases, and inflammatory cytokines [8]. The ADM is a collagen-abundant ECM that plays a substantial function in wound healing. Many collagen-based ADM scaffolds are manufactured by the decellularizing animal, and human skin and tissues, such as bovine collagen-derived Integra®, swine small intestine submucosa-derived Oasis®, and human placenta-derived Epifix® [5]. Porcine ADM is an ECM-rich collagen scaffold extensively employed in tissue engineering. It encourages wound healing because of its biodegradable and biocompatible nature [9].

Acellular dermal matrix (ADM) biomaterials can enhance vascular ingrowth and integrate with the host tissues. ADM products possess the ability to support vascular ingrowth and incorporate native tissues. This allows the host immune system to access the repair site, which increases resistance to infection by controlling necrotic debris and bacteria that contribute to the development of chronic wounds [10]. Porcine ADM is an outstanding substitute for human ADM because of its surplus tissue source, cost-effectiveness and structural resemblances to collagen in humans [11]. Porcine ADM possesses immunogenic epitopes which cause host rejection and encapsulation. The chemical cross-linking process is used to decrease the immunogenicity of porcine ADM. The cross-linking process decreases the immunogenicity of the ADM by chemically covering its antigenic epitope, it decreases the scaffold deprivation by matrix metalloprotease and cell infiltration essential for matrix remodeling [12]. The ADM avoids infection by allowing immune cells to contact with necrotic debris and bacteria [10].

Decellularization technology is used to eradicate the antigenicity of the xenogeneic tissue while retaining the components of ECM. Decellularization techniques are classified based on physical, chemical, or enzymatic approaches by adding detergents, such as Triton-X100, sodium dodecyl sulfate (SDS), and sodium deoxycholate [13]. Nevertheless, detergent residues elicit cytotoxicity in vivo and irreversible structural damage to ECM. In addition, decellularization by detergent is laborious [14]. Supercritical carbon dioxide ($SCCO_2$) extraction technology was an excellent alternative for the decellularization process [15,16]. The $SCCO_2$ technology can be employed to eradicate fat and cellular components from the fat cells and intracellular, functional cellular proteins while retaining the intact ECM structure without impairment. The $SCCO_2$ method is efficient in fat reduction and removal from porcine skin [16]. The histological and immunohistochemical (IHC) staining confirmed that there are no alterations and damage to ECM in $SCCO_2$ decellularization [15]. Carbon dioxide is the most commonly used supercritical fluid due to its suitable critical temperature (31.0 °C) for processing ECM. Carbon dioxide is an easily available ambient gas [16,17]. Moreover, the advantage of $SCCO_2$ decellularization technology is non-toxic, inexpensive, and eco-friendly [15,18]. The $SCCO_2$ process has few known disadvantages related to the processing of the tissues, except the fact that the $SCCO_2$ system is a costly and intricate apparatus working at high pressure.

The protective function, tensile strength and continuity of the skin are due to the presence of type I collagen. Throughout the maturation stage and skin remodeling, it is predicted that type III collagen declines compared to type I collagen to return to the ratio found in normal skin [19,20]. In ADM-treated animals type I collagen revealed a significant elevation in density compared to normal animals. In addition, a rise in total collagen in skin wounds was also observed in the presence of ADM [20].

In the present study, we decellularized the porcine hide employing $SCCO_2$ technology to produce ADM and examined the biocompatibility and mechanical strength in post-repair of full-thickness abdominal wall defects in the rabbit model.

2. Materials and Methods

2.1. Preparation of Acellular Dermal Matrix

Porcine hide was procured from Tissue Source, LLC (Lafayette, IN, USA). The fat layer of the hide was trimmed off and discarded. The remaining hide part was washed with phosphate-buffered saline (PBS). The hide was delicately sliced to 0.4–2 mm in thickness. The sliced hide was rolled along with polyethene gauze arranged in a tissue holder, which was then placed into a $SCCO_2$ vessel system (Helix SFE Version R3U, Applied Separations Inc (Allentown, PA, USA), with a cosolvent filled to 10% volume of the vessel with 75% ethanol. The operations of the $SCCO_2$ system were carried out in dynamic mode at 350 bar and 40 °C for 40 min to decellularize porcine hide to ADM. The flow rate of carbon dioxide is 0.3 liters per minute (LPM). Subsequently washed with acetic acid (1%), sodium hydroxide (0.1 N) and freeze-dried. The ADM is sterilized by γ-irradiation (25 kilo Gray), and the product is marketed as ABCcolla® Acellular Dermal Matrix in Taiwan.

2.2. Predicate Device

In the present study, the predicate device used is a popular commercial brand, recommended for hernia reconstruction by physicians. This predicate device is approved by US FDA and CE. The biological, chemical, physical and mechanical properties of this commercial brand are comparable to that of the $SCCO_2$-derived ADM, and therefore we used this predicate device (PRE) to compare in this study.

2.3. Hematoxylin and Eosin Staining

The native porcine hide and ADM were fixed in 4% buffered formaldehyde and paraffin and the sections were carried out using standard routine sectioning, followed by hematoxylin and eosin (H&E) for assessing the decellularization of the porcine hide. The stained native hide and ADM sections were placed on an Olympus bx53 microscope (Olympus, TX, USA) and photomicrographs were documented for assessment.

2.4. DAPI (4,6-Diamidino-2-phenylindole, Dihydrochloride) Staining

The 5 μm thickness of standard routine paraffin-embedded sections was cut, followed by dewaxing in xylene and graded alcohol series were used for rehydration and finally into water. Subsequently, the sections were stained with DAPI and photographs were recorded using a fluorescent microscope.

2.5. DNA Quantification

The native porcine hide and ADMs the standard genomic DNA were extracted employing a kit (Nautia Cat. NO.: NGTZ-S100, Nautiagene, Taipei City, Taiwan). The extracted DNA from the native porcine hide and ADMs DNA was measured employing PicoGreen dsDNa Quantitation Reagent and Kit (P-7589, Thermo Fisher Scientific, Tainan, Taiwan) at 260 nm in a spectrofluorometric microplate reader (BIO-TEK®Flx800, BioTek Instruments, Inc., Winooski, VT, USA). The residual DNA quantity and fragment size in the native porcine hide and ADM were analyzed by agarose gel electrophoresis.

2.6. Alpha-Gal (α-Gal) Staining

The native porcine hide and ADMs paraffin-embedded sections were carried out as stated in the former section, dewaxed, rehydrated and primary antibody alpha-Gal (α-gal) was added, incubated and developed by employing the standard immunohistochemistry staining Avidin-Biotin Complex (ABC) method, following Wu et al. [21]

2.7. Surgery for the Repair of Full-Thickness Abdominal Wall Defects

In this case, 24 New Zealand rabbits weighing greater than 2.5 kg (Male) were randomly allocated to two groups: the experimental ADM group ($n = 12$) and the predicate group (PRE) ($n = 12$), a well-known commercial brand. The animal study protocol was approved by Institutional Animal Care & Use Committee (Master Laboratory Co., Ltd.,

IACUC: 20T10-10). All animals overnight fasted before surgery, but the water was permitted. The animals were anaesthetized Zoletil and xylazine were injected into the intramuscularly at the dose of 10 mg/kg separately, and then anaesthetized with isoflurane 30 min before surgery. Before the surgery, the fur of the animal's abdomen was clipped with an electric animal shaver with aseptic techniques.

Using a sterile surgical technique, full-thickness abdominal wall defects with a size of 4×4 cm on both sides of the midline were carried out, by which the fascia, the underlying rectus abdominis muscle, and the peritoneum were resected. The experimental ADM (24 experimental ADM, one animal received 2) and PRE (24 predicate device, one animal received 2) (5×5 cm) were placed intraabdominal with 1 cm overlap and fixed tension-free to the abdominal wall with eight interrupted 2/0 polypropylene sutures. Next, the abdominal wall fascia was closed with a running 2/0 polypropylene suture. The subcutis and skin were closed with continuous resorbable 2/0 polyglactin sutures. The povidone-iodine was used to disinfect the wound. After surgery, the animals were given antibiotics intramuscularly once a day for seven days. At 2 and 8 weeks after implantation, the animal was sacrificed with humanity. The animal samples were collected for further biomechanical study.

2.8. Mechanical Testing

The mechanical testing was following the in vitro mechanical properties study. The instrument used is the HT 2402EC material testing machine (Hung Ta Instrument Co., Ltd., Taiwan, China) for the tensile strength test. The HT 2402EC capacity is 500 kgf with an accuracy of $\pm1\%$ and load accuracy of $\pm0.01\%$, with a capacity of 0.005–500 mm/min

2.9. Suture Retention Strength Test

The implanted ADM in the abdominal muscle of rabbit specimens were cut into a size of 25×50 mm^2. No #2 suture was used for the retention test, a suture needle was passed through at a distance of 10 mm from the side of the sample and tied the suture with a bowline knot. The suture was passed through a hole in a specimen and fixed on the hook which was attached to the grip of the tensile testing machine.

2.9.1. Tear Strength Test

The tear strength of the implanted ADM in the abdominal muscle of rabbit specimens was cut into a size of 25×50 mm^2, and a 10 mm slit was cut from the midline of the 2.5 cm edge towards the center of the specimen, which is suitable for clamping. Each specimen was clamped in the upper and lower grip of the tensile testing machine. The samples were then stretched at 25.4 mm/min. The test was stopped when the total displacement exceeded 30 mm (the specimen was fully stretched).

2.9.2. Burst Strength Test

The burst strength of the implanted ADM in the abdominal muscle of rabbit specimens was cut to a size of 50×50 mm^2, the surrounding muscle tissue was removed, and the specimens were clamped in the grip. The specimen was fixed into a sample holder, the clamping device could tightly clamp the four sides of the specimen with a diameter of 30 mm in the center tested. A diameter of 25 mm round rod bar fixture was load applied at a rate of 25.4 mm/min on the loading device. The test was stopped when the total displacement exceeded 30 mm or until complete material failure (load = 0 N).

2.9.3. Specimen Tissue Section Preparation and Staining

The raw specimens of each animal (implants on the left or right abdomen) were fixed in 10% buffered neutral formalin solution for 24 h and went through trimming, fixation, dehydration, embedding, and slicing for 3–4 μm slides. The slices were then stained with H&E staining and observed by a veterinary pathologist who was blinded to the experiment and scored by microscope observation.

Semi-quantitative scoring of histopathological lesions according to Shackelford et al. [22], the criteria of the severity grading system for all microscopic lesions were shown as follows: (i) Grade "minimal"—The tissue has undergone less than 10% of the tissue is involved. For hyperplasia, hypoplasia and/or atrophy lesions, the "minimal" grade is given when the affected tissue showed less than 10% increase or decrease in volume. (ii) Grade "mild"— The tissue has undergone 10–39% of the tissue involved. For hyperplasia, hypoplasia and/or atrophy lesions, the "mild" grade is given when the affected tissue has less than 10–39% increase or decrease in volume. (iii) Grade "moderate"—The tissue has undergone 40–79% of the tissue involved. For hyperplasia, hypoplasia and/or atrophy lesions, the "moderate" score is employed when the affected tissue has less than 40–79% increase or decrease in volume. (iv) Grade "marked"—The tissue has undergone 80–100% of the tissue involved. For hyperplasia, hypoplasia and/or atrophy lesions, the "marked" score is given when the affected tissue has experienced 80–100% increase or decrease in volume.

3. Results

3.1. H&E Staining of SCCO$_2$ Decellularized ADM

In native porcine hide, the nucleus is stained by hematoxylin displaying blue-purple color and eosin binds to the protein of cytoplasm giving pink color (Figure 1i,iii). The H&E staining showed no cells in the SCCO$_2$ decellularized ADM (Figure 1ii, iv). Therefore, the H&E staining shows ample decellularization of the porcine ADM as compared to normal hide.

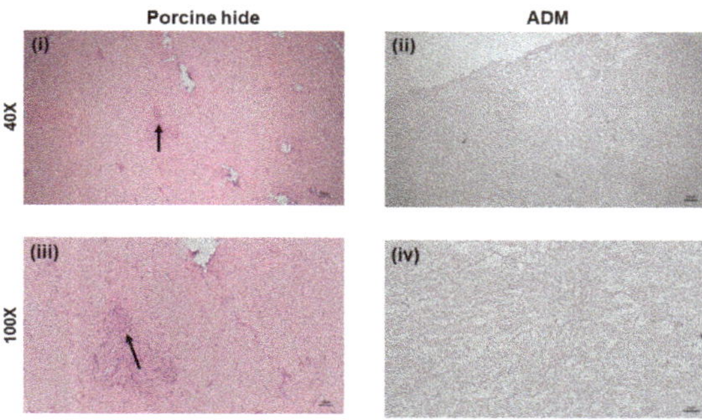

Figure 1. The characterization of ADM by H&E staining (**i,iii**) native porcine hide, and (**ii,iv**) SCCO$_2$ decellularized ADM. (**i,ii**) Scale bar: 100 µm; (**iii,iv**) Scale bar: 50 µm.

3.2. DAPI Staining of SCCO$_2$ Decellularized ADM

DAPI stains the nucleus of the cell and shows a bright blue color, under a fluorescent microscope. The DAPI staining of the SCCO$_2$-processed ADM displayed no nucleus demonstrating complete decellularization, whereas nucleus staining was visible in the native porcine hide (Figure 2i, iii). The results indicate that the SCCO$_2$-decellularization process can successfully remove the cells from the porcine hide (Figure 2ii, iv).

3.3. Residual DNA Content of SCCO$_2$ Decellularized ADM

The SCCO$_2$ decellularization of the porcine hide was confirmed by quantifying the content of DNA and agarose gel electrophoresis. For ideal and effective decellularization the criteria are that decellularized tissue must contain < 50 ng of dsDNA per mg of dry tissue. The SCCO$_2$ decellularized ADMs DNA quantification exhibited an average of 26.92 ng/mg of DNA (Figure 3A), which is below the acceptable level of 50 ng/mg residual DNA for medical implant devices based on Biological Evaluation of Medical Devices—Part 1 (ISO 2018). In addition, the SCCO$_2$ decellularized ADM by agarose gel electrophoresis showed no DNA band, whereas DNA bands were found in the native porcine hide (Figure 3B).

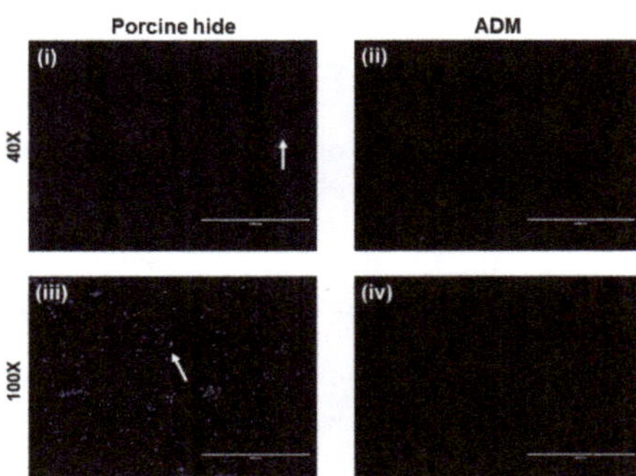

Figure 2. The characterization of ADM by DAPI staining (**i,iii**) native porcine hide, and (**ii,iv**) SCCO$_2$ decellularized ADM. (**i,ii**) Scale bar: 1000 μm; (**iii,iv**) Scale bar: 400 μm.

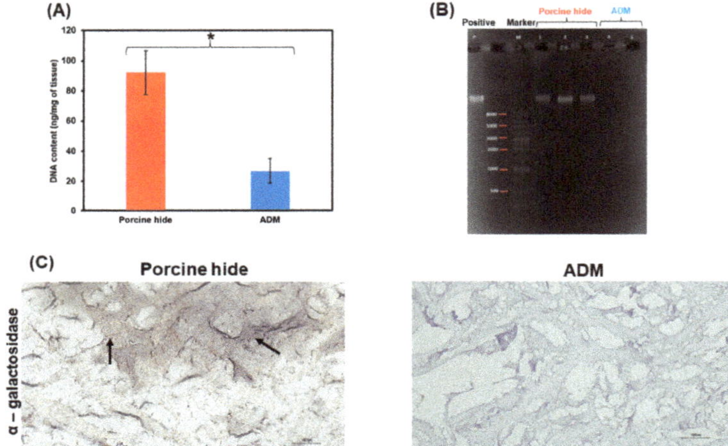

Figure 3. The characterization of ADM by DNA quantification and α-gal staining (**A**) DNA quantification from native porcine hide and SCCO$_2$ decellularized ADM. (**B**) Agarose gel electrophoresis of DNA from native porcine hide and SCCO$_2$ decellularized ADM. (**C**) α-gal staining. The data are expressed as mean ± SD ($n = 3$). * The differences between treatments with different letters are statistically significant ($p < 0.05$). (**C,ADM**) Scale bar: 100 μm.

3.4. Alpha-Gal Staining of SCCO$_2$ Decellularized ADM

The native porcine hide shows the presence of α-gal indicating live cells. α-gal is a catabolizing enzyme that breakdowns glycoproteins, glycolipids, and polysaccharides. Live cells show positive immunostaining of α-gal in the native porcine hide (Figure 3C). However, SCCO$_2$ decellularized ADM exhibited negative immunostaining for alpha-gal indicating complete decellularization in the ADM (Figure 3D).

3.5. The Mechanical Strength of SCCO$_2$ Decellularized ADM

The suture retention strength of SCCO$_2$ decellularized ADM after implantation was significantly higher (30.42 ± 1.23 N) than that in the PRE control (13.33 ± 5.05 N) at week 2. The suture retention strength of SCCO$_2$ decellularized ADM and PRE control after implantation were similar at week 8 (30.38 ± 2.31 N and 31.93 ± 3.05 N) (Figure 4A).

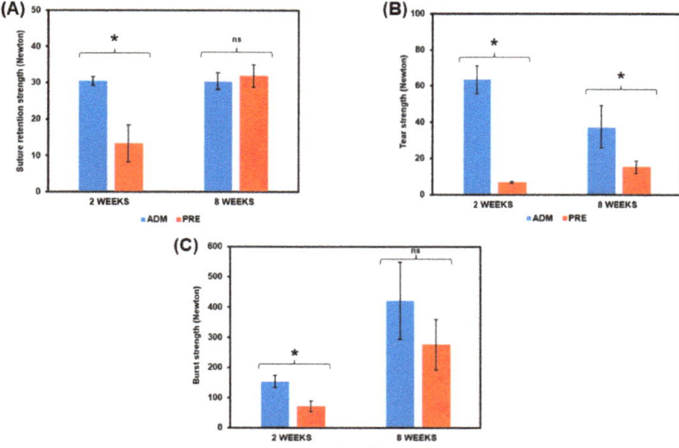

Figure 4. The mechanical strength of SCCO$_2$ decellularized ADM post 2 and 8 weeks of implantation. (**A**) Suture retention strength, (**B**) tear strength tests, and (C) burst strength tests. The data are expressed as mean ± SD. * $p < 0.05$ were considered statistically significant for the PRE group, ns—nonsignificant.

In the tear strength tests, the SCCO$_2$ decellularized ADM (63.45 ± 7.64 N and 37.34 ± 11.72 N) were significantly higher than the PRE control (6.83 ± 0.40 N and 15.27 ± 3.46 N) at week 2 and week 8 (Figure 4B).

In the burst strength test, the SCCO$_2$ decellularized ADM (153.92 ± 20.39 N) was significantly higher than the PRE control (71.77 ± 18.09 N) at week 2. However, there were no significant differences between the SCCO$_2$ decellularized ADM (421.50 ± 127.34 N) and the PRE control (276.42 ± 82.67 N) in strength at week 8 (Figure 4C).

3.6. Histological Evaluation of the Abdominal Wall Defect Model

The representative histological photomicrographs 2 weeks after implantation of the SCCO$_2$ decellularized ADM (Figure 5i,ii) and PRE control (Figure 5iii,iv) were shown in Figure 5. The representative histological photomicrographs 8 weeks after implantation of the SCCO$_2$ decellularized ADM (Figure 6i,ii) and PRE control (Figure 6iii,iv) were shown in Figure 6.

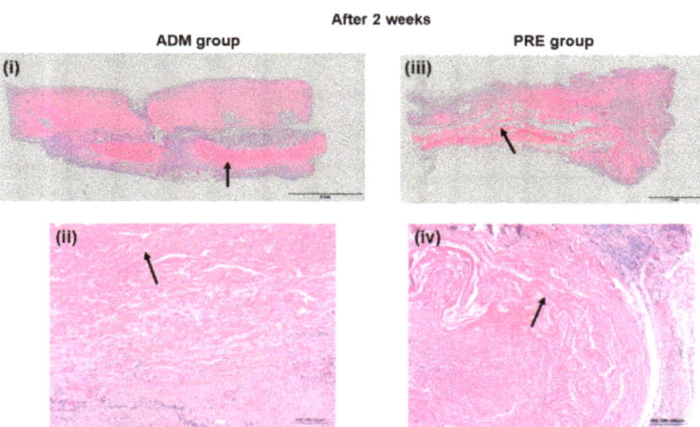

Figure 5. The histological evaluation of the abdominal wall defect model post 2 weeks. (**i,ii**) ADM group, and (**iii,iv**) PRE group. (**i,iii**) Scale bar: 5 mm; (**ii,iv**) Scale bar: 50 × 200.00 μm.

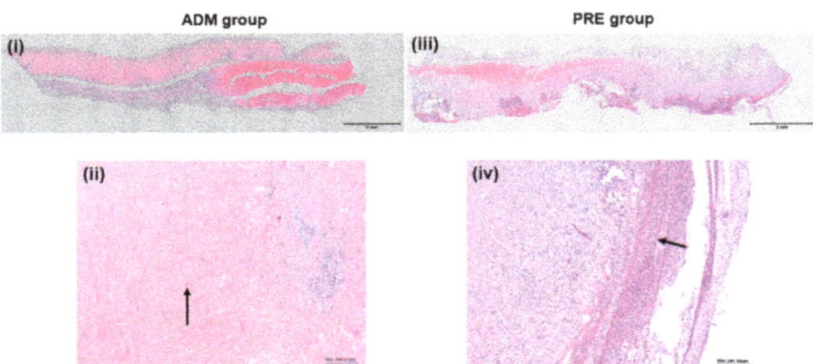

Figure 6. The histological evaluation of the abdominal wall defect model post 8 weeks. (**i,ii**) ADM group, and (**iii,iv**) PRE group. (**i,iii**) Scale bar: 5 mm; (**ii,iv**) Scale bar: 50×200.00 μm.

The histological scores of inflammation in the $SCCO_2$ decellularized ADM (2.83 ± 0.41 and 3.50 ± 0.55) showed similar levels to those in PRE control (3.17 ± 0.98 and 3.00 ± 0.89) at week 2 and week 8, respectively, indicating no significant difference in the clinical outcome after implantation between the $SCCO_2$ decellularized ADM and PRE control (Figure 7A,C). The histological scores of fibroblast and vessel formation in the $SCCO_2$ decellularized ADM and PRE control also showed similar levels at the observation time points. The scores of fibroblast formation in the $SCCO_2$ decellularized ADM (2.50 ± 0.55 and 2.00 ± 0), and (2.83 ± 0.41 and 2.67 ± 1.03) in PRE control at week 2 and week 8, respectively, indicating no significant changes between the $SCCO_2$ decellularized ADM and PRE control (Figure 7A,C). The scores of vessel formation in the $SCCO_2$ decellularized ADM (2.50 ± 0.55 and 2.17 ± 0.41) and (2.50 ± 0.55 and 2.67 ± 0.82) in PRE control at week 2 and week 8, respectively, indicating no significant changes between the $SCCO_2$ decellularized ADM and PRE control (Figure 7A,C).

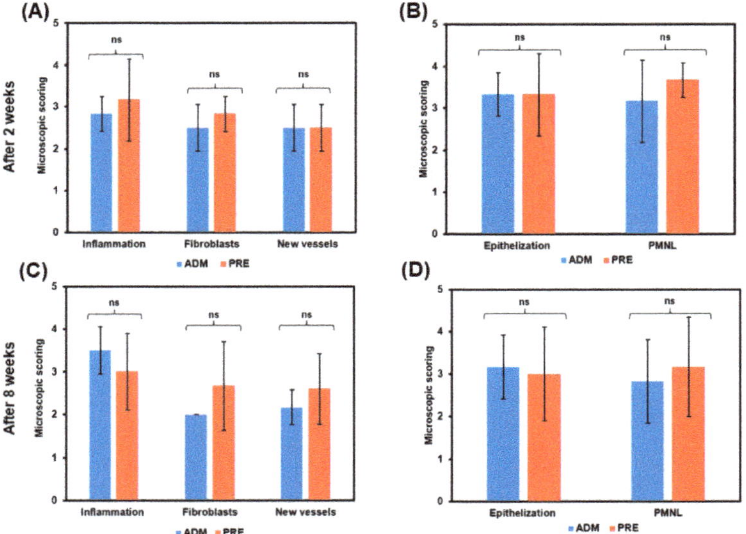

Figure 7. The histological scoring of the abdominal wall defect model post 2 and 8 weeks after implantation. At 2 weeks (**A**) inflammation, fibroblast and new vessels, (**B**) epithelization and PMNL. At 8 weeks (**C**) inflammation, fibroblast and new vessels, (**D**) epithelization and PMNL. The data are expressed as mean $\pm$ SD. ns—nonsignificant.

The scores of epithelization in the $SCCO_2$ decellularized ADM (3.33 ± 0.52 and 3.17 ± 0.75), and (3.33 ± 0.98 and 3 ± 1.1) in PRE control at week 2 and week 8, respectively, indicating no significant changes between the $SCCO_2$ decellularized ADM and PRE control (Figure 7B,D). The scores of PMNL in the $SCCO_2$ decellularized ADM (3.17 ± 0.98 and 2.83 ± 0.98), and (3.67 ± 0.41 and 3.17 ± 1.17) in PRE control at week 2 and week 8, respectively, indicating no significant changes between the $SCCO_2$ decellularized ADM and PRE control (Figure 7B,D).

4. Discussion

Today the budding topic of research is the development of various kinds of biomaterial for the restoration of tissue defects in the abdominal wall. For the past 20 years, the development has necessitated alterations to the numerous biomaterials in quest of a prosthetic material displaying ideal performance. However, there is no perfect biomaterial and strategy for tissue repair and regeneration processes so far. The new age of biomaterials has taken into account the features of the host tissue and its biology to ensure that the developed materials with proper characteristics are comparable to the host tissue. Therefore, engineers, biologists and pathologists play a vital role in the product research and development process. Various biomaterials are exposed to in vitro and in vivo biocompatibility studies, and pre-clinical animal performance studies and their mechanical strength before and after implantation are also tested [4]. We have developed and manufactured an innovative regenerative ADM using $SCCO_2$ technology and examined the biocompatibility and mechanical strength in post-repair of full-thickness abdominal wall defects in the rabbit model.

In the present study, the characterization of ADM was carried out to validate the complete decellularization, competent to be employed as ADM for repairing full-thickness abdominal wall defects. The ADM owns no residual cells as evidenced via H&E and DAPI staining. The $SCCO_2$ process completely removed the cells as evidenced by H&E staining. In addition, the $SCCO_2$ process eliminated the nucleus as evidenced by DAPI staining. The residual DNA quantification was found to be negligible and complied with the international standard for regulatory approval. Biomaterials accessible for the treatment of abdominal wall defects are essential to be tested by preclinical animal experimental studies. International Organization for Standardization have developed biocompatibility standards following ISO 10993-6, "Biological evaluation of medical devices "Tests for Local Effects after Implantation," epitomizes the features that are essential to be taken into consideration when executing implant studies. Among other features, ISO 10993-6 indicates the selection of species and the assessment of biological responses. Though ISO 10993-6 indicates various experimental animals such as rats, mice, and rabbits. Due to its size and easy handling, the rabbit is the animal of choice for implant testing. The New Zealand White rabbit models have offered a shred of outstanding evidence on the behavior of the various mesh implants in the abdominal wall and facilitated the progress of our understanding and predict the consequences that would transpire in human clinical practice [4,23]. Therefore, the current study examines the biocompatibility and mechanical strength in post-repair of full-thickness abdominal wall defects in the rabbit model further studies are needed for extrapolating to humans.

In the present study, ADM was characterized to validate the complete decellularization, α-gal negative in the ADM. A noteworthy difference was detected between the human and porcine dermis. The decellularized porcine ADM owns the immunogenic galactose-α-1,3-galactose epitope, while the galactose epitope was absent in the human dermis. Humans possess a cellular and humoral immune response to this epitope. Its occurrence generates the possibility of immune rejection [24,25]. Porcine ADM is treated to alleviate the galactose-α-1,3-galactose epitope, and this treatment has demonstrated efficacy in evading the immune response during the ADM is implanted in non-human primates. In vitro studies with human fibroblasts displayed higher numbers of human fibroblasts infiltrating deep into human ADM than in porcine ADM, this may be because of the increased density of collagen in the porcine dermis [26]. Therefore, the current study proved that $SCCO_2$ technology alleviated the galactose-α-1,3-galactose epitope offering biocompatibility and evading the immune response during the ADM implantation.

Hernia repair meshes need to be precisely designed for their uses. The perfect ADM used in the reinforced closure need to offer structural integrity during implantation and an ECM for regeneration. The implanted ADM remodels both mechanically and biologically. Commercially around 14 biologically derived ADMs are available for hernia repair in clinical use [27,28]. Furthermore, the tissue-derived ADMs showed no satisfactory positive responses, in the integration to the host [29]. The current study with the $SCCO_2$-derived ADM had an excellent integration to the host tissue remodel both mechanically and biologically in the post-repair of full-thickness abdominal wall defects in the rabbit model.

In the case of big incisional hernias, a multifaceted situation of biological and mechanical signaling and remodeling can impact the remodeling of ADM employed for repair. In addition, inflammation is linked to high protease activity, which degrades ADM [30]. ADM-ECM scaffolds implanted in large incisional hernias aid physiologically pertinent to loading and evading repetition, against the occurrence of protease activity. The inflammation and swelling were significantly decreased in gross observations in porcine ADM-reinforcement at 2, 4, and 6 weeks with minimal remodeling at 4 and 6 weeks after the repair [30]. The current study with the $SCCO_2$-derived ADM is not degraded and no inflammation was observed even after 8 weeks in post-repair of full-thickness abdominal wall defects in the rabbit model.

In Förstemann's abdominal wall model, the average minimum ultimate tensile strength of the porcine ADM was at least 25 times higher than the average loading in a human abdominal wall [31,32]. The absolute tensile strength of the porcine ADM was lower than the surrounding fascia, which is vital in avoiding hernia reappearance. Comparable results in tensile strength decreased in porcine ADM were observed in primates [29]. Furthermore, the average Young's modulus of the porcine ADM and fascia at 6 weeks was comparable. Young's modulus is a measure of the stress required to distort a material. In the abdominal wall defect model, the ADM and abdominal wall interact for a complete closure repair, both the ADM and the peritoneum-fascia complex will work together to counterattack stress and loads in the abdominal wall, totaling the capability of the repair to endure protruding and hernia recurrence in implanting collagen-based ADM in vivo [30]. The current study with the $SCCO_2$-derived ADM had excellent mechanical strength so that the ADM and abdominal wall interact for a complete closure repair in post-repair of full-thickness abdominal wall defects in the rabbit model.

Implanted porcine ADMs histological evaluation confirmed remodeling into fascia-like tissue, without any adhesions to the internal organs. Moderate revascularization occurs on an average of 2 weeks after implantation of porcine ADM and is increased vascularization in 4 weeks. Cellular repopulation took a similar trend in implanted porcine ADM. In the porcine ADM or peritoneum/fascia complex no inflammation and scarring were observed with the functional repair of a herniated abdominal wall [30]. The synthetic mesh and cross-linked porcine ADM are encapsulated with scar tissue and are not entirely integrated into the host tissue [33,34]. The host resident's abdominal wall inflammatory cells and the collagen synthesis of fibroblastic cells will respond to the ADM used for hernia repair. The ADM will modulate the integrity and mechanical properties of the repair over time [30]. The present study with the $SCCO_2$-derived ADM had a minimal infiltration of host cells to ADM for revascularization and to maintain the mechanical strength of the reconstructed abdominal wall in post-repair of full-thickness abdominal wall defects in the rabbit model.

The generally employed bioprosthetic in ventral hernia repair is the human ADM (AlloDerm; Life-Cell Corp., Branchburg, NJ, USA). However, it has a major disadvantage of propensity to stretch after implantation, leading to protruding of the repair site [35,36]. The novel xenogeneic non-crosslinked porcine ADM (Strattice; LifeCell) is a good substitute, due to its negligible bulge rate and the porcine dermis is similar to the human dermis. The temperature regulation of porcine is by the subcutaneous fat instead of hair, which is similar to humans. In addition, similar collagen arrangement and structure between porcine and human dermis, even though porcine collagen is tightly packed and possesses a smaller amount of elastin [26,37]. However, the present study with the $SCCO_2$-derived ADM had no stretching and protrusion of the repair site was observed that maintain the mechanical strength of the reconstructed abdominal wall in post-repair of full-thickness abdominal wall defects.

In the present study, we produced an innovative porcine ADM by using $SCCO_2$ extraction technology. Our results proposed that the $SCCO_2$ method can be successfully used to produce decellularized ADM with excellent mechanical strength, a suitable dermal substitute for the reconstruction of full-thickness abdominal wall defects.

Author Contributions: Conceptualization, D.-J.H. and Y.-N.L.; methodology, Y.-L.C., Y.-J.C., S.P. and K.-C.Y.; supervision, D.-J.H. and Y.-N.L.; writing—original draft, S.P. and D.-J.H.; writing—review and editing, S.P. and D.-J.H. All authors have read and agreed to the published version of the manuscript.

Funding: This research received no external funding.

Institutional Review Board Statement: The animal study protocol was approved by Institutional Animal Care & Use Committee (Master Laboratory Co., Ltd., IACUC: 20T10-10).

Informed Consent Statement: Not applicable.

Data Availability Statement: Not applicable.

Conflicts of Interest: The authors declare no conflict of interest.

References

1. Kingsnorth, A.; LeBlanc, K. Hernias: Inguinal and incisional. *Lancet* **2003**, *362*, 1561. [CrossRef]
2. Luijendijk, R.W.; Hop, W.C.; van del Tol, M.P.; de Lange, D.C.; Braaksma, M.M.; Ijzermans, J.N.; Boelhouwer, R.U.; de Vriers, B.C.; Salu, M.K.; Wereldsma, J.C.; et al. A comparison of suture repair with mesh repair for incisional hernia. *N. Engl. J. Med.* **2000**, *343*, 392. [CrossRef] [PubMed]
3. Ventral Hernia Working Group; Breuing, K.; Butler, C.; Ferzoso, S.; Franz, M.; Hultman, C.S.; Kilbridge, J.F.; Rosen, M.; Silverman, R.P.; Vargo, D. Incisional ventral hernias: Review of the literature and recommendations regarding the grading and technique or repair. *Surgery* **2010**, *148*, 544. [CrossRef] [PubMed]
4. Bellón, J.M.; Rodríguez, M.; Pérez-Köhler, B.; Pérez-López, P.; Pascual, G. The New Zealand White Rabbit as a Model for Preclinical Studies Addressing Tissue Repair at the Level of the Abdominal Wall. *Tissue Eng. Part C Methods* **2017**, *23*, 863–880. [CrossRef] [PubMed]
5. Dickinson, L.E.; Gerecht, S. Engineered Biopolymeric Scaffolds for Chronic Wound Healing. *Front. Physiol.* **2016**, *7*, 341. [CrossRef] [PubMed]
6. Eweida, A.M.; Marei, M.K. Naturally Occurring Extracellular Matrix Scaffolds for Dermal Regeneration: Do They Really Need Cells? *BioMed Res. Int.* **2015**, *2015*, 839694. [CrossRef]
7. Gelse, K.; Poschl, E.; Aigner, T. Collagens—Structure, function, and biosynthesis. *Adv. Drug Deliv. Rev.* **2003**, *55*, 1531–1546. [CrossRef] [PubMed]
8. Wiegand, C.; Schonfelder, U.; Abel, M.; Ruth, P.; Kaatz, M.; Hipler, U.C. Protease and pro-inflammatory cytokine concentrations are elevated in chronic compared to acute wounds and can be modulated by collagen type I in vitro. *Arch. Dermatol. Res.* **2010**, *302*, 419–428. [CrossRef] [PubMed]
9. Chattopadhyay, S.; Raines, R.T. Review collagen-based biomaterials for wound healing. *Biopolymers* **2014**, *101*, 821–833. [CrossRef]
10. Silverman, R.P. Acellular dermal matrix in abdominal wall reconstruction. *Aesthet. Surg. J.* **2011**, *31* (Suppl. 7), 24S–29S. [CrossRef]
11. Gowda, A.U.; Chang, S.M.; Chopra, K.; Matthews, J.A.; Sabino, J.; Stromberg, J.A.; Zahiri, H.R.; Pinczewski, J.; Holton, L.H., 3rd; Silverman, R.P.; et al. Porcine acellular dermal matrix (PADM) vascularises after exposure in open necrotic wounds seen after complex hernia repair. *Int. Wound J.* **2016**, *13*, 972–976. [CrossRef]
12. Carlson, T.L.; Lee, K.W.; Pierce, L.M. Effect of cross-linked and non-cross-linked acellular dermal matrices on the expression of mediators involved in wound healing and matrix remodeling. *Plast. Reconstr. Surg.* **2013**, *131*, 697–705. [CrossRef] [PubMed]
13. Gilbert, T.W.; Sellaro, T.L.; Badylak, S.F. Decellularization of tissues and organs. *Biomaterials* **2006**, *27*, 3675–3683. [CrossRef] [PubMed]
14. Flynn, L.; Semple, J.L.; Woodhouse, K.A. Decellularized placental matrices for adipose tissue engineering. *J. Biomed. Mater. Res. A* **2006**, *79*, 359–369. [CrossRef] [PubMed]

15. Wang, J.K.; Luo, B.; Guneta, V.; Li, L.; Foo, S.E.M.; Dai, Y.; Tan, T.T.Y.; Tan, N.S.; Choong, C.; Wong, M.T.C. Supercritical carbon dioxide extracted extracellular matrix material from adipose tissue. *Mater. Sci. Eng. C Mater. Biol. Appl.* **2017**, *75*, 349–358. [CrossRef] [PubMed]
16. Vaquero, E.M.; Beltrán, S.; Sanz, M.T. Extraction of fat from pigskin with supercritical carbon dioxide. *J. Supercrit. Fluids* **2006**, *37*, 142–150. [CrossRef]
17. Jenkins, C.L.; Raines, R.T. Insights on the conformational stability of collagen. *Nat. Prod. Rep.* **2002**, *19*, 49–59.
18. Chou, P.R.; Lin, Y.N.; Wu, S.H.; Lin, S.D.; Srinivasan, P.; Hsieh, D.J.; Huang, S.H. Supercritical Carbon Dioxide-decellularized Porcine Acellular Dermal Matrix combined with Autologous Adipose-derived Stem Cells: Its Role in Accelerated Diabetic Wound Healing. *Int. J. Med. Sci.* **2020**, *17*, 354–367. [CrossRef]
19. Liu, S.; Zhang, H.; Zhang, X.; Lu, W.; Huang, X.; Xie, H.; Zhou, J.; Wang, W.; Zhang, Y.; Liu, Y.; et al. Synergistic angiogenesis promoting effects of extracellular matrix scaffolds and adipose-derived stem cells during wound repair. *Tissue Eng. Part A* **2011**, *17*, 725–739. [CrossRef]
20. Carvalho-Júnior, J.D.C.; Zanata, F.; Aloise, A.C.; Ferreira, L.M. Acellular dermal matrix in skin wound healing in rabbits— Histological and histomorphometric analyses. *Clinics* **2021**, *76*, e2066. [CrossRef]
21. Wu, C.C.; Tarng, Y.W.; Hsu, D.Z.; Srinivasan, P.; Yeh, Y.C.; Lai, Y.P.; Hsieh, D.J. Supercritical carbon dioxide decellularized porcine cartilage graft with PRP attenuated OA progression and regenerated articular cartilage in ACLT-induced OA rats. *J. Tissue Eng. Regen. Med.* **2021**, *15*, 1118–1130. [CrossRef] [PubMed]
22. Shackelford, C.; Long, G.; Wolf, J.; Okerberg, C.; Herbert, R. Qualitative and quantitative analysis of nonneoplastic lesions in toxicology studies. *Toxicol. Pathol.* **2002**, *30*, 93–96. [CrossRef] [PubMed]
23. Jardelino, C.; Takamori, E.R.; Hermida, L.F.; Lenharo, A.; Castro-Silva, I.I.; Granjeiro, J.M. Porcine peritoneum as source of biocompatible collagen in mice. *Acta Cir. Bras.* **2010**, *25*, 332. [CrossRef] [PubMed]
24. McPherson, T.B.; Liang, H.; Record, R.D.; Badylak, S.F. Galalpha(1,3)Gal epitope in porcine small intestinal submucosa. *Tissue Eng.* **2000**, *6*, 233–239. [CrossRef] [PubMed]
25. Daly, K.A.; Stewart-Akers, A.M.; Hara, H.; Ezzelarab, M.; Long, C.; Cordero, K.; Johnson, S.A.; Ayares, D.; Cooper, D.K.; Badylak, S.F. Effect of the alphaGal epitope on the response to small intestinal submucosa extracellular matrix in a nonhuman primate model. *Tissue Eng. Part A* **2009**, *15*, 3877–3888. [CrossRef]
26. Campbell, K.T.; Burns, N.K.; Rios, C.N.; Mathur, A.B.; Butler, C.E. Human versus non-cross-linked porcine acellular dermal matrix used for ventral hernia repair: Comparison of in vivo fibrovascular remodelling and mechanical repair strength. *Plast. Reconstr. Surg.* **2011**, *127*, 2321–2332. [CrossRef]
27. Smart, N.J.; Marshall, M.; Daniels, I.R. Biological meshes: A review of their use in abdominal wall hernia repairs. *Surgeon* **2012**, *10*, 159–171. [CrossRef]
28. Meintjes, J.; Yan, S.; Zhou, L.; Zheng, S.; Zheng, M. Synthetic, biological and composite scaffolds for abdominal wall reconstruction. *Expert Rev. Med. Devices* **2011**, *8*, 275–288. [CrossRef]
29. Sandor, M.; Xu, H.; Connor, J.; Lombardi, J.; Harper, J.R.; Silverman, R.P.; McQuillan, D.J. Host response to implanted porcine-derived biologic materials in a primate model of abdominal wall repair. *Tissue Eng. Part A* **2008**, *14*, 2021–2031. [CrossRef]
30. Monteiro, G.A.; Rodriguez, N.L.; Delossantos, A.I.; Wagner, C.T. Short-term in vivo biological and mechanical remodeling of porcine acellular dermal matrices. *J. Tissue Eng.* **2013**, *4*, 2041731413490182. [CrossRef]
31. Förstemann, T.; Trzewik, J.; Holste, J.; Batke, B.; Konerding, M.A.; Wolloscheck, T.; Hartung, C. Forces and deformations of the abdominal wall—A mechanical and geometrical approach to the linea alba. *J. Biomech.* **2011**, *44*, 600–606. [CrossRef] [PubMed]
32. Konerding, M.A.; Bohn, M.; Wolloscheck, T.; Batke, B.; Holste, J.L.; Wohlert, S.; Trzewik, J.; Förstemann, T.; Hartung, C. Maximum forces acting on the abdominal wall: Experimental validation of a theoretical modeling in a human cadaver study. *Med. Eng. Phys.* **2011**, *33*, 789–792. [CrossRef] [PubMed]
33. Burns, N.K.; Jaffari, M.V.; Rios, C.N.; Mathur, A.B.; Butler, C.E. Non-cross-linked porcine acellular dermal matrices for abdominal wall reconstruction. *Plast. Reconstr. Surg.* **2010**, *125*, 167–176. [CrossRef] [PubMed]
34. Butler, C.E.; Burns, N.K.; Campbell, K.T.; Mathur, A.B.; Jaffari, M.V.; Rios, C.N. Comparison of cross-linked and non-cross-linked porcine acellular dermal matrices for ventral hernia repair. *J. Am. Coll. Surg.* **2010**, *211*, 368–376. [CrossRef] [PubMed]
35. Sacks, J.M.; Butler, C.E. Outcomes of complex abdominal wall reconstruction with bioprosthetic mesh in cancer patients. *Plast. Reconstr. Surg.* **2008**, *121*, 39.
36. Glasberg, S.B.; D'Amico, R.A. Use of regenerative human acellular tissue (AlloDerm) to reconstruct the abdominal wall following pedicle TRAM flap breast reconstruction surgery. *Plast. Reconstr. Surg.* **2006**, *118*, 8–15. [CrossRef]
37. Ge, L.; Zheng, S.; Wei, H. Comparison of histological structure and biocompatibility between human acellular dermal matrix (ADM) and porcine ADM. *Burns* **2009**, *35*, 46–50. [CrossRef]

Article

Inactivation of *Clostridium* Spores in Honey with Supercritical CO_2 and in Combination with Essential Oils

Alejandro Dacal-Gutiérrez [1], Diego F. Tirado [2] and Lourdes Calvo [1,*]

[1] Department of Chemical and Materials Engineering, School of Chemical Sciences, Universidad Complutense de Madrid, Av. Complutense s/n, 28040 Madrid, Spain
[2] Dirección Académica, Universidad Nacional de Colombia, Sede de La Paz, La Paz 202017, Colombia
* Correspondence: lcalvo@ucm.es

Abstract: The presence of tens of *Clostridium botulinum* spores per gram of honey can cause infantile botulism. Thermal treatment is insufficient to inactivate these resistant forms. This study explored the effectiveness of supercritical CO_2 ($scCO_2$) on its own and combined with lemon (LEO), clove (CLEO), and cinnamon (CEO) essential oils on the inactivation of *Clostridium sporogenes* (CECT 553) as a surrogate of *Clostridium botulinum*. In water, the degree of inactivation at 10 MPa after 60 min increased with the increasing temperature, reducing the population by 90% at 40 °C and by 99.7% at 80 °C. In contrast, when applied to honey, $scCO_2$ did not inactivate *Clostridium* spores satisfactorily at temperatures below 70 °C, which was related to the protective effect of honey. Meanwhile, $scCO_2$ modified with CEO (<0.4% mass) improved the inactivation degree, with a 1.3-log reduction achieved at 60 °C. With this same mixture, a reduction of 3.7 logs was accomplished in a derivative with 70% moisture. Honey was very sensitive to the temperature of the applied CO_2. The obtained product could be used as a novel food, food ingredient, cosmetic, or medicine.

Keywords: high-pressure carbon dioxide; lemon essential oil; clove essential oil; cinnamon essential oil; *Clostridium sporogenes*

Citation: Dacal-Gutiérrez, A.; Tirado, D.F.; Calvo, L. Inactivation of *Clostridium* Spores in Honey with Supercritical CO_2 and in Combination with Essential Oils. *Processes* **2022**, *10*, 2232. https://doi.org/10.3390/pr10112232

Academic Editors: Maria Angela A. Meireles, Ádina L. Santana and Grazielle Nathia Neves

Received: 23 September 2022
Accepted: 28 October 2022
Published: 31 October 2022

Publisher's Note: MDPI stays neutral with regard to jurisdictional claims in published maps and institutional affiliations.

1. Introduction

Clostridium species are anaerobic spore-forming gram-positive bacilli that are widely distributed in nature, both in soil and dust, in addition to sweet, salty, and residual aqueous bodies [1]. They are highly resistant to extreme temperatures, ultraviolet (UV) radiation, water scarcity, and physical and chemical agents, and, therefore, to the environment [2]. In particular, the ability of *Clostridium botulinum* to cause disease is linked to its ability to survive in adverse environmental conditions through spore formation, rapid growth, and toxin production. Thus, this bacterium can result in botulism, a disease contracted by food poisoning, due to the ingestion of its spores or toxins (type A and B). Botulism is mainly characterized by descending paralysis that leads to death by cardiorespiratory failure [3].

Honey is known as a tasty and nutritious sweetener; it is usually considered a natural, healthy, and clean product. The low pH (typically 3.9), low water activity (a_w 0.5–0.6), and high sugar concentration of honey prevent the growth and survival of many types of bacteria. In addition to its osmolarity, the antimicrobial effect of honey is mainly due to the generation of hydrogen peroxide from the oxidation of glucose by glucose oxidase [4]. Thus, the European Commission (EC) Regulation No. 2073/2005 [5] on microbiological criteria for foodstuffs considers honey a ready-to-eat food during its marketed shelf-life, in which no limits are described for any microorganism, toxin, or metabolites other than *Listeria monocytogenes*, and only if honey is intended for infants and under special medical purposes. However, honey may be a natural reservoir of *Clostridium* species [6].

Grabowski and Klein [7] reported that *C. botulinum* spores were identified in up to 62% of honey samples although spore counts varied widely, from 36 to 60 spores per gram in the most contaminated samples to less than 1 spore per gram in the others. Primary sources of this biocontamination in honey may be pollen, the bee itself, soil, water, air, or dust, which are natural sources that are very difficult to control [7]. Secondary sources are closely connected with the hygiene of processing, handling, and storage of honey [8]. Because of this, honey has been identified as a possible cause of infantile botulism [7].

Clostridial spores are ingested and then germinate and multiplicate to start toxic genesis in babies because of their poorly developed (anaerobic) intestinal flora. Accordingly, health authorities in several countries have ordered or advised the use of labels on honey packaging alerting to the danger of feeding infants under one year of age. Related to known outbreaks of infant botulism, the number of *Clostridium* spores is usually in the order of 1 to 80 spores per gram [7].

Conventional processing of commercial honey involves indirect heating around 60 °C–65 °C for about 25 min–30 min to destroy yeasts that are responsible for honey fermentation when the moisture content is high and storage temperature is favourable. This treatment is also intended to dissolve crystals of glucose as it exists in a supersaturated state, to avoid the total or partial crystallization of honey in storage, allowing it to remain liquid for longer [9]. Inconveniently, *Clostridium* spores are not inactivated and certain starch-digesting enzymes (invertase, amylase, glucose oxidase) are destroyed [10]. Heating also causes the formation of hydroxymethylfurfural (HMF), a substance characteristic of heated or old honey that can even have detrimental effects (mutagenic, genotoxic, organotoxic, and enzyme inhibition) [11]. Moreover, thermal processing reduces the aromatic richness, as volatile components are lost at high temperatures. Similarly, if not carried out appropriately, this processing results in a loss of diversity in flavours, as it produces a darkening in the tonality of the honey, a situation that is attributed to a partial caramelization of the sugars [12].

Other preservation methods are sought to produce honey with superior quality and free from *Clostridium* spores. Al-Ghamdi et al. [13] studied the effect of high-pressure thermal treatments on the inactivation of nonproteolytic spores type E *Clostridium botulinum*. High pressures (300 MPa to 600 MPa) and elevated temperatures (80 °C to 100 °C) were tested in four low-acid foods. In the said work, processing at 90 °C and 600 MPa resulted in inactivation below the detection limit after 5 min. Traditional thermal processing of spores at 90 °C for 10 min, on the other hand, did not result in an estimated 6-log reduction [13]. According to these studies, high-energy treatments are required to eliminate *Clostridium* spores, which is a difficult task to achieve without damaging the sensitive quality of the honey.

Meanwhile, the inactivation of *Clostridium sporogenes* spores in honey was studied with the application of ultraviolet-C (UV-C) [14]. In the aforementioned work, a maximum reduction of 2.5 Log_{10} colony-forming units g^{-1} (CFU g^{-1}) was observed after an 18 J mL^{-1} treatment. The effect of UV-C on some quality parameters of honey, such as HMF, pH, and colour, was also assessed by the authors, who claimed that UV-C light induced changes in most of these parameters.

Another option is the application of supercritical carbon dioxide ($scCO_2$) [15]. The $scCO_2$ conservation method presents some fundamental advantages related to the mild conditions employed, particularly because it allows processing to be carried out at a much lower temperature than those used in conventional pasteurization and sterilization techniques [16]. In the $scCO_2$ technique, food is exposed to pressurized carbon dioxide (CO_2) for a certain amount of time in a batch, semi-batch, or continuous manner [17]. In addition, CO_2 is inert, non-toxic, accessible, and affordable. Under ambient conditions, it is a gas, so it leaves no residue on the treated product and it is a Generally Recognized as Safe (GRAS) solvent [18].

The $scCO_2$ treatment has been mainly explored for the inactivation of pathogens in liquid food samples [15,17] and less so on viscous or solid foodstuffs. To the best of our knowledge, the reported cases of the use of $scCO_2$ to inactivate bacterial spores in these latter products are the inactivation of natural biocontamination of mesophilic and thermophilic spores in cocoa powder [19] and the bioburden, mostly *Bacillus cereus* spores, in paprika [20], *Alicyclobacillus acidoterrestris* in apple cream [21], and *Geobacillus stearothermophilus* in cheese [22]. In this previous research, the inactivation efficiency of the scCO$_2$ treatment was limited by the high resistance of these forms in media of low a_w. More recently, Zambon et al. [23,24] explored the feasibility to apply scCO$_2$ to dry and increase the microbial safety of strawberries, finding that the process was efficient against different strains of pathogens (*Escherichia coli*, *Salmonella* spp., and *L. monocytogenes*). Studies have also established that essential oils (EOs) have sporicidal effects in the range of 1% to 3% in mass fraction [25]. Our research group successfully used oregano EO in combination with scCO$_2$ to inactivate *Bacillus cereus* spores in paprika [26]. González-Alonso et al. [27] also inactivated *E. coli* inoculated in raw poultry meat after treatment with herbal EOs in combination with scCO$_2$. Other works reporting the application of scCO$_2$ to improve food shelf-life and safety by inactivating spores can be found in a recent review [28].

With the growing number of consumers seeking natural products from medicinal plants for the well-being of people, new initiatives are already trying to boost all the benefits of honey with different plants, fruits, and EOs. Latin American entrepreneurs are already commercializing 100% pure honey with EOs [29]. According to the companies themselves, EOs are added to honey to boost the benefits that it already has. To this end, EOs of mint, eucalyptus, ginger, cinnamon, and lemon have been added. This type of initiative increases the spectrum of the possible final market in which the product could be purchased, since besides food use (honey with EOs is mainly used for infusions), the final product can be used as a cosmetic and/or ancestral medicine [30].

Therefore, the current study aimed to evaluate the use of pure scCO$_2$, and scCO$_2$ in combination with several EOs to reduce the count of *Clostridium* spores in honey. Because of its morphological and genetic similarity and its nontoxigenicity, *C. sporogenes* was used as a surrogate for *C. botulinum* [31]. Moreover, it has previously been considered an adequate substitute since *C. sporogenes* spores are more resistant than those of *C. botulinum* to heat [32]. The impact of the most relevant operational variables of the technology was investigated according to previous studies in solid or low a_w matrices and within the range of the technology itself [33]. A minimum reduction of one order of magnitude in the *Clostridium* count was sought based on the contamination typically found in honey. In addition, this treatment's impact on the quality of the honey was analysed. This is the first time that the inactivation of *Clostridium* spores by high-pressure CO$_2$ has been reported both in aqueous media and in complex food matrices such as honey.

2. Materials and Methods

2.1. Raw Materials

Lemon EO (LEO; CAS No. 8008-56-8; Ref. W262528, lot No. 13503DE), clove EO, *Eugenia* spp. (CLEO; CAS No. 8000-34-8; Ref. C8392, lot No. 018K1137), and cinnamon EO, Ceylon type (CEO; CAS No. 8015-91-6; Ref. W229105, lot No. 05211MA), all from Sigma-Aldrich (Spain) were stored in a refrigerator until assayed. Commercial honey (Granja San Francisco, Spain) was also used. The honey label indicated that it was thousand-flower honey from a mixture of both native and non-native European honey. Its containers were opened in a laminar-flow chamber to ensure they were kept sterile and free of enzyme activity. After opening, the containers were carefully closed and refrigerated. The CO$_2$ (purity >99.95%, <7 vpm H$_2$, <10 vpm O$_2$, <5 vpm THC, <2 vpm CO, <25 vpm N$_2$) was supplied by Carburos Metálicos (Spain). All reagents and culture media were supplied by Sigma-Aldrich (Spain) and used as received.

2.2. Bacteria

C. sporogenes was used as a surrogate for *C. botulinum* [31]. The strain used was the CECT 553 (NCIMB 8053; ATCC 7955) from the Spanish Type Culture Collection (Universidad de Valencia, Spain).

2.3. Culture Medium

The reinforced clostridial medium (RCM) specific culture recommended by the strain supplier was used. A volume of 1 L of RCM was composed of 10 g of meat extract, 10 g of peptone, 5 g of glucose, 5 g of NaCl, 3 g of yeast extract, 3 g of sodium acetate, 1 g of raffinose, and 0.5 g of cysteine. The optimal pH was 7.1 and Na_2CO_3 (0.66 M) was used for its correction.

The lyophilized strain was dissolved in approximately 1 mL of fresh culture medium and then transferred to a small volume of fresh medium that was incubated anaerobically at 37 °C for 24 h to recover the bacteria. Once recovered, the revived strain was transferred to 100 mL of fresh medium and allowed to grow for 48 h at 37 °C.

2.4. Generation and Concentration of Spores

Aliquots were extracted from the live culture in a growing medium after 48 h of incubation. The aliquots were inoculated in the sporulation medium and impoverished at a volumetric ratio of 1 to 10. The composition of 1 L of sporulation medium was 50 g of peptone trypticase, 5 g of peptone, and 1 g of sodium thioglycolate. The inoculated sporulation medium was then incubated for 7 days at 30 °C. After this time, the presence of spores was confirmed by differential staining and subsequent microscopic observation. The spores were centrifuged at $5000 \times g$ for 15 min and washed with sterile water. The supernatant was discarded. The sediment was washed with distilled water and the process was repeated until it was centrifuged three times. The final sediment was diluted in 50 mL of distilled water and stored at 4 °C.

The order of magnitude of the spore suspension was measured by counting the CFU in Petri dishes per deep seeding in solid RCM. To verify that the plate count did not include vegetative forms and that only spores had been sown, a parallel count was carried out in which the dilutions were introduced to a water bath at 80 °C for 15 min. The initial spore concentration in the suspension was of the order of 10^6 CFU mL^{-1}.

2.5. Sample Preparation

For inactivation tests of spores in honey, samples had to be previously contaminated. Since the resulting contamination was intended to be maximal, but not capable of altering the properties of the honey, a small amount of the spore water suspension in high concentration was added. Thus, for each 50 g of honey, 5 mL of suspension was added, and the sample was shaken until a homogeneous mixture was observed. The obtained concentrations ranged from 10^4 CFU mL^{-1} to 10^5 CFU mL^{-1}. These concentrations were much higher than those of *Clostridium* spores usually found in honey, but they ensured precision in counts and analyses. The resulting honey was made more fluid by the addition of water; thus, if the water content was initially between 18% and 20% in mass fraction, the newly prepared samples had a water content between 25% and 27% in mass fraction. To prepare diluted honey samples, more water was added to the already contaminated honey to increase moisture levels until the desired moisture content was determined, as determined by weight.

2.6. Experimental Installation for the scCO$_2$ Treatment

CO_2 was supplied in liquid form. It passed through a thermally controlled bath (Selecta, Frigiterm-30, Spain) and was cooled to −10 °C before it reached the diaphragm pump (Milroyal D; Dosapro Milton Roy, Spain). The cooling of CO_2 prevented cavitation during pressurization. The pressurized CO_2 was preheated in a spiral inside of a heating jacket before entering a 50 mL capacity 316 ss vessel (Autoclave Engineers, MicroClave™, Series 401A-8067, USA). The temperature in the vessel was controlled with an external heating jacket and read by a type K thermocouple (± 1 °C) located inside the vessel, and the pressure was read at the outlet of the vessel with a Bourdon gauge with an accuracy of ± 0.2 MPa. Stirring was achieved by an impeller with a speed that could vary between 50 rpm and 500 rpm. A pre-vessel was located upstream from the main vessel for tests in which the CO_2 was mixed with the EO. The CO_2 pressure and flow rate were controlled by the combined action of a back-pressure regulator (BPR, 26-1761-24-161, TESCOM Europe, Germany) and a pump flow regulator, respectively. The installation had a rupture disk at 38 MPa to avoid overpressure. The total mass and flow rate of CO_2 were measured using a mass flow meter (M-10 SLPM-D, Alicat Scientific, Tucson, AZ, USA) connected to the outlet. The scheme of the $scCO_2$ apparatus is shown in a previous work of our research group [34].

2.7. Method for the Inactivation of Clostridium Spores by scCO₂

To begin a $scCO_2$ treatment, the vessel was loaded with the raw material. For the tests with pure water, 15 mL of suspension was introduced. For assays with honey and diluted honey, 15 g of the sample was loaded. After the closures were adjusted, the heating jacket was connected and heated to the operating temperature. Then, the CO_2 container was opened, and the pump was turned on to reach the working pressure. When the EO was used, a piece of cotton soaked with about 0.5 g was placed in the pre-vessel. When the CO_2 passed through the pre-vessel, the EO was solubilized by the CO_2, and the mixture passed into the vessel containing the sample. Then, the BPR valve was opened to provide a continuous flow of CO_2 over the sample at approximately 1 g min^{-1}. The CO_2 (or its mixture with the EO) entered from the bottom of the vessel and passed over the sample. The stirrer in the vessel was set to the desired rotation (300 rpm for spore suspension; 60 rpm for honey and diluted honey samples). The mixing was carried out both to facilitate the contact of the sample with the CO_2 stream inside the container and to guarantee a homogeneous mixture. There was no honey carry-over during processing. When the end of the operating time was reached, the pump was turned off, the CO_2 supply was cut off, and depressurization began very slowly to avoid dragging or freezing the sample. Typically, the time required for the operating temperature and pressure to be reached was 5 min, and the depressurization time was 12 min. Once depressurized and separated from the equipment, the vessel containing the sample was taken to the sterile laminar-flow chamber for further analyses.

2.8. Thermal Treatments

To study the separate effect of temperature on spore inactivation, several non-CO_2 tests were carried out. In these experiments, the operating procedure was the same, except that there was no CO_2 flow in the vessel. Instead, the spore suspension or the contaminated honey was kept inside the vessel at the desired temperature for the same time as in the comparative tests with CO_2.

2.9. Microbial Analysis

In a laminar-flow chamber, 1 mL of spore suspension or 1 g of honey or diluted honey was taken to prepare serial dilutions in the RCM to obtain the concentration of spores in the sample by counting the CFU in Petri dishes. The degree of inactivation of the *Clostridium* spores was expressed as logarithm (log) reduction, which is the logarithm of the count after treatment (N) divided by the initial count (N_0) before each test.

2.10. Quality Analysis of the Honey

The impact of the $scCO_2$ treatment on honey quality variables such as HMF content, diastase index, and pH was analysed.

2.10.1. Determination of Hydroxymethylfurfural (HMF)

The increase of HMF in honey and diluted honey samples was determined according to the method proposed by the International Honey Commission [35]. This method is based on the absorbance of HMF at 284 nm using a UV-Vis spectrophotometer (MRC, model UV-1800, Tel-Aviv, Israel). To avoid the interference of other components at this wavelength, the difference between the absorbances of a clear aqueous honey solution and the same solution after the addition of bisulphite was determined. The HMF content was calculated after subtraction of the background absorbance at 336 nm as shown in Equation (1):

$$\text{HMF}\left(\text{mg kg honey}^{-1}\right) = \frac{(A_{284} - A_{336}) \times 149.7 \times 5 \times D}{W} \tag{1}$$

where A_{284} and A_{336} are absorbances at 284 nm and 336 nm, respectively; D is the final volume of the sample solution divided by 10; 5 is a theoretical nominal sample weight; W is the weight in g of the honey sample; and 149.7 is a constant [35].

2.10.2. Determination of the Diastase Activity

The diastase activity determination was conducted by an enzymatic-spectrophotometric method according to the International Honey Commission [35]. The principle of the method is based on the fact that the enzymes in the sample act on a starch standard solution that is capable of developing colour, with iodine, in a defined range of intensity, under standard conditions. The decrease in the blue colour was measured photometrically at 660 nm. A plot of absorbance against time was used to determine the time (t_x) required to reach the specified absorbance, 0.235. The results were expressed as a diastase number (DN, also known as Schade or Goethe units), which was calculated as 300 divided by t_x. DN is defined as the amount of enzyme which converts 10 mg of starch to the prescribed endpoint in one hour at 40 °C under the conditions of the test.

2.10.3. Measurement of pH

The pH was measured during the CO_2 treatment using reactive strips (McolorpHast, sensitivity of 0.3 pH unit, Merck, Spain) introduced into the high-pressure vessel along with the sample, where the pH was in two ranges: 2.5−4.5 and 4.0−7.0.

2.11. Data Analysis

The experiments were replicated three times. Microbial analyses were performed in duplicate for each replicate ($n = 3 \times 2$). Means and standard deviations were computed for all data. The maximum standard deviation in the readings of the *Clostridium* degree of inactivation was 0.3-log cycles. The statistical significance of the factors on the *Clostridium* inactivation degree was analysed using the ANOVA General Linear Models tool of STATGRAPHICS XVIII at a 5% significance level ($p \leq 0.05$). For this purpose, the following factors were used as categorical variables: (a) the type of treatment: thermal, scCO$_2$, or scCO$_2$ plus CEO; and (b) the matrix: water, honey, or derivatives. Temperature (data from Figures 1–3) and water content (data from Figure 4) were included as quantitative factors.

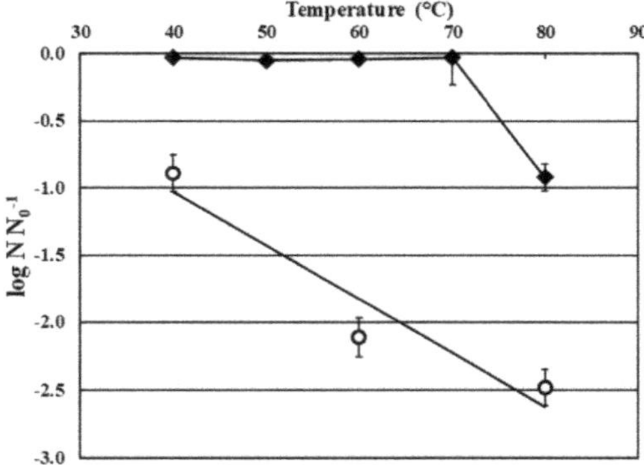

Figure 1. Inactivation of *Clostridium* spores suspended in water using scCO$_2$ (○) and comparison with heat at atmospheric pressure (♦). Conditions: P = 10 MPa; Q = 1 g min^{-1}; operation time = 60 min; stirring rate = 300 rpm. The solid line is the linear fitting.

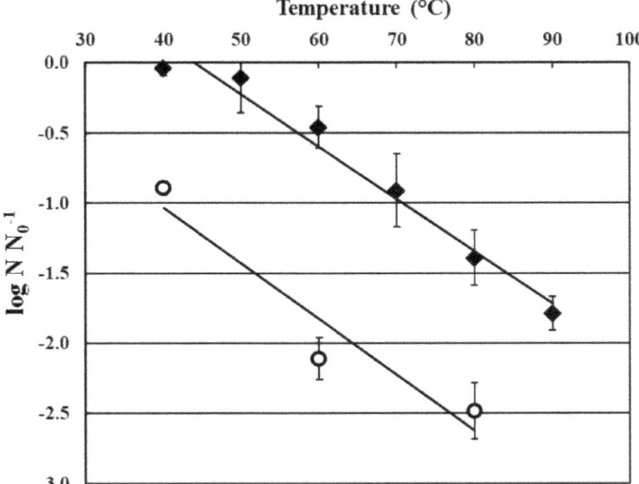

Figure 2. Effect of temperature on the inactivation of *Clostridium* spores using scCO$_2$ in aqueous suspension at 300 rpm (○) and biocontaminated honey at 60 rpm (♦). Conditions: P = 10 MPa; Q = 1 g min^{-1}; operation time = 60 min. The solid line is the linear fitting.

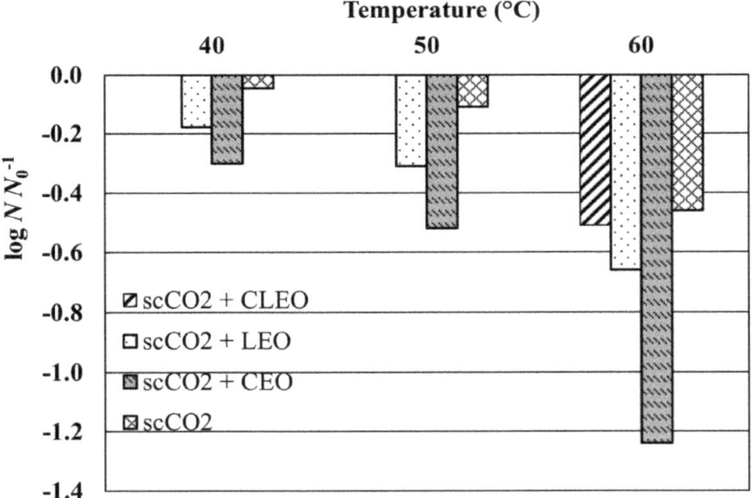

Figure 3. Inactivation of *Clostridium* spores in honey using scCO$_2$ alone and scCO$_2$ combined with lemon essential oil (LEO), clove essential oil (CLEO), and cinnamon essential oil (CEO). Conditions: P = 10 MPa; Q = 1 g min^{-1}; stirring rate = 60 rpm; operation time = 60 min.

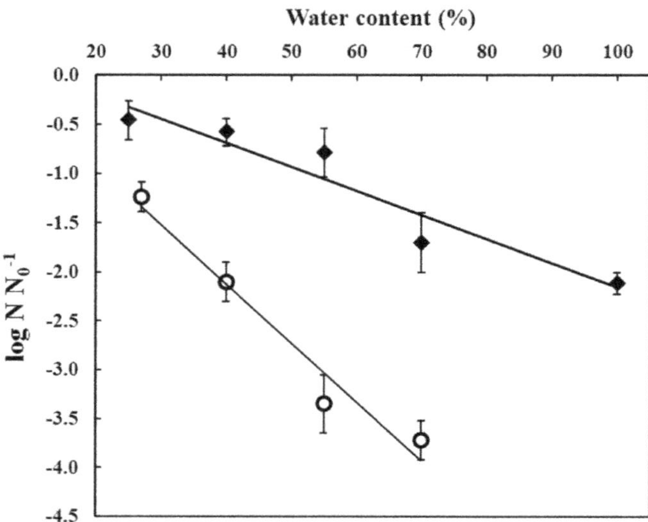

Figure 4. Effect of the initial water content on the inactivation of *Clostridium* spores in honey using scCO$_2$ (♦) and scCO$_2$ saturated with CEO (○). Conditions: P = 10 MPa; T = 60 °C; Q = 1 g min^{-1}; stirring rate = 60 rpm; operation time = 60 min. The solid line is the linear fitting.

3. Results and Discussions

First, the efficacy of the scCO$_2$ method on the inactivation of *Clostridium* spores suspended in water was analysed. Subsequently, the inactivation of these spores in honey and diluted honey samples was studied. The previous study on the water made it possible to establish the possible effects of the food matrix on the effectiveness of the treatment. The main variables that affected the operation and the influence that honey had on the degree of inactivation were studied. The pressure varied between 10 MPa and 30 MPa. The temperature ranged between 40 °C and 80 °C. Up to 60 °C, the treatment could be considered a low-temperature conservation method, but higher values were explored to reach a superior degree of inactivation and to be able to compare with the conventional heat treatment. Agitation at different speeds and increased CO$_2$ flow rates were explored to promote the transfer of CO$_2$ to water and viscous honey. Treatment times in the order of several hours were studied to evaluate their impact on the clostridial spores' survival. Furthermore, the use of scCO$_2$ saturated with EOs was investigated. Both CO$_2$ and EOs do not present a risk of toxicity and their combination could lead to a non-thermal treatment of honey for the elimination of its microbial contamination. The impact of the increasing amounts of water in the honey was intended to prove the importance of the a_w of the food in the inactivation of the bacterial spores. In the treated honey, the quality parameters were analysed to determine the feasibility of the implementation of this technology.

3.1. Inactivation of Clostridium Spores in Aqueous Suspension by scCO$_2$ and Comparison with Heating at Atmospheric Pressure

Figure 1 shows the results of the inactivation of *Clostridium* spores in an aqueous solution using scCO$_2$ and its comparison with heat at atmospheric pressure after 60 min of treatment. The pressure was set at 10 MPa and the solution was stirred at 300 rpm. With the scCO$_2$ process, at 40 °C, a significant ($p \leq 0.05$) degree of inactivation was observed; about 90% of the spore population with respect to the initial population was inactivated.

As the temperature increased (see Figure 1), the degree of inactivation was higher, reducing the population by 99.7% at 80 °C, which corresponded to a 2.5-log reduction. In contrast, in tests performed with heat at atmospheric pressure, without the passage of CO$_2$, only a small decrease in the total population was observed, less than 1-log reduction at 80 °C. Therefore, it appeared that CO_2 had sporicidal effects on its own. This fact was previously published by other authors [36–38]. They reported that $scCO_2$ had a lethal effect because in media with high a_w it could promote the activation and germination of spores, reducing their tolerance to heat [39].

3.2. Effect of Operation Conditions on the Inactivation of Clostridium Spores in Honey by scCO$_2$

For these assays, biocontaminated honey with *Clostridium* spores was used as a raw material to determine the buffering effects of the food itself on the process. The pressure was set at 10 MPa and the treatment time at 60 min, as in the water tests, unless otherwise specified. However, the stirring speed was reduced to 60 rpm because it was observed that a high stirring speed produced a dense foam.

3.2.1. Effect of Temperature

Similar to the aqueous suspension, the degree of inactivation of *Clostridium* spores increased with the temperature, as can be seen in Figure 2. However, clear differences could be observed compared with the evolution observed in the aqueous suspension. The lethal effect of $scCO_2$ was much lower in honey than in the aqueous suspension, as the degree of inactivation did not achieve a 2-log reduction, even after reaching 80 °C. This was probably due to the protective effect of honey.

A previous study [40] reported that the presence of salts and sugars in the media reduced the CO_2 effectiveness on the inactivation of *G. stearothermophilus* spores. This negative impact was proportional to their concentration, and therefore to a lower a_w. Additionally, on the inactivation of *A. acidoterrestris* in apple cream, the effectiveness of the $scCO_2$ in this medium was lower than that obtained in liquid juice [21]. The apple cream of the study was much more viscous than the juice and this resulted in a less effective level of contact with CO_2.

3.2.2. Effect of Pressure

The effect of the $scCO_2$ treatment was studied using a pressure of 30 MPa to compare with the results already obtained at 10 MPa. However, no important differences were obtained. A reduction of less than 0.05-log cycles was found in both cases at 40 °C. This was in line with other results obtained on the inactivation of spores in viscous [21] or solid [19,20] food products. It was widely demonstrated that a hydrostatic pressure of the range used had no effect on spores. Specifically, a recent paper reviewed all of the research undertaken on the inactivation of *Clostridium* spores in low-acid foods under high-pressure conditions in the range of 345 MPa–900 MPa [41]. Modest results were reported (<3-log reduction) even in combination with temperatures of the order of 80 °C–86 °C after 15 min–16 min of treatment. Therefore, it is unlikely that there was any significant effect of pressure related to the $scCO_2$ treatment in honey. Since no significant difference was found between 10 MPa and 30 MPa ($p > 0.05$), for economic and safety reasons, the lowest pressure (10 MPa) was used for the following tests.

3.2.3. Effect of the CO$_2$ Flow Rate

The effectiveness of increasing the CO_2 flow rate was tested to improve the contact between the CO_2 and honey, which had a high viscosity. These tests had good results in the inactivation of *A. acidoterrestris* in apple cream [21], which was also very viscous. However, no differences ($p > 0.05$) were found at 10 MPa and 60 °C between the resulting inactivation using $1\ \text{g min}^{-1}$ or $4\ \text{g min}^{-1}$ after 60 min (not shown). On the contrary, the introduction of CO_2 at a higher flow rate produced a greater amount of foam, and therefore, it was necessary to spend more rest time after the treatment to allow the CO_2 to escape from the honey. Specifically, by using $1\ \text{g min}^{-1}$, a rest time close to 2 h was needed for a partial recovery (only one layer of foam remained on the surface), and more than 3 h for its total recovery. When using $4\ \text{g min}^{-1}$, times longer than 6 h were needed for the partial recovery, and around 24 h to return to the original state. Therefore, since no significant difference was found between the highest and the lowest CO_2 flow rate, to reduce the CO_2 consumption and the pumping costs, 1 g min^{-1} was set for the rest of the tests.

3.2.4. Effect of Treatment Time

Operating at 10 MPa, 60 °C, 60 rpm, and 1 g min^{-1}, it was attempted to increase the treatment time to 4 h and consequently the contact with the scCO$_2$. However, this method did not have any significant benefit ($p > 0.05$) on the inactivation of *Clostridium* spores in honey in comparison with the results obtained from a treatment time of 60 min. Specifically, at 60 min, the inactivation degree reached -0.5 ± 0.2 logs, while after 240 min, it was -0.4 ± 0.1 logs.

3.3. Inactivation of Clostridium Spores via scCO$_2$ + EOs

Since the treatment with scCO$_2$ was not effective enough to achieve a 1-log reduction of spores at a mild temperature, the addition of EOs was tested.

In a previous study by our research group [26], it was demonstrated that it was possible to reduce *B. cereus* spores in paprika to 3-log cycles using scCO$_2$ mixed with oregano EO. For this reason, a similar treatment was considered for honey. Three EOs were chosen from the list of EOs that are effective as antimicrobials [42] and whose organoleptic properties were more compatible with honey: lemon (LEO), clove (CLEO), and cinnamon (CEO). For these treatments, the CO_2 was passed through the pre-vessel where the EO was placed before contact with the samples. Conditions were set in the pre-vessel at 10 MPa and 40 °C to improve the solubilization of the EO in the CO_2 because the solubility of EO is better at higher supercritical solvent densities. Under these conditions, the EO solubility in CO_2 was in the order of 0.3–0.4% in mass fraction [43]. Figure 3 compares the results obtained after honey treatment with scCO$_2$ combined with the mentioned EOs and with scCO$_2$ on its own at temperatures between 40 °C and 60 °C.

As can be seen in Figure 3, the CEO was the only EO in which significant ($p \leq 0.05$) improvement was found. The explanation for the different behaviour of the three EOs could lie in the different composition and richness of the main constituents, as demonstrated by Bagheri et al. [44]. In this same work, CEO (Cassia-Aliksir) is also one of the best against *C. tyrobutiricum*. The degree of efficacy achieved with this EO was better than that obtained with scCO$_2$ for the three explored temperatures. Thus, by using scCO$_2$ with CEO at 60 °C, a reduction of more than 1.3-log cycles (94%) in the *Clostridium* spore count in honey was achieved. A previous study discussed the mechanism of action of supercritical CO_2 on spores. With flow cytometry analysis, Rao et al. [45] probed that the permeability of the inner membrane and the cortex was increased. Moreover, the electron microscopy images showed clear evidence of damages to the external structure and morphology changes. Consequently, the pyridine-2,6-dicarboxylic acid (DPA) of the CO_2-treated spores was released to the medium. The presence of moisture would increase the fluidity of the membranes, facilitating the attack of CO_2. In parallel, scanning electron microscope (SEM) observations showed that exposure to high concentrations of EO resulted in damage to the spore coat of *Bacillus subtilis* [25]. Thus, it is logical to find that both agents combined cause a greater sporicidal effect.

Finally, it is highly likely that under these conditions in which *Clostridium* spores were greatly inactivated, other pathogens in the vegetative form were also inactivated due to their lower resistance to the scCO$_2$ treatment [46,47].

3.4. Impact of Initial Product Moisture

As indicated above, honey is a product with a low a_w, which may protect the spores and make scCO$_2$ inactivation difficult. Previous research on the microbial inactivation of solid foods, such as herbs and spices [48], cocoa [19], and paprika [20], have shown that CO_2 under pressure was sporicidal only in the presence of some water. These results were directly related to the role of water in the germination of spores in vegetative forms that are much less resistant [49].

For this reason, the effect of increasing the proportion of water in honey on the inactivation of *Clostridium* spores was analysed. Honey samples with different water content were treated with $scCO_2$ on their own and modified with the CEO. The percentages of moisture investigated were between 25% in mass fraction with respect to the sample (the minimal amount reached after *Clostridium* spore contamination) and 100% in mass fraction with respect to the sample (corresponding to the suspension of spores in water). Honey mead typically contains around 70% water. Please note that mead also contains alcohol and other components that could change these results.

As can be seen in Figure 4, the degree of inactivation of *Clostridium* spores was progressively augmented as the moisture of the samples was increased ($p \leq 0.05$), while operating at 60 °C. A higher degree of inactivation in the treatment with $scCO_2$ + CEO was obtained than in the treatment with $scCO_2$ alone ($p \leq 0.05$). For example, by using the CO_2 mixed with the CEO, a 1.2 ± 0.2-log reduction was reached when the water content in honey was 27%; meanwhile, 2.1 ± 0.2-log cycles (equivalent to a 99% reduction in the total population of spores) was found with an increase of the water content up to 40%. The linear adjustment shown in Figure 4 with the solid line predicts one log reduction in honey with 20% water, i.e., native. During the inactivation of *Clostridium* spores in the honey with 70% water, using $scCO_2$ modified with the CEO, a 3.7 ± 0.2-log reduction was reached, which would be equivalent in practice to obtaining a sterile product.

3.5. Effect of scCO₂ Treatment on Honey Quality

The increase in HMF content in honey and the decrease in the diastase activity are parallel processes to the degradation of vitamins, proteins, enzymes, and flavour of this product [11]. HMF is generated by the decomposition of fructose in acid conditions. It occurs naturally in most honey and increases rapidly with heat treatment. Therefore, it can be used as an indicator of heating and storage time [50]. The diastase activity is a measurement of the enzyme content in honey, so it is used as a quality parameter because of the sensitivity of enzymes to heat [9]. For this reason, the European legislation [5] sets a minimum DN of 8 for diastase activity and a maximum HMF content value of 40 mg HMF kg honey^{-1}, excluding honey produced in tropical areas, for which the highest level of HMF allowed is 80 mg HMF kg honey^{-1}.

This part of the study aimed to examine the HMF and diastase contents in honey samples with and without $scCO_2$ treatment to obtain objective information regarding its quality, and to study how the $scCO_2$ process affected these variables. The values of these parameters were initially determined in honey as purchased and are shown in Table 1.

Table 1. Hydroxymethylfurfural and diastase content in untreated honey and honey subjected to the $scCO_2$ treatment at 10 MPa and 60 min at increasing temperatures.

Treatment	Diastase Number (DN)	mg HMF kg Honey^{-1}
Untreated honey	13.4 ± 1.1 [a]	30.5 ± 2.5 [a]
$scCO_2$ at 45 °C	10.9 ± 0.2 [b]	33.1 ± 2.0 [a]
$scCO_2$ at 60 °C	7.2 ± 0.1 [c]	40.7 ± 1.9 [b]

Different lowercase letters within the same column represent statistically significant differences at a 5% significance.

Values of 13.4 for DN and 30.5 mg HMF kg honey^{-1} were found in the starting honey. These values were already low for DN and high for HMF, which showed that the honey had already been treated with heat. Table 1 also shows HMF values and DN in honey samples treated with $scCO_2$. The HMF content in honey samples treated with $scCO_2$ at 45 °C had no statistically significant differences ($p > 0.05$) with the HMF content of untreated honey samples. However, a significant ($p \leq 0.05$) change in the HMF content in samples treated by $scCO_2$ at 60 °C was found, with an increase of 33% with respect to the initial value. This also resulted in a darkening of the honey, which was noticeable to the naked eye at temperatures higher than 60 °C.

In contrast, the diastase index was more affected by the $scCO_2$ treatment. For example, the DN of honey subjected to $scCO_2$ at 45 °C was reduced by 18% of its initial value, and by 46% using $scCO_2$ at 60 °C, with statistically significant differences ($p \leq 0.05$) among all groups. This was because $scCO_2$ had a damaging impact on enzymes. The application of $scCO_2$ has been previously explored as an effective non-thermal technique to inactivate harmful enzymes in liquid and solid food systems. Structural, morphological, and electrophoretic behaviour changes in the enzymes have been detected after $scCO_2$ contact [51].

The dissolution of CO_2 in water generates carbonic acid (H_2CO_3), which reduces the pH. This reduction in acidity has synergistic effects on the inactivation of spores through the use of $scCO_2$. Haas et al. [48] claimed that the efficacy of $scCO_2$ for the inactivation of *C. sporogenes* in thioglycolate broth (pH = 5.5) at 5.5 MPa and 70 °C was substantially greater (7-log) if the medium was acidified to pH 2.5–3.0. In the same way, Casas and Calvo [52] showed that the inactivation of *B. cereus* spores in a phosphate-buffered solution (pH = 7) was null in contrast to that achieved in pure water, where at 30 MPa and 70 °C, the spore population was reduced by four orders of magnitude. The pH of the water under these conditions was found to be 3.9 inside the vessel.

Table 2 shows the variation of the pH in the vessel due to the $scCO_2$ treatment in honey, diluted honey, and the spore suspension. In the aqueous suspension, the pH was significantly ($p \leq 0.05$) lowered after $scCO_2$ treatment due to the better solubilization of CO_2. In contrast, the pH in pure honey was not significantly ($p > 0.05$) altered by the $scCO_2$ treatment since it was already low and the free water (a_w) for CO_2 dissolution was low. The naturally low pH of the honey could be one of the causes of the lower inactivation of *Clostridium* spores in this medium, along with the other protecting factors earlier mentioned. In contrast, in honey samples with added water, the pH was slightly reduced ($p \leq 0.05$).

Table 2. Variation of pH due to $scCO_2$ treatment.

Sample	Initial pH	Final pH
Honey	3.9 [a]	4.1 [a]
Diluted honey (40% water)	4.1 [a]	3.8 [b]
Diluted honey (55% water)	4.7 [a]	4.2 [b]
Diluted honey (70% water)	4.8 [a]	4.3 [b]
Spore suspension	5.5 [a]	4.3 [b]

Different lowercase letters within the same row represent statistically significant differences at a 5% significance. The average standard deviation in the readings was ±0.1.

The honey aroma is rich in volatile compounds such as alcohols, ketones, aldehydes, acids, terpenes, hydrocarbons, benzene, and furan derivatives [53]; many studies showed good solubility of flavours and fragrances in $scCO_2$ [54] although the pressure was usually higher than that used by us. Still, we did not observe any extract in the vessel placed after the BPR where depressurization to room conditions occurred, even after many runs.

By adding increasing quantities of the EO to the honey, our research group identified that the level of smell that was perceptible but pleasant was of the order of 200 ppm, while it was excessive at or above 400 ppm and even unpleasant when it was higher than 800 ppm. However, given the relatively high solubility of the EO in $scCO_2$ [43], it would be possible to remove all or part of the EO with an extra passage of pure CO_2. This was done successfully in an earlier study to eliminate the remaining odour of oregano EO from paprika [26]. In the aforementioned work, 20 min of continuous CO_2 passage was enough to render the odour in the paprika imperceptible to a panel of six trained people. Nevertheless, more experiments are needed to confirm this hypothesis.

4. Conclusions

Among several parameters, temperature and initial moisture content of the honey derivatives were variables with the greatest impact on the efficacy of the *Clostridium* spores' inactivation. Pressure and/or treatment time were less influential. The smallest reduction of *Clostridium* spore count (i.e., 1-log reduction) was obtained in an aqueous suspension with $scCO_2$ at 10 MPa and 40 °C after 60 min. In comparison, this minimum inactivation degree in pure honey was not achieved until 70 °C due to its low water content, naturally low pH, other protective effects associated with the presence of nutrients (sugar), and its high viscosity. However, adding CEO to the CO_2 (0.3–0.4% in mass fraction) significantly increased the effectiveness of inactivation, so the temperature to achieve a 1-log reduction was about 55 °C. Therefore, this combined method could be an alternative to traditional thermal treatments for the inactivation of honey microflora causing infant botulism. However, this method should be validated with *C. botulinum* spores before its implementation.

This treatment, if carried out at temperatures below 60 °C, did not cause significant increases in the HMF content of the honey, nor changes in its pH or colour. However, it significantly reduced the enzymatic activity of the honey due to the inactivation capacity of free enzymes caused by the $scCO_2$. The sensory impact of the addition of EOs may be acceptable for consumers based on flavour compatibility, but it could also be controlled with an extraction step of pure CO_2.

Better clostridial spore inactivation degrees were found in honey with a high-water content, as it could be mead. $scCO_2$ combined with the addition of CEO caused an inactivation degree of nearly 4-log cycles of the *Clostridium* spores at 60 °C, which opens the possibility of applying the process for the sterilization of other thermolabile liquid food products or those with high moisture content. In liquid products, this could be done by pumping them simultaneously with the $scCO_2$, providing contact time in a holding tube. This operating method would allow continuous treatment with higher capacity and lower operating costs [34].

The introduction of the CEO in combination with the $scCO_2$ left a noticeable odour in the honey. However, recipes for flavouring honey with herbs, edible flowers, and spices can be easily found on the Internet and are already marketed. Thus, it is highly possible that consumers would readily accept the aroma and flavour imparted by the CEO. Indeed, all the research team members of our group found the aroma and flavour to be very pleasant. Therefore, the product obtained in this work could have a wide range of uses, including as a food ingredient (e.g., for infusions and tea); in cosmetics (mainly for use on the skin); and in traditional medicine, since honey with CEO is used ancestrally in Latin America against colds, joint pains, indigestion, and other ailments [30]. Furthermore, its possible use as a novel food (since it includes an EO) should be authorized according to the EU regulation.

Author Contributions: Conceptualization, L.C.; Data curation, A.D.-G., D.F.T. and L.C.; Formal analysis, A.D.-G., D.F.T. and L.C.; Funding acquisition, L.C.; Investigation, A.D.-G.; Methodology, L.C.; Project administration, L.C.; Re-sources, L.C.; Software, D.F.T. and L.C.; Supervision, L.C.; Validation, D.F.T. and L.C.; Visualization, D.F.T. and L.C.; Writing—original draft, A.D.-G., D.F.T. and L.C.; Writing—review & editing, D.F.T. and L.C. All authors have read and agreed to the published version of the manuscript.

Funding: This research was financed by the Ministry of Science and Innovation of the Government of Spain through the project Ref. CTQ2009-14676/PPQ and Universidad Nacional de Colombia (project reference 55144).

Institutional Review Board Statement: Not applicable.

Informed Consent Statement: Not applicable.

Data Availability Statement: The data that support the findings of this study are available from the corresponding author, upon reasonable request.

Conflicts of Interest: The authors declare no competing interests.

References

1. Harris, A. *Clostridium botulinum*. In *Encyclopedia of Food and Health*; Elsevier: Amsterdam, The Netherlands, 2016; pp. 141–145.
2. Smith, T.J.; Hill, K.K.; Raphael, B.H. Historical and current perspectives on *Clostridium botulinum* diversity. *Res. Microbiol.* **2015**, *166*, 290–302. [CrossRef] [PubMed]
3. Long, S.S. *Clostridium botulinum* (Botulism). In *Principles and Practice of Pediatric Infectious Diseases*; Elsevier: Amsterdam, The Netherlands, 2018; pp. 999–1006.
4. Snowdon, J.A.; Cliver, D.O. Microorganisms in honey. *Int. J. Food Microbiol.* **1996**, *31*, 1–26. [CrossRef]
5. The Council of The European Union. Council Directive 2001/110/EC of 20 December 2001 relating to honey. *Off. J. Eur. Union* **2001**, *10*, 47–52.
6. Maikanov, B.; Mustafina, R.; Auteleyeva, L.; Wiśniewski, J.; Anusz, K.; Grenda, T.; Kwiatek, K.; Goldsztejn, M.; Grabczak, M. *Clostridium botulinum* and *Clostridium perfringens* occurrence in Kazakh honey samples. *Toxins* **2019**, *11*, 472. [CrossRef] [PubMed]
7. Grabowski, N.T.; Klein, G. Microbiology and foodborne pathogens in honey. *Crit. Rev. Food Sci. Nutr.* **2017**, *57*, 1852–1862. [CrossRef]
8. Klutse, C.K.; Larbi, D.A.; Adotey, D.K.; Serfor-Armah, Y. Assessment of effect of post-harvest treatment on microbial quality of honey from parts of Ghana. *Radiat. Phys. Chem.* **2021**, *182*, 109368. [CrossRef]
9. Subramanian, R.; Hebbar, H.U.; Rastogi, N.K. Processing of honey: A review. *Int. J. Food Prop.* **2007**, *10*, 127–143. [CrossRef]
10. White, J.W.; Kushnir, I.; Subers, M. Effect of storage and processing temperatures on honey quality. *Food Technol.* **1964**, *18*, 153–156.
11. Shapla, U.M.; Solayman, M.; Alam, N.; Khalil, M.I.; Gan, S.H. 5-Hydroxymethylfurfural (HMF) levels in honey and other food products: Effects on bees and human health. *Chem. Cent. J.* **2018**, *12*, 35. [CrossRef]
12. Wang, H.; Cao, X.; Han, T.; Pei, H.; Ren, H.; Stead, S. A novel methodology for real-time identification of the botanical origins and adulteration of honey by rapid evaporative ionization mass spectrometry. *Food Control* **2019**, *106*, 106753. [CrossRef]
13. Al-Ghamdi, S.; Sonar, C.R.; Patel, J.; Albahr, Z.; Sablani, S.S. High pressure-assisted thermal sterilization of low-acid fruit and vegetable purees: Microbial safety, nutrient, quality, and packaging evaluation. *Food Control* **2020**, *114*, 107233. [CrossRef]
14. Roig-Sagués, A.X.; Gervilla, R.; Pixner, S.; Terán-Peñafiel, T.; Hernández-Herrero, M.M. Bactericidal effect of ultraviolet-C treatments applied to honey. *LWT Food Sci. Technol.* **2018**, *89*, 566–571. [CrossRef]
15. Perrut, M. Sterilization and virus inactivation by supercritical fluids (a review). *J. Supercrit. Fluid.* **2012**, *66*, 359–371. [CrossRef]
16. Silva, E.K.; Alvarenga, V.O.; Bargas, M.A.; Sant'Ana, A.S.; Meireles, M.A.A. Non-thermal microbial inactivation by using supercritical carbon dioxide: Synergic effect of process parameters. *J. Supercrit. Fluid.* **2018**, *139*, 97–104. [CrossRef]
17. Zhang, J.; Davis, T.A.; Matthews, M.A.; Drews, M.J.; LaBerge, M.; An, Y.H. Sterilization using high-pressure carbon dioxide. *J. Supercrit. Fluid.* **2006**, *38*, 354–372. [CrossRef]
18. Brunner, G. Supercritical fluids: Technology and application to food processing. *J. Food Eng.* **2005**, *67*, 21–33. [CrossRef]
19. Calvo, L.; Muguerza, B.; Cienfuegos-Jovellanos, E. Microbial inactivation and butter extraction in a cocoa derivative using high pressure CO_2. *J. Supercrit. Fluid.* **2007**, *42*, 80–87. [CrossRef]
20. Calvo, L.; Torres, E. Microbial inactivation of paprika using high-pressure CO_2. *J. Supercrit. Fluid.* **2010**, *52*, 134–141. [CrossRef]
21. Casas, J.; Valverde, M.T.; Marín-Iniesta, F.; Calvo, L. Inactivation of *Alicyclobacillus acidoterrestris* spores by high pressure CO_2 in apple cream. *Int. J. Food Microbiol.* **2012**, *156*, 18–24. [CrossRef]
22. Sikin, A.M.; Walkling-Ribeiro, M.; Rizvi, S.S.H. Synergistic effect of supercritical carbon dioxide and peracetic acid on microbial inactivation in shredded mozzarella-type cheese and its storage stability at ambient temperature. *Food Control* **2016**, *70*, 174–182. [CrossRef]
23. Zambon, A.; Zulli, R.; Boldrin, F.; Spilimbergo, S. Microbial inactivation and drying of strawberry slices by supercritical CO_2. *J. Supercrit. Fluid.* **2022**, *180*, 105430. [CrossRef]
24. Zambon, A.; Facco, P.; Morbiato, G.; Toffoletto, M.; Poloniato, G.; Sut, S.; Andrigo, P.; Dall'Acqua, S.; de Bernard, M.; Spilimbergo, S. Promoting the preservation of strawberry by supercritical CO_2 drying. *Food Chem.* **2022**, *397*, 133789. [CrossRef] [PubMed]
25. Lawrence, H.A.; Palombo, E.A. Activity of essential oils against *Bacillus subtilis* spores. *J. Microbiol. Biotechnol.* **2009**, *19*, 1590–1595. [CrossRef] [PubMed]
26. Casas, J.; Tello, J.; Gatto, F.; Calvo, L. Microbial inactivation of paprika using oregano essential oil combined with high-pressure CO_2. *J. Supercrit. Fluid.* **2016**, *116*, 57–61. [CrossRef]
27. González-Alonso, V.; Cappelletti, M.; Bertolini, F.M.; Lomolino, G.; Zambon, A.; Spilimbergo, S. Microbial inactivation of raw chicken meat by supercritical carbon dioxide treatment alone and in combination with fresh culinary herbs. *Poult. Sci.* **2020**, *99*, 536–545. [CrossRef]
28. Hart, A.; Anumudu, C.; Onyeaka, H.; Miri, T. Application of supercritical fluid carbon dioxide in improving food shelf-life and safety by inactivating spores: A review. *J. Food Sci. Technol.* **2022**, *59*, 417–428. [CrossRef]
29. Sweet Gold Honey with Essential Oil. Available online: https://sweetgoldcr.com/aceites-esenciales/ (accessed on 24 July 2022).
30. Repretel Giros Honey with Essential Oil—Sweet Gold. Available online: https://www.youtube.com/watch?v=lS4GH89e4uA (accessed on 24 July 2022).
31. Bradbury, M.; Greenfield, P.; Midgley, D.; Li, D.; Tran-Dinh, N.; Vriesekoop, F.; Brown, J.L. Draft genome sequence of *Clostridium sporogenes* PA 3679, the common nontoxigenic surrogate for proteolytic *Clostridium botulinum*. *J. Bacteriol.* **2012**, *194*, 1631–1632. [CrossRef]

32. Shao, Y.; Ramaswamy, H.S. *Clostridium sporogenes*-ATCC 7955 spore destruction kinetics in milk under high pressure and elevated temperature treatment conditions. *Food Bioprocess Technol.* **2011**, *4*, 458–468. [CrossRef]
33. Ribeiro, N.; Soares, G.C.; Santos-Rosales, V.; Concheiro, A.; Alvarez-Lorenzo, C.; García-González, C.A.; Oliveira, A.L. A new era for sterilization based on supercritical CO_2 technology. *J. Biomed. Mater. Res. Part B Appl. Biomater.* **2020**, *108*, 399–428. [CrossRef]
34. Martín-Muñoz, D.; Tirado, D.F.; Calvo, L. Inactivation of *Legionella* in aqueous media by high-pressure carbon dioxide. *J. Supercrit. Fluid.* **2022**, *180*, 105431. [CrossRef]
35. International Honey Commission. Harmonised Methods of the International Honey Commission. 2009. Available online: https://www.ihc-platform.net/ihcmethods2009.pdf (accessed on 24 July 2022).
36. Ballestra, P.; Cuq, J. Influence of pressurized carbon dioxide on the thermal inactivation of bacterial and fungal spores. *LWT Food Sci. Technol.* **1998**, *31*, 84–88. [CrossRef]
37. Spilimbergo, S.; Bertucco, A.; Lauro, F.M.; Bertoloni, G. Inactivation of *Bacillus subtilis* spores by supercritical CO_2 treatment. *Innov. Food Sci. Emerg. Technol.* **2003**, *4*, 161–165. [CrossRef]
38. Watanabe, T.; Furukawa, S.; Hirata, J.; Koyama, T.; Ogihara, H.; Yamasaki, M. Inactivation of *Geobacillus stearothermophilus* spores by high-pressure carbon dioxide treatment. *Appl. Environ. Microbiol.* **2003**, *69*, 7124–7129. [CrossRef] [PubMed]
39. Watanabe, T.; Furukawa, S.; Tai, T.; Hirata, J.; Narisawa, N.; Ogihara, H.; Yamasaki, M. High pressure carbon dioxide decreases the heat tolerance of the bacterial spores. *Food Sci. Technol. Res.* **2003**, *9*, 342–344. [CrossRef]
40. Furukawa, S.; Watanabe, T.; Koyama, T.; Hirata, J.; Narisawa, N.; Ogihara, H.; Yamasaki, M. Inactivation of food poisoning bacteria and *Geobacillus stearothermophilus* spores by high pressure carbon dioxide treatment. *Food Control* **2009**, *20*, 53–58. [CrossRef]
41. Evelyn; Silva, F.V.M. Heat assisted HPP for the inactivation of bacteria, moulds and yeasts spores in foods: Log reductions and mathematical models. *Trends Food Sci. Technol.* **2019**, *88*, 143–156. [CrossRef]
42. Tariq, S.; Wani, S.; Rasool, W.; Shafi, K.; Bhat, M.A.; Prabhakar, A.; Shalla, A.H.; Rather, M.A. A comprehensive review of the antibacterial, antifungal and antiviral potential of essential oils and their chemical constituents against drug-resistant microbial pathogens. *Microb. Pathog.* **2019**, *134*, 103580. [CrossRef]
43. Antonie, P.; Pereira, C.G. Solubility of functional compounds in supercritical CO_2: Data evaluation and modelling. *J. Food Eng.* **2019**, *245*, 131–138. [CrossRef]
44. Bagheri, L.; Khodaei, N.; Salmieri, S.; Karboune, S.; Lacroix, M. Correlation between chemical composition and antimicrobial properties of essential oils against most common food pathogens and spoilers: In-vitro efficacy and predictive modelling. *Microb. Pathog.* **2020**, *147*, 104212. [CrossRef]
45. Rao, L.; Zhao, F.; Wang, Y.; Chen, F.; Hu, X.; Liao, X. Investigating the inactivation mechanism of *Bacillus subtilis spores* by high pressure CO_2. *Front. Microbiol.* **2016**, *7*, 1411. [CrossRef]
46. Damar, S.; Balaban, M.O. Review of dense phase CO_2 technology: Microbial and enzyme inactivation, and effects on food quality. *J. Food Sci.* **2006**, *71*, 1–11. [CrossRef]
47. García-González, L.; Geeraerd, A.H.; Spilimbergo, S.; Elst, K.; Van Ginneken, L.; Debevere, J.; Van Impe, J.F.; Devlieghere, F. High pressure carbon dioxide inactivation of microorganisms in foods: The past, the present and the future. *Int. J. Food Microbiol.* **2007**, *117*, 1–28. [CrossRef] [PubMed]
48. Haas, G.J.; Prescott, H.E.; Dudley, E.; Dik, R.; Hintlian, C.; Keane, L. Inactivation of microorganisms by carbon dioxide under pressure. *J. Food Saf.* **1989**, *9*, 253–265. [CrossRef]
49. Setlow, P. Spore germination. *Curr. Opin. Microbiol.* **2003**, *6*, 550–556. [CrossRef] [PubMed]
50. White, J. The role of HMF and diastase assays in honey quality evaluation. *Bee World* **1994**, *75*, 104–117. [CrossRef]
51. Hu, W.; Zhou, L.; Xu, Z.; Zhang, Y.; Liao, X. Enzyme inactivation in food processing using high pressure carbon dioxide technology. *Crit. Rev. Food Sci. Nutr.* **2013**, *53*, 145–161. [CrossRef]
52. Casas, J.; Calvo, L. Inactivation of spores by high pressure CO_2. In Proceedings of the 6th International Symposium on High Pressure Process Technology, Belgrade, Serbia, 8 September 2013.
53. Kaškoniene, V.; Venskutonis, P.R.; Čeksteryte, V. Composition of volatile compounds of honey of various floral origin and beebread collected in Lithuania. *Food Chem.* **2008**, *111*, 988–997. [CrossRef]
54. Capuzzo, A.; Maffei, M.E.; Occhipinti, A. Supercritical fluid extraction of plant flavors and fragrances. *Molecules* **2013**, *18*, 7194–7238. [CrossRef]

Article

Optimization of Subcritical Fluid Extraction for Total Saponins from *Hedera nepalensis* Leaves Using Response Surface Methodology and Evaluation of Its Potential Antimicrobial Activity

Hoang Thanh Duong [1], Ly Hai Trieu [1], Do Thi Thuy Linh [1], Le Xuan Duy [1], Le Quang Thao [2], Le Van Minh [1], Nguyen Tuan Hiep [1,*] and Nguyen Minh Khoi [1,3,*]

[1] National Institute of Medicinal Materials, Hanoi 100000, Vietnam; htduong210195@gmail.com (H.T.D.); lhtrieu12csh@gmail.com (L.H.T.); dothithuylinhht@gmail.com (D.T.T.L.); leduy.hust@gmail.com (L.X.D.); lvminh05@gmail.com (L.V.M.)
[2] National Institute of Drug Quality Control, Hanoi 100000, Vietnam; thaolq@nidqc.gov.vn
[3] Department of Medical and Traditional Pharmacology, University of Medicine and Pharmacy, Hanoi 100000, Vietnam
* Correspondence: nguyentuanhiep@nimm.org.vn (N.T.H.); khoinm@nimm.org.vn (N.M.K.)

Citation: Duong, H.T.; Trieu, L.H.; Linh, D.T.T.; Duy, L.X.; Thao, L.Q.; Minh, L.V.; Hiep, N.T.; Khoi, N.M. Optimization of Subcritical Fluid Extraction for Total Saponins from *Hedera nepalensis* Leaves Using Response Surface Methodology and Evaluation of Its Potential Antimicrobial Activity. *Processes* **2022**, *10*, 1268. https://doi.org/10.3390/pr10071268

Academic Editors: Irena Zizovic, Maria Angela A. Meireles, Grazielle Nathia Neves and Ádina L. Santana

Received: 16 May 2022
Accepted: 22 June 2022
Published: 27 June 2022

Abstract: (1) Background: *Hedera nepalensis* (Araliaceae) is a recognized medicinal plant founded in Asia that has been reported to work in antioxidant, antifungal, antimicrobial, and antitumor capacities. (2) Methods: The subcritical fluid extraction of saponin from *Hedera nepalensis* leaves and the optimum of the extraction process based on yield of saponin contents (by calculating the hederacoside C contents in dried *Hedera nepalensis* leaves) are examined by response surface methodology (RSM). Furthermore, the antimicrobial activity of the extract is tested for potential drug applications in the future. (3) Results: Based upon RSM data, the following parameters are optimal: extraction time of 3 min, extraction temperature of 150 °C, and a sample/solvent ratio of 1:55 g/mL. Under such circumstances, the achieved yield of saponin is 1.879%. Moreover, the extracts inhibit the growth of some bacterial strains (*Streptococcus pneumoniae*, *Streptococcus pyogenes*, *Haemophilus influenza*) at a moderate to strong level with inhibition zone diameter values ranging from 12.63 to 19.50 mm. (4) Conclusions: The development of such a model provides a robust experimental process for optimizing the extraction factors of saponin contents from *Hedera nepalensis* extract using subcritical fluid extraction and RSM. Moreover, the current work reveals that saponin extracts of *Hedera nepalensis* leaves exhibit a potential antimicrobial activity, which can be used as scientific evidence for further study.

Keywords: *Hedera nepalensis*; subcritical fluid extraction; response surface methodology; antimicrobial

1. Introduction

Medicinal herbs have traditionally been used to both prevent and treat a variety of ailments. Several attempts have been made to investigate medicinal flora, and pharmacists continue to investigate the value of them across the world. *Hedera nepalensis* (Araliaceae) is a recognized medicinal plant found in Asia, mainly in Japan, Afghanistan, and the Himalayas [1,2]. According to past research, *Hedera nepalensis* contains a diverse array of natural substances, such as triterpene saponins, flavonoids, steroids, tannins, terpenoids, and phenolic compounds [3]. Particularly, triterpene saponins, with hederacoside C and alpha-hederin as major constituents, have been reported to work in antioxidant, antifungal, antimicrobial, and antitumor capacities [4–6].

A vast collection of methods can be applied for recovering bioactive compounds from natural resources, which include maceration extraction, microwave-assisted extraction, ultrasound-assisted extraction, and others [7]. Numerous downsides were found when the usual extraction methods were applied to obtain these compounds, such as time and solvent consumption, as well as tedious and limited selectivity and/or extraction yields. Recently, subcritical fluid extraction has developed quickly as a conventional substitute for traditional extraction methods. The extraction of a liquid under pressure benefits from the enhanced solubility that happens when the solvent temperature rises. Higher temperatures improve the ability of solvents to be reconstituted for the analytes. Furthermore, raising the temperature of the solvent causes a reduction in viscosity, allowing for improved sample matrix penetration [8]. This method has been extensively utilized in recent years to extract bioactive compounds from natural materials [9–12], especially saponins [13–16]. To the best of our knowledge, there have been no publications regarding the extraction of saponins from *Hedera nepalensis* leaves by pressurized liquid extraction.

In recent years, microbial infections have increased to a great extent, and resistance to antimicrobial drugs will put the health of millions of people at risk. For mainly years, various plants have been used for daily remedies based only on traditional medicine, without adequate scientific research. With the advancement of technology in science and medicine, natural products of plants may provide a new source of antibacterial substances that might have a major impact on infectious diseases and overall community health [17,18].

The aim of our research is to optimize, by means of an experimental design using a Box–Behnken model in response surface methodology analysis, the process for extraction of saponin-rich extract from *Hedera nepalensis*. Such optimization with the proposed mathematical models will be able to properly predict the performance of the system by evaluating changes in extraction aspects such as temperature and time [19]. Moreover, the activity of these extracts against a wide range of microorganisms will be evaluated to determine the range of activity of the extract and to provide information about the antimicrobial potential of *Hedera nepalensis* extracts against infection agents.

2. Materials and Methods

2.1. Plant Materials

Leaves of *Hedera nepalensis* were harvested in Ha Giang province, Vietnam. Plant identification was carried out by the National Institute of Medicinal Material (Hanoi, Vietnam). The collected leaves were cleaned, oven-dried at 55 °C, and ground. They were stored under dry and dark conditions at room temperature. Moisture content (7.21%) was determined before further experiments.

2.2. Accelerated Solvent Extraction (ASE) Procedure

The pressured liquid extraction was performed using an ASE 350 System (Dionex, Sunnyvale, CA, USA) with a stainless-steel extraction cell. About 2 g of *Hedera nepalensis* sample was placed into an extraction cell after being uniformly mixed with the similar weight of diatomaceous earth. To avoid the powder from penetrating into the extraction bottle, a frit and a filter (Dionex) were positioned at the end of the cell. The method was as follows: solvent ethanol 50%, constant pressure 1600 psi, and other parameters (volume of used solvent, temperature, and dynamic extraction time) were chosen by performing the initial experiments (not reported here). The extract was evaporated to dryness using a rotary evaporator at 50 °C. The extract was dried using a rotary evaporator set to 50 °C. The dried extract was then diluted in 50 mL of methanol and filtered through a 0.45 μm filter for HPLC analysis.

2.3. HPLC Analysis

The saponin analysis of the extracts was conducted using HPLC as per our reported procedure [20]. Applied to this analysis, hederacoside C, a major saponin in *Hedera nepalensis*, was used as the marker for quality control of the products. The Shimadzu SPD-20A system (C18 column; 250 mm × 4.6 mm, 5 μm) (Shimadzu Co., Ltd., Kyoto, Japan) was used for HPLC analysis. A mixture of acetonitrile–0.02% phosphoric acid solution was used as the mobile phase. The composition of the mobile phase was: 0–25 min, 20–60%

acetonitrile; 25–30 min, 60–100% acetonitrile. Other running conditions included the detection wavelength (210 nm), the flow rate (1 mL/min), the injection volume (20 mL), and the column temperature (25 °C).

The percentage of hederacoside C in the portion of *Hedera nepalensis* leaves taken was calculated as follows:

$$X(\%) = \frac{C \times V \times P}{m \times (100 - W) \times 10}$$

C: the concentration of analyzed compound in the sample solution from the calibration curve equation (μg/mL); V: volume of the sample solution (mL); m: weight of *Hedera nepalensis* leaves taken to prepare the sample solution (mg); W: the moisture of *Hedera nepalensis* dried leaves (%); P: purity of standard (%).

2.4. Experimental Design and Statistical Analytic

Response surface methodology (RSM) is the pattern of the design and analysis of testing, modeling procedures, and optimization approaches that utilize experimental data to obtain an approximated operational correlation between a response target and a set of proposed variables. Specifically, RSM is an empirical method created to find the best response within the individual variations among the parameters. RSM is a statistical-based technique and is a potent experimental design instrument that acknowledges the execution of whole systems [21].

Box–Behnken is a spherical design, with a centrally positioned point and middle points at the edges of the cube; it does not contain any points on the outermost corners of the cube. Application of this design was used for optimization of numerous extraction procedures, and the number of experiments was appropriately selected [22].

Design-Expert 11 (State-Ease Inc., Minneapolis, MN, USA) was used to generate the experimental designs, a statistical analysis, and regression model. In this experiment, a three-level-three-factor BBD was employed to establish the best combination of extraction variables to produce saponin-rich extract from *Hedera nepalensis*. Volume of used solvent, temperature, and dynamic extraction time were optimized by means of an experimental design. Initial experiments (not reported here) were performed to choose the experimental area for each parameter. The temperature range was confined to the 200 °C maximum system operating temperature. The range of the selected factors is reported, and three independent variables, namely volume (A), temperature (B), and time (C), were chosen, as shown in Table 1. The particle size was kept below 1.4 mm to make it easier for experimental work [23].

Table 1. Independent process variables, range and levels used for Box–Behnken design.

Independent Variable	Factors	Coded Levels		
		−1	0	+1
Volume used (mL)	A	80	100	120
Extraction temperature (°C)	B	140	170	200
Extraction time (min)	C	3	6	9

$$Y = \beta_0 + \sum_{i=1}^{3} \beta_i X_i + \sum_{i=1}^{3} \beta_{ii} X_i^2 + \sum_{i=1}^{2} \sum_{j=1}^{3} \beta_{ij} X_i X_j$$

Y is the response function (yield of saponin content (%)), β_0 is the constant, and β_i, β_{ii}, and β_{ij} are the coefficients of the linear, quadratic, and cross-product terms, respectively. Accordingly, X_i and X_j are levels of the independent variables.

The analysis of variance (ANOVA) tables were created, and the effect and regression coefficients of individual linear, quadratic, and interaction terms were defined. Then, optimal conditions were counted from the final model and verified by an actual experiment attempt.

2.5. Antimicrobial Assay

2.5.1. Preparation of Test Organism Cultures

Bacterial strains: The antibacterial effectiveness of the extracts was evaluated using six bacterial strains that cause respiratory illnesses. Three strains of Gram-positive (*Staphylococcus aureus* ATCC® 25923, *Streptococcus pneumoniae* ATCC® 49619 and *Streptococcus pyogenes* ATCC® 12344) and three strains of Gram-negative (*Haemophilus influenza* ATCC® 49247, *Klebsiella pneumoniae* sub sp. *pneumoniae* ATCC® 13883 and *Pseudomonas aeruginosa* ATCC® 27853) bacteria were obtained from The Global Bioresource Center (American Type Culture Collection).

2.5.2. Inoculums Preparation

The agar plates were incubated at 37 °C and the colonies formed on them were counted after 24 h. A tryptic soy agar plate (TSA) was used for *S. aureus*, *S. pyogenes*, *K. pneumoniae*, and *P. aeruginosa*; a blood agar plate (BAP) was used for *S. pneumoniae*; and a chocolate agar plate (CAP) was used for *H. influenzae*. Using a spectrophotometer, the bacterial growth was collected using 5 mL of sterile saline water, and the wavelength was tweaked at 580 nm and diluted such that the turbidity value of the standard of 0.5 MFU (McFarland Units) corresponded to a culture density of about 1.5×10^8 cells/mL.

2.5.3. Qualitative Antibacterial Activity by Disc Diffusion Assay

Antibacterial activity of the extract was performed using the disc diffusion method. Dimethyl sulfoxide (DMSO) was used to reconstitute the extracts to reach a concentration of 200 mg/mL (final DMSO concentration not more than 5% *v/v*). After that, 60 μL of these test samples were loaded over preformed wells (8 mm in diameter) on top of the agar medium suitable for each strain. Then, 20 μg/disc of amoxicillin (Sigma) or cefotaxime (Sigma) were used as positive controls, and were loaded into filter paper discs (6 mm in diameter). For negative control, DMSO was used at a final concentration. The plates were incubated at 37 °C for 24 h. The inhibition zone surrounding the disc was measured by a ruler and was considered to be an indication of antibacterial activity: the larger the zone of inhibition, the more potent the bioactivity.

2.5.4. Quantitative Antibacterial Activity by Minimum Inhibitory Concentration

The in vitro antibacterial activity of optimum extract of *Hedera nepalensis* was found by determination of minimum inhibitory concentration (MIC). *S. pneumoniae*, *S. pyogenes*, and *H. influenzae* were investigated. The MIC value of the extract was settled as the lowest concentration that fully inhibited bacterial growth after 24 h of incubation at 37 °C. The MIC value was determined by a two-fold serial dilution technique in suitable culture media. The MIC was tested in the concentration range of 0.078–5 mg/mL. Then, 1 μL of bacterial suspension was added and cultured for 24 h at 37 °C with a density of ~1×10^4 CFU/mL to the surface of the agar plate. The agar plates were incubated at 37 °C and detected for counts of colonies growing on agar plates after 24 h. The MIC is the lowest concentration of antimicrobial agents, which stopped the visible growth of bacteria on the agar plate. Antibiotics (amoxicillin) at 0.0312–32 μg/mL and DMSO, a dissolving solvent, were used as positive controls and negative control, respectively.

3. Results and Discussion

3.1. Optimization of Extraction Using RSM

BBD designs were used in this study to explore the influence of independent factors, such as extraction temperature (140–200 °C), extraction time (3–9 min), and volume of used solvent (80–120 mL) of the *Hedera nepalensis* extracts by ASE technique. To analyze the merged effects of these factors, experiments were executed for different patterns of the parameters using statistically designed experiments (Table 2).

Table 2. Experimental results for the response value.

Std	Run	Factor A: Volume (mL)	Factor B: Temperature (°C)	Factor C: Time (min)	Response: Yield of Saponin Contents (%)
4	1	120	200	6	0.836
5	2	80	170	3	1.244
14	3	100	170	6	1.420
11	4	100	140	9	1.423
3	5	80	200	6	0.458
2	6	120	140	6	1.466
1	7	80	140	6	1.638
8	8	120	170	9	1.437
12	9	100	200	9	0.448
10	10	100	200	3	0.672
15	11	100	170	6	1.702
13	12	100	170	6	1.621
9	13	100	140	3	1.865
6	14	120	170	3	1.860
7	15	80	170	9	1.757
16	16	100	170	6	1.757
17	17	100	170	6	1.679

Table 2 shows that saponin contents from extracts varied from 0.448% to 1.865%. By applying multiple regression analysis methods, the predicted response for the extraction yield of saponin for *Hedera nepalensis* extract can be obtained and given as the second-order polynomial equation in Equation (1):

$$Y = +1.64 - 0.06 \times A - 0.50 \times B - 0.07 \times C - 0.03 \times A^2 - 0.50 \times B^2 - 0.03 \times C^2 + 0.14 \times A \times B - 0.23 \times A \times C + 0.05 \times B \times C \tag{1}$$

where Y is the predicted response and A, B, and C are the test variables: volume of used solvent (mL), temperature (°C), and time (min), respectively. The F-test was employed to determine the statistical significance of Equation (1), and the analysis of variance (ANOVA) for the response surface quadratic model is presented in Table 3.

Table 3. ANOVA for quadratic model results.

Response 1: Yield.

Source	Sum of Squares	df	Mean Square	F-Value	p-Value	
Model	3.46	9	0.3841	17.22	0.0006	significant
A-Volume	0.0316	1	0.0316	1.41	0.2731	
B-Temperature	1.98	1	1.98	88.66	<0.0001	
C-Time	0.0414	1	0.0414	1.86	0.2154	
AB	0.0758	1	0.0758	3.40	0.1078	
AC	0.2191	1	0.2191	9.82	0.0165	
BC	0.0118	1	0.0118	0.5292	0.4906	
A^2	0.0043	1	0.0043	0.1916	0.6747	
B^2	1.07	1	1.07	48.03	0.0002	
C^2	0.0036	1	0.0036	0.1615	0.6997	

Table 3. *Cont.*

Response 1: Yield.

Source	Sum of Squares	df	Mean Square	F-Value	p-Value	
Residual	0.1561	7	0.0223			
Lack of Fit	0.0884	3	0.0295	1.74	0.2969	Not significant
Pure Error	0.0677	4	0.0169			
Cor Total	3.61	16				
	Fit Statistics					
Std. Dev.	0.1493		R^2		0.9568	
Mean	1.37		Adjusted R^2		0.9012	
C.V.%	10.90		Predicted R^2		0.5794	
			Adeq Precision		13.4584	

The regression model's analysis of variance (ANOVA) reveals that it is very significant, as evidenced by the Fisher's F-test with a very low probability value [(Pmodel > F) = 0.0006]. The determination coefficient R^2 and the multiple correlation coefficients R can be used to assess the model's quality. The R^2 of 0.9568 reasonably settled with the adjusted R^2 of 0.9012 (both > 0.8), which showed that the model had a strong connection between the experimental data and data anticipated by the model. The closer the R (multiple correlation coefficient) values are to one, the stronger the correlation between the experimental and projected values. The value of R^2 (0.9568) suggests a good relationship between the experimental and anticipated response levels. Table 3 also shows that the linear coefficients (B) and quadratic coefficients (AC; B^2) were statistically significant with p-values < 0.05. Therefore, B, AC, and B^2 were variables that affected saponin extraction efficiency. The lack-of-fit test assesses the model's inability to reflect data in the experimental domain at locations not included in the regression. The non-significant result of lack-of-fit (>0.05) indicates that the quadratic model is statistically significant for the response, and thus may be employed in additional investigations.

The 3D response surface and 2D contour line were used to characterize the impacts of independent variables and their synergy with the yield of the saponin-rich extract. In the response surface and contour plots, extraction yield was obtained along with two continuous variables while the other variable was fixed constant at its zero level (center value of the testing ranges). It was clear that extraction yield was sensitive to minor alterations of the test variables. These graphs were drawn by imposing two other variables at their zero level, which are shown in Figure 1.

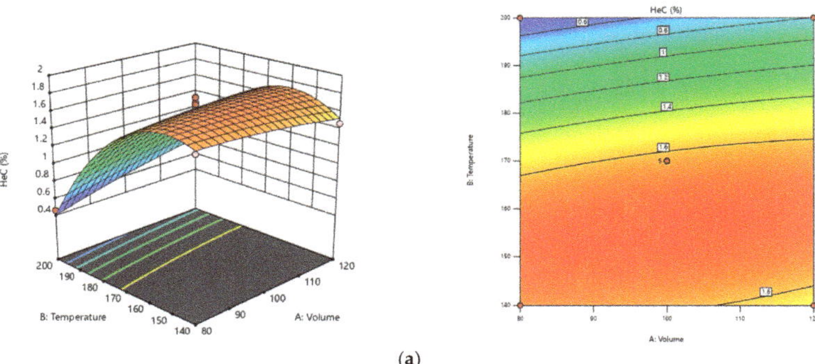

(**a**)

Figure 1. *Cont.*

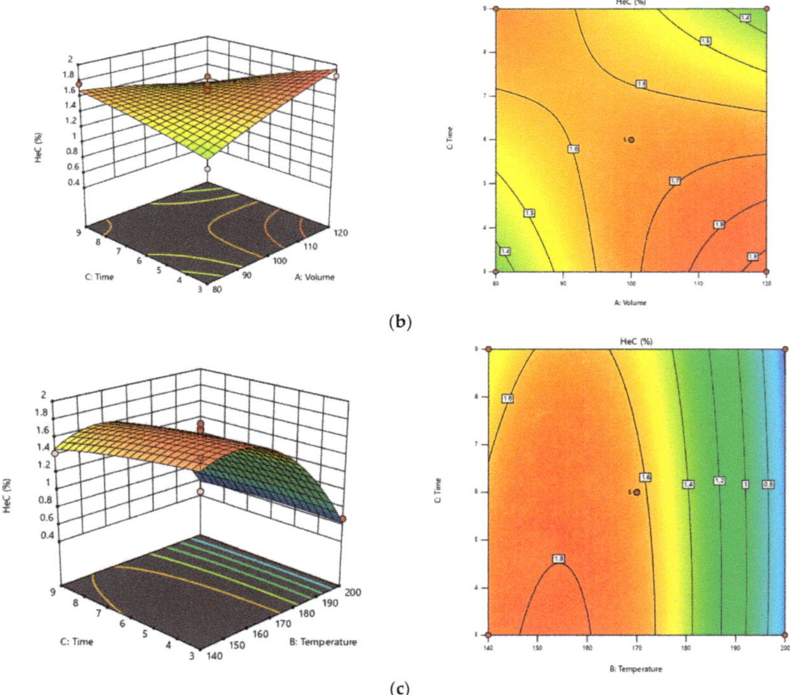

Figure 1. Response surface and contour plots for factors influencing yields of saponin extraction include: (**a**) temperature and volume; (**b**) volume and time; and (**c**) temperature and time.

The extraction yield of saponin compounds was better when the extraction temperature increased from 140 to 170 °C, but noticeably decreased when the temperature was prolonged. Because molecular mobility is accelerated with the rising temperature, these higher temperatures resulted in greater extraction efficiency. Furthermore, at higher temperatures, not only did the solvent's dissolving capability improve, so did the decrease in surface tension and solvent viscosity, which enhanced the mass transfer rate and, as a result, the availability of bioactive chemicals for extraction [24]. Although higher temperatures have a significant beneficial impact on extraction yield, they cannot be raised indefinitely. When the temperature exceeds a certain point, saponin degradation may occur in the thermal processes, reducing extraction effectiveness. This finding is consistent with earlier research demonstrating that saponin is a thermolabile substance and that high temperatures can reduce saponin extraction efficiency [25].

Figure 1b,c indicate a growing trend for saponin content when the extraction period is raised to 3 min, and a minor reduction when the duration is increased to 9 min. In this case, a longer extraction period resulted in a higher saponin percentage yield. This might be because the solute and solvent interacted with each other for a longer time. A longer contact duration enhanced mass transfer in the system. Excessive extraction time, on the other hand, is unnecessary because the solvent and sample would be in full equilibrium after a given time, based on Fick's second diffusion law. By then, the efficiency of extraction procedure would have slowed [26].

Figure 1a,b depict the interaction between extraction yield and solvent ratio. The extraction yield was observed to improve when the solvent increases. However, statistical analysis revealed that the solvent (80–120 mL) had little substantial influence on the overall saponin concentration of the extracts (Table 2). The increased concentration gradients obtained with higher solvent-to-sample ratios aid in the motion of saponins from the interior of the material to the surface and eventually into the solvent phase, allowing less saponin to remain trapped inside the sample [27].

By examining the maximum created by the X and Y coordinates, the optimal values of the variables may be determined. Responses were temperature range of 150 °C, extraction time of 3 min, and a solvent level of 110 mL. After a batch of tripled experiments, the efficiency of extraction was 1.879% ± 1.343%, which was close to the prediction of the model (1.899%). These points were found within the experimental ranges, suggesting that these analytical approaches might be utilized to find the best conditions.

It is common practice to validate the fitted model to ensure that it gives a sufficient approximation of the real system. The residuals from least squares are critical for determining the appropriateness of a model. A normality assumption check was performed by creating normal probability plots of the residuals (Figure 2a). As the residual plots resembled a straight line, the normality conditions were met. The plots of residuals vs. expected responses are shown in Figure 2b. In general, the residuals spread randomly over the display, meaning that the variances of the original observations of these figures are constant for all values of extract yield. Figure 2c depicts the relationship between the actual and anticipated values of saponin yield, proving that the model was appropriate based on the minimum residuals and the residuals' close relationship to the diagonal line. As a result, the empirical models may be used to describe the extraction yield and bioaccessibility of total saponin by the response surface.

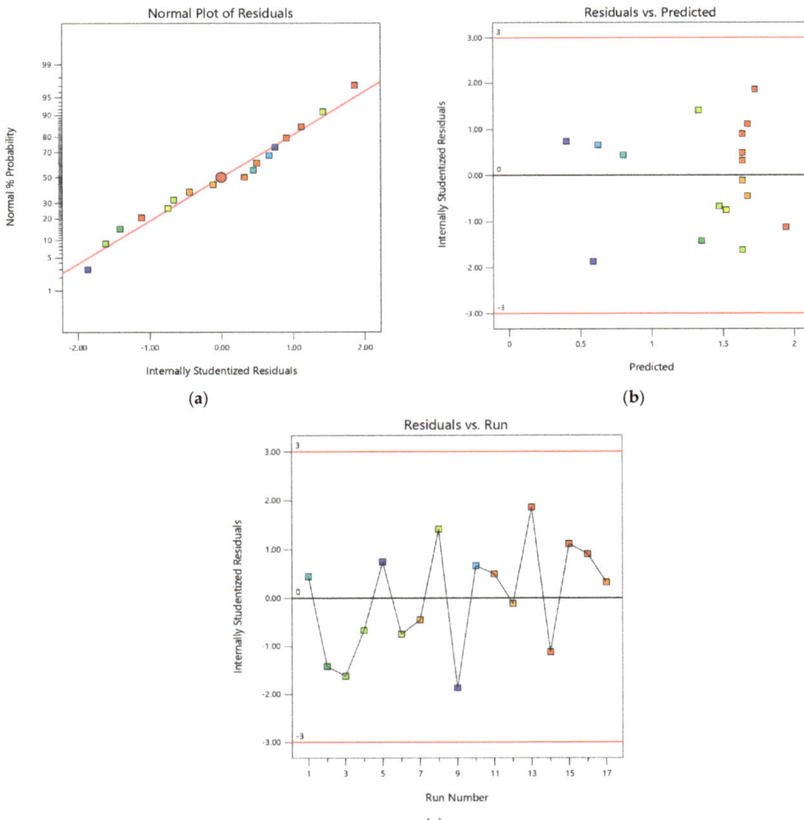

Figure 2. (**a**) The normal probability plot of residuals, (**b**) plot of internally studentized residuals versus experimental runs, and (**c**) plot of predicted and actual values.

Extraction is the first part of the process of isolating the active components from natural material and usually needs plenty of work to achieve significant success. For example, research on *Hedera nepalensis* using convenient methods such as maceration or percolation will take at least 24 h, or even days [5,28]. Thus, the process of study will take more time simply for the extraction and preparation of the sample process. Recently, with the help of modern technology such as ASE and RSM, the extraction method is considerably more rapid and can be calculated to accomplish the desired target under acceptable conditions. As in the experiments conducted above, the time needed to finish the extraction method was below ten minutes for each sample, a significant decrease in extraction times compared to conventional extraction methods reported in the literature. The results of our experiments are in agreement with the outcome of previous studies, where the production of bioactive compounds in medicinal herbs was carried out using similar techniques: the pressurized liquid method achieved a reasonable extraction efficiency in less time using less solvent [29–31]. Additionally, the natural products business is very interested in optimizing the extraction of natural plants to develop higher-quality goods. Response surface methodology (RSM) is a powerful tool for studying the interactions between variables and improving processes or products in which several variables might affect the outputs. The RSM model developed in this study enables the optimal parameters to be applied not only to laboratory tests, but it also has the potential to be applied at an industrial scale [32,33].

3.2. Antibacterial Activity

The antimicrobial activity of the extracts from *H. nepalensis* leaves was investigated by the disk diffusion method to determine the antibacterial ability of the extracts against some bacterial strains associated with respiratory diseases (Table 4). The results show that the extract of *H. nepalensis* leaves inhibited the growth of 3/6 types of bacteria, such as *S. pyogenes*, *S. pneumonia*, and *H. influenza*. Moreover, the extract inhibited the growth of three bacterial strains at a moderate to strong level with inhibition zone diameter values ranging from 12.63 to 19.50 mm.

Table 4. Antibacterial activity of the extracts from *H. nepalensis* leaves on several strains of bacteria related to respiratory diseases.

Microorganism		Zone of Inhibition (mm)			
		Control	*Hedera nepalensis* Extract	Amoxicillin	Cefotaxime
Gram (+)	*S. aureus*	-	-	29.23 ± 0.03	ND
	S. pneumoniae	-	19.33 ± 0.09	45.60 ± 0.06	ND
	S. pyogenes	-	18.60 ± 0.06	32.10 ± 0.06	ND
Gram (−)	*H. influenzae*	-	12.63 ± 0.19	15.83 ± 0.09	ND
	K. pneumoniae	-	-	-	24.10 ± 0.10
	P. aeruginosa	-	-	-	10.60 ± 0.10

Note: "-" has no antimicrobial activity; the inhibition zone diameter, including the diameter of the paper discs, is 6 mm for the antibiotics, and the agar well is 8 mm for the test samples. ND: not determined, $n = 3$, Mean ± SEM.

Furthermore, the MIC values provided evidence for the inhibitory ability at low concentrations of the extract (Table 5). The lower the MIC, the more sensitivity of bacteria react to tested extract. The MIC value of the extract was 5 mg/mL for *S. pneumoniae*, while it was greater than 5 mg/mL for the remaining strains. Overall, among the bacteria tested, the *Gram* (−) group was less sensitive than the *Gram* (+) group to the extract from *H. nepalensis* leaves. This might be related to changes in bacterial characteristics, such as cell walls, the proportion of peptidoglycan, and the type of cross-linking effect of bacterial activity [34].

Table 5. MIC values of the extracts from *H. nepalensis* leaves for several strains of bacteria related to respiratory diseases.

Microorganism		MIC	
		Hedera nepalensis **Extract (mg/mL)**	**Amoxicillin (µg/mL)**
Gram (+)	*S. pneumoniae*	5	0.0312
	S. pyogenes	>5	<0.0312
Gram (−)	*H. influenzae*	>5	2

Antibiotic resistance continues to be a concern in a number of developing and industrialized nations, presenting a substantial threat to the global health sector [35]. Due to the ineffectiveness of currently available antimicrobials for treating infectious disorders, many researchers have turned their attention to natural products as potential sources of novel bioactive chemicals [36].

In this study, the pharmacological activity of the *Hedera nepalensis* plant was confirmed by the antimicrobial activity of the extract sample that showed activity against *S. pyogenes*, *S. pneumonia*, and *H. influenza*. These results are in agreement with previous reports [1,3,37]. The antibacterial activity of the plant could be due to the presence of saponins [38,39], with active compounds such as hederacoside C in particular [40,41]. While the extract showed a significant antibacterial activity against a variety of tested bacteria, this sample only had a negligible amount of antibacterial activity against the test bacteria, as determined by their MIC values. The reason for the different performance of antibacterial activity for the sample may be related to the use of crude extracts. However, further research is necessary to determine their efficiency in suppressing the development of parasites, viruses, and/or fungus.

4. Conclusions

Accelerated solvent extraction is a sustainable and effective technique for extracting *Hedera nepalensis* leaves. The ASE followed by the RSM model is a practical method for enriching and observing saponin concentrations in *Hedera nepalensis* leaves. In the natural product extraction process, mathematical tools and models that can explain and estimate experimental data from the extraction process would be very valuable. Additionally, the current work reveals that crude extracts of *Hedera nepalensis* leaves exhibit a potential antimicrobial activity, hence providing a scientific validation of traditional techniques and supporting scientific data in favor of in vitro research.

Author Contributions: Conceptualization, N.T.H.; Formal analysis, D.T.T.L. and L.X.D.; Investigation, H.T.D. and L.H.T.; Project administration, N.M.K.; Supervision, N.T.H.; Validation, L.Q.T. and L.V.M.; Writing—original draft, H.T.D.; Writing—review and editing, L.H.T. and L.X.D. All authors have read and agreed to the published version of the manuscript.

Funding: This research received no external funding.

Conflicts of Interest: The authors declare no conflict of interest.

Abbreviations

Accelerated solvent extraction	ASE
American Type Culture Collection	ATCC
Analysis of variance	ANOVA
Blood agar plate	BAP
Box–Behnken design	BBD
Chocolate agar plate	CAP
Colony forming unit	CFU
Dimethyl sulfoxide	DMSO
Haemophilus influenza	*H. influenza*

High-performance liquid chromatography	HPLC
Klebsiella pneumoniae	*K. pneumoniae*
McFarland units	MFU
Minimum inhibitory concentration	MIC
Pseudomonas aeruginosa	*P. aeruginosa*
Response surface methodology	RSM
Standard error of mean	SEM
Staphylococcus aureus	*S. aureus*
Streptococcus pneumoniae	*S. pneumoniae*
Streptococcus pyogenes	*S. pyogenes*
Subcritical water extraction	SWE
Tryptic soy agar	TSA

References

1. Ahmad, B.; Munir, N.; Bashir, S.; Azam, S.; Khan, I.; Ayub, M. Biological screening of *Hedera nepalensis*. *J. Med. Plants Res.* **2012**, *6*, 5250–5257.
2. Hiep, N.T.; Nhung, P.T.H.; Nguyen, N.H.; Duong, H.T.; Huyen, P.T.; Thom, V.T.; Khoi, N.M.; Long, D.D. Identification, preliminary genetic, and biochemical analyses the Hedera plants which naturally distribute in Vietnam. *Pharmacogn. Res.* **2020**, *12*, 450.
3. Romman, M. *Comparative Pharmacological and Biological Evaluation of the Stem and Leaves of Hedera Nepalensis from District Malakand Khyber Pakhtunkhwa, Pakistan*; Islamia College Peshawar: Khyber Pakhtunkhwa, Pakistan, 2016.
4. Bibi, Y.; Naeem, J.; Zahara, K.; Arshad, M.; Qayyum, A. In Vitro antimicrobial assessment of selected plant extracts from pakistan. *Iran. J. Sci. Technol. Trans. A Sci.* **2018**, *42*, 267–272. [CrossRef]
5. Romman, M.; Jan, S.; Hamayun, M.; Ahmad, I.; Wali, S. In Vitro antioxidant activity of stem and leaves extracts of Hedera nepalensis K. Koch. *Int. J. Biosci. IJB* **2015**, *7*, 19–24.
6. Lutsenko, Y.; Bylka, W.; Matlawska, I.; Darmohray, R. Hedera helix as a medicinal plant. *Herba Pol.* **2010**, *56*, 83–96.
7. Zhang, Q.-W.; Lin, L.-G.; Ye, W.-C. Techniques for extraction and isolation of natural products: A comprehensive review. *Chin. Med.* **2018**, *13*, 1–26. [CrossRef]
8. Richter, B.E.; Jones, B.A.; Ezzell, J.L.; Porter, N.L.; Avdalovic, N.; Pohl, C. Accelerated solvent extraction: A technique for sample preparation. *Anal. Chem.* **1996**, *68*, 1033–1039. [CrossRef]
9. Sun, H.; Ge, X.; Lv, Y.; Wang, A. Application of accelerated solvent extraction in the analysis of organic contaminants, bioactive and nutritional compounds in food and feed. *J. Chromatogr. A* **2012**, *1237*, 1–23. [CrossRef]
10. Mottaleb, M.A.; Sarker, S.D. Accelerated solvent extraction for natural products isolation. In *Natural Products Isolation*; Springer: Berlin/Heidelberg, Germany, 2012; pp. 75–87.
11. Brachet, A.; Rudaz, S.; Mateus, L.; Christen, P.; Veuthey, J.L. Optimisation of accelerated solvent extraction of cocaine and benzoylecgonine from coca leaves. *J. Sep. Sci.* **2001**, *24*, 865–873. [CrossRef]
12. Hiep, N.T.; Duong, H.T.; Anh, D.T.; Hoai Nguyen, N.; Thai, D.Q.; Linh, D.T.T.; Anh, V.T.H.; Khoi, N.M. Subcritical Water Extraction of Epigallocatechin Gallate from Camellia sinensis and Optimization Study Using Response Surface Methodology. *Processes* **2020**, *8*, 1028. [CrossRef]
13. Lee, H.K.; Koh, H.L.; Ong, E.S.; Woo, S.O. Determination of ginsenosides in medicinal plants and health supplements by pressurized liquid extraction (PLE) with reversed phase high performance liquid chromatography. *J. Sep. Sci.* **2002**, *25*, 160–166. [CrossRef]
14. Ong, E.S.; Len, S.M. Pressurized hot water extraction of berberine, baicalein and glycyrrhizin in medicinal plants. *Anal. Chim. Acta* **2003**, *482*, 81–89. [CrossRef]
15. Güçlü-Üstündağ, Ö.; Balsevich, J.; Mazza, G. Pressurized low polarity water extraction of saponins from cow cockle seed. *J. Food Eng.* **2007**, *80*, 619–630. [CrossRef]
16. Gil-Ramirez, A.; Salas-Veizaga, D.M.; Grey, C.; Karlsson, E.N.; Rodriguez-Meizoso, I.; Linares-Pastén, J.A. Integrated process for sequential extraction of saponins, xylan and cellulose from quinoa stalks (*Chenopodium quinoa* Willd.). *Ind. Crops Prod.* **2018**, *121*, 54–65. [CrossRef]
17. Elisha, I.L.; Botha, F.S.; McGaw, L.J.; Eloff, J.N. The antibacterial activity of extracts of nine plant species with good activity against *Escherichia coli* against five other bacteria and cytotoxicity of extracts. *BMC Complement. Altern. Med.* **2017**, *17*, 1–10. [CrossRef] [PubMed]
18. Andersson, D.I.; Hughes, D. Persistence of antibiotic resistance in bacterial populations. *FEMS Microbiol. Rev.* **2011**, *35*, 901–911. [CrossRef]
19. Riswanto, F.D.O.; Rohman, A.; Pramono, S.; Martono, S. Application of response surface methodology as mathematical and statistical tools in natural product research. *J. Appl. Pharm. Sci* **2019**, *9*, 125–133.
20. Do Thi, T.L.; Hoang, T.D.; Nguyen, T.H.; Pham, T.H.; Nguyen, M.K.; Dinh, D.L. Simultaneous Quantification of Hederacoside C and α-hederin in Hedera Nepalensis K. Koch Using HPLC-UV. *VNU J. Sci. Med. Pharm. Sci.* **2020**, *36*, 17–23.
21. Venter, G. *Non-Dimensional Response Surfaces for Structural Optimization with Uncertainty*; University of Florida: Gainesville, FL, USA, 1998.

22. Kumar, A.; Prasad, B.; Mishra, I. Process parametric study for ethene carboxylic acid removal onto powder activated carbon using Box-Behnken design. *Chem. Eng. Technol. Ind. Chem. Plant Equip. Process Eng. Biotechnol.* **2007**, *30*, 932–937. [CrossRef]

23. Mokhtar, W.N.A.W.; Bakar, W.A.W.A.; Ali, R.; Kadir, A.A.A. Deep desulfurization of model diesel by extraction with N, N-dimethylformamide: Optimization by Box–Behnken design. *J. Taiwan Inst. Chem. Eng.* **2014**, *45*, 1542–1548. [CrossRef]

24. Chen, Y.; Xie, M.-Y.; Gong, X.-F. Microwave-assisted extraction used for the isolation of total triterpenoid saponins from *Ganoderma atrum. J. Food Eng.* **2007**, *81*, 162–170. [CrossRef]

25. Shi, J.; Arunasalam, K.; Yeung, D.; Kakuda, Y.; Mittal, G.; Jiang, Y. Saponins from edible legumes: Chemistry, processing, and health benefits. *J. Med. Food* **2004**, *7*, 67–78. [CrossRef] [PubMed]

26. Tan, M.; Tan, C.; Ho, C. Effects of extraction solvent system, time and temperature on total phenolic content of henna (*Lawsonia inermis*) stems. *Int. Food Res. J.* **2013**, *20*, 3117.

27. He, J.; Wu, Z.Y.; Zhang, S.; Zhou, Y.; Zhao, F.; Peng, Z.Q.; Hu, Z.W. Optimization of microwave-assisted extraction of tea saponin and its application on cleaning of historic silks. *J. Surfactants Deterg.* **2014**, *17*, 919–928. [CrossRef]

28. Ullah, S.; Abbasi, M.; Raza, M.; Khan, S.; Muhammad, B.; Rehman, A.; Mughal, M. Antibacterial activity of some selected plants of Swat valley. *Biosci. Res.* **2011**, *8*, 15–18.

29. Li, W.; Zhao, L.-C.; Sun, Y.-S.; Lei, F.-J.; Wang, Z.; Gui, X.-B.; Wang, H. Optimization of pressurized liquid extraction of three major acetophenones from *Cynanchum bungei* using a box-behnken design. *Int. J. Mol. Sci.* **2012**, *13*, 14533–14544. [CrossRef]

30. Choi, M.P.; Chan, K.K.; Leung, H.W.; Huie, C.W. Pressurized liquid extraction of active ingredients (ginsenosides) from medicinal plants using non-ionic surfactant solutions. *J. Chromatogr. A* **2003**, *983*, 153–162. [CrossRef]

31. Kim, S.H.; Kim, H.K.; Yang, E.S.; Lee, K.Y.; Du Kim, S.; Kim, Y.C.; Sung, S.H. Optimization of pressurized liquid extraction for spicatoside A in *Liriope platyphylla. Sep. Purif. Technol.* **2010**, *71*, 168–172. [CrossRef]

32. Alzorqi, I.; Singh, A.; Manickam, S.; Al-Qrimli, H.F. Optimization of ultrasound assisted extraction (UAE) of β-D-glucan polysaccharides from *Ganoderma lucidum* for prospective scale-up. *Resour. Effic. Technol.* **2017**, *3*, 46–54. [CrossRef]

33. Chan, C.-H.; See, T.-Y.; Yusoff, R.; Ngoh, G.-C.; Kow, K.-W. Extraction of bioactives from *Orthosiphon stamineus* using microwave and ultrasound-assisted techniques: Process optimization and scale up. *Food Chem.* **2017**, *221*, 1382–1387. [CrossRef]

34. Green, D.W. The bacterial cell wall as a source of antibacterial targets. *Expert Opin. Ther. Targets* **2002**, *6*, 1–20. [CrossRef] [PubMed]

35. Ovais, M.; Zia, N.; Khalil, A.T.; Ayaz, M.; Khalil, A.; Ahmad, I. Nanoantibiotics: Recent developments and future prospects. *Front. Clin. Drug Res. Anti. Infect.* **2019**, *5*, 158–174.

36. Manandhar, S.; Luitel, S.; Dahal, R.K. In Vitro antimicrobial activity of some medicinal plants against human pathogenic bacteria. *J. Trop. Med.* **2019**, *5*, 158. [CrossRef] [PubMed]

37. Uddin, G.; Khan, A.A.; Alamzeb, M.; Ali, S.; Alam, M.; Rauf, A.; Ullah, W. Biological screening of ethyl acetate extract of Hedera nepalensis stem. *Afr. J. Pharm. Pharmacol.* **2012**, *6*, 2934–2937. [CrossRef]

38. Kavya, N.; Adil, L.; Senthilkumar, P. A Review on Saponin Biosynthesis and its Transcriptomic Resources in Medicinal Plants. *Plant Mol. Biol. Report.* **2021**, *39*, 833–840. [CrossRef]

39. Fialová, S.B.; Rendeková, K.; Mučaji, P.; Nagy, M.; Slobodníková, L. Antibacterial Activity of Medicinal Plants and Their Constituents in the Context of Skin and Wound Infections, Considering European Legislation and Folk Medicine—A Review. *Int. J. Mol. Sci.* **2021**, *22*, 10746. [CrossRef]

40. Akhtar, M.; Shaukat, A.; Zahoor, A.; Chen, Y.; Wang, Y.; Yang, M.; Guo, M.; Deng, G. Hederacoside-C Inhibition of Staphylococcus aureus-Induced Mastitis via TLR2 & TLR4 and Their Downstream Signaling NF-κB and MAPKs Pathways In Vivo and In Vitro. *Inflammation* **2020**, *43*, 579–594.

41. Schmidt, S.; Heimesaat, M.; Fischer, A.; Bereswill, S.; Melzig, M. Saponins increase susceptibility of vancomycin-resistant enterococci to antibiotic compounds. *Eur. J. Microbiol. Immunol.* **2014**, *4*, 204–212. [CrossRef]

Article

Production of Oil and Phenolic-Rich Extracts from *Mauritia flexuosa* L.f. Using Sequential Supercritical and Conventional Solvent Extraction: Experimental and Economic Evaluation [†]

Ivan Best [1,2,*], Zaina Cartagena-Gonzales [1], Oscar Arana-Copa [1], Luis Olivera-Montenegro [1] and Giovani Zabot [3]

[1] Grupo de Ciencia, Tecnología e Innovación en Alimentos, Universidad San Ignacio de Loyola, Lima 15024, Peru; zaina.cartagena@usil.pe (Z.C.-G.); oscar.arana@usil.pe (O.A.-C.); lolivera@usil.edu.pe (L.O.-M.)

[2] Instituto de Ciencias de los Alimentos y Nutrición (ICAN-USIL), Universidad San Ignacio de Loyola, Lima 15024, Peru

[3] Laboratory of Agroindustrial Processes Engineering (LAPE), Federal University of Santa Maria (UFSM), Cachoeira do Sul 97105-900, Brazil; giovani.zabot@ufsm.br

* Correspondence: ibest@usil.edu.pe; Tel.: +51-1-3171000

[†] This paper is an extended version of paper published in the international conference: The 2nd International Electronic Conference on Foods 2021-Future Foods and Food Technologies for a Sustainable World, sciforum-048831, https://sciforum.net/paper/view/10988.

Citation: Best, I.; Cartagena-Gonzales, Z.; Arana-Copa, O.; Olivera-Montenegro, L.; Zabot, G. Production of Oil and Phenolic-Rich Extracts from *Mauritia flexuosa* L.f. Using Sequential Supercritical and Conventional Solvent Extraction: Experimental and Economic Evaluation. *Processes* **2022**, *10*, 459. https://doi.org/10.3390/pr10030459

Academic Editors: Maria Angela A. Meireles, Ádina L. Santana and Grazielle Nathia Neves

Received: 22 January 2022
Accepted: 18 February 2022
Published: 24 February 2022

Publisher's Note: MDPI stays neutral with regard to jurisdictional claims in published maps and institutional affiliations.

Abstract: *Mauritia flexuosa* L.f. is a palm from the Amazon. Pulp and oil are extracted from its fruits, with a high content of bioactive compounds. This study presents the economic evaluation of two extraction processes: (a) Conventional solvent extraction (CSE) with 80% ethanol for the recovery of phenolic-rich extracts; and (b) Supercritical fluid extraction (SFE) followed by CSE to obtain oil and phenolic-rich extracts. The objective of this study was to compare the feasibility of both extraction processes. The economic evaluation and the sensitivity study were evaluated using the SuperPro Designer 9.0® software at an extraction volume of 2000 L. Similar global extraction yields were obtained for both processes; however, 8.4 and 2.4 times more total polyphenol and flavonoid content were extracted, respectively, using SFE+CSE. Cost of manufacturing (COM) was higher in SFE+CSE compared to CSE, USD 193.38/kg and USD 126.47/kg, respectively; however, in the first process, two by-products were obtained. The sensitivity study showed that the cost of the raw material was the factor that had the highest impact on COM in both extraction processes. SFE+CSE was the most economically viable process for obtaining bioactive compounds on an industrial scale from *M. flexuosa* L.f.

Keywords: *Mauritia flexuosa* L.f.; conventional solvent extraction; supercritical fluid extraction; phenolic compounds; economic analysis

1. Introduction

Mauritia flexuosa L.f. is a palm from the South American Amazon and it is distributed in Peru, Bolivia, Brazil, Colombia, Ecuador, Venezuela and Guyana [1]. The fruit of *M. flexuosa* is considered a functional food due to its high content of phenolic compounds, carotenoids, essential fatty acids, vitamin E (tocopherols) and dietary fiber [2–4]. Moreover, from the pulp, 20–30% (wt.) of oil can be extracted [5], which contains 89.81% and 10.19% of unsaturated and saturated fatty acids, respectively, as well as a high content of β-carotene (911.4 mg/kg) and tocopherol (800 mg/kg) [6,7]. Oleic acid, a monounsaturated fatty acid, is the most abundant (89.81%) compound in the oil, followed by palmitic acid and linoleic acid [7,8].

The phenolic compounds extracted from *M. flexuosa* have anti-inflammatory, antioxidant, and antimicrobial properties [1–4,9], important for the prevention of chronic or non-chronic diseases, which have great potential in the food, pharmaceutical and cosmetic industries for the development of new products as colorants, flavorings, additives, antimicrobials and antioxidants [10]. On the other hand, *M. flexuosa* oil could be used in the cosmetic industry for the treatment of skin and hair due to its high content of carotenoids and vitamin C, which are related to its high antioxidant activity [11]. Interestingly, *M. flexuosa* oil also has antimicrobial activity [8,12] with potential application in the food industry.

Currently, the COVID-19 pandemic has increased the size of the global nutraceutical market, corresponding to $ 417.66 billion in 2020, which is projected to grow at a compound annual growth rate (CAGR) of 8.9% from 2020 to 2028. Within this market, the segments that have shown the highest growth are dietary supplements and functional foods [13]. *M. flexuosa* oil and its bioactive compounds also have a high nutraceutical value. It is estimated that there are approximately 4000 vegetable species from which oil can be extracted [14]. There is not much information regarding the market value of oil extracted from tropical fruits such as *M. flexuosa*. However, omega-6 and omega-9 fatty acids obtained from the hydrolysis *of M. flexuosa* oil represent important products with high added value in the cosmetic, food and pharmaceutical industries [7].

For the recovery of phenolic-rich extracts from the pulp of *M. flexuosa* and fat-soluble compounds from the oil of this fruit, solid–liquid extraction [2–4] and supercritical adsorption in columns packed with γ-alumina [5] were used, respectively. In some of these processes, large volumes of petroleum-derived solvents are required, as well as a long extraction time, which could reduce the quality of the bioactive compounds obtained [15]. One strategy for reducing costs without affecting the quality of these products is the intensification of processes that allow efficient use of energy and capital, improving the techno-economic parameters [16].

Using the concept of biorefinery and through the sequential integration of green extraction processes, the yield and recovery of bioactive compounds can be increased on an industrial scale, to be used as functional foods, as well as in the food, pharmaceutical and cosmetic industry [17]. However, before starting up an industrial-scale biorefinery, it is important to know all the production costs that would be associated with its implementation.

A previous study based on the fruits of *M. flexuosa* shows that the sequential use of a supercritical and a conventional solvent extraction, compared to conventional solvent extraction alone, makes it possible to obtain two by-products with high nutraceutical and commercial value: oil and phenolic-rich extracts [18]. Therefore, the objective of this study was to carry out an economic evaluation and sensitivity study of the by-products generated through the two extraction processes: a single-stage process with conventional solvent extraction and a two-stage sequential process using supercritical and conventional solvent extraction, at an extraction volume of 2000 L.

2. Materials and Methods

2.1. Sample Preparation

Fruits of *M. flexuosa* of the "Shambo" morphotype, acquired in October 2018 in the "Veinte de Enero" Community of the Marañon River, Iquitos Region, Peru (latitude: 4°39′19.5″ S, longitude: 73°49′27.9″ W), were used in this study. The fruits were selected from their sanity and ripening stage, and washed in water containing 25 ppm of sodium hypochlorite. Then, the pulp was obtained, which was lyophilized for subsequent assays as previously described [3].

2.2. Single-Stage Process and Two-Stage Sequential Process

A supercritical CO_2 extraction equipment (Top Industrie, Vaux-le-Pénil, France) was used to obtain oil from *M. flexuosa* on a laboratory scale. The optimized conditions using SFE to maximize oil extraction were: pressure, 2×10^7 Pa; extraction temperature, 42 °C and CO_2 flow rate, 42 g CO_2/min. For each extraction, the 50 mL extraction vessel was filled with approximately 50 g of lyophilized *M. flexuosa* pulp.

Phenol-rich extracts under previously optimized conditions were obtained from SFE defatted pulp or freeze-dried pulp of *M. flexuosa* on a laboratory scale, as previously described [3]. During this study, two extraction processes were evaluated: (a) Single-stage process by conventional solvent extraction (CSE) for obtaining phenolic-rich extracts and (b) Two-stage sequential process using supercritical and conventional solvent extraction (SFE+CSE) for the recovery of oil and phenolic-rich extracts.

The global extraction yields (GEY) for both extraction processes were calculated as the ratio between the total mass of extract and the mass of raw material loaded in the extractor on a dry weight (dw) basis [19].

2.3. Total Phenolic and Flavonoid Content

Total phenolics were extracted using a modified Folin–Ciocalteau method as described in Best et al. [3]. Briefly, 750 µL of 0.2 N Folin–Ciocalteau reagent was added to 100 µL of extract and allowed to react for 5 min. Then, 750 µL of a 7.5% sodium carbonate solution was added to the mixture and it was incubated in a water bath at 40 °C for 30 min. After this time, the absorbance at 725 nm was recorded. Total phenolics were expressed in µg of gallic acid equivalents per g of sample (µg GAE/g).

Total flavonoids were measured using the aluminum chloride colorimetric method as previously described [3]. First, 75 µL of a 5% $NaNO_2$ solution was added to 100 µL of extract and kept for 5 min at 25 °C. Then, 150 µL of a 10% $AlCl_3.6H_2O$ solution was added to the mixture and it was incubated for 5 min. Subsequently, 500 µL of 1 M NaOH was added and it was left to react for 15 min at room temperature. After this time, the absorbance at 510 nm was read. Total flavonoids were expressed as µg of catechin equivalents per g of sample (µg CE/g).

2.4. Process Simulation Model

SuperPro Designer 9.0® software was used to perform the simulations of the CSE and SFE+CSE. Direct costs (buildings, yard improvement, electrical facilities, insulation, instrumentation, installation, etc.) and indirect costs (administration rates, engineering, and construction, insurance, human resources for administration, cleaning services, etc.) were also estimated by the simulator, and both are considered in the economic evaluation.

The input parameters and simulation conditions for the single-stage process (CSE) and sequential two-stage process (SFE+CSE) are shown in Tables 1 and 2.

Table 1. Experimental data used to simulate the single-stage process by conventional solvent extraction (CSE).

Parameter	Value
Lyophilized pulp—1st step	
Lyophilized yield	36.17 g/100 g whole fruit
Lyophilization temperature	−50 °C
Time	3–4 h
Pressure	≥50 Pa
Conventional solvent extraction (80% ethanol)—2nd step	
Extraction yield	87.3 g ground and lyophilized pulp/100 g lyophilized pulp
Temperature	30 °C
Time	1 h
S/F	10 m ethanol/1 g lyophilized pulp
Lyophilized extract—3rd step	
Lyophilized yield	85.44 g lyophilized extract/100 g lyophilized pulp
Lyophilization temperature	−50 °C
Time	3–4 h
Pressure	≥50 Pa

S/F: mass ratio of solvent to feed.

Table 2. Experimental data used to simulate the sequential two-stage process using supercritical and conventional solvent extraction (SFE+CSE).

Parameter	Value
Lyophilized pulp—1st step	
Lyophilized yield	36.17 g/100 g whole fruit
Lyophilization temperature	$-50\,^\circ$C
Time	3–4 h
Pressure	$\geq$50 Pa
Oil extract—2nd step	
Extraction yield	38.85 g oil/100 g lyophilized pulp
Temperature	80 $^\circ$C
Extraction time	1 h
Pressure	2×10^7 Pa
CO_2 flow rate	42 g CO_2/min
Lyophilized extract–3rd step	
Lyophilized yield	47.418 g lyophilized extract/100 g lyophilized pulp
Lyophilization temperature	$-50\,^\circ$C
Extraction time	3–4 h
Pressure	$\geq$50 Pa

In Figure 1, the flowsheets of the CSE and SFE+CSE are shown. During the scale-up process, it was observed that the yield obtained on an industrial scale can increase the extraction yield compared to the laboratory scale, under the same processing conditions for each technology (pressure, temperature, extraction time, density) [20].

Both extraction processes include a wash tank (P-0/WSH-101), a disinfection tank (P-02/WSH-103), a rinse tank (P-03/WSH-104), a maturing kettle (P-04/V-104), a bleaching kettle (P-05/V-105), a pulping machine (P-06/SR-101), a packaging machine (P-07/SL-101), a freezer (P-08/FT-101), a lyophilizer (P-09/V-103), a plate filter (P-10/GR-101), and a sieving machine (P-11/VSCR-101). In the CSE, the lyophilizate was placed in an extraction tank (P-12/MSX-101), then it was centrifuged (P-13/DS-101, filtered by plates (P-14/NFD-10), evaporated (P-15/EV-101), packed (P-16/SL-102), frozen (P-17/FT-102), and lyophilized (P-18/V-101) until obtaining phenolic-rich extracts.

On the other hand, in the SFE+CSE, the lyophilizate was placed in the supercritical CO_2 equipment (P-12/V-102, P-13/V-106, P-14/G-101, P-15/MX-101, P-16/EC-101) for oil separation, then the aqueous phase was pumped (P-17/PM-101) into the extraction tank (P-18/MSX-101), then it was centrifuged (P-19/DS-101), plate filtered (P-20/NFD-101), evaporated (P-21/EV-101), packed (P-22/SL-102), frozen (P-23/FT-102), and lyophilized (P-24/V-101) to obtain the second by-product of this process: phenolic-rich extracts.

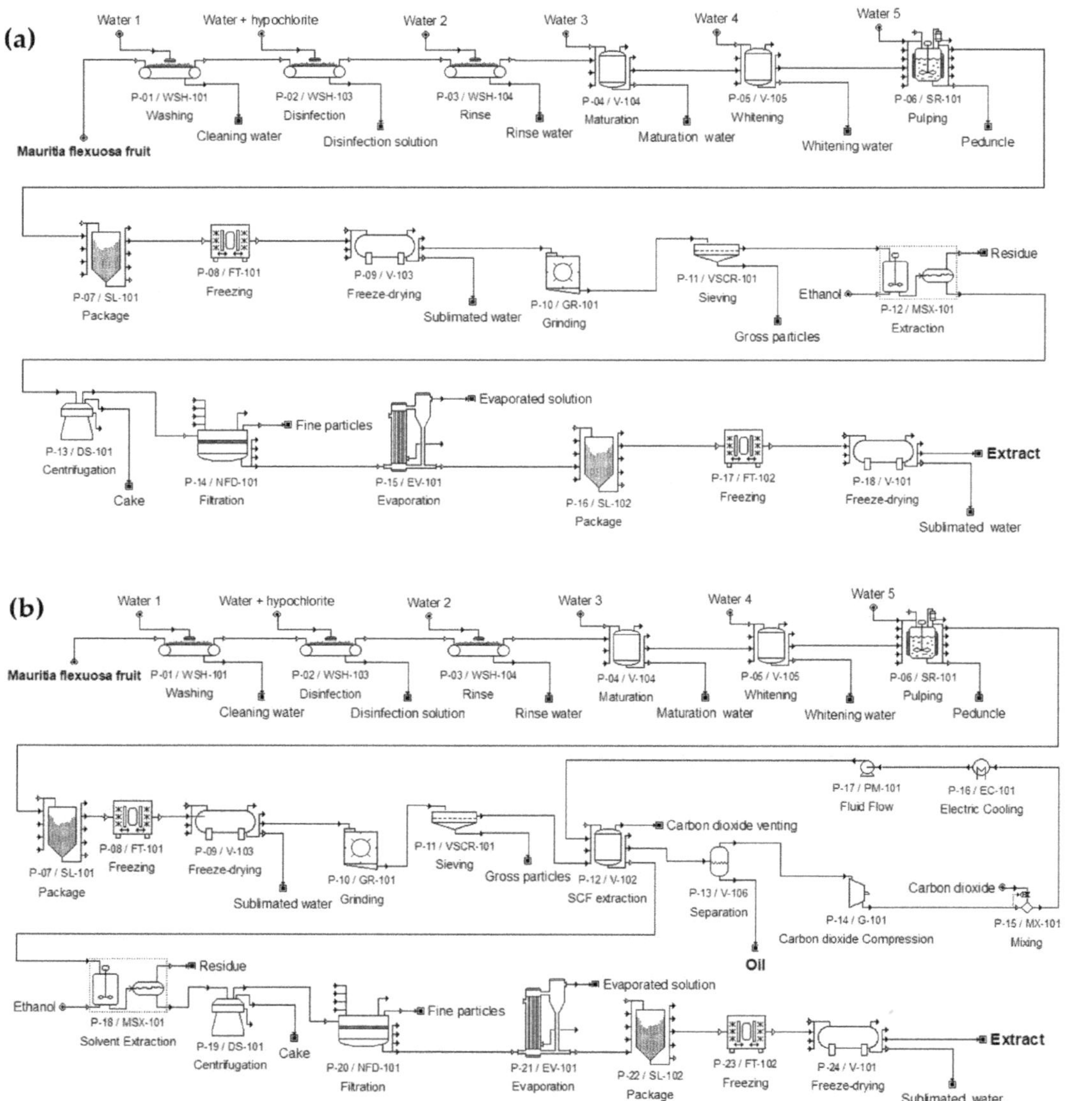

Figure 1. Flowsheets of the (**a**) single-stage process by conventional solvent extraction (CSE) and (**b**) sequential two-stage process using supercritical and conventional solvent extraction (SFE+CSE), designed using SuperPro Designer 9.0® software. Source: Ref. [18], reproduced with permission from Best et al., The 2nd International Electronic Conference on Foods 2021-Future Foods and Food Technologies for a Sustainable World, sciforum-048831; published by MDPI, 2021.

2.5. Economic Evaluation

The cost of the extraction plants for the CSE and SFE+CSE was calculated using past quotes from vendors and previous reports [21]. In some cases, the quotes and detailed specifications of the equipment were of different capacities than those required [22]. Equation

(1) was used to obtain the cost of each large-scale equipment based on the quote obtained for small-scale equipment.

$$C_1 = C_2 \left(\frac{Q_1}{Q_2}\right)^n \tag{1}$$

where C_1 is the cost of the equipment with capacity Q_1, C_2 is the known base cost for equipment with capacity Q_2, and n is a constant depending on the equipment type. The values of n were collected from the literature [23–26]. The cost of the supercritical fluid equipment was calculated according to [27]. Unit base cost and n values used for the extraction plants for the CSE and SFE+CSE are shown in Table 3.

Table 3. Base costs for each equipment composing the extraction plants.

Equipment	N [a]	Unit Base Cost (USD)	CSE Plant		SFE+CSE Plant	
			Number of Equipment	Total Base Cost (USD)	Number of Equipment	Total Base Cost (USD)
Sorting machine [b]	0.89	3900.00	1	1,824,167.05	1	1,824,167.05
Immersion washer (washed) [b]	0.53	3937.74	2	306,391.73	2	306,391.73
Immersion washer (rinsed) [b]	0.53	3937.74	2	306,391.73	2	306,391.73
Rinse tank [b]	0.53	4000.00	4	622,472.23	4	622,472.23
Water boiler (maturation) [b]	0.59	2500.00	4	588,843.66	4	588,843.66
Water boiler (bleached) [b]	0.59	2500.00	4	588,843.66	4	588,843.66
Automatic pulper [b]	0.60	1895.73	5	598,062.38	5	598,062.38
Automatic packaging machine [b]	0.60	1650.00	1	104,107.96	1	104,107.96
Freezing tunnel [b]	0.63	2500.00	1	194,061.78	1	194,061.78
Lyophilizer [b]	0.65	20,000.00	1	1,782,501,88	1	1,782,501,88
Roller mill [b]	0.91	5700.00	1	3,061,081.24	1	3,061,081.24
Industrial sieve [b]	0.91	1700.00	1	912,954.05	1	912,954.05
Extraction tank [b]	0.82	2000.00	1	576,806.30		
		1500.00			1	432,604.73
Supercritical CO$_2$ equipment [c]	0.60	2,520,106.41	-	-	1	159,007,964.56
Centrifuge [b]	0.71	7000.00	7	5,665,644.11	4	3,237,510.92
Plate filter [b]	0.66	1500.00	1	143,248.89	-	-
		700.00	-	-	1	66,849.48
Evaporator [b]	0.59	10,000.00	6	706,612.39	3	353,306.19
Lyophilizer [b]	0.65	15,000.00	1	1,336,876.41	-	-
		10,000.00	-	-	1	891,250.94
Conveyor belts [b]	0.89	769.00	24	8,632,519.77	24	8,632,519.77
Centrifugal pump [b]	0.55	900.00	5	201,007.62	5	201,007.62
Total		-	-	28,152,594.81	-	183,712,893.54

[a] n constant depending on equipment type based on references [23–26]. [b] Direct quotation. [c] Calculated based on [27].

For both extraction processes, the scale-up was carried out for a vessel with a volume of 2000 L. To perform the simulations, process operation of three daily shifts for 330 days per year was considered, corresponding an annual operation for 7920 h. For each batch, two tons of *M. flexuosa* were processed in both CSE and SFE+CSE.

The cost of raw material (*M. flexuosa*) was quoted as USD 15.63/kg (direct quotation of wholesale market, Lima, Peru in 2021). The commercialization of phenolic-rich extracts obtained by the CSE and SFE+CSE was estimated at USD 100.00/kg and USD 180.00/kg, respectively. The commercialization of oil was estimated at USD 314.47/L. The other input information is shown in Table 4.

Table 4. Input economic parameters used in SuperPro Designer 9.0® software.

Pameter	Value
Fixed Capital Investment (FCI)	
CSE plant [a]	USD 28,152,594.00
SFE+CSE plant [a]	USD 183,712,894,00
Depreciation rate [b]	10%/year
Maintenance rate [b]	6%/year
Project lifetime	25 years
Inflation	4%/year
Low NPV interest	7%
Depreciation period	25 years
Loan period for equipment	12 years
Loan interest for equipment	7%/year
Loan	100%
Cost of operational labor (COL)	
Wage (with administration and benefits) [c]	USD 4.91/h
Number of workers per shift	8
Operational time	7920 h/year
Cost of Raw Material (CRM)	
Mauritia flexuosa L.f. [a]	15.63 USD/kg
Industrial CO_2 [a]	0.033 USD/kg
Ethanol 80% [a]	0.53 USD/kg
Cost of utilities (COU)	
Electricity	0.1183 USD/kW.h
Steam	12 USD/ton
Water	1.63 USD/ton

[a] Based on local quotations. [b] Calculated based on reference [23]. [c] Based on reference [28].

Experimental data obtained at fixed operating conditions were used as input for the model. The cost of manufacturing (COM) for the production of phenolic-rich extracts by the CSE, as well as the production of oil and phenolic-rich extracts by SFE+CSE, was determined as the sum of three main components: direct costs, fixed costs, and general expenses. COM was estimated according to a methodology proposed elsewhere [26], in which the three main components are estimated in terms of four major costs: fixed capital investment (FCI), cost of raw material (CRM), cost of operational labor (COL), and cost of utilities (CUT). The FCI is related to expenses involved in the implementation of the production plant. CRM considers the cost of the raw material, including the costs of the extraction solvents. COL is related to the number of operators required to perform all stages of extraction. CUT considers electricity requirements, steam, and treated water for the process.

2.6. Sensitivity Study

The simulation was carried out considering an industrial scale at an extraction volume of 2000 L. The value of COM was simulated in CSE and SFE+CSE, considering six different scenarios: (1) Normal or real value of COM; (2) Plant at 50% the cost; (3) *M. flexuosa* at 50% the cost; (4) Ethanol 50% recycled; (5) Extract lyophilized 50% more expensive; and (6) Merging scenarios 2–5.

In addition to COM, to carry out the sensitivity study, the gross margin (GM), return over the investment (ROI), payback time (PBT), internal rate of return (IRR), and net present value (NPV) at 7% interest were also simulated considering the above-mentioned selling prices of oil and phenolic-rich extracts.

2.7. Statistical Analysis

The results were expressed as mean ± standard deviation (SD), and analyzed using Statistical Package for the Social Sciences (SPSS v26.0, IBM, Chicago, IL, USA). Differences between groups were evaluated using the Mann–Whitney U test, at a significance level of $p < 0.05$.

3. Results and Discussion

3.1. Experimental Results

As shown in Table 5, in the CSE, the global extraction yield was 13.84 g extract/100 g *M. flexuosa* pulp (dry basis), while in the SFE+CSE, the global extraction yield was 44.5 g oil/100 g *M. flexuosa* pulp (dry basis) and 13.84 g extract/100 g *M. flexuosa* pulp (dry basis). These results are in agreement with previous studies showing an extraction yield of 8.04% for phenolic-rich extracts obtained from the pulp of *M. flexuosa* defatted by Sohxlet [29] and an oil extraction yield between 23.5 to 41.1 g oil/100 g *M. flexuosa* using SFE-CO_2 [30]. A study carried out in Brazil showed that the annual productivity of pulp and oil from *M. flexuosa* was 0.79 ± 0.23 t/ha and 57.5 ± 17.0 kg/ha, respectively [31]. Currently, the average cost of *M. flexuosa* oil in Peru is USD 314.47 L; however, to increase its productivity, due to its high antioxidant potential, it is necessary to extract it on an industrial scale.

Table 5. Global extraction yields, total phenolics, and total flavonoids in the extracts obtained by the single-stage process by conventional solvent extraction (CSE) and the sequential two-stage process using supercritical and conventional solvent extraction (SFE+CSE).

	CSE	SFE+CSE	Reference
Global extraction yield (%)	13.84% (extract)	44.85% (oil) 13.8% (extract)	-
Total phenolics (µg GAE/g extract)	3423.94 ± 24.93	28800.95 ± 1180.37 *	Best et al. [3]
Total flavonoids (µg CE/g extract)	165.34 ± 4.11	390.82 ± 21.11 *	Best et al. [3]

CSE: Conventional solvent extraction; SFE: Supercritical fluid extraction; GAE: Gallic acid equivalents; CE: Catechin equivalent. * Mann–Whitney's U test, $p < 0.05$.

In the SFE+CSE, the total content of polyphenols and flavonoids were 8.4- and 2.4-fold higher, respectively, compared to the CSE ($p < 0.01$). The levels of total polyphenols obtained by the CSE were similar to those reported by previous studies in methanol extracts from *M. flexuosa* pulp [9,32]. However, these levels were significantly lower compared to those found from the pulp defatted by the SFE+CSE. This last method allows to concentrate the content of phenolic compounds, and therefore to increase the activity and the market value of this phenolic-rich extracts from *M. flexuosa*. Regarding the total flavonoid levels, the content obtained by both extraction methods in the present study was lower than that reported by previous studies [9,32].

3.2. Economic Evaluation of the Extraction Processes

The total investment for CSE and SFE+CSE was USD 28,152,594.00 and USD 183,712,894.00, respectively (Table 4). These differences in the cost of the total investment are due to the use of a supercritical fluid equipment in the SFE+CSE, which makes it possible to obtain two by-products: oil and phenolic-rich extracts. For the CSE, the productivity was 731.1 tons extract/year, while for the SFE+CSE, the productivity was 335.9 tons oil/year and 405.8 tons extract/year (Figures 2 and 3).

In the sensitivity study, for both extraction processes, no variation was observed in the productivity of oil and/or phenolic-rich extracts among all the evaluated scenarios. The differences on the COM and productivities of the by-products obtained by both extraction processes were related to the input data for the simulation, extraction yields, and total investment cost of each process.

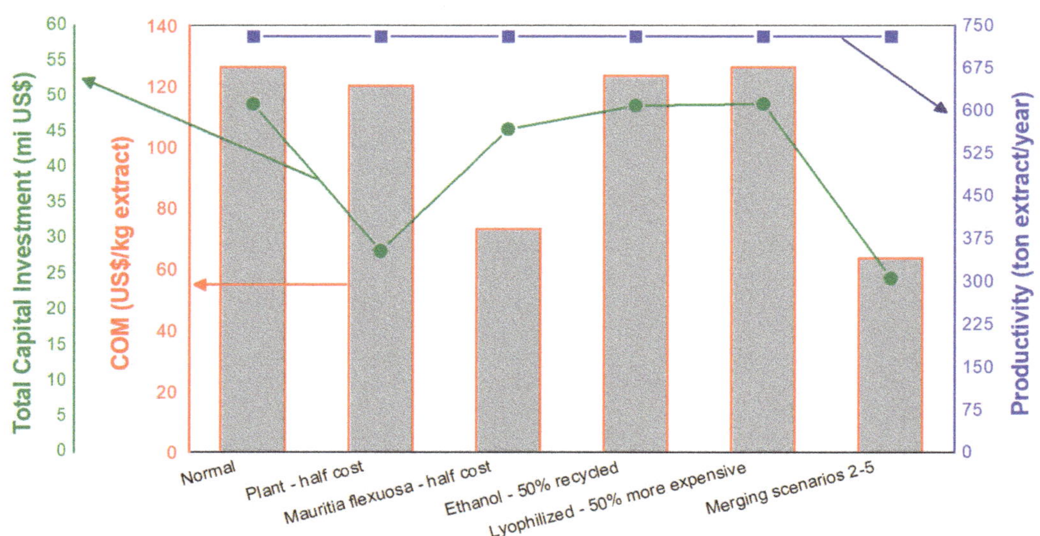

Figure 2. COM, productivity, and total capital investment to produce phenolic-rich extracts by the single-stage process using conventional solvent extraction (CSE).

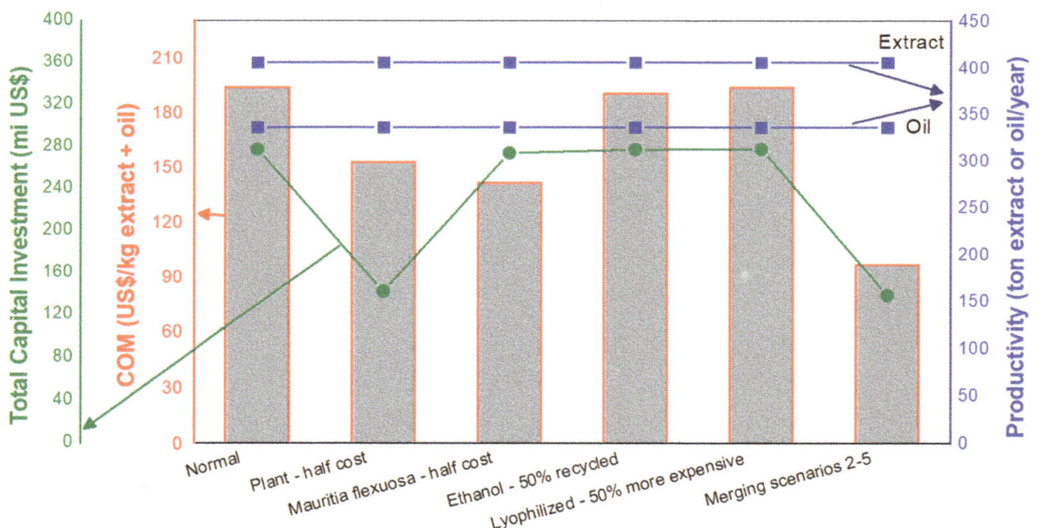

Figure 3. COM, productivity, and total capital investment to produce oil and phenolic-rich extracts by the sequential two-stage process using supercritical and conventional solvent extraction (SFE+CSE).

Figure 4 shows the contribution of the main cost factors (CRM, COL, FCI and CUT) on COM for each extraction process. For CSE and SFE+CSE, the CRM and FCI were the components that presented the highest contribution to COM. CRM corresponded to 89.22% and 57.02% of the COM in the CSE and SFE+CSE, respectively; while the FCI represented 10.10% and 42.48% of the COM in the CSE and SFE+CSE, respectively. Other costs such as CUT and COL had a lesser influence on the COM, together representing 0.69%

and 0.51% in the CSE and SFE+CSE, respectively. As the production capacity of a plant increases, the CUT and COL increase and decrease, respectively [33]. In the present study, the CUT and COL values were not very significant, since different production capacities were not evaluated.

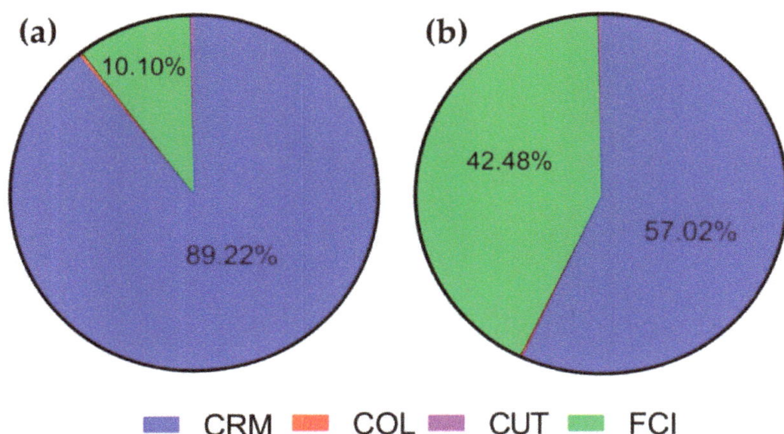

Figure 4. Contribution of each component (CRM, CUT, COL, and FCI) on the COM for bioactive compounds of *M. flexuosa* obtained by (**a**) Conventional solvent extraction (CSE) and (**b**) Supercritical fluid and conventional solvent extraction (SFE+CSE).

These results suggest that the acquisition cost of the raw material in both extraction processes exerted a strong influence on the COM. The cost of fresh *M. flexuosa* corresponded to 96.52% of the CRM, while the cost of supercritical CO_2 and ethanol represented only 3.48%, and fresh *M. flexuosa* obtained at a more affordable cost could notably decrease the COM [34]. Moreover, the FCI was the cost that had the second highest impact on the COM; however, an increase in the extraction time from 930 min to 1145 min in the CSE and SFE+CSE, respectively, decreased the contribution of the CRM and increased the impact of the FCI on the COM due to an increase in the operation time and use of supercritical CO_2 equipment, as described by Rosa et al. [35] for the extraction of clove bud oil and ginger oleoresin using supercritical fluid technology.

3.3. Sensitivity Study and Comparison between Extraction Methods

As shown in Table 6, when the CSE was used considering six different scenarios, the COM of one kg of phenolic-rich extracts ranged between USD 63.63 and USD 126.47. The main influence on the COM was the CRM, which has an impact of approximately 80–90%. This occurs because *M. flexuosa* is a biomass of great importance in the Peruvian market, as well as in the rest of South America. When it is possible to purchase this raw material at half cost, the parameters of return indicate the feasibility of the process. For example, the GM, ROI, and IRR were 26.94%, 33.46% and 42.42%, respectively. The PBT was 2.99 years with an NPV of USD 123,274,000.00. The best scenario for processing *M. flexuosa* by the CSE was achieved when merging scenarios 2–5, decreasing the value of COM by approximately two times.

Table 6. Project indices of the single-stage process by conventional solvent extraction (CSE).

Scenario	Condition	COM (USD/kg)	GM (%)	ROI (%)	PBT (Year)	IRR (%)	NPV (USD) (at 7% Interest)
1	Normal	126.47	NA	NA	NA	NA	NA
2	Plant—half cost	120.09	NA	NA	NA	NA	NA
3	*M. flexuosa*—half cost	73.06	26.94	33.46	2.99	42.42	123,274,000.00
4	Ethanol—50% recycled	123.44	NA	NA	NA	NA	NA
5	Lyophilized extract—50% more expensive	126.47	15.68	27.97	3.57	34.30	103,938,000.00
6	Merging scenarios 2–5	63.63	57.58	161.53	0.62	324.14	417,069,000.00

NA: Not applicable; COM: Cost of manufacturing; GM: Gross margin; ROI: Return on investment; PBT: Payback time; IRR: Internal rate of return after taxes; NPV: Net present value at 7%. Source: Ref. [18], reproduced with permission from Best et al., The 2nd International Electronic Conference on Foods 2021-Future Foods and Food Technologies for a Sustainable World, sciforum-048831; published by MDPI, 2021.

In the same trend, when the SFE+CSE was carried out, two by-products were obtained in each batch. COM of one kg of oil + phenolic-rich extracts ranged from USD 96.31 to USD 193.38. Overall, the values of COM were a bit higher in SFE+CSE than CSE because the SFE equipment increased the FCI cost significantly. This itemized cost contributed to approximately 50% of the total cost of these by-products. In this process, both CRM and FCI had a significant impact on the COM. Despite the values of COM being high, all scenarios for SFE+CSE presented positive returns on the initial capital and operational investment. The best scenario for processing *M. flexuosa* by SFE+CSE was achieved by merging scenarios 2–5, which decreased the COM by two times. In this scenario, the GM, ROI and IRR were 73.34%, 92.91% and 152.58%, respectively. The PBT was 1.08 years, with an NPV of USD 1,294,690,000.00 (Table 7).

Table 7. Project indices of the sequential two-stage process using supercritical and conventional solvent extraction (SFE+CSE).

Scenario	Condition	COM (USD/kg)	GM (%)	ROI (%)	PBT (Year)	IRR (%)	NPV (USD) (at 7% Interest)
1	Normal	193.38	19.73	15.54	6.43	16.48	193,979,000.00
2	Plant—half cost	152.30	36.78	35.38	2.83	45.55	416,182,000.00
3	*M. flexuosa*—half cost	140.75	41.57	24.29	4.12	28.83	457,940,000.00
4	Ethanol—50% recycled	190.01	21.13	16.10	6.21	17.27	210,877,000.00
5	Lyophilized extract—50% more expensive	193.38	46.94	35.13	2.85	45.08	800,105,000.00
6	Merging scenarios 2–5	96.31	73.34	92.91	1.08	152.58	1,294,690,000.00

COM: Cost of manufacturing; GM: Gross margin; ROI: Return on investment; PBT: Payback time; IRR: Internal rate of return after taxes; NPV: Net present value at 7%. Source: Ref. [18], reproduced with permission from Best et al., The 2nd International Electronic Conference on Foods 2021-Future Foods and Food Technologies for a Sustainable World, sciforum-048831; published by MDPI, 2021.

As shown in Tables 6 and 7, the COM calculated for the by-products of both extraction processes was lower than the sale price when scenarios 2 to 5 were merged, which suggests that both extraction processes are profitable under those conditions.

In 2021, the global market for plant extracts was USD 30.8 billion, forecast to reach USD 55.3 billion by 2026, with a CAGR of 6.0%. Within this market, phytomedicines, herbal extracts, essential oils and flavors are in greater demand. In the current scenario, due to the COVID-19 pandemic, the food and pharmaceutical industries have increased the consumption of plant extracts that enhance human immunity. However, the confinement and the increase in infections due to the emergence of new variants of SARS-CoV-2, has limited the supply and transportation of raw materials from extracts of natural products, which, added to its high demand, will increase the cost of extracts from natural products during the global spread of this virus [36].

It is estimated that the COM of natural extracts ranges from USD 3.00/kg to USD 5000.00/kg [16]. In the present study, in both extraction processes, the calculated COM value was within the range for commercial extracts. Interestingly, in both cases, the COM value was lower than the commercial value of the extracts obtained by CSE and SFE+CSE (wholesale market, Lima, Peru in 2021). Raw materials can represent up to 80% of COM when supercritical fluid extraction is used [16]. According to Osorio-Tobón et al. [37], raw materials, despite their high variability in cost, are generally the components with the highest contribution to COM. In the present study, when scenarios 2 to 5 were merged, the COM was 1.5 higher in the SFE+CSE compared to the CSE; however, using the former process, two by-products were recovered. In the latter scenario, our COM for SFE+CSE was similar to the COM obtained for the extraction of essential oil and curcuminoid-rich extracts from *Curcuma longa* L using the supercritical fluid extraction, pressurized liquid extraction, and supercritical antisolvent processes [37]. Moreover, in the present study, the COM for phenolic-rich extracts using CSE was similar or even lower than the COM for extraction of phenolic compounds from the pulp of *Euterpe edulis* [38].

GM evaluates the short-term benefits of the extraction process [21], with a higher GM indicating that the project is more feasible, because this indicator represents the percentage of every dollar of a product sold that the company will retain as gross profit [21,39]. In general, for both extraction processes, the GM decreased as the COM increased, with a higher GM being observed when scenarios 2 to 5 were merged. For CSE, the GM was positive for a COM of USD 73.06. Similar results were found by Galviz-Quezada et al. [33], who obtained positive GM values for the extraction of phenolic compounds from iraca at selling prices higher than USD 100/kg. On the other hand, for SFE+CSE, positive GM values were obtained under all of the evaluated scenarios. However, in the most favorable scenario, the GM was significantly higher than that reported for the production of essential oil and curcuminoid-rich extracts from *C. longa* L. at a raw material cost of USD 7.27/kg and USD 1.59/kg [37].

Another important parameter for evaluating the performance of extraction processes is the ROI, where the higher the value, the more attractive the project [40]. However, for a project to be feasible, a minimum value of ROI between 10% to 15% is acceptable [21]. Similar to GM, in both extraction processes, ROI decreased as COM increased, reaching its highest value in both extraction processes when scenarios 2 to 5 were merged. When comparing both processes, it was observed that in the case of the CSE, the ROI was more significantly influenced by the costs of the raw materials and the extract, while for SFE+CSE, the costs of the plant and the extract had a higher impact on ROI, due to the higher level of investment in equipment necessary to carry out this process. For SFE+CSE, an ROI > 10% was achieved for a COM of USD 193.38, indicating that the maximum sale price for the production of oil and phenolic-rich extracts could be achieved using this process.

PBT is also an important parameter in the sensitivity study, making it possible to evaluate the time until the initial investment has been paid back. It is estimated that the shorter the PBT, the faster the initial investment will be recovered; however, this depends on the type of company and the investors [33]. For small and large plants, the PBT should be between 2 to 3 years and 7 to 10 years, respectively [21]. In the case of SFE+CSE, a time between 1.08 to 6.43 years was obtained when the COM was in the range from USD 96.31 to USD 193.38, indicating the feasibility of the process under all of the evaluated scenarios. When scenarios 2 to 5 were merged, SFE+CSE had a value of PBT that was 1.74 times higher than that of CSE, due to the higher level investment in equipment in SFE+CSE. A similar behavior was observed previously [36], with a PBT value in the range from 3.25 to 4.71 years in order to obtain two by-products. Similarly, PBT values ranging from 0.60 to 3.02 years were reported for obtaining oil from Sucupira Branca *(Pterodon emarginatus)* seeds by SFE with an extractor capacity of 30 L [41].

The IRR is another parameter used to evaluate the profitability of a project; similar to the ROI, the higher the IRR, the more desirable the project [33]. Internal rate of return (IRR) is an important parameter for assessing the profitability of a process, as it accounts for factors such as plant income, capital investment, and time value of money [42]. In general, for both extraction processes, the IRR increased when the COM decreased, reaching its highest value when scenarios 2 through 5 were merged. Similar to the ROI, in the CSE, it was observed that the costs of the raw materials and the extract had the highest impact on the COM, while in the SFE+CSE, the costs of the plant and the extract had a greater influence on the value of the COM. Therefore, in the most favorable scenario, the SFE+CSE showed an IRR 2.12 times lower than that of CSE. However, the IRR value in the SFE+CSE was similar to or even higher than that reported in other studies [37,43].

Finally, NPV assesses the present value of all future cash flows generated by a project, including the initial capital investment, making it possible to establish which projects are able to generate the most profit [33]. A project can be considered feasible if the NPV is positive after generally assuming an interest rate of 7% [37]. In the CSE, only scenarios 3, 5 and 6 present a positive NPV value, which would depend on the 50% reduction in the cost of raw material and the sale of the extract at prices that were 50% more expensive. For SFE+CSE, all of the evaluated scenarios presented a positive NPV value, which indicates that all of them are feasible. Similar to IRR, for both extraction processes, when scenarios 2 to 5 were merged, the NPV reached its highest value. In the latter scenario, SFE+CSE had an NPV value 3.10 times than that of CSE. However, to perform a more adequate economic evaluation of the cost of production of oil and phenolic-rich extracts of *M. flexuosa* using the extraction methods evaluated in the current study, as previously described [34], other factors such as raw material characteristics and seasonality, market size, product demand, and costs related to product quality control, packaging, and distribution must also be considered.

4. Conclusions

In the sensitivity study, the scenario with the greatest individual impact on the economic parameters was the reduction in the cost of raw materials by 50%. In this scenario, in the CSE and SFE+CSE, the COM decreased by 1.7 and 1.4 times, respectively. However, in both extraction processes, the COM reached its lowest value when scenarios 2 to 5 were merged, decreasing the COM value approximately two times in both extraction processes. Comparing both extraction processes, SFE+CSE was the most profitable economic process, because it made it possible to obtain two value-added by-products, oil and phenolic-rich extracts, with high nutraceutical value and desirable profit potential.

Author Contributions: Conceptualization, I.B. and L.O.-M.; methodology and investigation, I.B., L.O.-M., Z.C.-G., O.A.-C. and G.Z.; validation, I.B., L.O.-M. and G.Z.; writing—original draft preparation, I.B., Z.C.-G., O.A.-C. and G.Z.; writing—review and editing, I.B., L.O.-M. and G.Z.; supervision, I.B.; funding acquisition, I.B. All authors have read and agreed to the published version of the manuscript.

Funding: This research was funded by the National Fund for Scientific, Technological Development and Technological Innovation (FONDECYT) of the National Council of Science, Technology and Technological Innovation (CONCYTEC) of Peru, Contract 007-2018-FONDECYT-BM, and Universidad San Ignacio de Loyola.

Data Availability Statement: The data presented in this study are available on request from the corresponding author.

Acknowledgments: We thank the staff of the ICAN-USIL of the Universidad San Ignacio de Loyola for their valuable assistance in obtaining and processing the *M. flexuosa* samples.

Conflicts of Interest: The authors declare no conflict of interest.

Nomenclature

CAGR	Compound annual growth rate
CE	Catechin equivalents
COL	Cost of operational labor
COM	Cost of manufacturing
COU	Cost of utilities
CRM	Cost of raw material
CSE	Conventional solvent extraction
CUT	Cost of utilities
FCI	Fixed capital investment
FSE	Fluid supercritical extraction
GAE	Gallic acid equivalents
GEY	Global extraction yield
GM	Gross margin
IRR	Internal rate of return after taxes
NPV	Net present value at 7%
PBT	Payback time
ROI	Return on investment
S/F	Mass ratio of solvent to feed

References

1. Pereira Freire, J.A.; Barros, K.B.N.T.; Lima, L.K.F.; Martins, J.M.; de Araújo, Y.C.; da Silva Oliveira, G.L.; de Souza Aquino, J.; Ferreira, P.M.P. Phytochemistry Profile, Nutritional Properties and Pharmacological Activities of *Mauritia flexuosa*. *J. Food Sci.* **2016**, *81*, R2611–R2622. [CrossRef]
2. Cândido, T.L.N.; Silva, M.R.; Agostini-Costa, T.S. Bioactive compounds and antioxidant capacity of buriti (*Mauritia flexuosa L.f.*) from the Cerrado and Amazon biomes. *Food Chem.* **2015**, *177*, 313–319. [CrossRef] [PubMed]
3. Best, I.; Casimiro-Gonzales, S.; Portugal, A.; Olivera-Montenegro, L.; Aguilar, L.; Muñoz, A.M.; Ramos-Escudero, F. Phytochemical Screening and DPPH Radical Scavenging Activity of Three Morphotypes of *Mauritia flexuosa* L.f. from Peru, and Thermal Stability of a Milk-Based Beverage Enriched with Carotenoids from These Fruits. *Heliyon* **2020**, *6*, e05209. [CrossRef] [PubMed]
4. Abreu-Naranjo, R.; Paredes-Moreta, J.G.; Granda-Albuja, G.; Iturralde, G.; González-Paramás, A.M.; Alvarez-Suarez, J.M. Bioactive Compounds, Phenolic Profile, Antioxidant Capacity and Effectiveness against Lipid Peroxidation of Cell Membranes of *Mauritia flexuosa* L. Fruit Extracts from Three Biomes in the Ecuadorian Amazon. *Heliyon* **2020**, *6*, e05211. [CrossRef] [PubMed]
5. Cunha, M.A.E.; Neves, R.F.; Souza, J.N.S.; França, L.F.; Araújo, M.E.; Brunner, G.; Machado, N.T. Supercritical Adsorption of Buriti Oil (*Mauritia flexuosa* Mart.) in γ-Alumina: A Methodology for the Enriching of Anti-Oxidants. *J. Supercrit. Fluids* **2012**, *66*, 181–191. [CrossRef]
6. De Aquino, J.S.; Soares, J.K.B.; Magnani, M.; Stamford, T.C.M.; de Mascarenhas, R.J.; Tavares, R.L.; Stamford, T.L.M. Effects of Dietary Brazilian Palm Oil (*Mauritia flexuosa* L.) on Cholesterol Profile and Vitamin A and E Status of Rats. *Molecules* **2015**, *20*, 9054–9070. [CrossRef]
7. Pérez, M.M.; Gonçalves, E.C.S.; Salgado, J.C.S.; de Rocha, M.S.; de Almeida, P.Z.; Vici, A.C.; da Infante, J.C.; Guisán, J.M.; Rocha-Martin, J.; Pessela, B.C.; et al. Production of Omegas-6 and 9 from the Hydrolysis of Açaí and Buriti Oils by Lipase Immobilized on a Hydrophobic Support. *Molecules* **2018**, *23*, E3015. [CrossRef]
8. Faustino Pereira, Y.; do Socorro Costa, M.; Relison Tintino, S.; Esmeraldo Rocha, J.; Fernandes Galvão Rodrigues, F.; de Sá Barreto Feitosa, M.K.; de Menezes, I.R.A.; Douglas Melo Coutinho, H.; da Costa, J.G.M.; de Sousa, E.O. Modulation of the Antibiotic Activity by the *Mauritia flexuosa* (Buriti) Fixed Oil against Methicillin-Resistant Staphylococcus Aureus (MRSA) and Other Multidrug-Resistant (MDR) Bacterial Strains. *Pathogens* **2018**, *7*, E98. [CrossRef]
9. Koolen, H.H.F.; da Silva, F.M.A.; Gozzo, F.C.; de Souza, A.Q.L.; de Souza, A.D.L. Antioxidant, Antimicrobial Activities and Characterization of Phenolic Compounds from Buriti (*Mauritia flexuosa* L.f.) by UPLC–ESI-MS/MS. *Food Res. Int.* **2013**, *51*, 467–473. [CrossRef]
10. Resende, L.M.; Franca, A.S.; Oliveira, L.S. Buriti (*Mauritia flexuosa* L.f.) Fruit by-Products Flours: Evaluation as Source of Dietary Fibers and Natural Antioxidants. *Food Chem.* **2019**, *270*, 53–60. [CrossRef]
11. Pereira, G.S.; Freitas, P.M.; Basso, S.L.; de Moreira Araújo, P.; Rodrigues dos Santos, R.; Conde, C.F.; Pereira, E.F.D.; Haverroth, M.; Amaral, A.M.F. Quality Control of the Buriti oil (*Mauritia flexuosa* L.f.) for Use in 3-Phase Oil Formulation for Skin Hydration. *Int. J. Phytocosmet. Nat. Ingred.* **2018**, *5*, 1. [CrossRef]
12. Nobre, C.B.; de Sousa, E.O.; de Lima Silva, J.M.F.; Melo Coutinho, H.D.; da Costa, J.G.M. Chemical Composition and Antibacterial Activity of Fixed Oils of *Mauritia flexuosa* and *Orbignya speciosa* Associated with Aminoglycosides. *Eur. J. Integr. Med.* **2018**, *23*, 84–89. [CrossRef]

13. Nutraceuticals Market Size, Share & Trends Analysis Report by Product (Dietary Supplements, Functional Food, Functional Beverages), by Region (North America, Europe, APAC, CSA, MEA), and Segment Forecasts, 2021–2030. Available online: https://www.grandviewresearch.com/industry-analysis/nutraceuticals-market (accessed on 15 January 2022).
14. Santori, G.; Di Nicola, G.; Moglie, M.; Polonara, F. A Review Analyzing the Industrial Biodiesel Production Practice Starting from Vegetable Oil Refining. *Appl. Energy* **2012**, *92*, 109–132. [CrossRef]
15. More, P.R.; Arya, S.S. Intensification of Bio-Actives Extraction from Pomegranate Peel Using Pulsed Ultrasound: Effect of Factors, Correlation, Optimization and Antioxidant Bioactivities. *Ultrason. Sonochem.* **2020**, *72*, 105423. [CrossRef]
16. Zabot, G.L.; Moraes, M.N.; Carvalho, P.I.N.; Meireles, M.A.A. New Proposal for Extracting Rosemary Compounds: Process Intensification and Economic Evaluation. *Ind. Crops Prod.* **2015**, *77*, 758–771. [CrossRef]
17. Martínez, J.; de Aguiar, A.C.; da Machado, A.P.F.; Barrales, F.M.; Viganó, J.; dos Santos, P. 2.51—Process Integration and Intensification. In *Comprehensive Foodomics*; Cifuentes, A., Ed.; Elsevier: Oxford, UK, 2021; pp. 786–807. ISBN 978-0-12-816396-2.
18. Best, I.; Olivera-Montenegro, L.; Cartagena-Gonzales, Z.; Arana-Copa, O.; Zabot, G. Techno-Economic evaluation of the production of oil and phenolic-rich extracts from *Mauritia flexuosa* L.f. using sequential supercritical and conventional solvent extraction. In Proceedings of the 2nd International Electronic Conference on Foods—"Future Foods and Food Technologies for a Sustainable World", online, 15–30 October 2021; MDPI: Basel, Switzerland, 2021.
19. Veggi, P.C.; Cavalcanti, R.N.; Meireles, M.A.A. Production of Phenolic-Rich Extracts from Brazilian Plants Using Supercritical and Subcritical Fluid Extraction: Experimental Data and Economic Evaluation. *J. Food Eng.* **2014**, *131*, 96–109. [CrossRef]
20. Farías-Campomanes, A.M.; Rostagno, M.A.; Meireles, M.A.A. Production of Polyphenol Extracts from Grape Bagasse Using Supercritical Fluids: Yield, Extract Composition and Economic Evaluation. *J. Supercrit. Fluids* **2013**, *77*, 70–78. [CrossRef]
21. Ochoa, S.; Durango-Zuleta, M.M.; Felipe Osorio-Tobón, J. Techno-Economic Evaluation of the Extraction of Anthocyanins from Purple Yam (*Dioscorea Alata*) Using Ultrasound-Assisted Extraction and Conventional Extraction Processes. *Food Bioprod. Process.* **2020**, *122*, 111–123. [CrossRef]
22. De Aguiar, A.C.; Osorio-Tobón, J.F.; Viganó, J.; Martínez, J. Economic Evaluation of Supercritical Fluid and Pressurized Liquid Extraction to Obtain Phytonutrients from Biquinho Pepper: Analysis of Single and Sequential-Stage Processes. *J. Supercrit. Fluids* **2020**, *165*, 104935. [CrossRef]
23. Peters, M.; Timmerhaus, K.; West, R.; Peters, M. *Plant Design and Economics for Chemical Engineers*; McGraw-Hill: New York, NY, USA, 2002; ISBN 978-0-07-239266-1.
24. Smith, R. *Chemical Process Design and Integration*; Wiley: Chichester, UK; Hoboken, NJ, USA, 2005; ISBN 978-0-471-48680-0.
25. Green, D.W.; Perry, R.H. *Perry's Chemical Engineers' Handbook*, 8th ed.; McGraw-Hill Education: Berkshire, UK, 2008; ISBN 978-0-07-142294-9.
26. Turton, R. *Analysis, Synthesis and Design of Chemical Processes*; Pearson Education: Upper Saddle River, NJ, USA, 2012.
27. Rocha-Uribe, J.A.; Novelo-Pérez, J.I.; Araceli Ruiz-Mercado, C. Cost Estimation for CO_2 Supercritical Extraction Systems and Manufacturing Cost for Habanero Chili. *J. Supercrit. Fluids* **2014**, *93*, 38–41. [CrossRef]
28. Lachos-Perez, D.; Buller, L.S.; Sganzerla, W.G.; Ody, L.P.; Zabot, G.L.; Forster-Carneiro, T. Sequential Hydrothermal Process for Production of Flavanones and Sugars from Orange Peel: An Economic Assessment. *Biofuels Bioprod. Bioref.* **2021**, *15*, 202–217. [CrossRef]
29. Nobre, C.B.; Sousa, E.O.; Camilo, C.J.; Machado, J.F.; Silva, J.M.F.L.; Filho, J.R.; Coutinho, H.D.M.; Costa, J.G.M. Antioxidative Effect and Phytochemical Profile of Natural Products from the Fruits of "Babaçu" (*Orbignia speciose*) and "Buriti" (*Mauritia flexuosa*). *Food Chem. Toxicol.* **2018**, *121*, 423–429. [CrossRef] [PubMed]
30. Chañi-Paucar, L.O.; Yali, E.T.; Maceda Santivañez, J.C.; Garcia, D.A.; Jonher, J.C.F.; Angela, A.; Meireles, M. Supercritical Fluid Extraction from Aguaje (*Mauritia flexuosa*) Pulp: Overall Yield, Kinetic, Fatty Acid Profile, and Qualitative Phytochemical Profile. *Open Food Sci. J.* **2021**, *13*, 1–11. [CrossRef]
31. Barbosa, R.I.; Lima, A.D.; Mourão Junior, M. Biometria de frutos do buriti (*Mauritia flexuosa* L.F.—Arecaceae): Produção de polpa e óleo em uma área de savana em Roraima. *Embrapa Amaz. Orient.-Artig. Periódico Indexado* **2010**, *5*, 71–85.
32. Pereira-Freire, J.A.; da Oliveira, G.L.S.; Lima, L.K.F.; Ramos, C.L.S.; Arcanjo-Medeiros, S.R.; de Lima, A.C.S.; Teixeira, S.A.; de Oliveira, G.A.L.; Nunes, N.M.F.; Amorim, V.R.; et al. In Vitro and Ex Vivo Chemopreventive Action of *Mauritia flexuosa* Products. *Evid. Based Complement. Alternat. Med.* **2018**, *2018*, e2051279. [CrossRef] [PubMed]
33. Galviz-Quezada, A.; Ochoa-Aristizábal, A.M.; Arias Zabala, M.E.; Ochoa, S.; Osorio-Tobón, J.F. Valorization of Iraca (*Carludovica palmata*, Ruiz & Pav.) Infructescence by Ultrasound-Assisted Extraction: An Economic Evaluation. *Food Bioprod. Process.* **2019**, *118*, 91–102. [CrossRef]
34. De Aguiar, A.C.; Osorio-Tobón, J.F.; Silva, L.P.S.; Barbero, G.F.; Martínez, J. Economic Analysis of Oleoresin Production from Malagueta Peppers (*Capsicum frutescens*) by Supercritical Fluid Extraction. *J. Supercrit. Fluids* **2018**, *133*, 86–93. [CrossRef]
35. Rosa, P.T.V.; Meireles, M.A.A. Rapid Estimation of the Manufacturing Cost of Extracts Obtained by Supercritical Fluid Extraction. *J. Food Eng.* **2005**, *67*, 235–240. [CrossRef]
36. Plant Extracts Market by Type (Phytomedicines & Herbal Extracts, Spices, Essential Oils, Flavors & fragrances), Application (Pharmaceutical & Dietary Supplements, Food & Beverages, Cosmetics), Sources, and Region—Forecast to 2026. Available online: https://www.marketsandmarkets.com/Market-Reports/plant-extracts-market-942.html (accessed on 15 January 2022).
37. Osorio-Tobón, J.F.; Carvalho, P.I.N.; Rostagno, M.A.; Meireles, M.A.A. Process Integration for Turmeric Products Extraction Using Supercritical Fluids and Pressurized Liquids: Economic Evaluation. *Food Bioprod. Process.* **2016**, *98*, 227–235. [CrossRef]

38. Vieira, G.S.; Cavalcanti, R.N.; Meireles, M.A.A.; Hubinger, M.D. Chemical and Economic Evaluation of Natural Antioxidant Extracts Obtained by Ultrasound-Assisted and Agitated Bed Extraction from Jussara Pulp (*Euterpe edulis*). *J. Food Eng.* **2013**, *119*, 196–204. [CrossRef]
39. Náthia-Neves, G.; Vardanega, R.; Meireles, M.A.A. Extraction of Natural Blue Colorant from *Genipa americana* L. Using Green Technologies: Techno-Economic Evaluation. *Food Bioprod. Process.* **2019**, *114*, 132–143. [CrossRef]
40. Vlysidis, A.; Binns, M.; Webb, C.; Theodoropoulos, C. A Techno-Economic Analysis of Biodiesel Biorefineries: Assessment of Integrated Designs for the Co-Production of Fuels and Chemicals. *Energy* **2011**, *36*, 4671–4683. [CrossRef]
41. Chañi-Paucar, L.O.; Johner, J.C.F.; Zabot, G.L.; Meireles, M.A.A. Technical and Economic Evaluation of Supercritical CO_2 Extraction of Oil from Sucupira Branca Seeds. *J. Supercrit. Fluids* **2022**, *181*, 105494. [CrossRef]
42. Kurambhatti, C.; Kumar, D.; Rausch, K.D.; Tumbleson, M.E.; Singh, V. Improving Technical and Economic Feasibility of Water Based Anthocyanin Recovery from Purple Corn Using Staged Extraction Approach. *Ind. Crops Prod.* **2020**, *158*, 112976. [CrossRef]
43. Johner, J.; Hatami, T.; Zabot, G.; Meireles, M.A. Kinetic Behavior and Economic Evaluation of Supercritical Fluid Extraction of Oil from Pequi (*Caryocar brasiliense*) for Various Grinding Times and Solvent Flow Rates. *J. Supercrit. Fluids* **2018**, *140*, 188–195. [CrossRef]

Article ·

Supercritical Technology-Based Date Sugar Powder Production: Process Modeling and Simulation

Hooralain Bushnaq [1], Rambabu Krishnamoorthy [1,2,*], Mohammad Abu-Zahra [1,3], Shadi W. Hasan [1,4], Hanifa Taher [1], Suliman Yousef Alomar [5], Naushad Ahmad [6] and Fawzi Banat [1,*]

1   Department of Chemical Engineering, Khalifa University of Science and Technology, Abu Dhabi 127788, United Arab Emirates; 100052972@ku.ac.ae (H.B.); mohammad.abuzahra@ku.ac.ae (M.A.-Z.); shadi.hasan@ku.ac.ae (S.W.H.); hanifa.alblooshi@ku.ac.ae (H.T.)
2   Tata Consultancy Services, Bletchley, Milton Keynes, Buckinghamshire MK3 5JL, UK
3   Research and Innovation Center on CO_2 and Hydrogen (RICH Center), Khalifa University of Science and Technology, Abu Dhabi 127788, United Arab Emirates
4   Center for Membranes and Advanced Water Technology (CMAT), Khalifa University of Science and Technology, Abu Dhabi 127788, United Arab Emirates
5   Zoology Department, King Saud University, Riyadh 11451, Saudi Arabia; syalomar@ksu.edu.sa
6   Chemistry Department, College of Science, King Saud University, Riyadh 11451, Saudi Arabia; anaushad@ksu.edu.sa
*   Correspondence: rambabu.krishnamoorthy@ku.ac.ae (R.K.); fawzi.banat@ku.ac.ae (F.B.)

Abstract: Date palm fruits (*Phoenix dactylifera*) contain high levels of fructose and glucose sugars. These natural sugar forms are healthy, nutritional and easily assimilate into human metabolism. The successful production of soluble date sugar powder from nutritious date fruits would result in a new food product that could replace the commercial refined sugar. In this work, a novel process technology based on the supercritical extraction of sugar components from date pulp was modeled and simulated using Aspen Plus software. The process model consisted of three main steps that were individually simulated for their optimal working conditions as follows: (a) freeze-drying of the date pulp at $-42\,°C$ and 0.0001 bar; (b) supercritical extraction of the sugar components using a 6.77 wt.% water mixed CO_2 solvent system at a pressure of 308 bar, temperature of $65\,°C$, and CO_2 flow rate of 31,000 kg/h; and (c) spray-drying of the extract using 40 wt.% Gum Arabic as the carrier agent and air as drying medium at $150\,°C$. The overall production yield of the process showed an extraction efficiency of 99.1% for the recovery of total reducing sugars from the date fruit. The solubility of the as-produced date sugar powder was improved by the process selectivity, elimination of insoluble fiber contents, and the addition of Gum Arabic. The solubility of the final date sugar product was estimated as 0.89 g/g water.

Keywords: date sugar production; freeze-drying; supercritical extraction; spray-drying; process simulation; aspen plus

Citation: Bushnaq, H.; Krishnamoorthy, R.; Abu-Zahra, M.; Hasan, S.W.; Taher, H.; Alomar, S.Y.; Ahmad, N.; Banat, F. Supercritical Technology-Based Date Sugar Powder Production: Process Modeling and Simulation. *Processes* **2022**, *10*, 257. https://doi.org/10.3390/pr10020257

Academic Editors: Maria Angela A. Meireles, Ádina L. Santana and Grazielle Nathia Neves

Received: 5 January 2022
Accepted: 18 January 2022
Published: 27 January 2022

Publisher's Note: MDPI stays neutral with regard to jurisdictional claims in published maps and institutional affiliations.

1. Introduction

Recently, the increasing awareness of the health effects of refined sugar has shifted consumers' focus towards natural and organic sugars. This has provided a tremendous scope for the global organic sugar market, which is expected to grow by 14% between 2019 and 2027 for a net worth of USD 4500 million by 2027 [1]. To meet the growing demands for organic sugar, a wide variety of plant sources and their products are being analyzed for the development of natural fruit-sugar products that benefit human health. Among them, date palm (*Phoenix dactylifera*) is a promising plant widely cultivated in the arid regions of the Persian Gulf and North Africa [2]. Date palm fruits contain high amounts of fructose and glucose sugars with very little sucrose content [3]. These natural sugar forms are simple, healthy, and assimilate easily into human metabolism. The fruit contains numerous health benefits, including nutritional, functional, and therapeutic values [4]. Date sugar is commercially produced in syrup form through a hot water extraction method. However, there are various drawbacks associated with this method, such as low extraction performance, high energy input, and degradation of thermolabile compounds [5].

Recovery of sugar components from the date fruit biomatrix is mainly hindered by the tough skin and gummy consistency of the fruit pulp. Moreover, the rigid cell membrane layers of the biomatrix retard the movement of solvent in and out of the plant cells. As a result, most of the valuable sugar components are lost with spent biomass in the conventional hot water extraction process, decreasing the extraction efficiency [6]. Thus, an innovative process technology is required to overcome these limitations and efficiently recover the sugar components from date fruits. Non-conventional techniques such as microwave-, ultrasonic-, supercritical-, enzymatic-, and magnetic-/electrical- assisted extraction are widely explored for the effective extraction of various bioactive compounds from different plants and their parts [7]. These advanced approaches are advantageous because they are less costly to operate, require a shorter treatment time, consume low amounts of energy, and do not produce any toxic byproducts.

Compared to other non-conventional extraction techniques, supercritical CO_2 (Sc-CO_2) extraction has gained increased popularity due to its sustainability, versatility, and absence of solvent contamination [8]. Due to the properties that can be readily modified with changes in pressure and temperature, supercritical fluids have a wide range of applications. Indeed, the properties of both liquid and gaseous matter are combined in the supercritical phase [9]. One of the notable benefits of supercritical solvent-based extraction is that the solvent can be fully separated from the desired solutes, thus ensuring product purity. Various studies have investigated the extraction of high-value components from fruits and vegetable tissues using supercritical fluids, specifically Sc-CO_2 [9–12]. However, usage of Sc-CO_2 for plant carbohydrates extraction is partly hampered by its low polarity, which limits the solubilization of the targeted compounds into Sc-CO_2. To overcome this problem, high-density polar solvents are used as a co-solvent to enhance the solubility of polar sugars in Sc-CO_2. Studies on carbohydrate solubility in Sc-CO_2 have shown that using polar co-solvents such as water significantly improves the solubility and separation of carbohydrates from mixtures at an optimal concentration of the co-solvent [13].

Further, a pre-drying step of the date pulp (prior to supercritical extraction) is essential for the maximum recovery of sugars and other nutritive analytes from the fruit matrix. Among the different pre-drying methods that could be applied to produce dehydrated food, freeze-drying is recognized as one of the best drying techniques to preserve food quality [14–16]. Freeze-drying, also known as lyophilization, is the process of removing water from the food by an initial freezing phase followed by a sublimation phase to remove the ice as vapor. The removal of water from food by sublimation protects the food product against the loss of essential components and damage caused by chemical reactions associated with the withdrawal or vaporization of liquid water [17].

A post-drying stage of the extract (obtained from the Sc-CO_2 extraction) to a solid powder form ensures product stability and longevity of the sugar. The surface area of the final date sugar product plays a vital role in determining its dissolution rate. Spray-drying is one of the most widely used solvent evaporation techniques in the food industry to obtain products with a large surface area [18,19]. However, the rubbery nature of the low molecular weight sugars in the extract (owing to their lower transition temperature) causes product instability and particle agglomeration during the drying process. Hence, a carrier agent is highly required in the post-drying step to overcome these limitations. Gum Arabic (GA) is a widely studied carrier agent due to its low viscosity and high solubility [18]. It is a natural prebiotic and is considered natural, edible, and a safe source of dietary fiber [20]. GA is a complex polysaccharide that benefits gut health, digestion, and cardiovascular health.

A careful evaluation of the scientific literature shows that freeze-drying and spray-drying are the best drying techniques for the pre- and post-drying stages of the main extraction [14,21]. Hence, the same methods were adopted in this study to develop a novel process technology based on Sc-CO$_2$ extraction for the enhanced recovery and production of soluble date sugar powder from the date fruit. Based on these literature findings, an Integrated Date Sugar Production Process (IDSPP) involving three main stages is conceptualized as follows: (i) Freeze-drying of date pulp, (ii) Sc-CO$_2$-assisted extraction of date syrup with water as a co-solvent [8], and (iii) spray-drying of the extract for the production of soluble powder date sugar.

This study focuses on the modeling and simulation of the IDSPP process using Aspen Steady-State Modeler software to establish the feasibility of the process and determine the optimum working conditions for the process to maximize the sugar yield of date fruits. The simulation results would provide the necessary process data for the new technology and the recommended values for various critical process parameters for the experimental development and realization of the IDSPP process. Continuous-model simulation was performed for each stage of the process as a discrete block to determine the optimal values of the critical operating parameters for the respective stage of the IDSPP plant.

2. Materials and Methods

2.1. IDSPP Flowsheet

The process flowsheet for the as-proposed IDSPP technology is shown in Figure 1. Freeze-drying, supercritical extraction, and spray-drying are the major unit operations in the process. Operating pressure, freezing temperature, and drying temperature are the main parameters examined for the freeze-drying (FD) stage to minimize the moisture content in the fruit. Regarding the supercritical extraction (SCE) stage, variables such as operating pressure, temperature, and solvent:co-solvent ratio were considered to optimize the extraction efficiency. For the final spray-drying (SD) stage, the mass flow rate and temperature of air (drying medium), and feed flow rate were studied to understand their effects on the size and solubility of the final date sugar product.

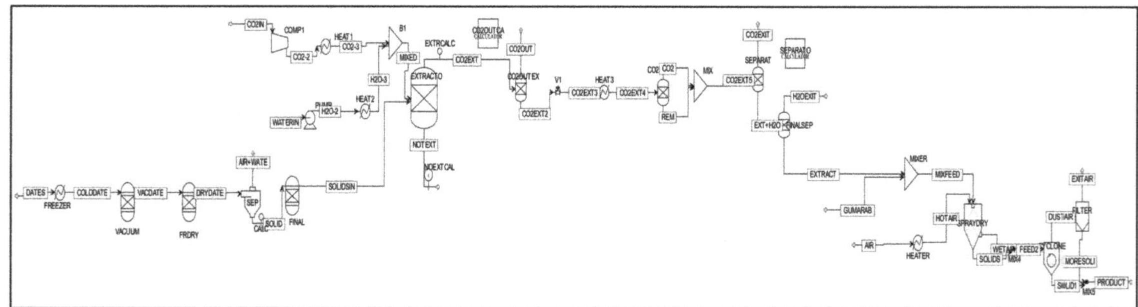

Figure 1. Integrated Date Sugar Production Process (IDSPP) flow diagram.

2.2. Materials

The lack of experimental data for the IDSPP at its design stage creates greater reliability on numerical simulation to provide useful information for the process. Commercial process simulation software, Aspen Plus V10 (2019), was used to solve the flowsheets encountered in this work. The software was used to understand the compositions of various phases and streams involved in each stage of the IDSPP. The simulations were run on a standalone personal computer (PC) using a Microsoft Windows 10 (2020) operating system.

2.3. Data

The essential properties and data for the Barhe variant of the date palm fruits were obtained from a study conducted by Rambabu et al., [2]. The properties (Table S1, Supplementary Materials) and structure of the GA (Figure S1, Supplementary Materials) were included in the Aspen database through the user interface [22]. The necessary user-defined functions (UDFs) pertaining to the solubility behavior of the analytes into the $Sc\text{-}CO_2$/water system (for the extraction stage) were developed using Phyton (V 3.9.9) scripting and patched to Aspen software.

2.4. Method

The simulation was performed through a model-based representation of chemical, physical, biological, and technical processes and unit operations in the software. The simulation prerequisites involved a detailed understanding of the chemical and physical properties of pure components and mixtures of various streams, as well as mathematical models that allowed stoichiometry-based process calculations. The simulator described processes in flow diagrams where the unit operations are positioned and connected by streams. The solver computed the necessary mass and energy balances for each node to find a stable operating point. The thermodynamic model suitable to present the behavior of each participating phase was selected after careful analysis of a wide range of scientific literature.

For the modeling, a basis of 100 kg/h of date fruit pulp, including the moisture content, was selected. All necessary components were found and defined in the Aspen database except for the date fruit, that was specified as a non-conventional solid. The date fruit was specified using the HCOALGEN enthalpy model and the DCHARIGT density model. Electrolyte Wizard was selected for the ice formation and drying of water, and the ice formation reaction was chosen and added to the chemistry profile to be used as an input to the units involving water phase change [23]. The Particle Size Distribution (PSD) was defined to describe the appearance of the non-conventional substream. A log-normal distribution function was used to describe the PSD with a geometric mean deviation of 27.5 mm, and a mass median diameter (D_{50}) value of 27.5 mm. The IDEAL property method was selected that accommodates both Raoult's law and Henry's law and is recommended for vaccum systems where gas is assumed to behave ideally.

The FD process was modeled using a Gibbs reactor which uses Gibbs free energy minimization with phase splitting to calculate the chemical equilibrium between any number of solid components and the fluid phases. The amount of sublimated water was calculated using a UDF that was developed using the results of İzli [14]. Accordingly, the Two Term model (as shown in Equation (1)) that best described the FD behavior was used for the freeze-dryer simulation.

$$MR = a_{FD} exp(-k_0 t) + b_{FD}\, exp(-k_1 t) \tag{1}$$

where MR is the moisture ratio (M_t/M_0), M_t represents the moisture at a specific time, t (g water/g dry solid); M_0 represents the initial moisture content (g water/g dry solid); k_0 and k_1 are the drying constants with values 0.002291 min^{-1} and 0.4768 min^{-1}, respectively; a_{FD} and b_{FD} are FD coefficients with values 0.835 and 0.4768, respectively. These values were experimentally estimated using various thin-layer drying models for the freeze-drying of dates [14].

For SCE stage modeling, the frozen dates were demonstrated as a mixture of specific components to simulate the extraction process and evaluate the exact yield of these components from the overall fruit matrix. Among the various represented components, glucose and fructose were highlighted as the targeted entities for the extraction by $Sc\text{-}CO_2$. The non-polar portion of the date fruit matrix was represented by fatty acids such as linoleic, palmitic, and oleic acids. The non-extractable material of the biomatrix was represented as cellulose due to its negligible solubility in $Sc\text{-}CO_2$ [12]. The phase equilibrium behavior of the glucose–fructose, $Sc\text{-}CO_2$, and water mixture was thermodynamically developed. The process simulation was based on the concept of stages in the extraction. Each stage was assumed as a component separator module at fixed temperature and pressure. The Soave-Redlich-Kwong (SRK) equation of state was used to calculate the thermodynamic properties. The SRK model is suitable for high-temperature and high-pressure extractions, such as supercritical conditioning [24]. Separation unit modules were used to manipulate the separation process of CO_2–water–glucose–fructose based on the solubility values at a given temperature and pressure. The separation was based on an empirical solubility equation (Equation (2)) proposed by Stahl and Schilz that was fed to manipulator block modules [25].

$$logy_{gf} = A + BP + CP^2 \tag{2}$$

where y_{gf} is the solubility of the glucose/fructose in the Sc-CO_2 (mol solute/mol CO_2); P is the total pressure of the system (bar); and A, B, C, are the empirical constants with values -4.3855, -6.2459×10^{-3}, and 2.0307×10^{-5} for glucose, and 4.5455, -1.2617×10^{-3}, and 1.1561×10^{-5} for fructose, respectively [25]. The solubilities of glucose and fructose in water were calculated according to the work of Saldaña et al., [26]. Additionally, Chrastil's equation [27] was used to model the solubility of the palmitic, oleic, and linoleic acids into Sc-CO_2. Equation (3) presents this model, which is based on the formation of acid–complex upon association of the solute and solvent molecules.

$$lnS = k \times ln\, \rho_s + \frac{a}{T} + b \tag{3}$$

where S is the solubility of the solute in the supercritical solvent (g/L); ρ_s is the density of the pure solvent (g/L); and k, a, and b are empirical constants. Further, the solubility of water in CO_2 was obtained through regression of experimental literature data [23] and is represented by Equation (4).

$$S_w = \left(1.66 \times 10^{-9}\right) + \left(7.36 \times 10^{-8} \times T\right) + \left(4.54 \times 10^{-8}P\right) + \left(1.22 \times 10^{-6} \times P \times T\right) \tag{4}$$

where S_w is the solubility of water in CO_2 (kg H_2O/kg CO_2), and T is the temperature (°C).

Modeling and simulation of the SD stage were performed through the built-in spray-dryer unit block model available in Aspen Plus. The necessary parameters for the spray-dryer model were specified for the drying process based on the IDEAL property method [29]. The properties for water, air, glucose, and fructose were defined with the conventional values available in the Aspen database. The properties of GA were explicitly included in the software. A rotary atomizer was chosen and modeled with a wheel diameter of 10 cm, four droplet intervals, four 30 mm blades, and a wheel speed of 18,000 rpm [19]. The characteristic droplet size distribution and nozzle exit velocity were calculated based on the geometry of the nozzle. The median droplet size (d_{50}, in μm) was calculated according to the model (Equation (5)) described by Masters [30].

$$d_{50} = \frac{K \times \dot{m}_{liq}^{p}}{N^q \times D_R^r \times (nh)^s} \times 10^4 \tag{5}$$

where n is the number of blades; m_{liq} is the mass flow rate of the feed (kg/h); N is the rotational speed (rpm); D_R is the diameter of the atomizer (m); and h is the height of the vanes (m). The constant K and exponents p, q, r, s were determined from the correlations for the vane liquid loading [30]. Subsequently, the mass flow ratio/vane (M_s) to the height of the vanes was determined as 3587.1 kg/h m. The validity limits were expressed in terms of the speed of the circumference of the disc (U) using Equation (6).

$$U = \pi \times \frac{N}{60} \times D_R \tag{6}$$

Thus, the speed of the circumference of the disc was evaluated as 94.25 m/s. As the values of the M_s and U values were in the same category, the design was considerably adequate according to the validity table [30]. The values of the constants and exponents were obtained from the same category. Based on these calculations and validations, the median droplet size (d_{50}) was estimated as 50.3 μm. Additionally, the critical moisture content for the drying process was evaluated as 2.67 kg water/kg dry solid (Equation (1)). The equilibrium moisture content was taken as 0.03 kg water/kg dry solid.

Table 1 summarizes the values for various parameters used for the simulation of each stage. The sensitivity analysis tool of the software was used to study the effect of different critical parameters on the final extracted amount of glucose and fructose.

Table 1. Temperature and pressure of each unit block used in the simulation model.

Unit	Temperature (°C)	Pressure (bar)
FREEZER	−42.0	1.01325
VACUUM	−42.0	0.00048
FRDRY	20.0	0.0001
SEP	20.0	0.0001
FINAL	20.0	1.01325
COMP1	783.9	308
HEAT1	90.0	308
PUMP	40.5	308
HEAT2	82.0	308
MIX1	65.0	308
EXTRACTO	65.0	308
CO2OUTEX	65.0	308
V1 VALVE	31.3	72
HEAT3	32.0	1
CO2DENS	32.0	1
MIX2	18.2	1
SEPARAT	18.2	1
FINALSEP	18.2	1
MIX	20.0	1
AIRHEAT	150.0	1
DRYER	150.0	1

3. Results and Discussions

Steady-state modeling and simulation of the IDSPP were conducted successfully using Aspen-Plus Software. Each sub-process was modeled and simulated separately. In addition, sensitivity analysis was performed for each block to identify the optimum operating conditions of the critical parameters for the respective stage.

3.1. Freeze-Drying

In the FD method, the biological qualities of the components, flavors, and colors of the food are usually preserved well [31]. For the analysis of IDSPP, vacuum-based FD was considered for the effective removal of moisture at a minimal drying time. The FD model accounted for the phase changes and water vapor diffusion within the date pulp. Modeling for the non-condensable species was accounted through the species transport model [32]. Technically, the feed sample was categorized into four layers, namely frozen front, vapor–ice interface, dry zone, and gas phase. A source term was used to adjust the distribution of gas-phase species due to water vapor generation [33].

Figure 2 depicts the process flow diagram of the FD process. Based on the simulation results, it was observed that the moisture content of the date pulp was reduced from 19.62% to 0.26%, calculated on a dry weight basis. The mass density of the freeze-dried dates was estimated as 1,639.05 kg/m^3. This relatively low-density value of the freeze-dried date pulp signified higher porosity and thus a potential for improved cell rupture in the SCE stage [32]. Further, sensitivity analysis was performed by varying the temperature and pressure for the FD to examine the effect on the moisture content and duty consumption. The results for the optimization studies of FD are displayed in Figure S2 (Supplementary Materials). A minimum heat duty of 13.23 kW at a temperature of −42 °C and pressure of 0.0001 bar was computed for the FD required for maximum moisture removal.

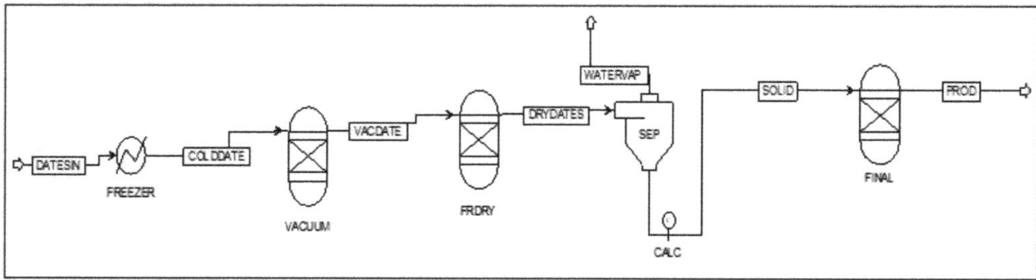

Figure 2. Freeze-drying process flow diagram.

3.2. Supercritical Extraction

An Aspen flowsheet for the SCE process is illustrated in Figure 3. For the proper design of the SCE stage, it was essential to employ appropriate mathematical representations. The SCE stage involved extraction from various solid natural matrices that were thermodynamically represented through solubility control, external mass transfer control, and/or internal mass transfer conventional models. Notably, three manipulators were used in the modeling of the SCE stage. The first one, "EXTRCALC" was used to estimate the quantity of the analytes to be extracted by the Sc-CO_2/water solvent system based on the UDF patched to the software. The UDF contained the necessary data and equations for the solubility behavior of glucose, fructose, and other essential components of date fruit into the solvent mixture. The second one, "CO2OUTCA" calculated the amount of CO_2 lost due to the opening of the extraction cell to fix the overall mass balance of the system. The third manipulator, "SEPARATO" was used to estimate the extracted mass lost with the exhaust stream of CO_2 gas.

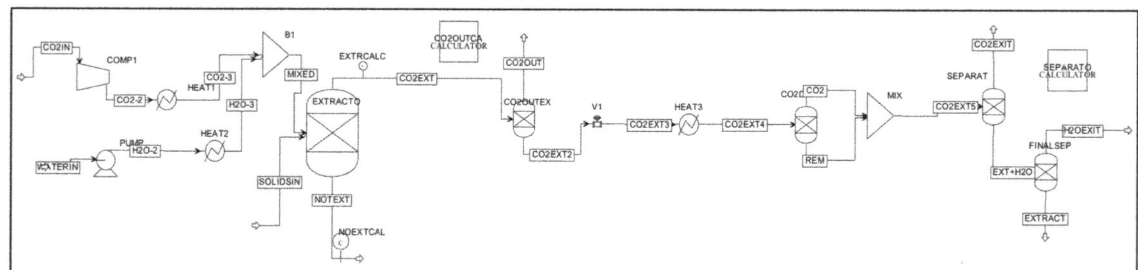

Figure 3. Aspen Plus Supercritical Extraction process flow diagram.

3.2.1. Operating Pressure

The extraction pressure of the SCE process is an important parameter that impacts the overall extraction efficiency [34]. Figure 4 shows the pressure effects on the extraction of reducing sugars from the freeze-dried date pulp. Simulation studies revealed that the extraction pressure had varying effects on the recovery of glucose and fructose sugars. For fructose sugar, the recovery efficiency decreased with incremental values in the operating pressure (Figure 4a). On the contrary, the extraction yield increased for glucose sugar with an increase in the operating pressure (Figure 4b). This opposing behavior can be ascribed to the different signs of the empirical coefficients in Equation (2) for glucose and fructose solubility in Sc-CO_2. Consequently, the trade-off optimization tool of Aspen software was utilized to estimate the optimum pressure required for the maximum extraction of the total reducing sugars. Results showed a maximum product flow (glucose + fructose) of 73.70 kg/h was achieved at an operating pressure of 308 bar.

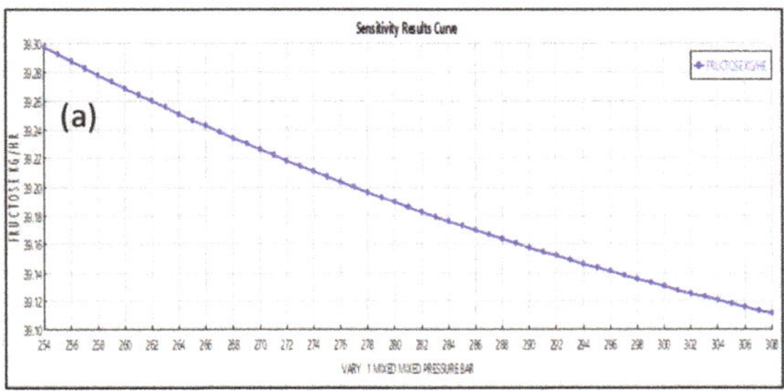
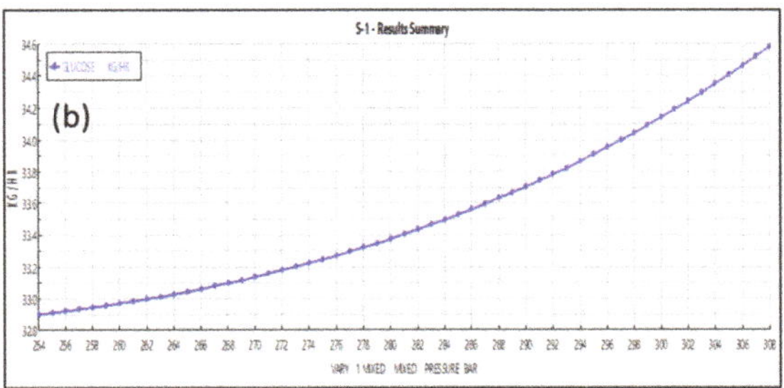

Figure 4. Effect of operating pressure on recovery of fruit sugars—(**a**) Fructose and (**b**) Glucose.

3.2.2. Extraction Temperature

Similar to pressure, the operating temperature of the SCE process is an imperative parameter that influences the efficiency of the separation process. Simulation studies showed that an increase in the extraction temperature had no significant effects on the recovery of reducing sugars from the fruit pulp. This insignificant temperature effect was anticipated as the solubility equation (Equation (2)) for the monosaccharide sugar molecules into $Sc\text{-}CO_2$ did not involve any temperature effects. However, the extraction of non-polar analytes such as linoleic acid, palmitic acid, and oleic acid by the solvent showed a significant increase with temperature for T > 70 °C (Figure S3, Supplementary Materials). The increase in the temperature enhanced the kinetic energy of the $Sc\text{-}CO_2$ fluid, which in turn improved the solvent diffusion and bioactive compound recovery [35]. However, it is worthy of specifying that higher extraction temperatures would cause the thermal degradation of the recovered sugar molecules, reducing the overall yield [36]. Based on the simulation results, an extraction temperature of 65 °C was recommended for the SCE stage.

3.2.3. Solvent Flow Rate and Composition

Results on the sensitivity analysis for the solvent:co-solvent ratio showed that a co-solvent concentration of 6.77 wt.% produced the maximum recovery of reducing sugars from the date pulp biomatrix. The simulation run resulted in an overall extraction yield of 99.1% for the reducing sugars with flow rates of 34.14 kg/h and 39.13 kg/h for glucose and fructose, respectively. Moreover, the sensitivity analysis showed that the total extracted quantity of glucose and fructose increased with the flow rate of $Sc\text{-}CO_2$, as shown in Figure 5. Almost all of the reducing sugars were extracted into the solvent system for an $Sc\text{-}CO_2$ flow rate of 31,000 kg/h. Further increases in the $Sc\text{-}CO_2$ flow rate had no significant effects on the sugar extraction. This quantity of $Sc\text{-}CO_2$ defines the solvent limit required for the maximum removal of targeted sugar components from the date palm fruit. Based on the optimization studies for the solvent:co-solvent ratio and the $Sc\text{-}CO_2$ flow rate, the optimum flow rate of water was estimated as 2100 kg/h for the maximum recovery of glucose and fructose components. Figure 6 shows the effect of the water flow rate on the extraction of reducing sugars from the fruit biomatrix. Similar to $Sc\text{-}CO_2$, the extraction yield increased with incremental levels of water flow rate. The improvement in the extraction of monosaccharides was due to the enhanced permeability and polarity of water molecules at the extraction conditions.

In general, the mass transfer mechanism for the recovery of plant bioactive compounds using supercritical technology involves four major steps [37]. Firstly, the analytes are desorbed from the biomatrix into the solvent film. Second, the analytes diffuse into the solvent, while in the third, the analytes dissolve into the solvent. The final step involves the bulk mass transfer of the solvent from the biomatrix to the external. Based on this mechanism, the increase in the extract yield with the flow rate of the solvent system signified that the external film transfer diffusion (second step) quantified by a diffusion coefficient, and the dissolution potential of sugars into the $Sc\text{-}CO_2$/water system (third step) quantified by a partitioning coefficient, played the limiting roles for the recovery of the targeted analytes [38].

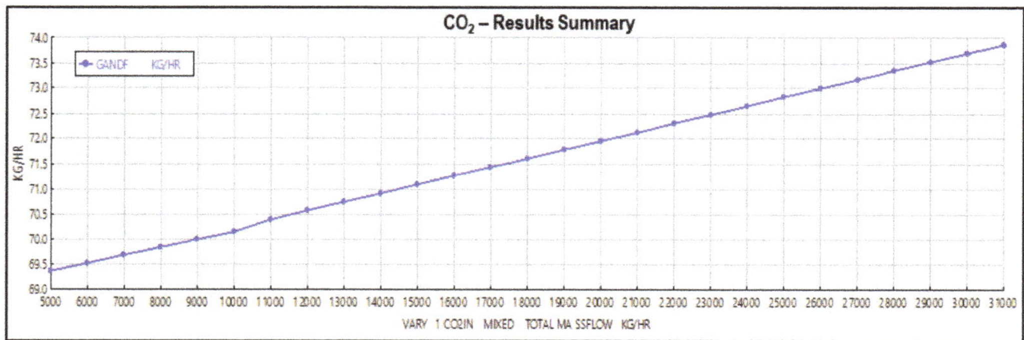

Figure 5. Effect of Sc-CO$_2$ flow rate on glucose and fructose recovery from date pulp.

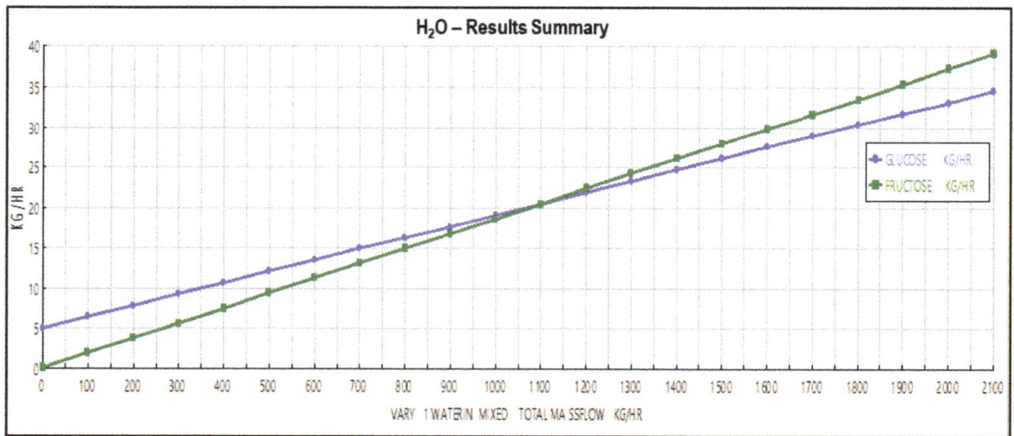

Figure 6. Effect of water flow rate on glucose and fructose recovery from date pulp.

3.3. Spray-Drying

In general, beneficial features such as high drying rate, shorter drying time, and broad working temperature make SD an ideal choice for drying of food products to powder form [32]. The complicated three-dimensional nature of spray-dryers often makes it difficult to describe their exact behavior using empirical models. In this work, Aspen's built-in simulator was used to characterize the SD process of the SCE extract (glucose and fructose solution) mixed with GA as the carrying agent. The model employed an Eulerian–Lagrangian approach to compute the particle motion paths, and heat and mass transfer phenomena occurring between the drying medium and the feed particles [32]. The choice of GA was to overcome the stickiness effect and increase the glass transition temperature of the mixture. Based on the previously reported experimental studies of SD for date fruit powder, the mass ratio of GA to SCE extract was fixed as 0.4 [18]. Figure 7 represents the process flow diagram for the SD stage of the IDSPP technology.

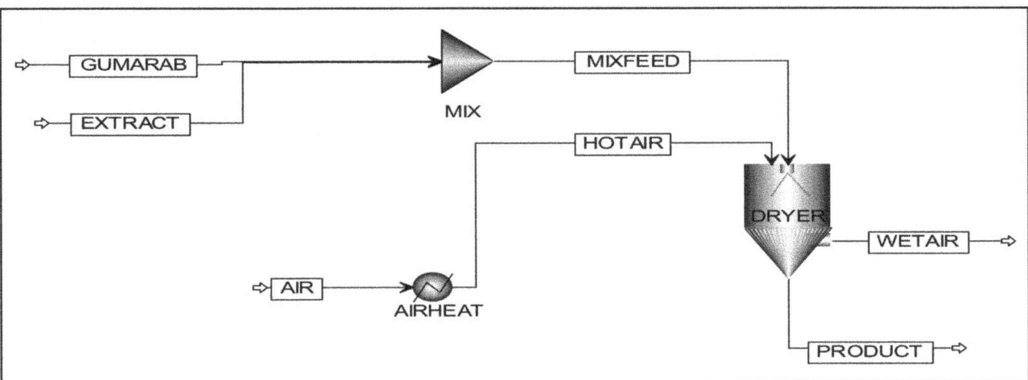

Figure 7. Aspen Plus Spray Drying process flow diagram.

Sensitivity analysis of the SD stage involved the study of the effect of different parameters on the final moisture content of the sugar powder. Results showed that the final moisture content of the powdered product decreased with an increase in the mass flow rate of the drying air (Figure S4, Supplementary Materials). Complete removal of excess moisture was observed at an airflow rate of 8000 kg/h resulting in powdered sugar with the equilibrium moisture content. High flow rates of the drying air reduced the droplet size due to their improved atomizing effect. This resulted in a finer particle size of the product with a larger surface area that facilitated the rapid and higher drying rate of the product with complete removal of moisture [30]. It is worth specifying that the blank simulation without GA addition resulted in low drying efficiency. This can be attributed to the possible stickiness of the feed extract, as reported in several other related experimental works [39].

Figure 8a shows the effect of the inlet air temperature on the moisture removal from the feed solution for the powdered product formation. It was observed that an increase in the air temperature resulted in improved moisture elimination. This was clearly due to the improved heat and mass transfer processes between the air and feed particles owing to the greater driving force for moisture evaporation [40]. Moreover, higher operating temperatures would eliminate insufficient drying of the particles, thereby reducing the probability of semi-wet/sticky solids formation that would typically stick to the wall and thus result in higher yield [39]. Simulation results showed the attainment of equilibrium moisture content at a minimum air inlet temperature of 150 °C for the formation of dry powdered products of date sugar. Further, the results for the dependency of product moisture content on the feed flow rate are shown in Figure 8b. It was evident that lower feed flow rates resulted in a relatively drier product than higher flow rates. In the SD stage, a higher feed rate may result in slower heat and mass transfer processes. An increase in feed flow rate reduces the contact time between the drying medium and the atomized particles leading to low heat transfer rates resulting in low moisture removal. In general, feed flow rate not only affects the moisture content and yield of the product but also significantly influences the product solubility and water activity [41]. Based on the simulation results, a feed flow rate of 435 kg/h was recommended for the SD stage of the IDSPP.

3.4. Product Solubility

Product solubility is a critical benchmark for food products, especially food-grade sugars, to be used as an additive in food and beverages. In general, several parameters of the product, such as composition, water activity, morphology, equilibrium moisture content, etc., affect the product solubility [42]. The solubility of the product (glucose and fructose mixture) in an aqueous medium was simulated over a temperature range from 0 to 100 °C and at a pressure of 1 bar based on the UNIQUAC (universal quasichemical) thermodynamic model. The solid–liquid phase equilibrium of the powdered date sugar in pure water was simulated using activity coefficients [43]. The solubility of the sugar product at a temperature of 25 °C and 1 bar was estimated as 0.89 g/g water. Comparing this value with the solubility of the table sugar (2.1 g/g water), it was observed that sucrose (in table sugar) had a higher solubility than the extracted date sugar. However, it is expected that experimental validation would result in higher solubility of the date sugar powder due to the good emulsification properties and high solubility of GA in aqueous solutions [22]. Figure 9 depicts the PSD for the final powder product obtained for the optimal operating conditions of the SD step. The mean particle size and density were estimated as 504 μm and 0.0023 kg/m^3, respectively, demonstrating the presence of significant levels of bioactive compounds in the product [42]. In addition, powders obtained using lower drying temperatures tend to possess a high percentage of smaller particles in the micro-range.

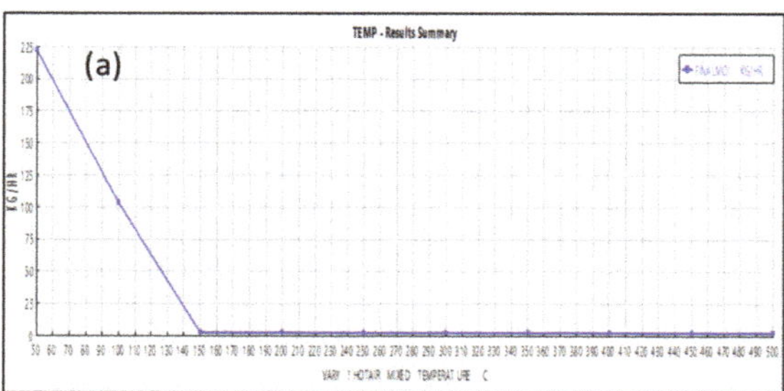
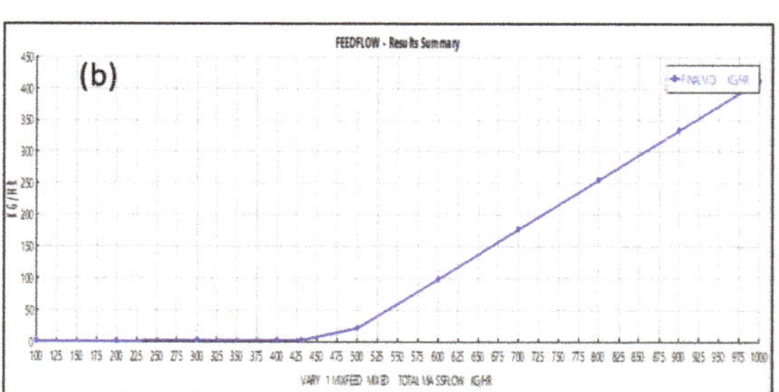

Figure 8. Effect of (**a**) inlet air temperature and (**b**) feed flow rate on the moisture content of the date sugar powder product.

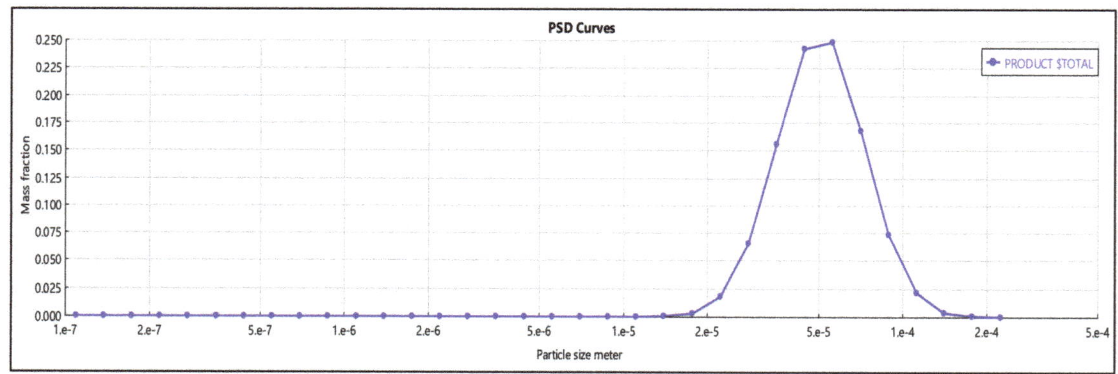

Figure 9. Particle size distribution of the spray-dried date sugar powder.

Experimental validation of IDSPP technology, with extensive analysis of extracts (including other vital analytes such as fibers, minerals, amino acids, etc.), determination of the extraction equilibrium and kinetics [44–47], and establishment of the necessary economics for production along with regulatory waste management approaches [48–50] would potentially realize effective production of date sugar that could ideally replace refined sugar.

4. Conclusions

In summary, this work conceptualized a novel IDSPP technology comprising of three main stages, namely freeze-drying, supercritical extraction, and spray-drying to process 100 kg/h of date fruit pulp to produce a soluble date sugar product. Stage-wise modeling and simulation studies were performed using Aspen Plus software with the sensitivity analysis to identify the optimum values of critical parameters for each stage. Freeze-drying studies revealed a minimum duty requirement of 13.23 kW at a temperature of -42 °C and pressure of 0.0001 bar to reduce the moisture content of the fruit pulp from 19.62% to 0.26%. The Sc-CO$_2$ assisted extraction resulted in a total sugar recovery of 99.1% with flow rates of 34.14 kg/h and 39.13 kg/h for glucose and fructose sugars, respectively. Optimization studies showed that a pressure of 308 bar, temperature of 65 °C, Sc-CO$_2$ flow rate of 31,000 kg/h, and Sc-CO$_2$: water ratio of 0.07 resulted in the maximum extraction efficiency. Further, spray-drying simulation of the date extract into a powdered product using Gum Arabic as the carrier agent showed a mean particle size of 50.4 μm for the date sugar powder with a solubility of 0.89 g/g water (at 25 and 1 bar) and density of 0.0023 kg/m^3. Sensitivity analysis revealed that the extraction pressure and solvent flow rates were the significant parameters for the extraction step while operating temperature had no salient effects on the recovery of the sugar. For the spray-drying process, drying air flow rate and its inlet temperature had positive effects, while feed flow rate negatively impacted moisture removal. Results showed that a mass flow rate of 8000 kg/h and an inlet temperature of 150 °C for the drying air was optimal to dry a feed stream mixture of date extract and Gum Arabic flowing at 435 kg/h containing 40 wt.% of Gum Arabic. Thus, this work confirmed the technical feasibility of IDSPP for the production of soluble date sugar powder and provided the necessary data with optimum operating conditions required for the process. Experimental validations and detailed characterizations of the product are further required to realize the outcomes of this study.

Supplementary Materials: The following supporting information can be downloaded at: https://www.mdpi.com/article/10.3390/pr10020257/s1, Figure S1: Chemical structure of Gum Arabic, Figure S2: Sensitivity analysis for FD: Temperature and pressure effects on the final moisture content and the net heat duty, Figure S3: Effect of temperature on the recovery of non-polar analytes of date fruit pulp, Figure S4: Effect of drying air flow rate on the moisture content of the date sugar powder product, Table S1: Properties of Gum Arabic used for the ASPEN simulation.

Author Contributions: Conceptualization, R.K. and F.B.; Data curation, H.B. and N.A.; Formal analysis, H.B., R.K. and S.W.H.; Investigation, H.B.; Methodology, R.K. and M.A.-Z.; Project administration, F.B.; Resources, S.W.H. and H.T.; Software, H.B. and R.K.; Supervision, F.B.; Validation, M.A.-Z., S.W.H., S.Y.A. and F.B.; Visualization, H.T.; Writing—original draft, H.B. and R.K.; Writing—review & editing, M.A.-Z., S.Y.A., N.A. and F.B. All authors have read and agreed to the published version of the manuscript.

Funding: This work was entirely funded and supported by the project grant CIRA-2019-028 under the Competitive Internal Research Award scheme of Khalifa University, UAE. The article processing charges were funded by grant RSP-2021/35, King Saud University, Riyadh, Saudi Arabia.

Institutional Review Board Statement: Not applicable.

Informed Consent Statement: Not applicable.

Data Availability Statement: Not applicable.

Acknowledgments: This work was entirely funded and supported by the project grant CIRA-2019-028 under the Competitive Internal Research Award scheme of Khalifa University, UAE. This project was supported by the Researchers Supporting Project Number (RSP-2021/35), King Saud University, Riyadh, Saudi Arabia.

Conflicts of Interest: The authors declare no conflict of interest.

Nomenclature

a	Empirical constant
A	Empirical constant
a_{FD}	Freeze Drying experimental coefficient
B	Empirical constant
b	Empirical constant
b_{FD}	Freeze Drying experimental coefficient
C	Empirical constant
D_5	Mass Median Diameter, mm
d_{50}	Median droplet size (μm)
D_R	Diameter of the atomizer (m)
FD	Freeze Drying
GA	Gum Arabic
h	Height of the vanes (m)
IDSPP	Integrated Date Sugar Production Process
k	Empirical constant; k_0–Drying constant (min^{-1});
k_0	Drying constant (min^{-1})
k_1	Drying constant (min^{-1})
M_0	Initial moisture content (g water/g dry solid)
m_{liq}	Mass flow rate of the feed (kg/h)
MR	Moisture ratio
M_s	Mass flow ratio/vane (kg/h)
M_t	Moisture at a specific time (g water/g dry solid)
n	Number of blades
N	Rotational speed (rpm)
P	Total pressure of the system (bar)
p	Vane liquid loading correlation
PSD	Particle Size Distribution
q	Vane liquid loading correlation
r	Vane liquid loading correlation
S	Solubility of the solute in the supercritical solvent (g/L)
s	Vane liquid loading correlation
Sc	CO_2–Supercritical Carbon dioxide
SCE	Supercritical Extraction
SD	Spray Drying
SRK	Soave-Redlich-Kwong
S_w	Solubility of the water in CO_2 (kg H_2O/kg CO_2)
T	Temperature ($^\circ$C)
U	Speed of the circumference of the disc (m/s)
UDF	User Defined Function
y_{gf}	Solubility of the glucose/fructose in the supercritical carbon dioxide (mol solute/mol CO_2)
ρ_s	Density of the pure solvent (g/L)

References

1. Voora, V.; Bermúdez, S.; Larrea, C. *Global Market Report: Sugar*; International Institute for Sustainable Development Geneva, Switzerland, 2020.
2. Rambabu, K.; Bharath, G.; Hai, A.; Banat, F.; Hasan, S.W.; Taher, H.; Zaid, H.F.M. Nutritional Quality and Physico-Chemical Characteristics of Selected Date Fruit Varieties of the United Arab Emirates. *Processes* **2020**, *8*, 256. [CrossRef]
3. Ahmed, I.A.; Ahmed, A.W.K.; Robinson, R.K. Chemical composition of date varieties as influenced by the stage of ripening. *Food Chem.* **1995**, *54*, 305–309. [CrossRef]
4. Alharbi, K.L.; Raman, J.; Shin, H.-J. Date Fruit and Seed in Nutricosmetics. *Cosmetics* **2021**, *8*, 59. [CrossRef]
5. Ameer, K.; Shahbaz, H.M.; Kwon, J.H. Green Extraction Methods for Polyphenols from Plant Matrices and Their Byproducts: A Review. *Compr. Rev. Food Sci. Food Saf.* **2017**, *16*, 295–315. [CrossRef] [PubMed]
6. Rambabu, K.; Edathil, A.A.; Nirmala, G.S.; Hasan, S.W.; Yousef, A.F.; Show, P.L.; Banat, F. Date-fruit syrup waste extract as a natural additive for soap production with enhanced antioxidant and antibacterial activity. *Environ. Technol. Innov.* **2020**, *20*, 101153. [CrossRef]

7.	Vallejo-Domínguez, D.; Rubio-Rosas, E.; Aguila-Almanza, E.; Hernández-Cocoletzi, H.; Ramos-Cassellis, M.E.; Luna-Guevara, M.L.; Rambabu, K.; Manickam, S.; Munawaroh, H.S.H.; Show, P.L. Ultrasound in the deproteinization process for chitin and chitosan production. *Ultrason. Sonochem.* **2021**, *72*, 105417. [CrossRef] [PubMed]
8.	Arumugham, T.; Rambabu, K.; Hasan, S.W.; Show, P.L.; Rinklebe, J.; Banat, F. Supercritical carbon dioxide extraction of plant phytochemicals for biological and environmental applications—A review. *Chemosphere* **2021**, *271*, 129525. [CrossRef] [PubMed]
9.	Al-Suod, H.; Ratiu, I.A.; Krakowska-Sieprawska, A.; Lahuta, L.; Górecki, R.; Buszewski, B. Supercritical fluid extraction in isolation of cyclitols and sugars from chamomile flowers. *J. Sep. Sci.* **2019**, *42*, 3243–3252. [CrossRef] [PubMed]
10.	Ratiu, I.-A.; Al-Suod, H.; Ligor, M.; Ligor, T.; Railean-Plugaru, V.; Buszewski, B. Complex investigation of extraction techniques applied for cyclitols and sugars isolation from different species of Solidago genus. *Electrophoresis* **2018**, *39*, 1966–1974. [CrossRef]
11.	Cavalcanti, R.N.; Albuquerque, C.L.C.; Meireles, M.A.A. Supercritical CO_2 extraction of cupuassu butter from defatted seed residue: Experimental data, mathematical modeling and cost of manufacturing. *Food Bioprod. Process.* **2016**, *97*, 48–62. [CrossRef]
12.	Mirofci, S. A supercritical fluid extraction process to obtain valuable compounds from Eruca sativa leaves. Masters Thesis, University of Padova, Padua, Italy, 2014.
13.	Dohrn, R.; Buenz, A.P. Solubility enhancement of carbohydrates in carbon dioxide (Part 1 and 2). *Int. Z. Lebensm.* **1995**, *46*, 30–31.
14.	İZLİ, G. Total phenolics, antioxidant capacity, colour and drying characteristics of date fruit dried with different methods. *Food Sci. Technol.* **2016**, *37*, 139–147. [CrossRef]
15.	Seerangurayar, T.; Manickavasagan, A.; Al-Ismaili, A.M.; Al-Mulla, Y.A. Effect of carrier agents on flowability and microstructural properties of foam-mat freeze dried date powder. *J. Food Eng.* **2017**, *215*, 33–43. [CrossRef]
16.	Franceschinis, L.; Salvatori, D.M.; Sosa, N.; Schebor, C. Physical and functional properties of blackberry freeze-and spray-dried powders. *Dry. Technol.* **2014**, *32*, 197–207. [CrossRef]
17.	Shishehgarha, F.; Makhlouf, J.; Ratti, C. Freeze-drying characteristics of strawberries. *Dry. Technol.* **2002**, *20*, 131–145. [CrossRef]
18.	Manickavasagan, A.; Thangavel, K.; Dev, S.R.S.; Delfiya, D.S.A.; Nambi, E.; Orsat, V.; Raghavan, G.S.V. Physicochemical characteristics of date powder produced in a pilot-scale spray dryer. *Dry. Technol.* **2015**, *33*, 1114–1123. [CrossRef]
19.	Minh, N.P.; Vo, T.T.; Tu, T.T.C.; Vi, M.N.T.; Diem, N.T.N.; Thanh, H.T.T. Spray Drying Parameters Affecting to Dried Powder from Palmyra Palm (*Borassus flabellifer*) Juice. *J. Pharm. Sci. Res.* **2019**, *11*, 1382–1387.
20.	Anderson, D.M.W. Evidence for the safety of gum arabic (*Acacia senegal* (L.) Willd.) as a food additive—A brief review. *Food Addit. Contam.* **1986**, *3*, 225–230. [CrossRef]
21.	Georgetti, S.R.; Casagrande, R.; Souza, C.R.F.; Oliveira, W.P.; Fonseca, M.J.V. Spray drying of the soybean extract: Effects on chemical properties and antioxidant activity. *LWT-Food Sci. Technol.* **2008**, *41*, 1521–1527. [CrossRef]
22.	Glicksman, M. Gum arabic (*Gum acacia*). In *Food Hydrocolloids*; CRC Press: Boca Raton, FL, USA, 2019; pp. 7–29.
23.	Rockstraw, D.A. ASPEN Plus in the Chemical Engineering Curriculum: Suitable Course Content and Teaching Methodology. *Chem. Eng. Educ.* **2005**, *39*, 68–75.
24.	Rostamian, H.; Lotfollahi, M.N. New functionality for energy parameter of Redlich-Kwong equation of state for density calculation of pure carbon dioxide and ethane in liquid, vapor and supercritical phases. *Period. Polytech. Chem. Eng.* **2016**, *60*, 93–97. [CrossRef]
25.	Stahl, E.; Schilz, W.; Schütz, E.; Willing, E. A quick method for the microanalytical evaluation of the dissolving power of supercritical gases. *Angew. Chem. Int. Ed. Engl.* **1978**, *17*, 731–738. [CrossRef]
26.	Saldaña, M.D.A.; Alvarez, V.H.; Haldar, A. Solubility and physical properties of sugars in pressurized water. *J. Chem. Thermodyn.* **2012**, *55*, 115–123. [CrossRef]
27.	Chrastil, J. Solubility of solids and liquids in supercritical gases. *J. Phys. Chem.* **1982**, *86*, 3016–3021. [CrossRef]
28.	King, A.D., Jr.; Coan, C.R. Solubility of water in compressed carbon dioxide, nitrous oxide, and ethane. Evidence for hydration of carbon dioxide and nitrous oxide in the gas phase. *J. Am. Chem. Soc.* **1971**, *93*, 1857–1862. [CrossRef]
29.	Petersen, L.N.; Poulsen, N.K.; Niemann, H.H.; Utzen, C.; Jørgensen, J.B. An experimentally validated simulation model for a four-stage spray dryer. *J. Process Control* **2017**, *57*, 50–65. [CrossRef]
30.	Masters, K. *Spray Drying Handbook*; Halstead Press: New York, NY, USA, 1985.
31.	Nakagawa, K.; Ochiai, T. A mathematical model of multi-dimensional freeze-drying for food products. *J. Food Eng.* **2015**, *161*, 55–67. [CrossRef]
32.	Malekjani, N.; Jafari, S.M. Simulation of food drying processes by Computational Fluid Dynamics (CFD); recent advances and approaches. *Trends Food Sci. Technol.* **2018**, *78*, 206–223. [CrossRef]
33.	Li, S.; Stawczyk, J.; Zbicinski, I. CFD model of apple atmospheric freeze drying at low temperature. *Dry. Technol.* **2007**, *25*, 1331–1339. [CrossRef]
34.	Malaman, F.S.; Moraes, L.A.B.; West, C.; Ferreira, N.J.; Oliveira, A.L. Supercritical fluid extracts from the Brazilian cherry (*Eugenia uniflora* L.): Relationship between the extracted compounds and the characteristic flavour intensity of the fruit. *Food Chem.* **2011**, *124*, 85–92. [CrossRef]
35.	Zhang, Q.; Zhao, M.; Xu, Q.; Ren, H.; Yin, J. Enhanced enzymatic hydrolysis of sorghum stalk by supercritical carbon dioxide and ultrasonic pretreatment. *Appl. Biochem. Biotechnol.* **2019**, *188*, 101–111. [CrossRef] [PubMed]
36.	Oladipupo Kareem, M.; Edathil, A.A.; Rambabu, K.; Bharath, G.; Banat, F.; Nirmala, G.S.S.; Sathiyanarayanan, K. Extraction, characterization and optimization of high quality bio-oil derived from waste date seeds. *Chem. Eng. Commun.* **2021**, *208*, 801–811. [CrossRef]

37. Islam, M.N.; Jo, Y.-T.; Jung, S.-K.; Park, J.-H. Thermodynamic and kinetic study for subcritical water extraction of PAHs. *J. Ind. Eng. Chem.* **2013**, *19*, 129–136. [CrossRef]
38. Khajenoori, M.; Asl, A.H.; Hormozi, F. Proposed models for subcritical water extraction of essential oils. *Chin. J. Chem. Eng.* **2009**, *17*, 359–365. [CrossRef]
39. Fazaeli, M.; Emam-Djomeh, Z.; Ashtari, A.K.; Omid, M. Effect of spray drying conditions and feed composition on the physical properties of black mulberry juice powder. *Food Bioprod. Process.* **2012**, *90*, 667–675. [CrossRef]
40. Rambabu, K.; Muruganandam, L.; Velu, S. CFD simulation for separation of carbon dioxide-methane mixture by pressure swing adsorption. *Int. J. Chem. Eng.* **2014**, *2014*, 402756. [CrossRef]
41. Cortés-Rojas, D.F.; Souza, C.R.F.; Oliveira, W.P. Optimization of spray drying conditions for production of *Bidens pilosa* L. dried extract. *Chem. Eng. Res. Des.* **2015**, *93*, 366–376. [CrossRef]
42. Lourenço, S.C.; Moldão-Martins, M.; Alves, V.D. Microencapsulation of pineapple peel extract by spray drying using maltodextrin, inulin, and arabic gum as wall matrices. *Foods* **2020**, *9*, 718. [CrossRef]
43. Zhao, H.; Li, J.; Wang, L.; Li, C.; Li, P. Thermodynamic investigation of 1,3,5-trioxane, methyl acrylate, methyl acetate, and water mixtures, in terms of NRTL and UNIQUAC models. *Ind. Eng. Chem. Res.* **2019**, *58*, 18378–18386. [CrossRef]
44. Ramu, A.G.; Telmenbayar, L.; Theerthagiri, J.; Yang, D.; Song, M.; Choi, D. Synthesis of a hierarchically structured Fe_3O_4—PEI nanocomposite for the highly sensitive electrochemical determination of bisphenol A in real samples. *New J. Chem.* **2020**, *44*, 18633–18645. [CrossRef]
45. Theerthagiri, J.; Lee, S.J.; Murthy, A.P.; Madhavan, J.; Choi, M.Y. Fundamental aspects and recent advances in transition metal nitrides as electrocatalysts for hydrogen evolution reaction: A review. *Curr. Opin. Solid State Mater. Sci.* **2020**, *24*, 100805. [CrossRef]
46. Madhavan, J.; Theerthagiri, J.; Balaji, D.; Sunitha, S.; Choi, M.Y.; Ashokkumar, M. Hybrid advanced oxidation processes involving ultrasound: An overview. *Molecules* **2019**, *24*, 3341. [CrossRef]
47. Senthil, R.A.; Priya, A.; Theerthagiri, J.; Selvi, A.; Nithyadharseni, P.; Madhavan, J. Facile synthesis of α-Fe_2O_3/WO_3 composite with an enhanced photocatalytic and photo-electrochemical performance. *Ionics* **2018**, *24*, 3673–3684. [CrossRef]
48. Theerthagiri, J.; Senthil, R.A.; Priya, A.; Madhavan, J.; Michael, R.J.V.; Ashokkumar, M. Photocatalytic and photoelectrochemical studies of visible-light active α-Fe_2O_3—gC_3N_4 nanocomposites. *Rsc Adv.* **2014**, *4*, 38222–38229. [CrossRef]
49. Theerthagiri, J.; Senthil, R.A.; Senthilkumar, B.; Polu, A.R.; Madhavan, J.; Ashokkumar, M. Recent advances in MoS_2 nanostructured materials for energy and environmental applications—A review. *J. Solid State Chem.* **2017**, *252*, 43–71. [CrossRef]
50. Theerthagiri, J.; Salla, S.; Senthil, R.A.; Nithyadharseni, P.; Madankumar, A.; Arunachalam, P.; Maiyalagan, T.; Kim, H.-S. A review on ZnO nanostructured materials: Energy, environmental and biological applications. *Nanotechnology* **2019**, *30*, 392001. [CrossRef]

Article

Fractional Factorial Design Study for the Extraction of Cannabinoids from CBD-Dominant Cannabis Flowers by Supercritical Carbon Dioxide

Sadia Qamar [1,*], Yady J. M. Torres [2], Harendra S. Parekh [1] and James Robert Falconer [1,*]

[1] School of Pharmacy, The University of Queensland, Brisbane, QLD 4102, Australia; h.parekh@uq.edu.au
[2] School of Clinical Sciences, Queensland University of Technology, Brisbane, QLD 4000, Australia; yadyjuliana.manriquetorres@qut.edu.au
* Correspondence: s.qamar@uq.edu.au (S.Q.); j.falconer@uq.edu.au (J.R.F.); Tel.: +61-7-3346-1852 (J.R.F.)

Citation: Qamar, S.; Torres, Y.J.M.; Parekh, H.S.; Falconer, J.R. Fractional Factorial Design Study for the Extraction of Cannabinoids from CBD-Dominant Cannabis Flowers by Supercritical Carbon Dioxide. *Processes* **2022**, *10*, 93. https://doi.org/10.3390/pr10010093

Academic Editors: Maria Angela A. Meireles, Ádina L. Santana and Grazielle Nathia Neves

Received: 22 November 2021
Accepted: 30 December 2021
Published: 3 January 2022

Publisher's Note: MDPI stays neutral with regard to jurisdictional claims in published maps and institutional affiliations.

Abstract: The optimization of the supercritical fluid extraction (SFE) of cannabinoids, using supercritical carbon dioxide (scCO$_2$), was investigated in a fractional factorial design study. It is hypothesized that four main parameters (temperature, pressure, dry flower weight, and extraction time) play an important role. Therefore, these parameters were screened at predetermined low, medium, and high relative levels. The density of scCO$_2$ was used as a factor for the extraction of cannabinoids by changing the pressure and temperature. The robustness of the mathematical model was also evaluated by regression analysis. The quantification of major (cannabidiol (CBD), cannabidiolic acid (CBDA), delta 9-tetrahydrocannabinol (Δ^9-THC), delta 8-tetrahydrocannabinol (Δ^8-THC), and delta 9-tetrahydrocannabinol acid (THCA-A)) and minor (cannabidivann (CBDV), tetrahydrocannabivann (THCV), cannabigerolic acid (CBG), cannabigerol (CBGA), cannabinol (CBN), and cannabichomere (CBC)) cannabinoids in the scCO$_2$ extract was performed by RP-HPLC analysis. From the model response, it was identified that long extraction time is a significant parameter to obtain a high yield of cannabinoids in the scCO$_2$ extract. Higher relative concentrations of CBD(A) (0.78 and 2.41% w/w, respectively) and THC(A) (0.084 and 0.048% w/w, respectively) were found when extraction was performed at high relative pressures and temperatures (250 bar and 45 °C). The higher yield of CBD(A) compared to THC(A) can be attributed to the extract being a CBD-dominant cannabis strain. The study revealed that conventional organic solvent extraction, e.g., ethanol gives a marginally higher yield of cannabinoids from the extract compared to scCO$_2$ extraction. However, scCO$_2$ extraction generates a cleaner (chlorophyll-free) and organic solvent-free extract, which requires less downstream processing, such as purification from waxes and chlorophyll.

Keywords: cannabis flowers; neutral cannabinoids (sp. *sativa*); supercritical extraction; supercritical carbon dioxide (scCO$_2$); SFE Nottingham unit; SFE Helix unit

1. Introduction

The extraction of cannabinoids from cannabis flowers is becoming increasingly popular, due to their potential therapeutic effects, medicinal benefits, and potential for utilization in patients' pain management [1–4]. Recently, the supercritical fluid extraction (SFE) technique has gained interest and has been used for various extraction applications [5–9]. In addition, SFE is also used to extract the bioactive components and essential oils on a pilot scale [10,11]. Due to the tunable properties of CO$_2$, it is used as the main solvent to extract these components. The density of CO$_2$ can be altered by making very minor changes to temperature and pressure [12]. SFE is considered a greener technique because of its low impact on the environment, and it does not require the use of hazardous organic solvents [13]. SFE has already been commercialized for the extraction of cannabis and related products. However, because of the lack of knowledge of how SFE affects the parameters, the interactions of the testing materials, and in-depth fluid dynamics, it is considered a black box design [14].

The factorial design is frequently used for the screening of different processing parameters in experiments, especially for the improvement of extraction efficiency and chromatographic separations; the most effective procedure is to evaluate the main contributing factors [15] because, theoretically, the number of experimental factors contributes to systematically resolving the issue. Additionally, a factorial design study significantly reduces the resources, time and effort needed to solve problems [16]. Therefore, factorial design studies are considered an efficient and information-rich technique for the analysis of analytical data and to obtain valid results [17,18].

In the reported literature studies, the extraction conditions were focused on high pressure and temperature [19–21]. In addition, these studies consider the density of CO_2 as a resulting parameter, with a change in pressure and temperature to extract the cannabinoids. However, a single factor cannot produce sufficient data to address all problems. By optimizing the main contributing parameters in SFE, the extractability and sensitivity of the process can be modified. Therefore, a careful study of SFE's tunable properties is needed to understand the effects of all participating parameters. Additionally, by optimizing the SCF extraction ability, a higher degree of freedom can be acquired, compared to the conventional techniques.

The SFE design of experimentation in any study depends on the various objectives, such as time consumption, cost-effectiveness, the feasibility of the experiment, and investigators' intention. There are three common screening stages to achieve the best results from any experimental design, including interpretation and data analysis, experimental trials, and the proper design of the software. For the screening of important SFE factors that affect the experiment, the most commonly used designs for experiments are the Plackett–Burman design, fractional factorial design, and the full factorial design [14]. This study aims to design a half-fractional factorial study to determine the best factors for the selective extraction of cannabinoids from cannabis.

2. Materials and Methods

2.1. Chemical and Reagents

The solvent (methanol, ethanol, and phosphoric acid) and cannabinoid reference standards, named as delta 9-tetrahydrocannabinol acid (THCA-A; Lot: FE12121601; with 99.18% purity), cannabichomere (CBC; lot: FE10011502; with 97.60% purity), delta 8-tetrahydrocannabinol (Δ^8-THC), delta 9-tetrahydrocannabinol (Δ^9-THC) with 97.66% purity (lot: FE1041701), cannabinol (CBN; lot: FE06131701; with 99.37% purity), cannabigerol (CBGA), cannabidiolic acid (CBDA; lot: FE12011601; with 98.3% purity), cannabigerolic acid (CBG; lot: FE06241604; with 98.98% purity), cannabidiol (CBD; lot FE08071702; with 99.66% purity), tetrahydrocannabivann (THCV), and cannabidivann (CBDV), were purchased from Sigma Aldrich Ltd. and Novachem Pty. Ltd., Victoria, Australia. Liquid-vapor CO_2 was supplied by BOC (Sydney, Australia). The dried cannabis flowers were a generous gift from PreveCeutical Medical Inc., Vancouver, BC, Canada.

2.2. Sample Preparation

The cannabis strain, "tower" (cannabidiol-dominant), with the *sativa* genotype was planted on 4 May 2017 under ideal growing conditions (12–18 h light exposure at 23 °C). The flowers of the cannabis sample were collected at the fluorescence stsgeand dried for 5 to 8 days at 20 °C. According to the certificate of analysis, the flower material had around 15% *w/w* of CBD and CBDA, whereas THC and THCA levels were around 0.517% *w/w*. After drying (with total moisture < 10%), samples were crushed to obtain a particle size < 2.7 mm. The samples were then pulverized in short pulses (to avoid warming) for two min (Breville coffee grinder, model BCG200), to further reduce particle size and increase the surface area, as well as for the efficiency of cannabinoid extraction.

2.3. SFE Equipment and Setup

The supercritical carbon dioxide (scCO$_2$) extraction of cannabis was performed using a Helix supercritical fluid extraction (SFE) unit (Applied Separations, Allentown, PA, USA) of the cannabis sample. The maximum sample-holding capacity of the reactor was 100 mL. The backpressure was directly regulated by a preconditioning chamber from the liquid-vapor CO$_2$ (liquid-vapor equilibrium) cylinder, as shown in Figure 1. The desired internal temperature was controlled through a heating jacket on the sample-holding chamber. The pulverized cannabis sample was placed in the bottom of the chamber. The CO$_2$ stream entered from the bottom inlet of the chamber and extraction was carried out according to the conditions listed in coming sections. The liquid-vapor CO$_2$ dissolved the matrix from the sample according to its density. After extraction, the CO$_2$ with the dissolved matrix entered the separation chamber, where the pressure was around 50 bar to avoid the throttling effect of dry ice. The extract was collected in the sample-collecting vessel, attached to the bottom of the separating chamber. The sample was washed with a continuous flow of CO$_2$ for 10 min, and dry ice was collected in the sample-collecting vessel. After the collection of 1st extract, the reactor was refilled again with fresh CO$_2$, and extraction was performed for 10 min to avoid the supersaturation of CO$_2$.

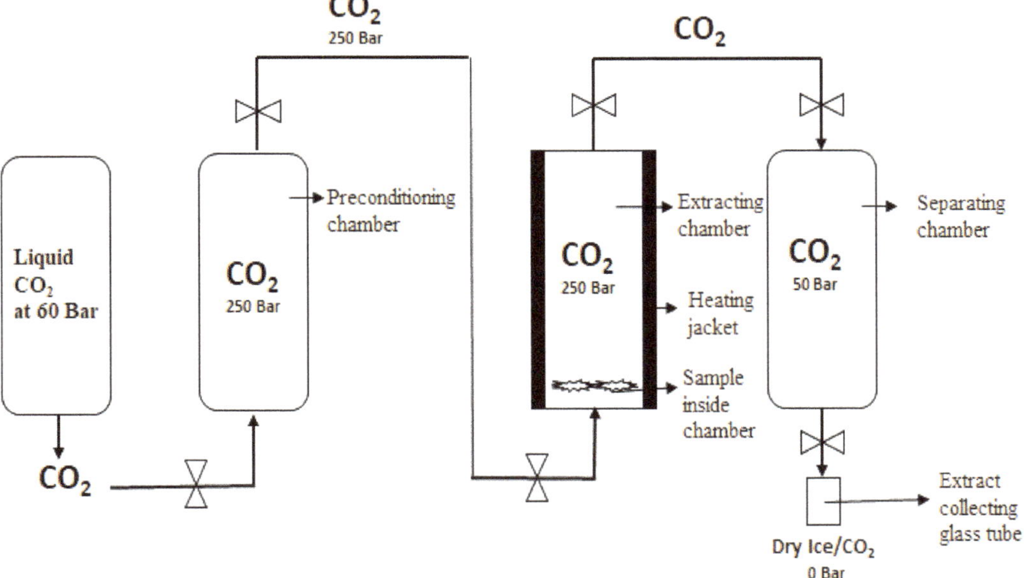

Figure 1. Schematic representation of the supercritical extraction Helix unit.

2.4. Parameters for the Factorial Design

The experimental domain factors of the fractional factorial study are shown in Table 1, including temperature (F1), pressure (F2), processing time (F3), and the amount of plant material (F4) at lower, central, and higher levels. To attain the various supercritical regions of CO$_2$ (scCO$_2$), the pressure and temperature ranges were employed from 150 to 250 bar and 35 to 45 °C, respectively. The processing time for the design of the experiment ranged from 30 to 180 min, and the quantity of plant material used was 0.5 to 1 g.

Table 1. Factors for the experimental factorial design study.

	Factor	Low (−)	Centre (0)	High (+)
F1	Temperature (°C)	35	39	45
F2	Pressure (bar)	150	200	250
F3	Processing time (h)	30	120	180
F4	Material (g)	0.5	0.6	1.0

2.5. Experimental Matrix

The factorial design of 2^{4-1} was employed to give an 8-run experimental plan with two center points, positioned at a medium level between the sets of low and high levels. Center points were used as a reference point to determine the factor–response linearity and experimental error. The experimental matrix is given in Table 2, together with the factors for F1, F2, F3 and F4. The relationship between pressure, temperature and density are also described in [12] and provided in Table 2.

Table 2. Experimental matrix for the 2^{4-1} design for the extraction.

Std. Order	Run. Order	Center. Pt.	Blocks	Temperature (°C) F1	Pressure (bar) F2	Time (Hour) F3	Sample Amount (g) F4 F1 × F2 × F3	Density of CO₂ Kg/m³
10	5	0	1	39	200	2.0	0.6	843
3	1	1	1	35	250	0.5	1.0	903
7	4	1	1	35	250	3.0	0.5	903
4	10	1	1	45	250	0.5	0.5	845
9	6	0	1	39	200	2.0	0.6	843
6	7	1	1	45	150	3.0	0.5	746
5	8	1	1	35	150	3.0	1.0	819
2	3	1	1	45	150	0.5	1.0	746
8	2	1	1	45	250	3.0	1.0	845
1	9	1	1	35	150	0.5	0.5	819

The central points are at rows 1 and 5.

2.6. Resolution

In this study, the resolution level of the half-fractional factorial design study plot for the cannabis sample was generated as 4. This demonstrates that the experimental matrix from F1 to F3 was considered a full factorial design. By using generators or statistical modeling, factor F4 was assessed. This was formed by multiplying the previous three-factor value; that is, F4 = F1 × F2 × F3

2.7. Regression Modeling

The multiple regression analysis for the selected independent factors, and between the responses of these factors, can be represented by Equation (1), as follows.

$$Y_j = \beta_0 + \beta_1 X_1 + \beta_2 X_2 + \beta_3 X_3 + \beta_4 X_1 X_2 X_3 + \beta_5 X_1 X_2 + \beta_6 X_1 X_3 + \beta_7 X_1 X_4 + \beta_8 X_2 X_3 + \beta_9 X_2 X_4 + \beta_{10} X_3 X_4 + \beta_{11} X_1 X_2 X_4 + \beta_{12} X_1 X_3 X_4 + \varepsilon \quad (1)$$

where

- ε = Experimental error term
- X_1 to X_4 = Variable effect
- β_0 = Coefficient constant of the average experimental response
- β_1 to β_3 = Variable main effects estimation
- β_4 to β_{12} = Variable interaction effects estimation
- Y = Experiment j response estimation.

2.8. Statistical Analysis of the Experimental Design

The cannabis experimental design (DOE) for data analysis and factorial runs was obtained with Minitab 17. The confidence levels of 90% ($p < 0.01$) and 95% ($p < 0.05$) were considered statistically significant.

2.9. Conventional Extraction

Organic solvent extraction was also performed to compare the results and the efficiency of $scCO_2$ extraction. The ethanolic extract of the cannabis ground material was obtained at a ratio of 1:10 (w/w). This mixture was sonicated for 15 min and stirred (with a magnetic stirrer) for 24 h in a dark, cold room at 4 °C to avoid any kind of degradation. This mixture was filtered using a Whatman filter paper and the obtained extract was dried using a nitrogen evaporator. After that, the extract was dissolved in 2 mL methanol and centrifuged at 10,000 rpm for 15 min. Further dilutions were performed to quantify the number of cannabinoids in the extract, using HPLC.

2.10. HPLC Quantification: Mobile Phase Elution Program

The reverse-phase HPLC quantification and separation of 11 cannabinoids was performed using a C18 chromatographic column (Shim-pack XR-ODSII, spherical silica particles, 2.2 µm particle size (Shimadzu Scientific) and Lab Solutions software or a Cannabis Analyzer (Shimadzu Scientific Instruments, NSW, Australia) were used to standardize the method.

The mobile phase A and B was a mixture of MilliQ water and methanol, with phosphoric acid (99.93/0.07% v/v). Both mobile phases were sonicated for 15 min and, after that, the pH of the mobile phases was monitored (the pH of mobile phase A was around 2.22 to 2.26, and B was around 2.43 to 2.48). The column oven temperature was attuned to 50 °C and the flow rate was 1.0 mL/min, to maintain the column pressure (~5400 to 5600 psi). The volume of injection was 10 µL and the total runtime was 45 min. Initially, mobile phase B (v/v) was at 65% for 1 min. Then, it gradually increased to 72% over 25 min; after that, it increased to 95% for 5 min. After maintaining these conditions for 2 min, the initial ratio of mobile phases was adjusted and the column was re-equilibrated for 12 min.

2.11. Standard Solution Preparation

The cannabinoid standard solution, with a concentration of 1000 µg/mL, was diluted in methanol to make 250 µg/mL as a stock solution. The calibration curve of mixed standards with 11 cannabinoids was prepared at a concentration range from 1.0 to 25.0 µg/mL in methanol, using the cannabinoids stock solution. All standards solutions were stored at −80 °C.

2.12. Sample Preparation for Cannabinoid Quantification

The obtained solvent-free $scCO_2$ extract was dissolved in 2 mL methanol, sonicated for 15 min, and centrifuged at 10,000 rpm for 10 min to prepare the primary extract. Then, 50 µL of this supernatant was transferred into a new centrifuged tube and made up to 1000 µL by adding 950 µL of methanol (20-times dilution, as a secondary extract). This extract was vortex-mixed at 1000 rpm for 1 min to dissolve the cannabinoids properly into the methanol. Then, 100 µL of this secondary extract was transferred into HPLC vials directly for HPLC quantification or diluted again to establish the area under the curve, respectively. All extract solutions were stored at −80 °C, until ready for testing.

3. Result and Discussion

In this study, a high yield and the selective extraction of 11 cannabinoids were achieved via the SFE method. The association between the SFE main parameters (pressure and temperature) and the experimental domain of cannabis (processing time and amount) were screened to obtain an effective extraction of cannabinoids. The mathematical modeling was performed via a 2^{4-1} factorial design study. This factorial design revealed the efficiency of the selected main parameters and their simultaneous effect on the $scCO_2$ extract. The selection of $scCO_2$ as an extraction solvent was due to its intrinsically non-polar behavior, because the chemical nature of cannabinoids is oily, and they easily dissolve in a non-polar medium [22].

3.1. Model Response for Extract Yield and Total Cannabinoids

The results of the half-fraction factorial design, with a resolution of 4 for the total extract yield from the original pulverized cannabis material, are represented in Tables 3 and 4. The percentage yield of the $scCO_2$ extract was studied using 0.5 to 1.0 g of the cannabis sample; because of sample sensitivity (having a high amount of CBA + CBDA), a low amount of cannabis was used to avoid the super-saturation of $scCO_2$ under experimentation.

Table 3. HPLC quantification for supercritical carbon dioxide extract yield, cannabidiol (CBD), cannabidiolic acid (CBDA), and total cannabinoids.

Std. Order	F1	F2	F3	F4	R1 Yield (mg/g)	% Yield with Respect to Sample Amount	R2 CBD (μg/mL) *	R2 % CBD (w/w)	R3 CBDA (μg/mL) *	R3 % CBDA (w/w)	R4 Total Cannabinoids (μg/mL) *	R4 Total Cannabinoids in $scCO_2$ Extract (w/w)
10	0	0	0	0	72.7	12.12	102.50 ± 0.02	0.41	220.95 ± 0.08	0.88	353.190 ± 0.08	1.413
3	−1	1	−1	1	98.0	9.80	60.88 ± 0.41	0.24	187.30 ± 0.11	0.75	268.53 ± 0.30	1.074
7	−1	1	1	−1	148.4	29.68	82.51 ± 0.06	0.33	171.00 ± 0.12	0.68	277.98 ± 0.18	1.112
4	1	1	−1	−1	132.0	26.40	63.19 ± 0.11	0.25	113.73 ± 0.46	0.45	188.65 ± 0.43	0.755
9	0	0	0	0	69.3	11.55	138.57 ± 0.23	0.55	225.7 ± 0.24	0.90	397.14 ± 0.36	1.588
6	1	−1	1	−1	90.7	18.14	166.50 ± 0.86	0.66	342.87 ± 1.66	1.37	556.91 ± 2.79	2.228
5	−1	−1	1	1	38.3	3.83	91.04 ± 0.09	0.36	121.12 ± 0.13	0.48	232.43 ± 0.22	0.930
2	1	−1	−1	1	47.3	4.73	73.18 ± 0.41	0.29	80.16 ± 0.36	0.32	171.18 ± 0.82	0.685
8	1	1	1	1	152	15.20	195.26 ± 0.52	0.78	603.37 ± 0.85	2.41	853.12 ± 1.42	3.412
1	−1	−1	−1	−1	38.8	7.76	52.55 ± 0.06	0.21	79.74 ± 0.04	0.64	145.19 ± 0.61	0.581

F1 = Temperature, F2 = pressure, F3 = processing time, F4 = material. The central points are at rows 1 and 5, * 20 times diluted extract.

Table 4. Estimated coefficient and effects for the processing of supercritical carbon dioxide ($scCO_2$) extract yield and total cannabinoids.

Term	Yield of $scCO_2$ Extract					Total Cannabinoids				
	Effect	Coefficient	SS	T Value	*p* Value	Effect	Coefficient	SS	T-Value	*p*-Value
Constant		93.19	18249	11.65	0.055		351.6	420,638	18.70	0.034
F1	24.63	12.31	1212	1.54	0.367	181.6	90.8	65,988	4.83	0.130
F2	78.83	39.41	12426	4.93	0.127	150.4	75.2	45,260	4.00	0.156
F3	28.33	14.16	1604	1.77	0.327	316.5	158.3	200,362	8.42	0.075
F4	−18.58	−9.29	690	−1.16	0.453	59.3	29.7	7042	1.58	0.360
F1F2	−5.82	−2.91	67	−0.36	0.778	6.4	3.2	82	0.17	0.893
F1F3	3.38	1.69	22	0.21	0.868	208.6	104.3	87,016	5.55	0.114
F1F4	6.88	3.44	94	0.43	0.742	80.0	40.0	12,808	4.53	0.280
Center points	-	−36.5	-	−2.04	0.290	-	−36.1			
Curvature	-	-	2130		0.290	-		2081	0.74	0.549
R^2 (%)			97.27					99.33		
R^2 adjusted (%)			75.44					93.99		

F1 = Temperature, F2 = pressure, F3 = processing time, F4 = material.

The effect of pressure and temperature during $scCO_2$ extraction on the yield of extract was also investigated because, previously, Span and Wagner [12] found that minor changes in pressure and temperature have a great influence on the density of $scCO_2$. The variation in the density of CO_2 on selected experimental conditions is shown in Table 2. The minor changes in the density of CO_2 also change its diffusivity, as a result of the solvation power of the CO_2 effects. Therefore, in this study, small variations in temperature and pressure were implemented to monitor the effect of $scCO_2$ density on experimentation.

From the results, it was found that the maximum yield of the $scCO_2$ extraction of cannabis was obtained from run 7 (around 148.4 mg/g, when the processing temperature was 35 °C, the pressure was 250 bar, processing time was 3 h, and sample amount was 500 mg), showing that low temperature and high pressure provided a higher amount of $scCO_2$ extract, whereas the lowest yield (38.3 mg/g) of $scCO_2$ extract was obtained at a low temperature and pressure (at runs 1 and 5) when the processing time and sample amount were higher. This shows that high pressure is a key parameter for obtaining a good yield of extract.

From the Pareto chart (Figure 2a), it can be seen that, at a 90% confidence level, all independent parameters and their interactions did not show any significant effect ($p > 0.09$) on the extract yield. A full regression model equation with respect to the extract yield (Equations (2) and (3)) and statistical analysis (Table 4) also represented that no independent and interaction parameters showed any profound effect. It is suggested that these findings were observed due to the selected low amount of cannabis loading material.

The results of the HPLC analysis of total cannabinoids in the $scCO_2$ extract show that standard order 8 provided a higher number of cannabinoids in the extract (853.12 µg/mL). From the Pareto charts (Figure 2d) and the estimated regression model (Equations (4) and (5)), it was revealed that high temperature, pressure, amount, and long processing time are the main contributing factors for ensuring the high quantity of cannabinoids in the extract. The regression model for total yield and cannabinoids are provided in Equations (2) and (3):

$$Y_{Total\ yield} = 93.19 + 12.31F1 + 39.41F2 + 14.16F3 - 9.29F4 - 2.91F1 \times F2 + 1.69F1 \times F3 + 3.44F1 \times F4 - 36.5\ Ct\ Pt \quad (2)$$

$$Y_{Total\ cannabinoids} = 351.6 + 90.8F1 + 75.2F2 + 158.3F3 + 29.7F4 + 3.2F1 \times F2 + 104.3F1 \times F3 + 40.0F1 \times F4 - 36.1\ Ct\ Pt \quad (3)$$

A simplified model for total yield and cannabinoids is presented in Equations (4) and (5):

$$Y_{Total\ yield} = 93.19 \quad (4)$$

$$Y_{Total\ cannabinoids\ in\ 50\mu L} = 351.6 + 158.3F3 \quad (5)$$

3.2. Model Response for CBD and CBDA

The relationship between selected experimental parameters and the extraction of CBD and CBDA was also determined by a half-fractional factorial design study. The goal of the study was to selectively extract the major cannabinoids CBD and CBDA from cannabis. The HPLC data for CBD and CBDA (presented in Table 3) showed that the standard order 8 yields a higher amount of these cannabinoids (195.26 and 603.37 µg/mL). As the original cannabis flower material was extracted without decarboxylation (the conversion of acidic cannabinoids into their neutral form), this resulted in a comparatively higher quantity of CBDA, compared to CBD. From the Pareto chart of CBD and CBDA (Figure 2b,c), it was found that a long processing time (F3) had a strong influence on the solubility of CBD and CBDA in $scCO_2$. Similarly, statistical analysis ($p \leq 0.09$) also showed that processing time strongly affected the extraction of CBD and CBDA, as shown in Table 5. Additionally, the two-way interaction between high pressure (250 bar) and a long processing time (3 h) demonstrated a profound effect on the extraction of CBDA. Previously, Perrotin-Brunel and van Roosmalen [23] also reported that high pressure (200 bar) increases the molar solubility of cannabinoids (CBD and THC) during $scCO_2$ extraction modeling.

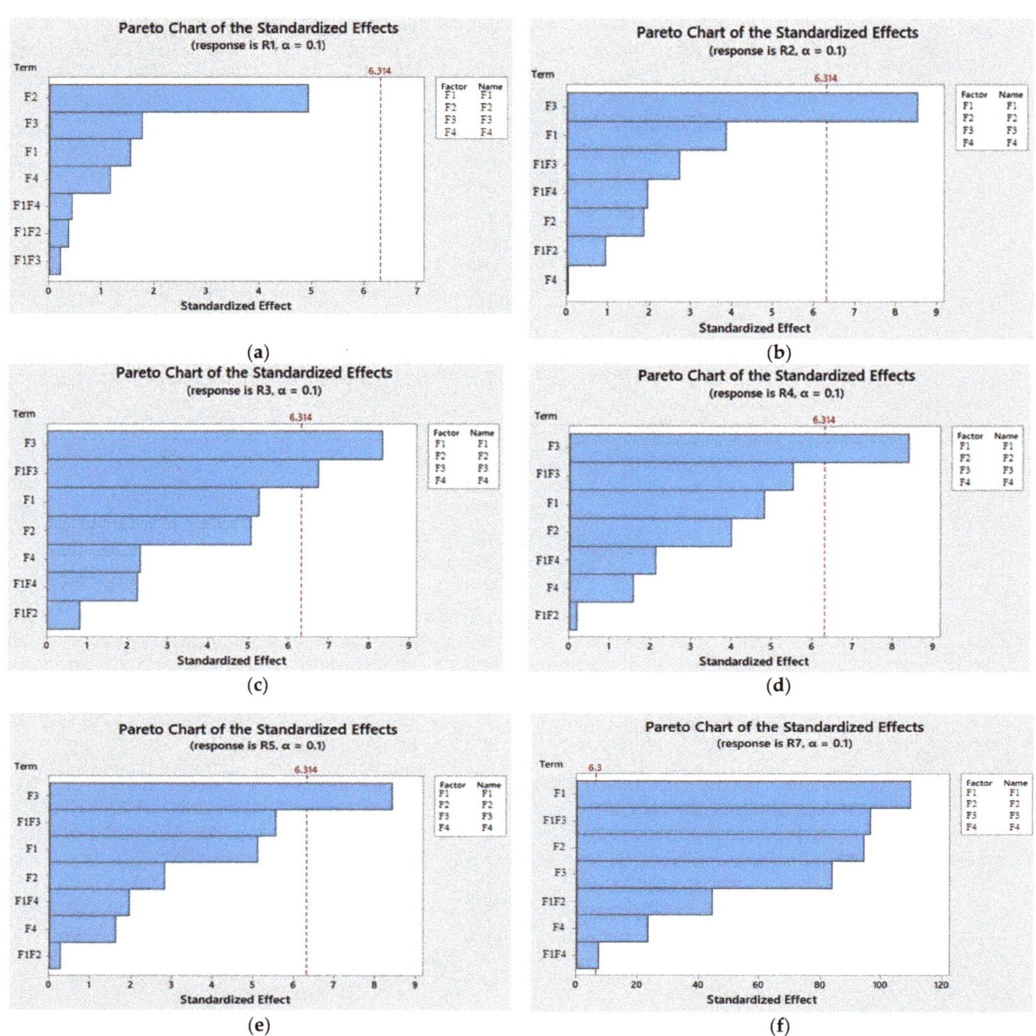

Figure 2. (**a**) Pareto chart of supercritical carbon dioxide (scCO$_2$) extract yield. (**b**) Pareto chart for the amount of cannabidiol (CBD) in supercritical carbon dioxide (scCO$_2$) extract. (**c**) Pareto chart for cannabidiolic acid (CBDA) in supercritical carbon dioxide (scCO$_2$) extract. (**d**) Pareto chart for the number of total cannabinoids in the supercritical carbon dioxide (scCO$_2$) extract yield. (**e**) Pareto chart for the amount of delta 9-tetrahydrocannabinol (D9 THC) in the supercritical carbon dioxide (scCO$_2$) extract yield. (**f**) Pareto chart for the amount of delta 9-tetrahydrocannabinolic acid (D9 THCA) in the supercritical carbon dioxide (scCO$_2$) extract yield.

The regression model for CBD and CBDA is presented in Equations (6) and (7):

$$Y_{CBD} = 105.15 + 19.39F1 + 9.33F2 + 42.70F3 - 0.06F4 - 4.64F1 \times F2 + 13.65F1 \times F3 + 9.74F1 \times F4 - 12.6\ Ct\ Pt \tag{6}$$

$$Y_{CBDA} = 219.2 + 65.8F1 + 63.3F2 + 104.0F3 + 28.7F4 + 10.2F1 \times F2 + 84.1F1 \times F3 + 28.0F1 \times F4 - 23.3\ Ct\ Pt \tag{7}$$

A simplified model for CBD and CBDA is provided in Equations (8) and (9):

$$Y_{CBD} = 105.15 + 42.70F3 \tag{8}$$

$$Y_{CBDA} = 219.2 + 104.0F3 + 84.1F1 \times F3 \tag{9}$$

Table 5. FFD statistical analysis: estimated coefficient and effects for cannabidiol (CBD) and cannabidiolic acid (CBDA).

Term	CBD					CBDA				
	Effect	Coefficient	SS	T-Value	*p*-Value	Effect	Coefficient	SS	T-Value	*p*-Value
Constant	-	105.15	20,966	21.04	0.030		219.2	224,333	17.56	0.036
F1	38.77	19.39	3006	3.88	0.161	131.6	65.8	34,620	5.27	0.119
F2	18.66	9.33	696	1.87	0.313	126.6	63.3	32,032	5.07	0.124
F3	85.40	42.70	14,584	8.55	0.074	208.0	104.0	86,554	8.33	0.076
F4	−0.11	−0.06	0	−0.01	0.993	57.5	28.7	6607	2.30	0.261
F1F2	−9.27	−4.64	172	−0.93	0.524	20.5	10.2	839	0.82	0.563
F1F3	27.31	13.65	1491	2.73	0.223	168.1	84.1	56,545	6.73	0.094
F1F4	19.49	9.74	759	1.95	0.302	56.0	28.0	6270	2.24	0.267
Center points	-	−12.6	-	−1.13	0.461	-	−23.3		−0.83	0.558
Curvature	-	-	2557	1.28	0.461	-	-	867	-	0.558
R^2 (%)			99.06					99.45		
R^2 adjusted (%)			91.51					95.02		

F1 = Temperature, F2 = pressure, F3 = processing time, F4 = material.

3.3. Model Response for THC and THCA

The effect of scCO$_2$ extraction at selected parameters on the solubility or extraction ability of Δ^9-THC, Δ^8-THC, and THCA was also investigated; THC is considered to be a major and psychoactive cannabinoid in cannabis.

From the obtained scCO$_2$ extraction data, it was found that the high temperature (45 °C) and pressure (250 bar) with 845 Kg/m^3 density of scCO$_2$ dissolved a higher amount of Δ^9-THC and THCA (Table 6). Previously, Rovetto and Aieta [20] identified that high pressure and temperature increased the percentage of THC and THCA in the scCO$_2$ extract. Similarly, in another study, it was also found that high pressure (200 bar) and temperature (52 °C) increased the solubility of THC (above 1×10^{-4}) in scCO$_2$ [24]. This study also suggested that at a low pressure of around 150 bar, the solubility of THC in scCO$_2$ decreases. These findings correlated with our study and confirmed that low temperature (35 °C) and pressure (150 bar) decreased the yield of Δ^9-THC in the extract. Conversely, in the case of THCA, it was suggested that the combination of low temperature and high pressure (run 2) produced the lowest amount. The low quantity of THC and THCA, compared to CBD and CBDA, in the extract was due to the nature of the plant material. However, no Δ^8-THC was found in the obtained extracts. This might be due to the absence of Δ^8-THC in the original plant material.

Table 6. HPLC quantification for delta 9-tetrahydrocannabinol (D9-THC), delta 9-tetrahydrocannabinol acid (D9-THCA), delta 8-tetrahydrocannabinol (D8-THC), cannabidivann (CBDV), and tetrahydrocannabivann (THCV) in the supercritical carbon dioxide extract.

Std. Order	F1	F2	F3	F4	R5 D9-THC (μg/mL) *	R5 % D9-THC (*w/w*)	R7 D9-THCA (μg/mL) *	R7 % D9-THCA (*w/w*)	R8 D8-THC (μg/mL) *	R9 CBDV (μg/mL) *	R9 % CBDV (*w/w*)	R10 THCV (μg/mL) * ^	R10 % THCV (*w/w*)
10	0	0	0	0	8.98 ± 0.01	0.036	3.59 ± 0.00	0.014	-	-	-	0.68±	0.003
3	−1	1	−1	1	6.04 ± 0.02	0.024	3.46 ± 0.00	0.014	-	-	-	0.53±	0.002
7	−1	1	1	−1	7.07 ± 0.01	0.028	2.46 ± 0.00	0.010	-	-	-	0.59±	0.002
4	1	1	−1	−1	4.31 ± 0.02	0.017	4.64 ± 0.01	0.019	-	-	-	0.40±	0.002
9	0	0	0	0	9.08 ± 0.02	0.036	3.65 ± 0.21	0.015	-	0.13 ± 0.00	0.001	0.69±	0.003
6	1	−1	1	−1	14.68 ± 0.09	0.059	6.08 ± 0.02	0.024	-	0.30 ± 0.00	0.001	0.85±	0.003
5	−1	−1	1	1	6.34 ± 0.02	0.025	1.31 ± 0.00	0.005	-	-	-	0.47±	0.002
2	1	−1	−1	1	4.78 ± 0.03	0.019	0.91 ± 0.00	0.004	-	-	-	0.50±	0.002
8	1	1	1	1	21.06 ± 0.02	0.084	11.92 ± 0.02	0.048	-	1.13 ± 0.00	0.004	0.76±	0.003
1	−1	−1	−1	−1	3.92 ± 0.01	0.016	1.17 ± 0.00	0.005	-	-	-	0.41±	0.002

F1 = Temperature, F2 = pressure, F3 = processing time, F4 = material. The central points are at rows 1 and 5, * 20 times diluted extract. ^: standard error mean is less than 0.00.

The statistical analysis of the module (shown in Table 7), Pareto chart (Figure 2e,f), and regression equation for the THC showed that only a long processing time was the main factor regarding obtaining a high yield from the extract. Conversely, in the case of THCA, it was suggested that all main parameters and their two-way interactions were contributing to producing a good yield in terms of extract (shown in Table 7). Currently, due to the lack of mathematical modeling study on the percentage yield, or the solubility of THC in the scCO$_2$, the findings of this work do not correlate with the literature.

Table 7. Estimated coefficient and effects for delta 9-tetrahydrocannabinol (D9-THC) and delta 9-tetrahydrocannabinol acid (D9-THCA).

Term	D9-THC Effect	D9-THC Coefficient	D9-THC SS	D9-THC T-Value	D9-THC *p*-Value	D9-THCA Effect	D9-THCA Coefficient	D9-THCA SS	D9-THCA T-Value	D9-THCA *p*-Value
Constant		8.776	259.66	18.46	0.034		3.9935	95.2478	231.51	0.003
F1	4.865	2.432	47.33	5.12	0.123	3.7885	1.8943	28.7055	109.81	0.006
F2	2.694	1.347	14.513	2.83	0.216	3.2565	1.6282	21.2096	94.39	0.007
F3	8.027	4.013	128.86	8.44	0.075	2.9005	1.4503	16.8258	84.07	0.008
F4	1.555	0.777	4.83	1.63	0.349	0.8105	0.4053	1.3138	23.49	0.027
F1F2	0.262	0.131	0.14	0.28	0.829	1.5310	0.7655	4.6879	44.38	0.014
F1F3	5.295	2.647	56.07	5.57	0.113	3.3290	1.6645	22.1645	96.49	0.007
F1F4	1.873	0.936	7.014	1.97	0.299	0.2430	0.1215	0.1181	7.04	0.090
Center points	-	−0.75	-	−0.71	0.608	-				
Curvature	-	-	0.903	-	0.608	-	−0.3730	0.2226	−9.67	0.066
R^2 (%)	99.31					100				
R^2 adjusted (%)	93.77					99.98				

F1 = Temperature, F2 = pressure, F3 = processing time, F4 = material.

The full regression model for THC and THCA are shown in the following equations (Equations (10)–(13)), showing the influence of statistical parameters. The factorial design study generated from SFE-processing conditions also revealed that processing time is one of the most significant parameters for THC, whereas, in the case of THCA, all main factors contributed effectively.

3.4. Regression Model for THC and THCA

Regression eqution for THC and THCA of FFD model are provided in Equations (10) and (11).

$$Y_{D9\text{-}THC} = 8.776 + 2.432F1 + 1.347F2 + 4.013F3 + 0.777F4 + 0.131F1 \times F2 + 2.647F1 \times F3 + 0.936F1 \times F4 - 0.75\,Ct\,Pt \tag{10}$$

$$Y_{D9\text{-}THCA} = 3.994 + 1.894F1 + 1.628F2 + 1.450F3 + 0.405F4 + 0.766F1 \times F2 + 1.665F1 \times F3 + 0.122F1 \times F4 - 0.373\,Ct\,Pt \tag{11}$$

3.5. A Simplified Model of THC and THCA

Regression equations for THC and THCA of simplified FFD model are provided in Equations (12) and (13).

$$Y_{D9\text{-}THC} = 8.776 + 4.013F3 \tag{12}$$

$$Y_{D9\text{-}THCA} = 3.994 + 1.894F1 + 1.628F2 + 1.450F3 + 0.405F4 + 0.766F1 \times F2 + 1.665F1 \times F3 + 0.122F1 \times F4 \tag{13}$$

3.6. Model Response for Other Selected Minor Cannabinoids

From the composition of other cannabinoids (presented in Table 8) in the extract, it can be seen that CBD, CBDA, THC and THCA were the main cannabinoids in the selected cannabis strain. Other micro-cannabinoids, such as CBDV, THCV, CBG, CBGA, CBN and CBC were found in trace amounts. The results have shown that the extracts obtained from runs 6 and 8 showed a considerable quantity of these micro-cannabinoids. However, another experimental run did not show a significant extraction rate. Similarly, conventional organic solvent extraction mirrored the findings regarding micro-cannabinoids.

Table 8. HPLC quantification for cannabigerol (CBGA), cannabigerolic acid (CBG), cannabinol (CBN), and cannabichomere (CBC) in the supercritical carbon dioxide extract.

Std. Order	F1	F2	F3	F4	R11 CBG (μg/mL) *	R11 % CBG (*w/w*)	R12 CBGA (μg/mL) *	R12 % CBGA (*w/w*)	R13 CBN (μg/mL) *	R13 % CBN (*w/w*)	R14 CBC (mg/mL)	R14 % CBC (*w/w*)
10	0	0	0	0	0.02 ± 0.00	0.000	0.99 ± 0.00	0.004	0.51 ± 0.00	0.002	5.96 ± 0.01	0.024
3	−1	1	−1	1	-	-	0.75 ± 0.00	0.003	0.33 ± 0.00	0.001	4.47 ± 0.00	0.018
7	−1	1	1	−1	-	-	0.68 ± 0.00	0.003	0.44 ± 0.00	0.002	7.59 ± 0.00	0.030
4	1	1	−1	−1	-	-	1.52 ± 0.00	0.006	0.99 ± 0.00	0.004	3.74 ± 0.01	0.015
9	0	0	0	0	0.02 ± 0.00	0.000	0.96 ± 0.00	0.004	0.74 ± 0.00	0.003	4.87 ± 0.00	0.019
6	1	−1	1	−1	3.58 ± 0.01	0.014	1.41 ± 0.00	0.006	0.78 ± 0.00	0.003	11.38 ± 0.00	0.045
5	−1	−1	1	1	-	-	0.41 ± 0.00	0.002	0.45 ± 0.00	0.002	4.87 ± 0.00	0.019
2	1	−1	−1	1	-	-	0.24 ± 0.00	0.001	0.28 ± 0.00	0.001	3.04 ± 0.00	0.012
8	1	1	1	1	1.20 ± 0.00	0.005	2.75 ± 0.00	0.011	0.59 ± 0.00	0.002	6.11 ± 0.00	0.024
1	−1	−1	−1	−1	-	-	0.41 ± 0.00	0.001	0.28 ± 0.00	0.001	2.28 ± 0.00	0.009

F1 = Temperature, F2 = pressure, F3 = processing time, F4 = material. The central points are at rows 1 and 5, * 20 times diluted extract.

3.7. Model Validation

After plotting the factorial design, it was identified that processing time was the main contributing factor that influences the extraction of cannabinoids. The prediction of the module for validation is represented in Table 9. The two center points were also used to determine module curvature response and repeatability. Additionally, a statistical module summary, such as the residual standard deviation (S.D.$_{res}$), the coefficient of determination (R^2), and the adjusted coefficient of determination (R^2_{adj}) of each cannabinoid were also verified, to determine the response of the module. However, due to the short supply of cannabis material, and the consideration of processing time only as a statistically significant influencing parameter, extraction at both low and high levels was not performed.

Table 9. Prediction for model validation.

Factor	Best Level	Minimized Level
Pressure	250 bar	100 bar
Temperature	45 °C	35 °C
Processing time	180 min	30 min
Raw material	1 g	0.5 g
Estimated total yield	93.19	93.19
Estimated yield of CBD	147.85	62.45
Estimated yield of CBDA	407.30	199.30
Estimated yield of D9-THC	12.79	4.76
Estimated yield of all 11 cannabinoids	509.90	193.30

3.8. Conventional Extraction

Cannabinoids from cannabis material are freely soluble in organic solvents. They can be easily extracted using alcohols with a 90% yield [25]. However, due to their toxic behavior and flammable properties, a promising alternative extraction technique is required. The results for the conventional extraction of cannabis are shown in Table 10.

Table 10. Conventional extraction of cannabinoids from cannabis plant material.

CBD % (w/w)	CBG % (w/w)	CBDA % (w/w)	CBN % (w/w)	D9-THC % (w/w)	CBC % (w/w)	D9-THCA % (w/w)
2.47 ± 0.064	0.07 ± 0.001	10.92 ± 0.282	0.17 ± 0.003	0.09 ± 0.002	0.14 ± 0.003	0.43 ± 0.01

The organic solvent extraction of cannabis material was performed to determine the efficiency of scCO$_2$ extraction. From the comparative study, it was observed that the extract obtained from organic solvent extraction had a relatively higher percentage of the main cannabinoid than with scCO$_2$ extraction. Previously, Omar et al. [26] also reported that the organic solvent extraction process of cannabinoids is more efficient than the scCO$_2$ extraction method. However, scCO$_2$ extraction provides a solvent-free extract with the advantage of selective extraction (it does not dissolve pigments, including chlorophyll). In terms of purity, scCO$_2$ also had a relatively high percentage of cannabinoids, compared to the conventional extract, and the process does not need an additional purification step.

The organic solvent extraction of cannabis leftover material (obtained after the factorial design experiment) was also performed (represented in Table 11), to determine the retained number of cannabinoids after scCO$_2$ extraction. It was found that samples obtained from the standard order run 8 demonstrated a very low quantity of cannabinoids, compared to other runs. Only a trace amount of CBDA was left after scCO$_2$ extraction, whereas a high number of cannabinoids were found in extracts obtained from run 1. The scCO$_2$ extraction for the standard order run 1 was performed at a low temperature (35 °C), pressure (150 bar), processing time (30 min), and amount (0.5 g). This run gives a very low number of cannabinoids with scCO$_2$ extraction.

Table 11. Conventional extraction of cannabinoids from the leftover cannabis sample (obtained after factorial design study) for cannabidiol (CBD) and cannabidiolic acid (CBDA), delta 9-tetrahydrocannabinol (D9-THC), delta 9-tetrahydrocannabinol acid (D9-THCA), and total cannabinoids.

Std. Order	F1	F2	F3	F4	CBD (µg/mL) *	CBD % (w/w)	CBDA (µg/mL) *	CBDA % (w/w)	D9-THC (µg/mL) *	D9-THCA (µg/mL) *	D9-THCA % (w/w)	Total Cannabinoids (µg/mL) *	Total Cannabinoids % (w/w)
10	0	0	0	0	-		6.43 ± 0.04	0.086	-	0.09	0.001	6.81 ± 0.05	0.091
3	−1	1	−1	1	1.51 ± 0.00	0.020	19.47 ± 0.12	0.260	-	0.54	0.007	22.40 ± 0.10	0.299
7	−1	1	1	−1	0.16 ± 0.00	0.002	7.87 ± 0.05	0.105	-	0.11	0.001	8.37 ± 0.01	0.112
4	1	1	−1	−1	0.973 ± 0.00	0.013	15.03 ± 0.04	0.200	-	0.42	0.006	16.96 ± 0.02	0.226
9	0	0	0	0	-		5.48 ± 0.02	0.073	-	0.02	0.000	5.79 ± 0.00	0.077
6	1	−1	1	−1	-		4.70 ± 0.00	0.063	-	0.05	0.001	5.01 ± 0.00	0.067
5	−1	−1	1	1	0.05 ± 0.00	0.001	9.78 ± 0.10	0.130	-	0.25	0.003	10.48 ± 0.05	0.140
2	1	−1	−1	1	-		7.95 ± 0.02	0.106	-	0.17	0.002	8.50 ± 0.03	0.113
8	1	1	1	1	-		0.67 ± 0.00	0.009	-	-	0	0.67 ± 0.00	0.009
1	−1	−1	−1	−1	3.05 ± 0.01	0.041	34.71 ± 0.14	0.463	-	0.94	0.013	40.03 ± 0.10	0.534

F1 = Temperature, F2 = pressure, F3 = processing time, F4 = material. The central points are at rows 1 and 5, * 20 times diluted extract.

Therefore, it was suggested that the collection of the low number of cannabinoids with scCO$_2$, compared to organic solvent extraction, might be due to the limitations of the instrument. It might be the case that during the separation of the extract from the Helix extraction chamber, a pressure effect (throttling effect) occurred that affected the yield of cannabinoids. Additionally, scCO$_2$ extract was collected in a glass chamber; as a result, volatile components from the extract evaporated, which may also have decreased the yield of cannabinoids.

4. Conclusions

The supercritical fluid extraction technique is growing in popularity as a solvent to selectively dissolve and extract non-polar components from raw materials. By applying the half-fractional factorial design study, the effectiveness of four main factors on the extraction of cannabinoids using scCO$_2$ was determined. The key finding identified from this study was that high pressure and temperature together generate the highest yield of cannabinoids, when using scCO$_2$ as the extraction solvent. In addition, a longer processing time was the most dominant factor aiding the dissolution of cannabinoids in scCO$_2$. From the mathematical model prediction, it was determined that the dry flower weight was a key factor to determine any associations between parameters. It was identified that the optimal temperature, pressure, and processing time were 250 bar, 45 °C, and 180 min, respectively. Due to the selection of a lesser amount of cannabis material for the factorial design study (low, medium, and high levels), the model response did not show a significant variation. Another possible reason for this was the density of CO$_2$, as the variation that was trialed was not adequate to produce appreciable interaction effects across the selected parameters. The quality of the model was also verified by regression analysis and selected central points. In addition, the study further identified that scCO$_2$ extraction at high pressure and temperature provided the highest yield of cannabinoids.

Author Contributions: Conceptualization, S.Q., J.R.F. and H.S.P.; methodology, S.Q., J.R.F. and H.S.P.; validation, S.Q., J.R.F., Y.J.M.T. and H.S.P.; formal analysis, S.Q., J.R.F., Y.J.M.T. and H.S.P.; writing—original draft preparation, S.Q. and J.R.F.; visualization, S.Q. and J.R.F.; project administration, S.Q., J.R.F., Y.J.M.T. and H.S.P. All authors have read and agreed to the published version of the manuscript.

Funding: This research received no external funding.

Institutional Review Board Statement: Not applicable.

Informed Consent Statement: Not applicable.

Data Availability Statement: Not applicable.

Acknowledgments: Sadia Qamar is a recipient of the Australian Government Research Training Program Scholarship from The University of Queensland, Brisbane, Australia. The authors also thank Andrew K. Whittaker of the Australian Institute for Bioengineering and Nanotechnology (AJBN) and Kristofer Thurecht at the Center for Advanced Imaging, the University of Queensland. Brisbane, QLD 4072 Australia, for their support and access to specialized equipment, including the CO_2 unit used in this research. The authors also acknowledge the support of the School of Pharmacy, The University of Queensland.

Conflicts of Interest: The authors declare no conflict of interest.

References

1. Schrot, R.J.; Hubbard, J.R. Cannabinoids: Medical implications. *Ann. Med.* **2016**, *48*, 128–141. [CrossRef]
2. Kuhathasan, N.; Dufort, A.; MacKillop, J.; Gottschalk, R.; Minuzzi, L.; Frey, B.N. The use of cannabinoids for sleep: A critical review on clinical trials. *Exp. Clin. Psychopharmacol.* **2019**, *27*, 383–401. [CrossRef] [PubMed]
3. Baron, E.P. Medicinal Properties of Cannabinoids, Terpenes, and Flavonoids in Cannabis, and Benefits in Migraine, Headache, and Pain: An Update on Current Evidence and Cannabis Science. *Headache J. Head Face Pain* **2018**, *58*, 1139–1186. [CrossRef] [PubMed]
4. Davis, M.P. Cannabinoids for Symptom Management and Cancer Therapy: The Evidence. *J. Natl. Compr. Cancer Netw.* **2016**, *14*, 915–922. [CrossRef] [PubMed]
5. Wrona, O.; Rafińska, K.; Możeński, C.; Buszewski, B. Supercritical Fluid Extraction of Bioactive Compounds from Plant Materials. *J. AOAC Int.* **2017**, *100*, 1624–1635. [CrossRef] [PubMed]
6. Bayrak, S.; Sökmen, M.; Aytaç, E.; Sökmen, A. Conventional and supercritical fluid extraction (SFE) of colchicine from Colchicum speciosum. *Ind. Crop. Prod.* **2019**, *128*, 80–84. [CrossRef]
7. Zoccali, M.; Donato, P.; Mondello, L. Recent advances in the coupling of carbon dioxide-based extraction and separation techniques. *TrAC Trends Anal. Chem.* **2019**, *116*, 158–165. [CrossRef]
8. Patil, P.D.; Patil, S.P.; Kelkar, R.K.; Patil, N.P.; Pise, P.V.; Nadar, S.S. Enzyme-assisted supercritical fluid extraction: An integral approach to extract bioactive compounds. *Trends Food Sci. Technol.* **2021**, *116*, 357–369. [CrossRef]
9. Saini, R.K.; Keum, Y.S. Carotenoid extraction methods: A review of recent developments. *Food Chem.* **2018**, *240*, 90–103. [CrossRef]
10. Ahangari, H.; King, J.W.; Ehsani, A.; Yousefi, M. Supercritical fluid extraction of seed oils–A short review of current trends. *Trends Food Sci. Technol.* **2021**, *111*, 249–260. [CrossRef]
11. Gallego, R.; Bueno, M.; Herrero, M. Sub-and supercritical fluid extraction of bioactive compounds from plants, food-by-products, seaweeds and microalgae—An update. *Trends Anal. Chem.* **2019**, *116*, 198–213. [CrossRef]
12. Span, R.; Wagner, W. A new equation of state for carbon dioxide covering the fluid region from the triple-point temperature to 1100 K at pressures up to 800 MPa. *J. Phys. Chem. Ref. Data* **1996**, *25*, 1509–1596. [CrossRef]
13. Płotka-Wasylka, J.; Rutkowska, M.; Owczarek, K.; Tobiszewski, M.; Namieśnik, J. Extraction with environmentally friendly solvents. *Trends Anal. Chem.* **2017**, *91*, 12–25. [CrossRef]
14. Sharif, K.; Rahman, M.; Azmir, J.; Mohamed, A.; Jahurul, M.; Sahena, F.; Zaidul, I.S.M. Experimental design of supercritical fluid extraction–A review. *J. Food Eng.* **2014**, *124*, 105–116. [CrossRef]
15. Hibbert, D.B. Experimental design in chromatography: A tutorial review. *J. Chromatogr. B* **2012**, *910*, 2–13. [CrossRef] [PubMed]
16. Elhalil, A.; Tounsadi, H.; Elmoubarki, R.; Mahjoubi, F.; Farnane, M.; Sadiq, M.; Abdennouria, M.; Qourzalb, S.; Barkaa, N. Factorial experimental design for the optimization of catalytic degradation of malachite green dye in aqueous solution by Fenton process. *Water Resour. Ind.* **2016**, *15*, 41–48. [CrossRef]
17. Kordkandi, S.A.; Forouzesh, M. Application of full factorial design for methylene blue dye removal using heat-activated persulfate oxidation. *J. Taiwan Inst. Chem. Eng.* **2014**, *45*, 2597–2604. [CrossRef]
18. Cornell, J.A. *A Primer on Experiments with Mixtures*; John Wiley & Sons: Hoboken, NJ, USA, 2011.
19. Grijó, D.R.; Osorio, I.A.V.; Cardozo-Filho, L. Supercritical Extraction Strategies Using CO_2 and Ethanol to Obtain Cannabinoid Compounds from Cannabis Hybrid Flowers. *J. CO_2 Util.* **2019**, *30*, 241–248. [CrossRef]
20. Rovetto, L.J.; Aieta, N.V. Supercritical carbon dioxide extraction of cannabinoids from *Cannabis sativa* L. *J. Supercrit. Fluids* **2017**, *129*, 16–27. [CrossRef]

21. Attard, T.M.; Bainier, C.; Reinaud, M.; Lanot, A.; McQueen-Mason, S.J.; Hunt, A.J. Utilisation of supercritical fluids for the effective extraction of waxes and Cannabidiol (CBD) from hemp wastes. *Ind. Crops Prod.* **2018**, *112*, 38–46. [CrossRef]
22. Hazekamp, A. The trouble with CBD oil. *Med. Cannabis Cannabinoids* **2018**, *1*, 65–72. [CrossRef] [PubMed]
23. Perrotin-Brunel, H.; van Roosmalen, M.; van Spronsen, J.; Verpoorte, R.; Peters, C.; Witkamp, G.J.I.-G. Supercritical fluid extraction of cannabis: Experiments and modelling of the process design. *J. Supercrit. Fluids* **2010**, *52*, 1–6.
24. Perrotin-Brunel, H.; Perez, P.C.; van Roosmalen, M.J.; van Spronsen, J.; Witkamp, G.-J.; Peters, C.J.J.T.J.o.S.F. Solubility of Δ9-tetrahydrocannabinol in supercritical carbon dioxide: Experiments and modeling. *J. Supercrit. Fluids* **2010**, *52*, 6–10. [CrossRef]
25. De Vita, D.; Madia, V.N.; Tudino, V.; Saccoliti, F.; De Leo, A.; Messore, A.; Roscilli, P.; Botto, A.; Pindinello, I.; Santilli, G.; et al. Comparison of different methods for the extraction of cannabinoids from cannabis. *Nat. Prod. Res.* **2019**, *34*, 2952–2958. [CrossRef] [PubMed]
26. Omar, J.; Olivares, M.; Alzaga, M.; Etxebarria, N. Optimisation and characterisation of marihuana extracts obtained by supercritical fluid extraction and focused ultrasound extraction and retention time locking GC-MS. *J. Sep. Sci.* **2013**, *36*, 1397–1404. [CrossRef]

 *processes*

Article

Ultrasound-Assisted Extraction of Semi-Defatted Unripe Genipap (*Genipa americana* L.): Selective Conditions for the Recovery of Natural Colorants

Grazielle Náthia-Neves [1,2,*], Ádina L. Santana [1,3], Juliane Viganó [1,4], Julian Martínez [1] and Maria Angela A. Meireles [1,*]

[1] School of Food Engineering, University of Campinas (UNICAMP), R. Monteiro Lobato 80, Campinas 13083-862, SP, Brazil; adina.santana@gmail.com (Á.L.S.); julianevigano@gmail.com (J.V.); julian@unicamp.br (J.M.)

[2] Department of Chemical Engineering and Environmental Technology, University of Valladolid, Prado de la Magdalena 5, 47011 Valladolid, Spain

[3] Food Science Institute, Kansas State University, 1530 N Mid Campus Drive, Manhattan, KS 66506, USA

[4] Department of Chemical Engineering, Institute of Environmental, Chemical and Pharmaceutical Sciences, Universidade Federal de São Paulo, R. São Nicolau 210, Diadema 09913-030, SP, Brazil

[*] Correspondence: grazinathia@yahoo.com.br (G.N.-N.); maameireles@lasefi.com (M.A.A.M.); Tel.: +55-19-3521-0100 (G.N.-N. & M.A.A.M.); Fax: +55-19-3521-4027 (G.N.-N. & M.A.A.M.)

Citation: Náthia-Neves, G.; Santana, Á.L.; Viganó, J.; Martínez, J.; Meireles, M.A.A. Ultrasound-Assisted Extraction of Semi-Defatted Unripe Genipap (*Genipa americana* L.): Selective Conditions for the Recovery of Natural Colorants. *Processes* **2021**, *10*, 1435. https://doi.org/10.3390/pr9081435

Academic Editor: Weize Wu

Received: 16 July 2021
Accepted: 17 August 2021
Published: 19 August 2021

Publisher's Note: MDPI stays neutral with regard to jurisdictional claims in published maps and institutional affiliations.

Abstract: Ultrasound-assisted extraction (UAE) of semi-defatted unripe genipap (SDG) using supercritical CO_2 was performed to enhance the recovery of natural colorant iridoids genipin and geniposide. There are currently few natural sources of iridoids, and their application as colorants is scarce. The UAE resulted in extracts with blue and green colors using water and ethanol, respectively. The highest global yield and genipin content was recovered with water, and the geniposide was significantly recovered with ethanol. With water at 450 W, the UAE raised the maximum global yield (25.50 g/100 g raw material). At 150 W and 7 min, the maximum content of genipin (121.7 mg/g extract) and geniposide (312 mg/g extract) was recovered. The total phenolic content (TPC) and antioxidant capacity with the oxygen reactive antioxidant capacity (ORAC) assay were also high in aqueous extracts. Ethanolic extracts showed high ferric-reducing ability antioxidant potential (FRAP) values. UAE showed an efficient and fast method to obtain different extracts' fractions from SDG, which have a wide spectrum of applications, especially as natural food colorants.

Keywords: ultrasound-assisted extraction; iridoids; genipin; geniposide; antioxidants

1. Introduction

Genipap (*Genipa americana* L.) is commonly used to produce jams, ice cream, soft drinks, liquor, and wine, consumed by the populations in the Brazilian Amazon. Unripe genipap contains genipin and geniposide (which are not present in the ripe fruits). These compounds have been recently associated with health benefits, such as antiviral [1] and anti-allergic [2] activities, as well as neuroprotection [3,4] and antidiabetics [5]. Genipin is responsible for the violet to the blue-dark complex coloring, and geniposide is attributed to the yellow-green complex [6].

Genipin can quickly react with amino acids in the presence of oxygen to produce a blue color complex [7]. Geniposide represents more than 70% of the iridoid content in the unripe genipap [8]. To obtain the blue-color complex, geniposide requires hydrolysis, often mediated by enzymes [8–10].

There is a need for natural colorants to replace the synthetic ones in food products such as beverages, desserts, gels, and confectionery. The color of food is often associated with the flavor, safety, and nutritional value of the product [11]. Therefore, these compounds' obtention from natural sources represents an important alternative to provide additives to the food industry. Indeed, the extraction of iridoids from genipap is common with organic solvents such as ethyl acetate and n-butanol or enzymatic hydrolysis with β-glucosidase [12].

Moreover, the low stability of natural colorants and other antioxidants in environmental conditions induces the need for rapid extraction processes of these target compounds, as well as the use of Generally Recognized as Safe (GRAS) solvents (water and ethanol). Recent studies recovered genipin and geniposide from crude unripe genipap using pressurized liquid extraction with ethanol [13] and enzyme-assisted extraction combined with high-pressure processing [14].

The use of supercritical fluid extraction (SFE) as a type of pre-treatment (by the recovering of oil from a plant matrix) carried out on the raw material before the extraction process of polar compounds has already been reported in the literature. SFE, for instance, has already been combined with other techniques such as pressurized liquid extraction, pressurized hot water extraction, low-pressure solvent extraction, and ultrasound-assisted extraction to obtain polyphenols, iridoids, flavonoids, and polar pigments [15–17]. Ultrasound-assisted extraction (UAE) has been shown as an efficient technique to recover bioactive compounds from plant matrices because of induced acoustic cavitation that disrupts cell walls and enhances the diffusion of target compounds into the solvent [18]. Considering extractions from by-product fractions, UAE was effective in recovering lycopene [19] and pectin [20] from tomato and phenolic compounds from lime peels [21].

Ramos-de-la-Peña et al. [22] used UAE of crude genipap coupled with enzyme-assisted extraction to enhance the obtaining of genipin. Strieder et al. [18] used UAE of crude genipap with the use of milk as a solvent to provide a blue-colored carrier system.

In the current work, we used, for the first time, UAE as an alternative process to recover iridoids from semi-defatted unripe genipap fruits (SDG), a genipap by-product derived from lipid extraction using supercritical carbon dioxide (SC-CO_2). The results obtained from this work are expected to support further development of processes to intensify the extraction and stability of natural colorants from genipap until the complete valorization of this plant for food and non-food applications.

2. Material and Methods

2.1. Sample Preparation

Crude unripe genipap fruits (*Genipa americana* L.) were provided by the company Sítio do Bello (Paraibuna, São Paulo, Brazil), transported and stored under freezing conditions (−18 °C) until drying in a freeze-dryer (LP1010, Liobrás, São Carlos, SP, Brazil) at 300 μm Hg for 96 h. The dried fruits (whole fruit with peel) were ground in a knife mill (Marconi, model MA-340, Piracicaba, Brazil), and the particle size distribution was determined in a vibratory system (Bertel, model 1868, Caieiras, Brazil) using sieves from 16 to 80 mesh (Tyler series, Wheeling, IL, USA). The mean particle diameter (dp) was 0.23 ± 0.03 mm, determined according to the method proposed by ASAE [23]. The ground genipap was semi-defatted (the nonpolar fraction of the raw material was reduced from 8 to 0.9 wt.%) by SFE using CO_2 as the solvent. The SFE process was performed according to the method described by Nathia-Neves et al. [15], which consists of continuous solvent flow through a fixed solid bed of sample particles placed inside the extraction column. The SFE conditions were at the temperature of 60 °C, the pressure of 300 bar, and the solvent to raw material ratio of 16 (*w/w*). The SDG was stored at −18 °C in the absence of light before further experiments.

The SDG was characterized according to moisture (method 920.151 from AOAC [24]): 13.9 ± 0.1 wt.% dry basis (d.b.), ash (method 923.03 from AOAC [24]): 4.4 ± 0.3 wt.% d.b., protein (method 970.22 from AOAC [24]): 10 ± 1 wt.% d.b., lipids (method 963.15 from AOAC [24]): 0.9 ± 0.3 wt.% d.b., and carbohydrates (calculated by difference): 84.7 wt.% d.b.

2.2. Reagents

Ethanol (99.0%, Dinâmica, Diadema, Brazil) and distilled water (Millipore, Bedford, USA) were used as solvents. The iridoids standards (genipin > 98% and geniposide > 98%)

were obtained from Sigma-Aldrich (Sao Paulo, Brazil). Acetonitrile (HPLC grade, J. T. Baker, Phillipsburg, NJ, USA), ultrapure water (Millipore, Bedford, MA, USA), and formic acid (Dinâmica, Diadema, Brazil) were used in the high-performance liquid chromatography (HPLC). The Folin–Ciocalteu reagent and gallic acid (Sigma-Aldrich, Sao Paulo, Brazil) were used for the total phenolic content analysis. The reagents used in the assays to measure antioxidant capacity were 6-hydroxy-2,5,7,8-tetramethylchromane-2-carboxylic acid (Trolox), ferric chloride ($FeCl_3$) 2,4,6-tris(2-pyridyl)-s-triazine (TPTZ), 2,2′-azobis(2-methylpropionamidine) dihydrochloride (APPH), and fluorescein, which were purchased from Sigma-Aldrich (Sao Paulo, Brazil).

2.3. Experimental

The UAE was performed with a 19 kHz ultrasonic probe (Unique, 800 W, Indaiatuba, Brazil). The effects of nominal power (150, 300, and 450 W), extraction time (1, 3, 5, and 7 min), and solvent (ethanol and water) were evaluated. Approximately 1.5 g of raw material was inserted in a 50 mL falcon tube containing 25 mL of solvent. The probe contact height with the medium was standardized to 25 mm. An ice bath was used to prevent the overheating of the extraction medium. The extracts were separated from the solid fraction using 0.22 mm Whatman paper filters and subsequently stored at $-18\,°C$ in the absence of light until further analyses.

2.4. Extract Evaluation
2.4.1. Global Yield

The global yield (X_0) of SDG extracts was determined by removing the solvent from the extract by evaporating a 5 mL aliquot in an oven (TE-395, Tecnal, Piracicaba, Brazil) at 100 °C for 24 h. The global yield for each UAE condition was calculated according to Equation (1):

$$X_0 = \left(\frac{m_{EXT}}{F_0} \right) \times 100 \qquad (1)$$

where X_0 (% (g extract/100 g SDG), dry basis) is the global extraction yield, m_{EXT} (g) is the mass of dried extract, and F_0 (g) is the mass of raw material used in UAE.

2.4.2. Total Phenolic Content (TPC)

The extracts' total phenolic content (TPC) was quantified using the Folin–Ciocalteu reagent [25]. Each extract was diluted in distilled water. Triplicates of the sample (20 µL) and Folin–Ciocalteu (20 µL) reagent were mixed and incubated at room temperature for 3 min. Afterwards, 20 µL of sodium carbonate solution and 140 µL of distilled water were added and the mixture was kept in the dark for 2 h at room temperature. The absorbance was read at 725 nm with the aid of a microplate reader (FLUOstar Omega, BMG LABTECH GmbH, Ortenberg, Germany). The gallic acid standard (0.02 to 0.16 mg/mL) was used, and the results were expressed as mg gallic acid equivalent (GAE) per g of SDG extract.

2.4.3. Iridoids Quantification by HPLC Analysis

The iridoids (genipin and geniposide) were quantified by HPLC coupled with a diode-array detector (HPLC-DAD, Waters, Alliance model E2695, Milford, CT, USA) system and a C18 column (Kinetex, 100 × 4.6 mm i.d.; 2.6 µm; Phenomenex, Torrance, CA, USA) at 35 °C. The iridoids were detected according to the method described by Náthia-Neves et al. [26]. The mobile phases consisted of water (A) and acetonitrile (B), both acidified with 0.1% formic acid (v/v). The elution gradient at 1.5 mL/min was performed as follows: 0 min: 99% (A), 9 min: 75% (A), 10 min: 99% (A), and 13 min: 99% (A). The calibration curves of the iridoids were obtained in the range of 0.1–1000 µg/mL for geniposide (R^2 = 0.9998) and 0.1–625 µg/mL for genipin (R^2 = 0.9998). Figure 1 shows the chromatograms of the standards and the extracts obtained from SDG. Each sample was injected in duplicate, and the results were expressed in mg of genipin or mg of geniposide per g of dried SDG extract.

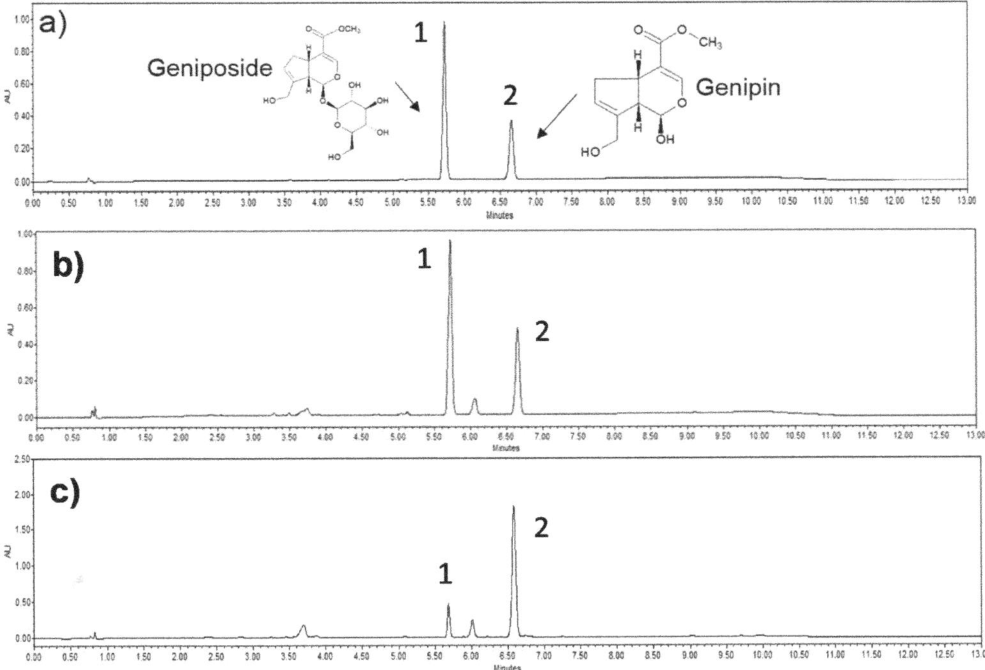

Figure 1. Representative HPLC/DAD chromatograms for genipin and geniposide recovered at 240 nm: (**a**) standard solution of genipin and geniposide, (**b**) ethanolic extract from SDG obtained at 150 W and 7 min, and (**c**) aqueous extract from SDG obtained at 150 W and 7 min. The retention times of geniposide and genipin were 5.73 and 6.65 min, respectively.

2.4.4. Antioxidant Capacity

The ferric-reducing ability of extracts was determined by the FRAP method [27]. The FRAP solution was prepared using 0.3 M acetate buffer (pH 3.6), 10 mM TPTZ in 40 mM HCl solution, and 20 mM $FeCl_3$ (10:1:1, $v/v/v$). Triplicates of the extract diluted in water (25 μL) were added to 175 μL of FRAP solution and kept in the dark for 30 min at 37 °C. The absorbance was measured at 595 nm using a microplate reader (a FLUOstar Omega BMG LABTECH GmbH, Ortenberg, Germany). The Trolox standard curve was obtained (0.01 to 0.06 mg/mL). Results were expressed in mg Trolox equivalent (TE) per g of SDG extract.

The oxygen radical absorbance capacity (ORAC) was measured by diluting the extracts and Trolox (5–25 μg/mL) in a 75 mM potassium phosphate buffer at pH 7.4 [28]. Pure potassium 75 mM phosphate buffer was used as a blank. Triplicates of the diluted sample, standard or blank (25 μL), followed by 150 μL of fluorescein working solution, were inserted into a black 96-well microplate. Afterward, the microplate was incubated at 37 °C for 15 min, and subsequently, 25 μL of AAPH solution was added to each well. The fluorescence decrease (excitation at 485 nm; emission at 510 nm) was measured for 100 min at 37 °C in the microplate reader. The results were expressed in mg of TE per g of SDG extract.

2.4.5. Color Analysis

The coloration of extracts was evaluated using a colorimeter (Hunter Associates Laboratory, Inc., Reston, VA, USA) equipped with a D65 light source and an angle of observation of 2° for all samples. The color was characterized with a CIELAB system to calculate the coordinates L^*, a^* and b^*, which were used to calculate the chroma (C^*) and hue angle (H), according to the Equations (2) and (3), respectively. The a^* represents differences in red (positive) and green (negative) colors. Positive and negative values of b^* represent yellow and blue colors, respectively. The C^* values denote the saturation or purity of color. The Hue angle (H^*) value denotes 0 for redness, 90 for yellowness, 180 for greenness, and 270 for blueness.

$$C* = \sqrt{a*^2 + b*^2} \tag{2}$$

$$H* = \arctan\left(\frac{b*}{a*}\right) \tag{3}$$

2.5. Statistical Evaluation

The effects of parameters studied were evaluated using a randomized full factorial design ($2 \times 3 \times 4$) for the solvent (water and ethanol), the nominal ultrasonic power (150, 350, and 450 W), and the extraction time (1, 3, 5, and 7 min), in duplicate (Table 1). Comparison of results within each column of Table 1 was carried out considering the different solvents used (water and ethanol). The effects of the parameters on global yield, genipin, geniposide, TPC, FRAP, and ORAC were evaluated by the analysis of variance (ANOVA) using Minitab 16® software (Minitab Inc., State College, PA, USA). Tukey's test's significant differences were evaluated with a 95% confidence level (p-value ≤ 0.05).

Table 1. Effect of the process parameters on extraction results.

Power (w)	Time (min)	X_0 (wt.%)	Genipin (mg/g Extract)	Geniposide (mg/g Extract)	TPC (mg GAE/g Extract)	FRAP (mg TE/g extract)	ORAC (mg TE/g Extract)
				Water Solvent			
150	1	14.9 ± 0.1 [e]	101 ± 1 [g]	52.30 ± 0.03 [h]	28.9 ± 0.1 [a]	5.5 ± 0.7 [a]	87 ± 6 [cdefg]
	3	19 ± 1 [d]	113 ± 1 [e]	57 ± 1 [h]	27 ± 2 [a]	5 ± 1 [a]	106 ± 13 [abcdef]
	5	19.8 ± 0.1 [cd]	120.8 ± 0.4 [ab]	23.1 ± 0.2 [ij]	19.3 ± 0.4 [b]	5.4 ± 0.2 [a]	89 ± 7 [bcdefg]
	7	22.1 ± 0.3 [bc]	121.7 ± 0.5 [a]	19.6 ± 0.3 [jk]	21 ± 1 [b]	5.05 ± 0.04 [a]	121 ± 12 [ab]
300	1	24.52 ± 0.1 [ab]	120 ± 1 [ab]	22.2 ± 0.5 [ijk]	22 ± 1 [b]	6 ± 1 [a]	113 ± 16 [abcd]
	3	25 ± 1 [a]	116 ± 1 [cd]	29 ± 1 [ij]	21.4 ± 0.4 [b]	4.4 ± 0.6 [a]	110 ± 9 [abcde]
	5	24.1 ± 0.2 [ab]	116.0 ± 0.4 [d]	20.3 ± 0.4 [ijk]	20.9 ± 0.3 [b]	5.06 ± 0.04 [a]	100 ± 3 [abcdefg]
	7	23.4 ± 0.2 [ab]	119 ± 1 [bc]	18.8 ± 0.4 [jk]	20 ± 2 [b]	6 ± 1 [a]	129 ± 8 [a]
450	1	25.5 ± 0.7 [a]	118.8 ± 0.5 [bc]	31.4 ± 0.6 [i]	21 ± 1 [b]	6 ± 1 [a]	110 ± 11 [abcde]
	3	25 ± 1 [a]	114.6 ± 0.5 [de]	7.01 ± 0.5 [l]	18 ± 1 [b]	5.1 ± 0.1 [a]	112 ± 14 [abcde]
	5	24.9 ± 0.2 [a]	110.3 ± 0.5 [f]	7.1 ± 0.1 [l]	17.9 ± 0.2 [b]	4.9 ± 0.1 [a]	119 ± 3 [abc]
	7	23 ± 1 [abc]	108.58 ± 0.04 [f]	11 ± 3 [kl]	21 ± 1 [b]	5.2 ± 0.5 [a]	126 ± 7 [a]
		X_0 (%)		**Ethanol Solvent**			
150	1	2.9 ± 0.7 [i]	71.1 ± 0.3 [l]	247 ± 5 [c]	8.7 ± 0.4 [c]	15 ± 5 [b]	73 ± 7 [g]
	3	2.1 ± 0.2 [i]	86.3 ± 0.2 [ij]	198 ± 5 [g]	10 ± 1 [c]	15 ± 3 [b]	72 ± 2 [g]
	5	3.1 ± 0.7 [i]	92 ± 1 [h]	242 ± 5 [c]	9 ± 1 [c]	17 ± 2 [b]	86 ± 13 [cdefg]
	7	3.4 ± 0.7 [i]	103.04 ± 0.05 [g]	312 ± 10 [a]	8 ± 1 [c]	12 ± 3 [b]	70 ± 19 [g]
300	1	6.9 ± 0.6 [h]	78.5 ± 0.7 [k]	225 ± 0.3 [ef]	9 ± 2 [c]	16 ± 2 [b]	89 ± 7 [bcdefg]
	3	8.3 ± 0.1 [gh]	80 ± 1 [k]	235 ± 2 [cde]	8.4 ± 0.3 [c]	16 ± 2 [b]	77 ± 8 [fg]
	5	9.1 ± 0.2 [fgh]	83 ± 1 [j]	243 ± 5 [c]	8 ± 1 [c]	16 ± 1 [b]	82 ± 7 [defg]
	7	9.7 ± 0.6 [fg]	81 ± 1 [k]	228 ± 11 [def]	7.8 ± 0.4 [c]	14.8 ± 0.2 [b]	92 ± 2 [bcdefg]
450	1	7.02 ± 0.05 [h]	85 ± 1 [ij]	260 ± 6 [b]	10 ± 1 [c]	15 ± 1 [b]	89.67 ± 0.04 [bcdefg]
	3	9 ± 1 [fg]	84 ± 1 [ij]	232 ± 12 [cde]	8.6 ± 0.3 [c]	16 ± 1 [b]	91 ± 3 [bcdefg]
	5	11.3 ± 0.2 [f]	85 ± 1 [ij]	223 ± 12 [ef]	7.6 ± 0.2 [c]	16 ± 1 [b]	80 ± 5 [efg]
	7	12 ± 1 [f]	87 ± 1 [i]	223 ± 5 [f]	7.5 ± 0.3 [c]	16.1 ± 0.1 [b]	87.2 ± 0.2 [cdefg]

X_0: global yield; TPC: total phenolic content; FRAP: ferric-reducing ability; ORAC: oxygen radical absorbance capacity. Results expressed on dry basis. Same superscript letters in the same column indicate no significant difference ($p < 0.05$).

3. Results and Discussion

The use of the SFE technique as a pre-treatment of the raw material allows the extraction of nonpolar compounds. By extracting these compounds with $SC\text{-}CO_2$, it is possible to obtain a matrix concentrated in polar compounds, such as phenolic compounds, iridoids, flavonoids, and polar pigments. This practice has been employed frequently in the literature to obtain a matrix more concentrated in polar compounds, making these compounds more efficient in processes that use water or ethanol as a solvent [15–17,29]. In addition to making polar compounds more accessible, the use of SFE combined with other techniques allows applying the biorefinery concept (by the full use of the raw material). In this work, the SFE removed 89% of the oil content from unripe genipap fruit, which was composed of palmitic acid (34 mg/mg extract), stearic acid (14 mg/mg extract), linoleic acid (276 mg/mg extract), and linolenic acid (48 mg/mg extract) [15].

3.1. Effect of the Process Parameters on Global Yield

The experimental X_0 obtained in all UAE conditions are presented in Table 1, and they range from 14.9 to 25.5 wt.% for aqueous extracts and from 2.1 to 12 wt.% for ethanolic extracts. It is worth highlighting that the raw material used in this work had already been subjected to another extraction process with $SC\text{-}CO_2$. An extract was obtained with a yield of 4.6% (composed mainly of linoleic acid) [15]. These results imply that after the extraction of the non-polar fraction, there are still many polar compounds to be extracted from the unripe genipap fruit.

Analysis of variance (ANOVA, $\alpha = 0.05$) showed that the interactions between time and power (p-value = 0.001), power and solvent (p-value = 0.005), and time and solvent (p-value = 0.008) significantly influenced the X_0. The plots of such interactions, showing each extraction variable's effect, are presented in Figure 2. It is important to mention that the dotted lines do not represent trends for non-numeric variables.

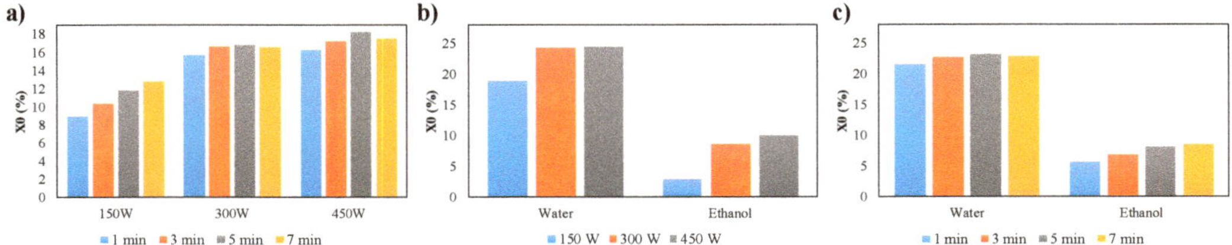

Figure 2. Effect of the process parameters on global yield: (**a**) time and power, (**b**) power and solvent, and (**c**) extraction time and solvent. Results expressed on a dry basis. Standard deviation by ANOVA = 0.6.

The efficiency of the extraction process depends on the nature of the compounds present in the raw material. The use of water, a solvent with higher polarity, allowed enhancing the X_0 compared with ethanol, as represented in Figure 2b,c. In other words, the use of water as the solvent increased X_0 by 4- to 8-folds at 150 W and 1- to 3-folds at 300 and 450 W (Table 1). Therefore, it can be concluded that water, especially for low ultrasound power, could access and solubilize a great number of molecules in the SDG, resulting in higher X_0. It is known that during the UAE, two mechanisms of mass transfer can occur: diffusion through the cell walls and washing out (rinsing) the cell contents once the walls are broken [30]. As the used material was dry and milled, once it is rehydrated, there is swelling during the steeping extraction stage, resulting in the fragmentation of the vegetal material, which increases the transfer of compounds from the raw material into the solvent [31].

The positive effect of ultrasonic power (Figure 2a,b) on the X_0 can be mainly attributed to the increase of energy density, which intensifies the phenomenon of acoustic cavitation [32]. This phenomenon can promote the rupture of plant cell walls, favoring the transport of the intracellular components present in the vegetable matrix into the bulk solvent, thus increasing the X_0 [33]. Similarly, the positive effect of extraction time (Figure 2a,c) on the X_0 is attributed to the required time to desorb and solubilize the extract molecules by the solvent. For extractions using water, the equilibrium time was reached at 3 min. While for ethanol, the X_0 increased until 7 min (Table 1). Therefore, once again showing the extraction efficiency of water as the solvent to obtain higher X_0 from SDG. On the other hand, the performance over X_0 is not always a relevant answer parameter for evaluating the extracts, since the concentration of target compounds and biological activity provide a potential application to the extract.

3.2. Effect of the Process Parameters on Iridoid Content

Among the evaluated UAE variables, the interactions between time and power (*p*-value = 0.001) and time and solvent (*p*-value = 0.001) significantly influenced the genipin recovery, as shown in Figure 3a,b, respectively. The extraction of geniposide was significantly influenced only by the interaction between solvent and power (*p*-value = 0.014, Figure 3c). It is worth mentioning that the plots presented in Figure 3 show the effect of UAE variables. Experimental results for the concentration of both target compounds obtained from SDG are presented in Table 1.

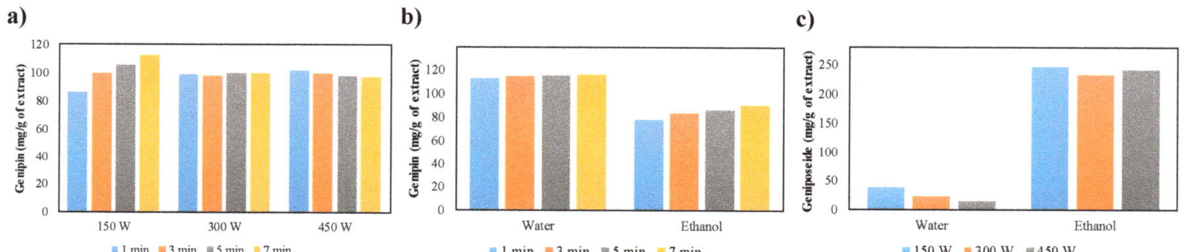

Figure 3. Effect of the process parameters on genipin and geniposide content: (**a**) extraction time and power on genipin content, (**b**) solvent and extraction time on genipin content, and (**c**) solvent and power on geniposide content. Results are expressed on a dry basis. Standard deviation by ANOVA for genipin = 0.6 and for geniposide = 5.5.

The highest genipin content (121.7 mg/g extract or 26.9 mg/SDG) was recovered at 150 W in 7 min using water as the solvent. This result is comparable with that extracted in our previous work from the whole unripe genipap without peel applying the low-pressure solvent extraction (LPSE) method with water at 40 °C and 1 bar for 25 min (233 mg/g extract) [34], and 81–196 mg/g unripe genipap obtained from crude genipap extracted with enzymes for 150 min [10]. It is worth mentioning that the extraction times used by these authors were much higher than the times used in this study. Considering that 7 min is a short time to extract bioactive compounds from a by-product plant matrix, the UAE was efficient in recovering such a target compound.

Regarding the ultrasonic power, different effects were observed. For instance, for water used as a solvent, as the power increased from 300 to 450 W and the extraction time increased, a reduction in the genipin content was observed. These findings are probably due to excessive cavitation and time exposed, which contributed to its degradation. Additionally, the water showed higher genipin content than ethanol, corroborating the results of X_0. A recent patent emphasized the use of water or other polar solvents for the obtaining of genipin-rich extracts. Indeed, non-polar organic solvents may be used as an alternative to water since their solubility is less than about 30% of the water solubility. Their polarity index varies from 0 to 5.0 [35]. However, the use of water as a solvent offers several benefits. Water is an environmentally friendly solvent, abundant, cheaper, non-toxic, and especially for extracts with food claim application, water is fully compatible.

On the other hand, the UAE with ethanol enhanced the geniposide content in ethanolic extracts (198–312 mg/g of extract), which was comparable to that from the mesocarp of crude unripe genipap (127 mg/g extract) obtained with pressurized ethanol [13]. Such finding indicates that the fruit water content may also impair the geniposide recovery in addition to the ultrasound's positive effect. The moisture content of the raw material used in the mentioned work was 80%, which leads to the formation of a hydroalcoholic complex inside the extraction system that probably decreased the geniposide recovery. The 14 wt.% moisture content of SDG explains the high recoveries of target compounds due to the reduced competition with the water bonded within the plant matrix. The enhancement of geniposide with ethanol was attributed to the affinity in such a solvent. The sugar moiety of the geniposide structure is another factor that explains its affinity to ethanol. The authors of [22,36,37] used enzymes such as pectinesterases and β-glucosidases to hydrolyze such sugar moiety to transform the glucose from geniposide to genipin. The cost of the process increased because of the use of enzymes. Besides, studies on the application of geniposide as a natural colorant should be supported.

From the findings presented in this work, it can be concluded that performing the SC-CO$_2$ extraction before the extraction of genipin and geniposides from unripe genipap fruit by UAE does not promote the degradation of these compounds. Thus, it is important to highlight that a sequential UAE could be applied to recover at least three extract fractions. A non-polar extract was obtained by SFE, a genipin-rich extract employing water as the solvent, and a second geniposide-rich extract obtained using ethanol as the solvent. Indeed, the sequential extraction process has been applied by several authors [16,38,39] to promote the best use of raw materials through their integral use.

3.3. Effect of the Process Parameters on Color

The color parameters of SDG extracts are presented in Table 2. The values of chroma (C^*) detected in the aqueous extracts ($C^* = 1.3$–3.3) for most extraction conditions indicated low saturation of colorant in the extracts, in comparison with the ethanolic ones ($C^* = 0.9$–6.8). For the extractions with ethanol, high ultrasonic powers (300 and 450 W) during 7 min increased the values of C^*, while the opposite effect was found in the remaining conditions (Table 2). The values of C^* found for the aqueous and ethanolic extracts of SDG were higher than those of heartwood extracted with a 0.1 M sodium hydroxide solution [40].

The lightness (L^*) parameter of ethanolic extracts was higher than those observed in the aqueous extracts. However, it was lower than those detected in the concentrated extracts from roselle flowers [41]. The increasing of ultrasonic powers enhanced the values of L^*, resulting in extracts with a more transparent aspect (Table 2).

Regarding the extractions with water, most extraction conditions were associated with color in the blue region (b^* negative, Table 2), similarly to those detected in the endocarp + seeds of genipap extracted with pressurized ethanol [13]. The hue angle (H) for most aqueous extracts (300–360°—blue region) was comparable with those detected in the extracts of *Spirulina platensis* [42].

The color region of ethanolic extracts was between green (negative values for a^*) and yellow (positive values for b^*). The increase in the power increased the values of b^* and decreased the hue angle (Table 2).

The blue color present in the aqueous extract is related to the presence of genipin, whereas the green color of the extract obtained with ethanol is mainly due to the presence of geniposide. The blue color of the aqueous extract (Table 2) shows that the genipin present in the raw material reacted with the proteins solubilized in the extract to form the blue pigment. However, HPLC analysis revealed that the extracts had a high content of genipin (ranged from 101 to 121.7 mg/g of extract), which indicates that not all genipin reacted with the proteins present in the extract to form the blue pigment. These results are promising because they allow obtaining an extract with natural blue pigments rich in genipin that could be used to manufacture food (candies, gums, dairy beverages) and pharmaceutical products.

Table 2. Color parameters for the SDG extracts.

Water						
Power (W)	**Time (min)**	L^*	C^*	H	a^*	b^*
150	1	1.91 ± 0.03	1.4 ± 0.1	333.5 ± 1.3	1.3 ± 0.1	−0.64 ± 0.05
	3	2.2 ± 0.4	1.5 ± 0.1	332.3± 1.7	1.4 ± 0.1	−0.72 ± 0.05
	5	2.02 ± 0.03	1.45 ± 0.01	324 ± 2	1.18 ± 0.03	−0.85 ± 0.03
	7	1.42 ± 0.14	1.3 ± 0.1	347 ± 12	1.25 ± 0.04	−0.3 ± 0.03
300	1	1.9 ± 0.1	1.7 ± 0.1	319 ± 6	1.3 ± 0.1	−1.1 ± 0.2
	3	4.3 ± 0.1	3.8 ± 0.1	106 ± 2	1.03 ± 0.09	−1.6 ± 0.1
	5	3.9 ± 0.9	3.2 ± 0.1	110 ± 2	1.01 ± 0.09	−1.5 ± 0.1
	7	3.0 ± 0.8	2.3 ± 0.3	312 ± 8	1.5 ± 0.1	−1.3 ± 0.4
450	1	2.1 ± 0.1	1.6 ± 0.1	333 ± 4	1.4 ± 0.02	−1.7 ± 0.1
	3	3.9 ± 0.2	2.7 ± 0.5	302 ± 2	1.4 ± 0.3	−2.3 ± 0.4
	5	3.5 ± 0.3	3.3 ± 0.6	301 ± 6	1.7 ± 0.2	−2.8 ± 0.7
	7	3.44 ± 0.04	2.6 ± 0.1	306 ± 1	1.51 ± 0.04	−2.1 ± 0.1
Ethanol						
Power (W)	**Time (min)**	L^*	C^*	H	a^*	b^*
150	1	4.0 ± 0.2	1.2 ± 0.2	110 ± 4	−0.4 ± 0.2	1.2 ± 0.2
	3	5.14 ± 0.02	1.0 ± 0.1	101 ± 3	−0.2 ± 0.1	1.0± 0.1
	5	4.7 ± 0.3	1.7 ± 0.3	114 ± 4	−0.7 ± 0.1	1.6 ± 0.3
	7	3.19 ± 0.01	0.9 ± 0.1	105 ± 3	−0.2 ± 0.1	1.9± 0.1
300	1	4.3 ± 0.2	3.2 ± 0.2	105 ± 2	−0.8 ± 0.1	3.1 ± 0.1
	3	4.4 ± 0.2	3.8 ± 0.2	107 ± 2	−0.9 ± 0.2	3.6 ± 0.3
	5	5.7 ± 0.3	4.1 ± 0.3	109 ± 3	−1.29 ± 0.07	3.9 ± 0.4
	7	5.03 ± 0.03	4.79 ± 0.03	104.1 ± 0.1	−1.17 ± 0.02	4.65± 0.02
450	1	5.6 ± 0.9	3.8 ± 0.5	110 ± 1	−1.3 ± 0.2	3.5 ± 0.4
	3	4.6 ± 1.1	2.6 ± 0.1	109 ± 5	−0.8 ± 0.2	2.4 ± 0.1
	5	6.6 ± 1.3	4.6 ± 0.5	107 ± 5	−1.0 ± 0.2	5.4 ± 0.3
	7	8.1 ± 0.9	6.8 ± 0.1	106.7 ± 0.4	−1.9 ± 0.1	6.5 ± 0.1

L^*: luminosity: black (L^* = 0) and white (L^* = 100); a^*: green color (−) and red color (+); b^*: blue color (−) and yellow color (+); C^*: croma; H^*: hue angle.

3.4. Effect of the Process Parameters on the Total Phenolic Content and Antioxidant Capacity

This work aimed to evaluate the extraction process conditions to obtain high amounts of genipin and geniposide. However, in addition to these iridoids, extracts obtained from unripe genipap fruit also present phenolic compounds and antioxidant capacity, measured in this work by FRAP and ORAC.

The interactions between time and solvent (p-value = 0.002) and power and solvent (p-value = 0.001) significantly influenced the TPC, as shown in Figure 4. The antioxidant capacities measured by FRAP (p-value = 0.0001) and ORAC (p-value = 0.0001) were significantly influenced only by the solvent, as shown in Figure 5. The TPC of extractions with water (17.9–28.9 mg GAE/g extract, Table 1) were higher than those detected in extractions with ethanol (7.5–10 mg GAE/g extract.). Similar effects of solvents were detected in the X0, genipin content, and ORAC assay (87–129 mg TE/g extract for water 70–92 mg Trolox TE/g extract for ethanol, Table 1). The opposite solvent effect was found in the geniposide recovered (7–57 mg/g extract for water and 198–312 mg/g extract for ethanol) and the FRAP assay (4.4–6 mg TE/g extract for water and 14.8–17 mg TE/g extract for ethanol).

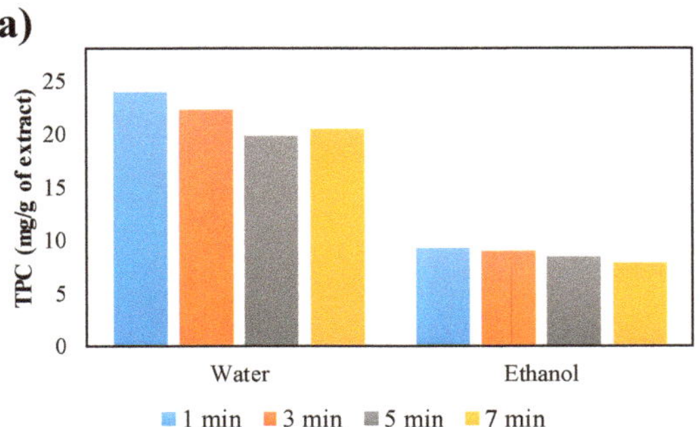

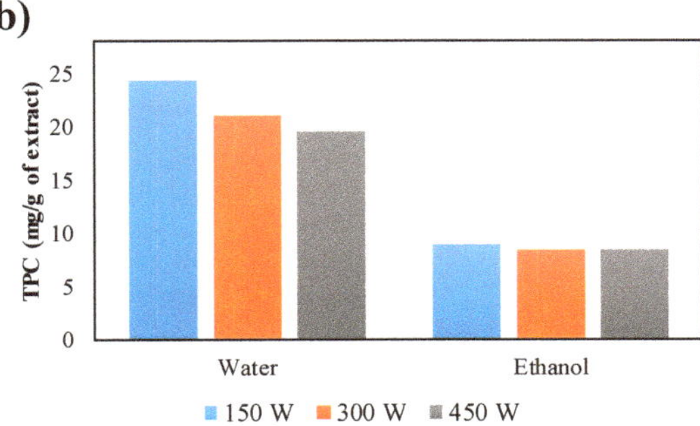

Figure 4. Effect of the process parameters on TPC: (**a**) solvent and extraction time and (**b**) solvent and power. Results expressed on a dry basis. Standard deviation by ANOVA for TPC = 0.9.

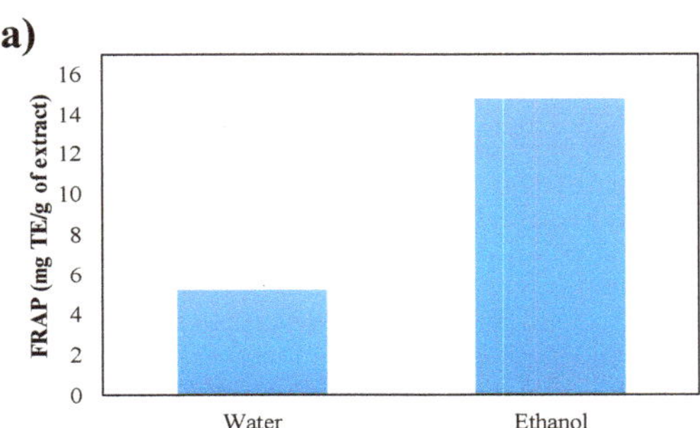

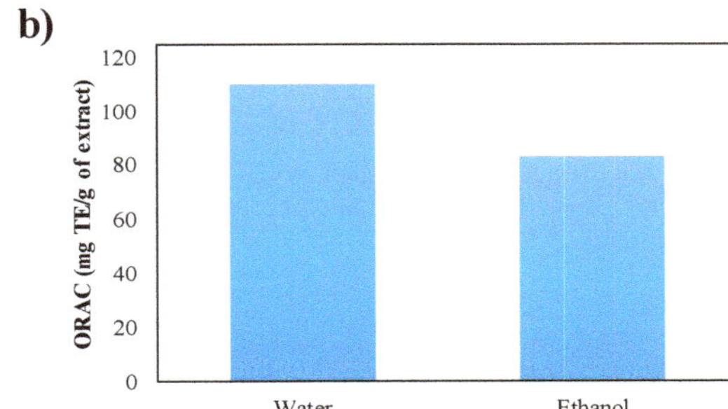

Figure 5. Effect of the solvent on antioxidant capacity: (**a**) effect of the solvent on FRAP and (**b**) effect of the solvent on ORAC. Results are expressed on a dry basis. Standard deviation by ANOVA for FRAP = 1 and for ORAC = 9.

The TPCs observed in the extracts were higher than those found in UAE extracts (4 mg GAE/g) and pressurized ethanol extracts (7–17 mg GAE/g of extract) [13,43], both from unripe crude genipap. For the FRAP, the antioxidant capacity of extracts was comparable with those obtained from the UAE of blue butterfly pea flower [44].

Based on the results provided in this section, it is possible to state that in addition to the color attributes, the SDG extracts have in their composition phenolic compounds, and the whole crude extract presents antioxidant capacities through ORAC and FRAP assays. These features reinforce the claim of the extract for food applications.

4. Conclusions

The UAE, up to 7 min, has shown as an efficient, fast, and environment-friendly method to recover natural antioxidant pigments from semi-defatted unripe genipap by-products, with results comparable to those obtained in the literature for crude unripe genipap.

Extractions with water preferentially extracted genipin, resulting in blue-colored extracts, whereas extraction with ethanol produced green-colored extracts, attributed to the enhancement in geniposide obtention. The TPC of the extracts was higher using water as the solvent, and the ultrasonic power presented a negative effect on this response.

The antioxidant capacity of the aqueous extracts was higher than that of the ethanolic extracts for the ORAC assay. In contrast, for the FRAP assay, the ethanolic extracts presented higher antioxidant capacity. The UAE was shown to be an effective technique to obtain extracts rich in iridoids. Our results are expected to valorize genipap by-products and further research the application of UAE extracts in food products and biological systems.

Author Contributions: Conceptualization: G.N.-N. and M.A.A.M.; Methodology: G.N.-N., Á.L.S., J.V. and J.M.; Writing—original draft preparation: G.N.-N. and Á.L.S.; Writing—review and editing: G.N.-N., Á.L.S., J.V., J.M. and M.A.A.M.; Supervision: M.A.A.M.; Project administration, M.A.A.M.; Funding Acquisition, M.A.A.M. All authors have read and agreed to the published version of the manuscript.

Funding: G. Náthia-Neves and Ádina L. Santana thank CAPES— This study was financed in part by the Coordenação de Aperfeiçoamento de Pessoal de Nível Superior—Brazil (Finance Code 001), for a Ph.D. and post-doctoral financial assistantships. M. Angela A. Meireles thanks CNPq for the productivity grant (309825/2020-2).

Institutional Review Board Statement: Not applicable.

Informed Consent Statement: Not applicable.

Conflicts of Interest: The authors confirm that this article's content has no conflict of interest.

References

1. Huang, A.-G.; Tan, X.-P.; Qu, S.-Y.; Wang, G.-X.; Zhu, B. Evaluation on the antiviral activity of genipin against white spot syndrome virus in crayfish. *Fish Shellfish Immunol.* **2019**, *93*, 380–386. [CrossRef]
2. Ko, J.-W.; Shin, N.-R.; Park, S.-H.; Cho, Y.-K.; Kim, J.-C.; Seo, C.-S.; Shin, I.-S. Genipin inhibits allergic responses in ovalbumin-induced asthmatic mice. *Int. Immunopharmacol.* **2017**, *53*, 49–55. [CrossRef] [PubMed]
3. Hughes, R.H.; Silva, V.A.; Ahmed, I.; Shreiber, D.I.; Morrison, B. Neuroprotection by genipin against reactive oxygen and reactive nitrogen species-mediated injury in organotypic hippocampal slice cultures. *Brain Res.* **2014**, *1543*, 308–314. [CrossRef] [PubMed]
4. Cai, L.; Li, R.; Tang, W.-j.; Meng, G.; Hu, X.-y.; Wu, T.N. Antidepressant-like effect of geniposide on chronic unpredictable mild stress-induced depressive rats by regulating the hypothalamus–pituitary–adrenal axis. *Eur. Neuropsychopharmacol.* **2015**, *25*, 1332–1341. [CrossRef]
5. Guo, L.X.; Liu, J.H.; Zheng, X.X.; Yin, Z.Y.; Kosaraju, J.; Tam, K.Y. Geniposide improves insulin production and reduces apoptosis in high glucose-induced glucotoxic insulinoma cells. *Eur. J. Pharm. Sci.* **2017**, *110*, 70–76. [CrossRef] [PubMed]
6. Náthia-Neves, G.; Meireles, M.A.A. Genipap: A New Perspective on Natural Colorants for the Food Industry. *Food Public Health* **2018**, *8*, 21–33. [CrossRef]
7. Yoo, J.S.; Kim, Y.J.; Kim, S.H.; Choi, S.H. Study on genipin: A new alternative natural crosslinking agent for fixing heterograft tissue. *Korean J. Thorac. Cardiovasc. Surg.* **2011**, *44*, 197–207. [CrossRef]
8. Bentes, A.d.S.; Mercadante, A.Z. Influence of the Stage of Ripeness on the Composition of Iridoids and Phenolic Compounds in Genipap (*Genipa americana* L.). *J. Agric. Food Chem.* **2014**, *62*, 10800–10808. [CrossRef]
9. Paik, Y.-S.; Lee, C.-M.; Cho, M.-H.; Hahn, T.-R. Physical Stability of the Blue Pigments Formed from Geniposide of Gardenia Fruits: Effects of pH, Temperature, and Light. *J. Agric. Food Chem.* **2001**, *49*, 430–432. [CrossRef]
10. Bellé, A.S.; Hackenhaar, C.R.; Spolidoro, L.S.; Rodrigues, E.; Klein, M.P.; Hertz, P.F. Efficient enzyme-assisted extraction of genipin from genipap (*Genipa americana* L.) and its application as a crosslinker for chitosan gels. *Food Chem.* **2018**, *246*, 266–274. [CrossRef]
11. Sigurdson, G.T.; Tang, P.; Giusti, M.M. Natural Colorants: Food Colorants from Natural Sources. *Annu. Rev. Food Sci. Technol.* **2017**, *8*, 261–280. [CrossRef]
12. Zhou, T.; Fan, G.; Hong, Z.; Chai, Y.; Wu, Y. Large-scale isolation and purification of geniposide from the fruit of Gardenia jasminoides Ellis by high-speed counter-current chromatography. *J. Chromatogr. A* **2005**, *1100*, 76–80. [CrossRef]
13. Náthia-Neves, G.; Tarone, A.G.; Tosi, M.M.; Maróstica Júnior, M.R.; Meireles, M.A.A. Extraction of bioactive compounds from genipap (Genipa americana L.) by pressurized ethanol: Iridoids, phenolic content and antioxidant activity. *Food Res. Int.* **2017**, *102*, 595–604. [CrossRef]
14. Ramos-de-la-Peña, A.M.; Montañez, J.C.; de la Luz Reyes-Vega, M.; Hendrickx, M.E.; Contreras-Esquivel, J.C. Recovery of genipin from genipap fruit by high pressure processing. *LWT Food Sci. Technol.* **2015**, *63*, 1347–1350. [CrossRef]
15. Náthia-Neves, G.; Vardanega, R.; Hatami, T.; Meireles, M.A.A. Process integration for recovering high added-value products from Genipa americana L.: Process optimization and economic evaluation. *J. Supercrit. Fluids* **2020**, *164*, 104897. [CrossRef]
16. Viganó, J.; Aguiar, A.C.; Moraes, D.R.; Jara, J.L.P.; Eberlin, M.N.; Cazarin, C.B.B.; Maróstica, M.R.; Martínez, J. Sequential high pressure extractions applied to recover piceatannol and scirpusin B from passion fruit bagasse. *Food Res. Int.* **2016**, *85*, 51–58. [CrossRef] [PubMed]
17. Lachos-Perez, D.; Baseggio, A.M.; Mayanga-Torres, P.C.; Maróstica, M.R.; Rostagno, M.A.; Martínez, J.; Forster-Carneiro, T. Subcritical water extraction of flavanones from defatted orange peel. *J. Supercrit. Fluids* **2018**, *138*, 7–16. [CrossRef]
18. Strieder, M.M.; Silva, E.K.; Meireles, M.A.A. Specific Energy: A New Approach to Ultrasound-assisted Extraction of Natural Colorants. *Probe* **2019**, *23*, 30.
19. Rahimi, S.; Mikani, M. Lycopene green ultrasound-assisted extraction using edible oil accompany with response surface methodology (RSM) optimization performance: Application in tomato processing wastes. *Microchem. J.* **2019**, *146*, 1033–1042. [CrossRef]

20. Grassino, A.N.; Brnčić, M.; Vikić-Topić, D.; Roca, S.; Dent, M.; Brnčić, S.R. Ultrasound assisted extraction and characterization of pectin from tomato waste. *Food Chem.* **2016**, *198*, 93–100. [CrossRef]
21. Rodsamran, P.; Sothornvit, R. Extraction of phenolic compounds from lime peel waste using ultrasonic-assisted and microwave-assisted extractions. *Food Biosci.* **2019**, *28*, 66–73. [CrossRef]
22. Ramos-De-La-Peña, A.M.; Renard, C.M.; Wicker, L.; Montañez, J.C.; García-Cerda, L.A.; Contreras-Esquivel, J.C. Environmental friendly cold-mechanical/sonic enzymatic assisted extraction of genipin from genipap (Genipa americana). *Ultrason. Sonochem.* **2014**, *21*, 43–49. [CrossRef]
23. MI ASAE. Method of Determining and Expressing Particle Size of Chopped Forage Material by Screening. 1998; pp. 562–564. Available online: https://standards.globalspec.com/std/373056/s424-1 (accessed on 2 July 2021).
24. AOAC. *Official Methods of Analysis of the Association of Official Analytical Chemistry*, 16th ed.; AOAC International: Gaithersburg, MD, USA, 1997; Volume 2, 850p.
25. Singleton, V.L.; Orthofer, R.; Lamuela-Raventós, R.M. Analysis of total phenols and other oxidation substrates and antioxidants by means of folin-ciocalteu reagent. In *Methods in Enzymology*; Academic Press: Cambridge, MA, USA, 1999; Volume 299, pp. 152–178.
26. Náthia-Neves, G.; Nogueira, G.; Vardanega, R.; Meireles, M.A.A. Identification and quantification of genipin and geniposide from Genipa americana L. by HPLC-DAD using a fused-core column. *Food Sci. Technol. Camp.* **2018**, *38*, 116–122. [CrossRef]
27. Benzie, I.F.F.; Strain, J.J. The Ferric Reducing Ability of Plasma (FRAP) as a Measure of "Antioxidant Power": The FRAP Assay. *Anal. Biochem.* **1996**, *239*, 70–76. [CrossRef] [PubMed]
28. Ou, B.; Hampsch-Woodill, M.; Prior, R.L. Development and Validation of an Improved Oxygen Radical Absorbance Capacity Assay Using Fluorescein as the Fluorescent Probe. *J. Agric. Food Chem.* **2001**, *49*, 4619–4626. [CrossRef] [PubMed]
29. Alcázar-Alay, S.C.; Osorio-Tobón, J.F.; Forster-Carneiro, T.; Meireles, M.A.A. Obtaining bixin from semi-defatted annatto seeds by a mechanical method and solvent extraction: Process integration and economic evaluation. *Food Res. Int.* **2017**, *99*, 393–402. [CrossRef] [PubMed]
30. Vinatoru, M. An overview of the ultrasonically assisted extraction of bioactive principles from herbs. *Ultrason. Sonochem.* **2001**, *8*, 303–313. [CrossRef]
31. Toma, M.; Vinatoru, M.; Paniwnyk, L.; Mason, T.J. Investigation of the effects of ultrasound on vegetal tissues during solvent extraction. *Ultrason. Sonochem.* **2001**, *8*, 137–142. [CrossRef]
32. Pereira, G.A.; Silva, E.K.; Peixoto Araujo, N.M.; Arruda, H.S.; Meireles, M.A.A.; Pastore, G.M. Obtaining a novel mucilage from mutamba seeds exploring different high-intensity ultrasound process conditions. *Ultrason. Sonochem.* **2019**, *55*, 332–340. [CrossRef]
33. Prado, J.M.; Veggi, P.C.; Náthia-Neves, G.; Meireles, M.A.A. Extraction Methods for Obtaining Natural Blue Colorants. *Curr. Anal. Chem.* **2018**, *14*, 1–28. [CrossRef]
34. Náthia-Neves, G.; Vardanega, R.; Meireles, M.A.A. Extraction of natural blue colorant from *Genipa americana* L. using green technologies: Techno-economic evaluation. *Food Bioprod. Process.* **2019**, *114*, 132–143. [CrossRef]
35. Wu, S.; Horn, G. Genipin-Rich Material and Its Use. U.S. Patent No. 61/556,441, 7 November 2012.
36. Xu, M.; Sun, Q.; Su, J.; Wang, J.; Xu, C.; Zhang, T.; Sun, Q. Microbial transformation of geniposide in Gardenia jasminoides Ellis into genipin by Penicillium nigricans. *Enzym. Microb. Technol.* **2008**, *42*, 440–444. [CrossRef]
37. Dong, Y.; Liu, L.; Bao, Y.; Hao, A.; Qin, Y.; Wen, Z.; Xiu, Z. Biotransformation of geniposide in Gardenia jasminoides to genipin by Trichoderma harzianum CGMCC 2979. *Chin. J. Catal.* **2014**, *35*, 1534–1546. [CrossRef]
38. Reyes-Giraldo, A.F.; Gutierrez-Montero, D.J.; Rojano, B.A.; Andrade-Mahecha, M.M.; Martínez-Correa, H.A. Sequential Extraction Process of Oil and Antioxidant Compounds from Chontaduro Epicarp. *J. Supercrit. Fluids* **2020**, *166*, 105022. [CrossRef]
39. Rebelatto, E.A.; Rodrigues, L.G.G.; Rudke, A.R.; Andrade, K.S.; Ferreira, S.R.S. Sequential green-based extraction processes applied to recover antioxidant extracts from pink pepper fruits. *J. Supercrit. Fluids* **2020**, *166*, 105034. [CrossRef]
40. Azilah Abdul Rahman, N.; Marsinah Tumin, S.; Tajuddin, R. Optimization of Ultrasonic Extraction Method of Natural Dyes from Xylocarpus Moluccensis. *Int. J. Biosci. Biochem. Bioinform.* **2013**, *3*, 53–55. [CrossRef]
41. Aryanti, N.; Nafiunisa, A.; Wardhani, D. Conventional and ultrasound-assisted extraction of anthocyanin from red and purple roselle (*Hibiscus sabdariffa* L.) calyces and characterisation of its anthocyanin powder. *Int. Food Res. J.* **2019**, *26*, 529–535.
42. İlter, I.; Akyıl, S.; Demirel, Z.; Koç, M.; Conk-Dalay, M.; Kaymak-Ertekin, F. Optimization of phycocyanin extraction from Spirulina platensis using different techniques. *J. Food Compos. Anal.* **2018**, *70*, 78–88. [CrossRef]
43. Porto, R.G.C.L.; Cardoso, B.V.S.; Barros, N.V.d.A.; Cunha, E.M.F.; Araújo, M.A.d.M.; Moreira-Araújo, R. Chemical Composition and Antioxidant Activity of *Genipa americana* L. (Jenipapo) of the Brazilian Cerrado. *J. Agric. Environ. Sci.* **2014**, *3*, 4. [CrossRef]
44. Mehmood, A.; Ishaq, M.; Zhao, L.; Yaqoob, S.; Safdar, B.; Nadeem, M.; Munir, M.; Wang, C. Impact of ultrasound and conventional extraction techniques on bioactive compounds and biological activities of blue butterfly pea flower (*Clitoria ternatea* L.). *Ultrason. Sonochem.* **2019**, *51*, 12–19. [CrossRef] [PubMed]

 *processes*

Review

Valorization of Cereal Byproducts with Supercritical Technology: The Case of Corn

Ádina L. Santana [1,2,*] and Maria Angela A. Meireles [2,*]

1   Grain Science and Industry Department, Kansas State University, 1301 N Mid Campus Drive, Manhattan, KS 66506, USA
2   School of Food Engineering, University of Campinas (UNICAMP), R. Monteiro Lobato 80, Campinas 13083-862, SP, Brazil
*   Correspondence: adina.santana@gmail.com or adina@ksu.edu (Á.L.S.); maameireles@lasefi.com (M.A.A.M.)

Abstract: Ethanol and starch are the main products generated after the processing of corn via dry grinding and wet milling, respectively. Milling generates byproducts including stover, condensed distillers' solubles, gluten meal, and the dried distillers' grains with solubles (DDGS), which are sources of valuable compounds for industry including lignin, oil, protein, carotenoids, and phenolic compounds. This manuscript reviews the current research scenario on the valorization of corn milling byproducts with supercritical technology, as well as the processing strategies and the challenges of reaching economic feasibility. The main products recently studied were biodiesel, biogas, microcapsules, and extracts of enriched nutrients. The pretreatment of solid byproducts for further hydrolysis to produce sugar oligomers and bioactive peptides is another recent strategy offered by supercritical technology to process corn milling byproducts. The patents invented to transform corn milling byproducts include oil fractionation, extraction of undesirable flavors, and synthesis of structured lipids and fermentable sugars. Process intensification via the integration of milling with equipment that operates with supercritical fluids was suggested to reduce processing costs and to generate novel products.

Keywords: yellow corn; biowaste; supercritical CO_2; biorefinery; process intensification; cost of manufacture

Citation: Santana, Á.L.; Meireles, M.A.A. Valorization of Cereal Byproducts with Supercritical Technology: The Case of Corn. *Processes* **2023**, *11*, 289. https://doi.org/10.3390/pr11010289

Academic Editor: Anet Režek Jambrak

Received: 29 December 2022
Revised: 6 January 2023
Accepted: 12 January 2023
Published: 16 January 2023

1. Introduction

Corn (*Zea mays* L.), also known as yellow or sweet corn, is a multipurpose crop used for many industrial applications including foods, fuels, cosmetics, and pharmaceuticals.

According to a recent report published by the United States Department of Agriculture, the three main producers of corn worldwide are the United States (353.84 million ton/year), China (274 million ton/year), and Brazil (126 million ton/year) [1].

Corn processing provides opportunities to generate multiple products including flour, ethanol, starch, protein, and syrups. Before processing, the grains are separated from the stover (leaves, stalks, cobs, and silk), which is an underutilized mixture of lignocellulosic material. The stover is produced at an equal proportion of grains processed, i.e., 1 kg of corn grain generates 1 kg of stover [2,3].

The mill is the main form of processing corn. The processes for milling corn include (a) dry milling, (b) dry grinding, and (c) wet milling.

According to Stanford and Keener [4], the term dry milling is erroneously used throughout the literature to describe both dry milling and dry grinding processes.

Dry milling consists of mechanical procedures to separate the germ, tip cap, and pericarp from the endosperm to produce flour, oil, grits, and corn meal. These are products commonly sold for human consumption and hominy feed, which is a byproduct sold for animal feeding [4,5].

Dry grinding is destined to produce ethanol via the fermentation of corn kernels. Wet milling differs from dry grinding in the utilization of huge amounts of water to fractionate corn kernels into starch, oil, protein, fiber, ethanol, and sweeteners [6,7].

The corn milling industry plays an important role in the corn supply chain. It starts from crop production and ends with feed and food industrial channels including retail markets, animal feed, food processors, cosmetics, and fuel, as well as the food supply for emergency government food aids [5]. In the United States, some companies that mill corn include Cargill [8], Bunge [9], Gavilon [10], and J-Six Enterprises [11].

The generation of underutilized fractions after crop processing is unavoidable. In addition, auto-oxidation of lipids, microbial growth in high moisture materials, and changes in enzymatic activity accelerates the decomposition of discarded fractions, and such difficulty in the reutilization of products turns into an environmental concern [12].

The undervalued fractions from dry grinding include the condensed distillers, and the wet or dried distiller's grains with solubles, whereas the byproducts derived from wet milling include spent germ meal, gluten feed, oil, and steep water. The stover is an undervalued fraction generated before corn grain processing. Even with the known utilization of some byproducts in animal feed [7,13,14], dry grinding and wet milling face a challenge in the valorization of byproducts.

Supercritical technology includes multiple separation methods that use solvents operating in the region of compressed liquid, and subcritical and supercritical state in which carbon dioxide (CO_2) is the most preferred solvent. Supercritical CO_2 (SC-CO_2) is a low-cost, non-flammable, and non-toxic solvent that shows high efficiency in extracting and fractionating low-polar products, including oils and waxes [15–17], to dry particles [18] and assist in chemical reactions [19]. As reviewed by Fărcaș and coworkers [20], SC-CO_2 provokes low oxidative and thermal impact in the products generated after the processing of undervalued cereal products.

In this context, supercritical technology emerged as a green approach to improve the processing of crops.

This review discusses the current opportunities offered by the application of supercritical technology to transform byproducts derived from dry grinding and wet milling into value-added products, including oil [21], wax [17], biofuels [22], biogas [23], and microparticles [18]. The patents survey and our opinions on the processing strategies to reduce costs and improve the quality of products were also included.

2. Dry Grinding

Corn is abundantly processed into ethanol via dry grinding because of the high conversion rates of corn starch.

Approximately 450 million bushels of corn are used for both dry grinding and wet milling to produce ethanol [24]. One bushel of corn is equivalent to 25.40 kg. Before grain processing, there is a generation of stover, a fraction consisting of leaves, cob, straw, and silk [25].

After harvesting corn, the stover is separated from the grains, and the grains are cleaned, stored, and tempered. The tempering increases the moisture of corn kernels to facilitate the degerming process, i.e., the separation of germ and bran coat [5].

The grains are ground to a fine powder, known as meal, which is mixed with water and heated at 85 °C. Afterward, the enzyme alpha-amylase is added, followed by heating at 110–150 °C for 1 h, to accelerate the conversion of starch into dextrose, resulting in a liquefied slurry (Figure 1).

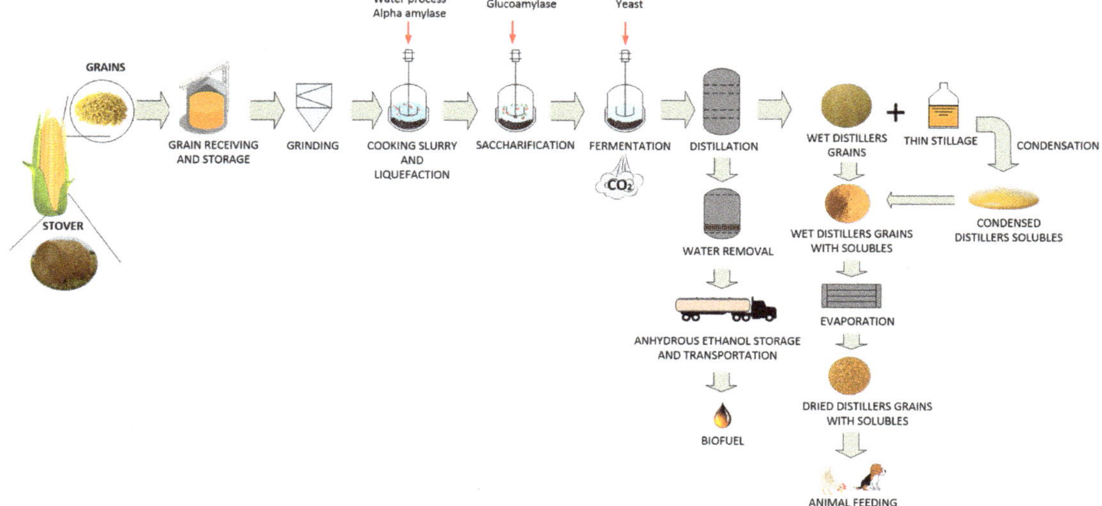

Figure 1. The steps involved in corn dry grinding.

The slurry is cooled to room temperature. Afterward, glucoamylase is added to saccharify the dextrose in the slurry into glucose. Glucose is the substrate for fermentation by yeast strains (*Saccharomyces cerevisiae*). The glucose fermentation results in two streams, one consisting of CO_2, and the other consisting of beer which is a slurry composed of up to 16% ethanol by volume [26].

The beer is pumped into distillation columns resulting in wet ethanol and the stillage. Wet ethanol is distilled up to 95.6% azeotropic point, and then transferred to the molecular sieve system to produce anhydrous ethanol. The stillage is a viscous liquid composed of 6–16% solids.

The stillage is dried, resulting in wet distillers' grains and thin stillage (TS). The storage and handling of wet distillers' grains (WDG) increase processing costs because they can be stored for up to 5–7 days and require the use of chemicals to inhibit microbial growth. In addition, the water in TS is evaporated resulting in the condensed distillers' solubles (CDS), a highly viscous syrup.

The drying of CDS demands high energy because of high syrup viscosity. Thus, to reduce the quantity of streams, one proposed solution was to mix the CDS with WDG and, subsequently, drying of the resulting mixture on the dried distillers' grains with solubles (DDGS), which has longer shelf life due to its low moisture (10–12%). Another possibility is to use the TS or the CDS in animal diets [27]. By the end of dry grinding, one bushel of corn can produce approximately 10.2 L ethanol and 8 kg DDGS [28,29].

DDGS is donated or sold to be used as an animal diet component or as fertilizer. The disadvantage of using DDGS as fertilizer is the excess of organic matter leading to soil pollution. In addition, there is a loss of nutrients from corn DDGS that could be reused. In 2019, dry grinding generated 22,591,477 tons of DDGS [30].

3. Wet Milling

The wet milling process separates the corn kernels into starch, oil, protein, ethanol, and fiber. Grains are received, separated from foreign materials, and stored. Afterward, the grains are inserted into a stainless-steel tank of up to 600 metric tons for steeping, mixed with an aqueous solution consisting of 0.12–0.20% sulfur dioxide at 50 °C for 24–48 h to prevent microbial growth and to get the germ easily separated from the endosperm.

Lactic acid produced by *Lactobacillus* bacteria in the steeping water induces the kernel's softening and enhances sulfur dioxide's sorption. Sulfur dioxide prevents putrefaction and helps to separate the starch from the protein because it acts as a reducing agent in breaking disulfide bonds within the protein, enhancing starch recovery [31].

The steeping process generates a stream consisting of steepwater which is a nutritive media composed of 5–10% solids including sugars and protein [6,32].

The steeped corn is ground and subsequently transformed via multiple separation steps. The oil is extracted from the germ via solvent extraction generating a byproduct known as germ meal (Figure 2). The mixture consisting of fiber, starch, and protein is inserted into hydrocyclones and washing screens to separate the fiber from the starch and protein (millstarch). The fiber is dried and pressed resulting in the corn gluten feed.

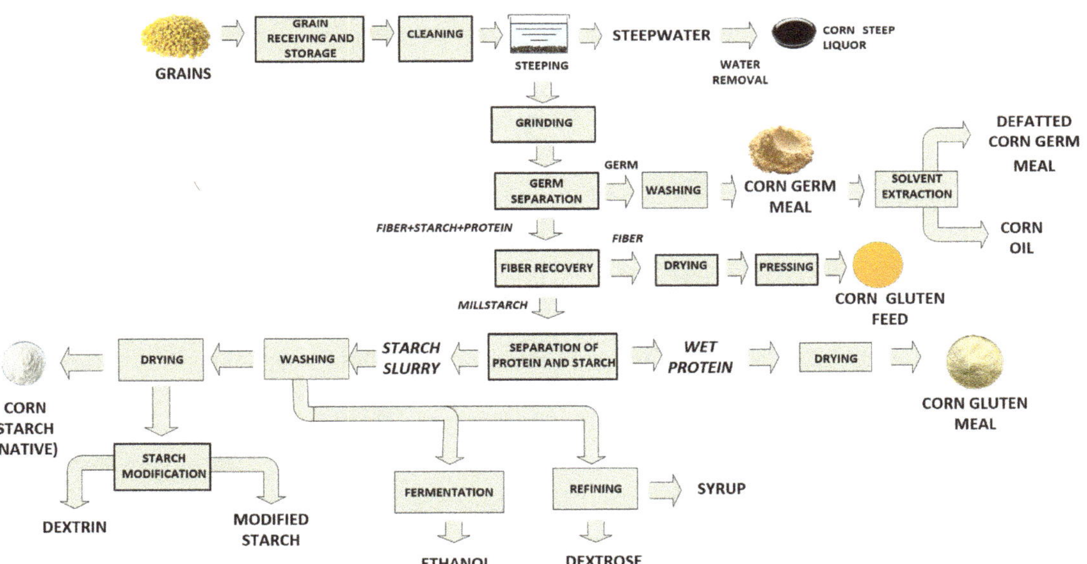

Figure 2. The steps involved in corn wet milling.

The millstarch is centrifuged to separate the starch and gluten meal. The gluten meal is dried, whereas the starch is washed several times to recover the granules by reducing soluble impurities and then, subsequently dried. The starch can be sold as such or used as a raw ingredient to produce ethanol via fermentation or to produce sweeteners via refining.

At the end of the wet milling process, one bushel of corn produces 6–9 L ethanol, 14 kg starch, 0.5–0.9 kg oil, 5–6.4 kg corn gluten feed, and 0.9–1.4 kg corn gluten meal [4]. In 2019, wet milling produced approximately 765,620 tons of corn germ meal, 3,468,292 tons of corn gluten feed, and 106,899 tons of gluten meal [30]. In addition, it is difficult in wet milling to reduce the water used for processing. For instance, 1.5–1.78 L of water processes 1 kg of corn. In addition, the amount of wastewater generated is equivalent to a medium-large city.

Furthermore, another concern faced by wet milling includes the environmental damage provoked by air pollution. Air pollution may be provoked by acid rain via sulfur dioxide emission from the steeping step, or particulate material and smoke from evaporators during drying. Aside from obtaining different products, wet milling demands high capital and operational costs. For instance, the cost of investment to construct a plant that processes 2542 tons of grains daily was estimated at USD 200–300 million [6].

4. Dry Grinding and Wet Milling Byproducts

4.1. Condensed Distillers' Solubles (CDS)

The condensed distillers' solubles (CDS) consist of a blend of corn and yeast after the fermentation step in dry grinding. Enzymatic hydrolysis transforms CDS protein into small peptides and amino acids with antioxidant potential against free radicals [33], and the potential to inhibit angiotensin I-converting enzyme (ACE) which is a protein that regulates the volume of fluids in the body thereby allowing an increase in blood pressure [34].

4.2. Dried Distillers' Grains with Solubles (DDGS)

The dried distiller's grains with solubles (DDGS) are a low-cost source of nutrients, including lipids, carotenoids, and protein. The nutrient composition supports studies on DDGS use in poultry, swine, and beef cattle diets [35].

According to Heuzé and coworkers [36], 63% of distillers' grains derived from corn are sold as dried products. Recently, Langemeier [37] estimated the average price of corn DDGS at USD 165/ton.

Xanthophylls are oxygenated carotenoids in corn DDGS that play a role in visual and cognitive development. Shin and coworkers [14] observed that lutein was the predominant xanthophyll found in corn DDGS, with a concentration of 4.4 to 9.3 times higher than zeaxanthin. In addition, the same authors detected tocopherols and tocotrienol, where gamma-tocopherol and gamma-tocotrienol were predominant in multiple corn DDGS samples.

Lutein is commercially extracted from marigold petals, costing USD 34.33/lb [38]. In this case, corn DDGS could be reused to decrease the costs of lutein.

Phenolic compounds were detected in corn DDGS, including vanillic, caffeic, p-coumaric, ferulic, and sinapic acids. The ferulic acid concentration detected in the DDGS was around three-fold higher than in crude corn [39]. Phenolic compounds have been extensively studied for their health benefits against cancer [40] and other illnesses.

In addition, corn DDGS contains low-cost oil, also known as distillers corn oil, from which antioxidants, tocopherols, and tocotrienols have been detected [14]. Based on this, some ethanol producers took the opportunity to extract oil mechanically from the WDG before drying and selling it as a secondary product. However, the semi-defatted corn DDGS distributed to animal feeding provoked problems attributed to the reduced energy value in the meal by 45 kcal/lb per % oil extracted [35,41].

4.3. Germ Meal

Germ meal is the low-fat solid fraction obtained after the hexane extraction of corn in wet milling. Germ meal is a source of oleic- and linoleic-fatty acids [21], carotenoids, and protein. Corn germ meal was studied mostly for animal feeding [7,13]. Zeaxanthin (989 µg/kg) and lutein (72 µg/kg) were also found in corn germ [42]. Recently, germ meal has been shown to be a source of protein with technological properties desirable for the industry [43].

4.4. Gluten Feed and Gluten Meal

Gluten feed is obtained after the separation of fiber from protein and starch. The ingestion of PROMITOR®, a commercial soluble fiber derived from corn, was studied by Whisner and coworkers [44], who observed the enhanced quality in the gut microbiome of female adolescents, allowing the increase of calcium absorption. In addition, Costabile and coworkers [45] showed that the ingestion of PROMITOR® with the probiotic *Lactobacillus rhamnosus* resulted in a positive synergy by decreasing inflammation, and by inhibiting the activity of interleukin 6 which is a proinflammatory cytokine [46].

Gluten meal is the protein byproduct obtained after separation from starch. Zein and glutelin are proteins of high amino acid content obtained from gluten meals. Zein is an alcohol-soluble product with 45–50% of protein which is not used directly for human consumption due to its hydrophobicity. The hydrophobic nature of zein redirected its application for the formulation of films and coatings [47]. Information on the utilization of corn glutelin in foods is scarce.

Peptides converted from corn gluten meal could in vitro scavenge free reactive oxygen species by increasing the levels of antioxidant enzymes [48,49].

4.5. Stover, Silk, and Steep Liquor

Corn stover, or straw, consists of plant parts left behind after grain harvesting including leaves, stalks, cobs, and silk. Corn stover is a lignocellulosic feedstock for energy production including polyol fuel via pyrolysis [50] and ethanol via enzymatic hydrolysis of pretreated stover [51]. Corn cob found a recent application as an activated carbon adsorbent for mercury removal [52].

Corn silk, also referred to as corn hair or Maydis stigma, is a byproduct of corn stigma and style used in Chinese medicine to treat multiple diseases including inflammation in the urinary tract [53]. The rich composition of phenolic compounds is strongly linked to corn silk's action against inflammation including maysin and caffeoylquinic acid derivatives [54]. Isolated maysin from corn silk showed in vitro effects against obesity by inhibiting the functionality of adipocytes [55].

Steep liquor is the waste generated after steeping in the wet milling process. Rodriguez-Lopez et al. [56] detected phenolic compounds and protein in corn steep liquor, which may serve as food nutrients. In addition, steep liquor was studied as a nutrient feedstock for fermentation [57].

Since it was administered at low dosages, corn's nutrients may contribute to agriculture as a growth stimulant and fertilizer [58].

The reviewed proximate composition and the bioactive compounds detected in corn milling byproducts are shown in Tables 1 and 2, respectively.

Table 1. The proximate composition of corn milling byproducts (%).

	Moisture	Protein	Fat	Ash	Carbohydrates							Reference
					Total *	Fiber	Starch	Glucan	Cellulose	Lignin	Xylan	
DDGS	10	27.5	10.5	4.1	47.9	32.8	5.2	21.2	16	-	8.2	[59,60]
TS	91.78	0.12	1.75	0.74	5.61	1.27	-	-	-	-	-	[61]
CDS	68.51	8.11	1.86	4.17	17.35	0.32	-	-	-	-	-	[62]
Stover	11.8	9.01	0	0	79.19	17.9	-	31.7	32.9–43.1	5.4–12.6–14.8	17.1	[17,63–65]
Gluten feed	10	20.3	3	7.4	59.3	32	15.5	-	-	-	-	[59]
Gluten meal	10	28.37	2.2	5.83	53.6	5.88–7.3	13.9		13.64		-	[59,66]
Germ meal	10	25.12	2.73	3.8	58.35	35.5	17.8	-	-	-	-	[13,59]
Steep liquor	50	16.4	2.3	1.5	29.8	1.8	5.7–14.8	-	-	-	-	[7,59,67]

* Total carbohydrates estimated by subtraction.

Table 2. The recently reviewed bioactive compounds identified in corn milling byproducts.

Substance	DDGS	Germ Meal	Gluten Feed Fiber	Gluten Meal	Thin Stillage	Stover	Reference
Carotenoids	Lutein: 101 µg/g	Lutein: 0.072 µg/g	Lutein: 5.24 µg/g	Lutein: 2.5–15 µg/g	Lutein: 31 µg/g	-	[42,61,68–70]
	Zeaxanthin: 69 µg/g	Zeaxanthin: 0.989 µg/g	Zeaxanthin: 6.61 µg/g	Zeaxanthin: 5–35 µg/g	Zeaxanthin: 32.5 µg/g Beta cryptoxanthin: 0.70 µg/g	-	
	Beta cryptoxanthin: 47 µg/g	Beta carotene: 0.16 µg/g	Beta-carotene:0.299 µg/g	Beta-carotene:1.22 µg/g	-	-	
	-	-	-	-	-	-	
Tocopherols	Alpha-tocopherol: 0.108 µg/g	Alpha-tocopherol: 31.74 µg/g	Alpha-tocopherol: 0.28 µg/g	Alpha-tocopherol: 0.18 µg/g	Alpha-tocopherol: 37.3 µg/g	-	[14,69]
	Gamma tocopherol: 0.069 µg/g	Gamma tocopherol: 535.59 µg/g	Gamma tocopherol: 9.78 µg/g	Gamma tocopherol: 13.96 µg/g	Gamma tocopherol: 143 µg/g		
	Delta tocopherol: 0.0182 µg/g	Delta tocopherol:17.38 µg/g	Delta tocopherol: 1.08 µg/g	Delta tocopherol: 0.97 µg/g	Delta tocopherol: 4.46 µg/g		
Tocotrienols	Alpha Tocotrienol: 0.0093 µg/g	Alpha Tocotrienol: 0.32 µg/g	Alpha Tocotrienol: 1.32 µg/g	Alpha Tocotrienol: Non identified	Alpha Tocotrienol: 29.6 µg/g		[14,69]
	Gamma tocotrienol: 0.014 µg/g	Gamma tocotrienol: 3.67 µg/g	Gamma tocotrienol: 6.61 µg/g	Gamma tocotrienol: 4.52 µg/g	Gamma tocotrienol: 38.5 µg/g		
		Delta tocotrienol: 0.24 µg/g	Delta tocotrienol: 0.54 µg/g	Delta tocotrienol: 0.50 µg/g	Delta tocotrienol 0.09 µg/g	-	
Phenolic compounds	Ferulic acid: 7010 µg/g	Ferulic acid: 9870—636,540 µg/g	Ferulic acid: 1020–4200 µg/g	-	-	Ferulic acid: 990.20–3515.50 µg/g-	[39,71–73]
	p-Coumaric acid: 530 µg/g	-	p-Coumaric acid: 60–340 µg/g	-	-	p-Coumaric acid: 2027.45–7307.75 µg/g	
	Caffeic acid: 770 µg/g	-	-	-	-	-	
	Vanillic acid: 500 µg/g	-	-	-	-	-	
	Sinapic acid: 8780 µg/g	-	-	-	-	-	
Phytosterols	Campesterol: 2916 µg/g	Campesterol: 87 µg/g	Campesterol: 201.1 µg/g	Campesterol: 87.2 µg/g	Beta sitosterol: 1467 µg/g	Campesterol: 226.4 µg/g	[17,61,68,69]
	Stigmasterol: 861 µg/g	Stigmasterol: 343.8 µg/g	Stigmasterol: 476.1 µg/g	Beta sitosterol: 1342.5 µg/g	Squalene: 168 µg/g	Stigmasterol: 319.6 µg/g	
	Sitosterol: 9066 µg/g	Beta sitosterol: 311.1 µg/g	Beta sitosterol: 2943.6 µg/g	-	Δ-5 avenasterol: 455 µg/g	Beta-sitosterol: 735.6 µg/g	
	Sitostanol: 4186 µg/g	-	-	-	-	-	

5. Supercritical Technology

Supercritical technology is a term that includes a wide range of separation methods that operate at the phase regions of compressed liquid, and subcritical and supercritical fluid. Biofuels, biopolymers, and extracts rich in antioxidants are among the value-added products generated via supercritical technology as opportunities to valorize corn fractions, reduce costs, and promote sustainability for the process.

The reusability, low cost, low toxicity, non-flammability, and moderate critical point (31.2 °C and 7.38 MPa) allow the use of CO_2 as the preferred supercritical fluid. In addition, SC-CO_2 is a carrier solvent that does not react with the extract, thereby avoiding contamination [15].

The components in the supercritical technology-based apparatus include a CO_2 storage tank, solvent reservoir, pump, heat and cooling devices, temperature and pressure indicators, valve assembly, flowmeter and high-pressure vessel or chamber (Figure 3(A.1)). Figure 3 shows a scheme of some apparatus for some techniques that use supercritical fluids. The configurations are modified according to the research method and goals.

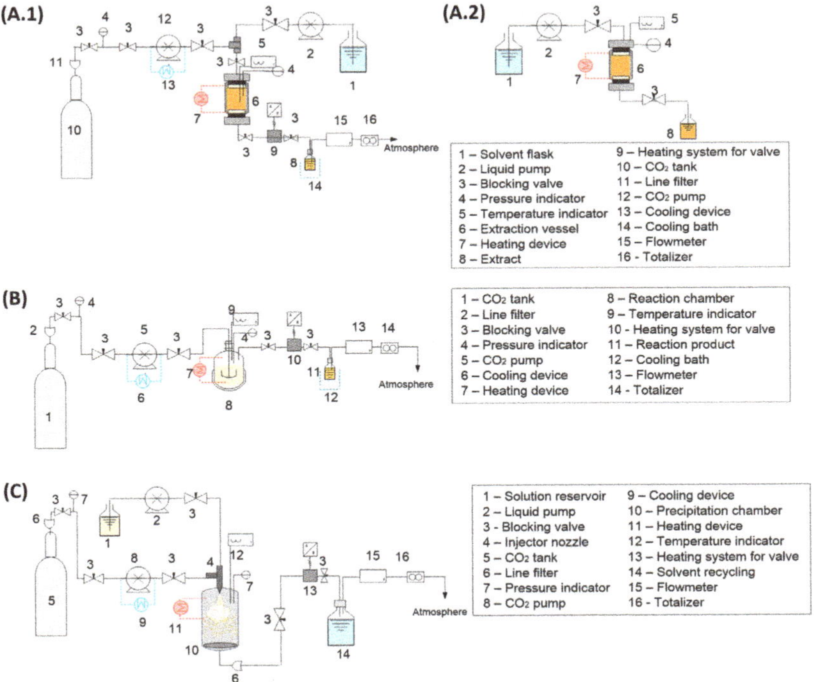

Figure 3. The schematic drawing of equipment designed for SFE with cosolvent (**A.1**), PLE (**A.2**), the chemical reaction in a batch reactor (**B**) and particle formation (**C**).

Details on costs and procedures to construct supercritical technology-based equipment are available elsewhere [74,75].

5.1. Processes

5.1.1. Supercritical Fluid Extraction (SFE) and Hydrolysis Assisted with SC-CO_2

Supercritical fluid extraction (SFE) has been used mostly to extract non-polar products including oils [21] and waxes [17] from corn milling byproducts.

Marinho and coworkers [21] observed that the SFE of oil from corn germ (45–85 °C, 15–25 MPa) was improved by including ethanol as a cosolvent in terms of the antioxidant potential detected in extracts. The fatty acids oleic and linoleic were predominant in the extracts.

Burlini and coworkers used SC-CO_2 to extract phytosterols from corn germ [71]. Previous studies used sequential extractions to obtain different products from crops which may be useful for further studies with corn. For instance, studies with turmeric optimized the extraction of essential oils with SC-CO_2, followed by the extraction of phenolic compounds and monosaccharides with pressurized solvents [76–78].

Espinosa-Pardo [43] and coworkers performed the extraction of corn germ by first extracting oil with SC-CO_2 (45 °C, 40 MPa, 5.6 g/min, without co-solvents) as a pretreatment strategy to prepare the raw material for protein extraction. As the authors mentioned, SFE provided a semi-defatted corn germ meal and an extract rich in phenolic compounds. In addition, the protein extracts obtained from semi-defatted corn germ meal possessed industry-interest technological properties, including foaming properties higher than 80%.

Even with proven applications in human health, there are no recent reports on the processing of corn silk with supercritical technology. Liu and coworkers [79] studied the SFE of flavonoids from corn silk using the Box-Behnken design combined with Response Surface Methodology. The authors did a sequential extraction, using pure carbon dioxide for 2 h, followed by extraction with ethanol [79].

Recently, Sun and coworkers [80] used SC-CO_2-assisted hydrolysis of corn cob and stalk as a pretreatment strategy to improve the release of oligosaccharides for further enzymatic hydrolysis. The authors hypothesized that the pretreatment with SC-CO_2 improved the utilization of lignocellulosic sugars. This behavior can be explained by the generation of carbonic acid due to the interaction between CO_2 and the water present in the raw material causing temporary acidification of the system [80].

In another study, supercritical CO_2 (120–170 °C) combined with ultrasound pretreated corn cob and stalk improved the sugar yield by 75% after enzymatic hydrolysis of pretreated raw material [81].

5.1.2. Subcritical Water-Assisted Hydrolysis

Water at subcritical conditions (100–374 °C, pressures up to 22.1 MPa) has been used as a green solvent to valorize lignocellulosic materials via bioactive compounds extraction or via the generation of fermentable sugars by hydrolysis [82]. The process using pressurized water at temperatures lower than 100 °C is known as pressurized hot water extraction [15]. Subcritical water is recommended for extraction in pseudo-continuous mode, which is performed via the collection of extracts after water flows across the extraction vessel (Figure 3(A.2)).

For the pretreatment of lignocellulosic materials via hydrolysis, subcritical water is injected in batch mode, i.e., the wet raw material is inserted in the vessel, followed by water injection as performed by Zhang and coworkers [83]. These authors pretreated corn zein with subcritical water in a batch reactor and observed positive improvements in the protein's solubility, thermal stability, and denaturation temperature.

According to Moraes and coworkers [84], extraction operated with supercritical technology at batch mode is not desirable in the industry because the pauses required between batches to remove the spent raw material and inserting a new material for processing increase operation times and reduce productivity.

5.1.3. Pressurized Liquid Extraction

Pressurized liquid extraction (PLE) is used to extract analytes from a solid or semi-solid matrix using compressed liquids below the subcritical state. Solvent mixtures containing water and ethanol were enough to intensify the extraction of bioactive compounds from underutilized vegetable products including phenolic compounds, and carotenoids.

Ethanol, water, and mixtures are the preferable solvents in PLE because they are generally considered as safe (GRAS). To the best of our knowledge, no recent studies on the PLE applied to yellow corn milling byproducts exist.

5.1.4. Chemical Reactions: Transesterification and Gasification

Transesterification or alcoholysis is the catalytic conversion of a mixture of triglycerides into fatty acid alkyl esters (FAAE), i.e., biodiesel. Supercritical CO_2 was shown to be a green alternative to assist biological (lipases) and non-biological catalysts in the rapid conversion of triglycerides in FAEE [85].

The supercritical state induces high diffusivity which accelerates the action of catalysts. In transesterification, SC-CO_2 is not consumed. After system depressurization, it is separated immediately from the reaction products and may be recycled [19].

Yadav and coworkers [22] studied the transesterification of corn oil, and methanol assisted with SC-CO_2 catalyzed with Nafion NR50. The highest conversion of triglycerides into fatty acid methyl esters was correlated with the highest oil-to-methanol molar ratio [22].

Gasification converts biomass into a mixture of gases. Supercritical water is used to gasify high-moisture biomass, promoting high gas yields and low residual chars and tars. Supercritical water gasification (SWG) includes multiple reactions, including oxidation, methanation, hydrolysis, and water gas shift [86].

The SWG of corn straw catalyzed by K_2CO_3 produced a high amount of hydrogen and zero carbon monoxide [23]. One reason hypothesized by the authors is the difficulty in the formation of carbon monoxide after lignin decomposition.

5.1.5. Micronization and Encapsulation

Supercritical technology is also used to improve the delivery of many active ingredients via micronization, encapsulation, and impregnation.

Micronization reduces the particle size of materials and improves solubility and the delivery of molecules. Encapsulation consists of incorporating the active ingredient with an encapsulating agent to enhance the shelf life of products. Multiple methods with supercritical fluids to precipitate particles via micronization and encapsulation include supercritical antisolvent fractionation (SAF), supercritical antisolvent (SAS), the rapid expansion of supercritical solutions, gas anti-solvent processes, and precipitation from gas-saturated solutions [87,88].

Rosa and coworkers [18] used zein as an encapsulant polymer to incorporate vitamin complexes after precipitation assisted with SC-CO_2 as antisolvent, generating microparticles with maximum incorporation of 13.74 mg riboflavin/g, 0.47 mg δ-tocopherol/g, and 14.57 mg β-carotene/g.

Palazzo and coworkers [89] used SC-CO_2 as co-solute to encapsulate the phenolic compound luteolin with zein incorporated with aqueous mixtures with ethanol and acetone, thus generating microparticles with a maximum encapsulation efficiency of 82%.

5.1.6. Impregnation and Extrusion

Corn starch is extensively used in research. Corn milling byproducts, including germ and the zein derived from gluten meal, are potential biopolymers for the formation of aerogels or alcogels to be dried with supercritical technology. In addition, formulations with such byproducts can be tested for printability in a 3D printer and subsequently incorporated with a drug of interest.

Liu and coworkers [90] studied the products generated after SC-CO_2-assisted impregnation of zein-based nanocomposites with cinnamon essential oil and suggested the products as long-term antibacterial delivery materials.

Dias and coworkers [91] incorporated beta-carotene in corn starch aerogels via SC-CO_2 impregnation (maximum yield of 0.96 mg beta-carotene/g aerogel) to enhance compound solubility in water. The poor solubility in water is related to the poor absorption of carotenoids in the intestine after consumption.

Extrusion of flour is highly performed to produce foods with distinct shapes and textures including snacks and pasta. Extrusion assisted with SC-CO_2 was also reviewed as a method to enhance the bioavailability of low-solubility drugs [92].

High-moisture corn fiber was extruded in a twin-screw extruder assisted with SC-CO$_2$. The total phenolic content in the products obtained in the extrusion assisted with CO$_2$ was statistically similar to the extrusion not assisted with SC-CO$_2$ for most of the conditions used [93]. The recently reviewed supercritical technology-based processes discussed in this review are summarized in Table 3.

Table 3. The supercritical-based processes used to valorize corn milling byproducts.

Process	Byproduct	Conditions	Reference
SFE	Germ	Solvent: CO$_2$ Cosolvent: acetone, ethanol, and hexane Temperature: 45–85 °C Pressure: 15–25 MPa CO$_2$ flow rate: 3 L/h Cosolvent flow rate: 0.1 mL/min	[21]
SFE	Germ	Solvent: CO$_2$ Temperature: 80 °C Pressure: 30 MPa CO$_2$ flow rate: 2.5 L/min Time: 10 min	[71]
SFE	Germ	Solvent: CO$_2$ Temperature: 45 °C Pressure: 40 MPa CO$_2$ flow rate: 5.6 g/min	[43]
SFE	Silk	Solvent: CO$_2$ Temperature: 40–60 °C Pressure: 25–45 MPa CO$_2$ flow rate: 20 L/h Cosolvent flow rate: 1.3 mL/g Time: 120 min	[79]
Hydrolysis assisted with SC-CO$_2$	Stover: cob and stalk	Solvent: CO$_2$ and water present in raw material Temperature: 40–70 °C Pressure: 35–45 MPa Time: Non identified	[80]
Hydrolysis assisted with SC-CO$_2$	Stover: cob and stalk treated with ultrasound	Solvent: CO$_2$ and water present in raw material Temperature: 170 °C Pressure 20 MPa Time: 30 min	[81]
Hydrolysis assisted with SW	Zein	Temperature:110–170 °C Pressure: 0.1–0.8 MPa Time: 20–120 min	[83]
Transesterification	Oil	Solvent: SC-CO$_2$ Temperature: 95 °C Pressure: 9.65 MPa Catalyst: Nafion NR50 Time: 240 min	[22]
SWG	Straw	Temperature: 450–550 °C Pressure: 41 MPa Catalyst: K$_2$CO$_3$ Time: 10–50 min	[23]
Encapsulation	Zein	Antisolvent: SC-CO$_2$ Solvent: Ethanol:water (94:6, *v/v*) Temperature: 40 °C Pressure: 7–16 MPa Solution flow rate: 0.02 g/mL CO$_2$ flow rate: 20–60 g/min	[18]

Table 3. *Cont.*

Process	Byproduct	Conditions	Reference
Encapsulation	Zein	Antisolvent: CO_2 Solvent: Aqueous mixtures with ethanol and acetone Temperature: 60 °C Pressure: 8.2–10 MPa Solution flow rate: non-identified CO_2 flow rate: non-identified	[89]
Impregnation	Zein	Temperature: 40 °C Pressure: 15 MPa Time: 60 min	[90]
Impregnation	Starch	Temperature: 40–60 °C Pressure: 15–30 MPa Number of cycles: 1–4 Depressurization rate: 0.25–2.61 MPa/min	[91]
Extrusion	Fiber	Temperature: 90–120 °C Pressure: non-identified Moisture: 30% CO_2 flow rate: 200 mL/min Screw speed: 100 g/min	[93]

5.2. Economic Evaluation

To understand the feasibility of plant-based biorefinery, it is necessary (a) to know the composition of the feedstock, (b) to understand how the present compounds interconnect together for process design and optimization, and (c) to understand the costs involved in processes that are reflected in the economic value of products and (d) the environmental impacts provoked.

The cost of manufacturing (COM) is used to evaluate the economic feasibility of a process [94]. The COM consists of three major types of costs: (a) general expenses (GE): costs which cover business maintenance and include the costs involved in management, sales administration, and research and development; (b) fixed costs (FC): costs that do not depend on the production, i.e., land cost, insurance, territorial taxes, and depreciation; (c) operating costs (OC): costs that are dependent on the production and consist of raw material, operational labor, waste treatment, and utility costs.

The three components of COM are calculated in terms of five types of costs, i.e., the cost of raw material (CRM), the fixed capital investment (FCI), the cost of utilities (CUT), the cost of operational labor (COL), and the cost of waste treatment (CWT).

The major challenge in the scale-up of processes conducted with supercritical technology is the economic profitability of the process that is determined by the return on investment and the payback period which is the time to recover the cost of investment applied in the process. Based on the characteristics of bench-scale SFE of *Artemisia annua*, Baldino and coworkers [95] scaled up a 0.05 L bench SFE plant (equivalent to USD 21,195.00–29,673.00) to 50 L (equivalent to USD 317,925.00) by considering (a) operation at optimized conditions (residence time, amount of raw material, temperature, pressure, and CO_2 flow rate), (b) 30% costs for automation, as commonly used in the industry and (c) equipment plants reported in the literature.

For the SFE-assisted bypressing oil from Baru seeds, the total cost including equipment for a 0.1 L unit was calculated as USD 213,327.00. The analysis of the sensitivity of this process considering a 40 L plant with the cost of raw material as USD20/kg, the return on investment for the conditions studied was estimated to range between 15.15 and 937.44%, with an internal rate of return between 11.95 and 401.33%, and a payback time between 0.11 and 6 years [96].

Works on the process simulation and economic evaluation of the utilization of corn milling byproducts with supercritical technology are scarce.

Rosa and coworkers [18], who evaluated the encapsulation of vitamins with zein with $SC\text{-}CO_2$ as an antisolvent, observed that the quantity of CO_2 used for processing strongly affected the costs because of the electricity demand of operating the CO_2 pump. Considering the scenario of excess CO_2 utilization, the COM calculated for the vitamin microcapsules was USD 0.2/capsule; this value is competitive with the costs of commercially available multivitamin capsules. However, considering a reduction in 50% of CO_2 consumption for encapsulation, the COM could be reduced by approximately 24%.

Attard and coworkers [17] estimated the economic scenario for the extraction of waxes from corn stover with supercritical CO_2 based on the biorefinery approach. In the wax extracted with SFE, multiple unsaturated fatty acids were found. The COM estimated for wax was initially at USD 94.20/kg wax but decreased to USD 4.83/kg wax considering the scenario of the highest efficiency, i.e., to reuse biomass as combustion feedstock for energy generation [17].

Studies on the economic evaluation of using corn milling byproducts or other underutilized products should carefully evaluate aspects including the raw material, demand, that the market the products are to be obtained in is focused, the country, the quality requirements, and the local fees, among other issues, in order to provide a realistic scenario to companies and stakeholders [76].

5.3. Patents Survey

Patents are exclusive rights granted by a country to inventors that contribute to stopping others from taking advantage of their ideas without permission. The patented processes that valorize corn milling byproducts with supercritical technology include extraction, fractionation, and chemical reactions.

The SFE at bench and at industrial scale was patented to deodorize corn starch by removing "cardboard- or cereal-like" off-flavors associated with the compounds hexanal, 2-heptanone, trimethylbenzene, nonanal, and BHT-aldehyde [97].

The bench scale equipment consisted of a pump, extraction vessel, oven, CO_2 regulating valve, and an extract collection vial. The solid starch to be purified is tightly packed in the extraction vessel and placed in the oven. The oven is closed and heated to the extraction temperature, and pre-heated CO_2 is pumped into the vessel with the CO_2-flow regulating valve closed, until the target processing conditions are reached. For the industrial scale, the inventors inserted starch slurry (20–50% solids) into a countercurrent extractor but found limitations attributed to starch gelatinization. The inventors suggested using ethanol (as the cosolvent) to overcome this limitation in isolated starches and starchy flours.

Another process with $SC\text{-}CO_2$ was patented to reuse corn germ to enhance the yield of oil up to approximately 20% using SFE equipment consisting of at least one extractor vessel, one heat source, one pressure generator (pump), and indicators of temperature and pressure [98].

The patent CN101077990A integrated SFE of corn germ in an extractor vessel followed by fractionation of extract into two separators that generate different fractions of vitamins and unsaturated fatty acids. The equipment also contained a device to clean CO_2 after extraction and further recycle CO_2 in the storage tank to decrease processing costs [99].

A process patented for the hydrolysis of lignocellulosic feedstock used a semi-batch mode by inserting corn stover in a small packed reactor that was processed with a supercritical ethanol-CO_2 mixture [100]. In another patent, NaOH (1%, v/v) followed by $SC\text{-}CO_2$ was used to pretreat corn straw to increase the release of polysaccharides to be converted to xylose via enzymatic hydrolysis [101].

In contrast, the CN102210356B patent introduces an enzymatic reaction of corn germ oil assisted with $SC\text{-}CO_2$ to synthesize structured lipids with medium-chain fatty acids by up to 32% [102]. Table 4 shows the conditions used on the patents searched for this review.

Table 4. The patents reviewed the reuse of corn byproducts with supercritical technology.

Material	Process	Conditions Used	Product	Patent Number	Location	Reference
Starch (solid and slurry at 20–50% solids)	Supercritical fluid extraction	Solvent: CO_2 and ethanol Temperature: 50–120 °C Pressure: >30 MPa Solvent to the raw material ratio: 1–10	Deodorized starch	US8216628B	United States	[97]
Germ meal	Supercritical fluid extraction	Solvent: CO_2 Temperature: 20–110 °C Pressure: 11–64 MPa Solvent to the raw material ratio: 2–30	Oil	US8603328B2	United States	[98]
Germ meal	Supercritical fluid extraction coupled with fractionation	Solvent: CO_2 Extractor: 40 °C and 8 MPa–20 MPa Separator 1: 35 °C and 8 MPa–6 MPa Separator 2: 35 °C and 6 MPa Flow rate: 20 L/h	Fractionated oils	CN101077990A	China	[99]
Stover	Supercritical hydrolysis	Solvent: water Temperature: 264 °C Pressure: 7.58 MPa Time: 20 min	Monosaccharides	US8282738B2	United States	[100]
Straw	Supercritical hydrolysis	Solvent: CO_2 and water/C1–C5 alcohol Temperature: 35–70 °C Pressure: Non identified Time: 60 min	Monosaccharides	CN112708647A	China	[101]
Germ oil	Enzymatic reaction	Enzymatic reaction Temperature: 45–65 °C Catalyst: Novozym 435 at 3–5% Pressure: 8–13 MPa Mixing speed: 120 r/min, Reaction time: 18–28 h	Structured lipid rich in caprylin	CN102210356B	China	[102]

6. Current Trends and Opportunities for Future Studies

Combining technologies is a trend for process intensification via the generation of multiple products with competitive costs. Based on the vast overview of possibilities offered by supercritical technology to process milling byproducts explored in this review, we propose a scenario to support further research on process optimization and feasibility.

In the first stage of this scenario, a storage tank that captures the CO_2 generated in the fermentation reservoirs in dry milling (Figure 4A) was included as a CO_2 reservoir for integrated equipment that operates extraction and particle formation integrated with CO_2 cleaning and recycling to reduce CO_2 emission to the atmosphere.

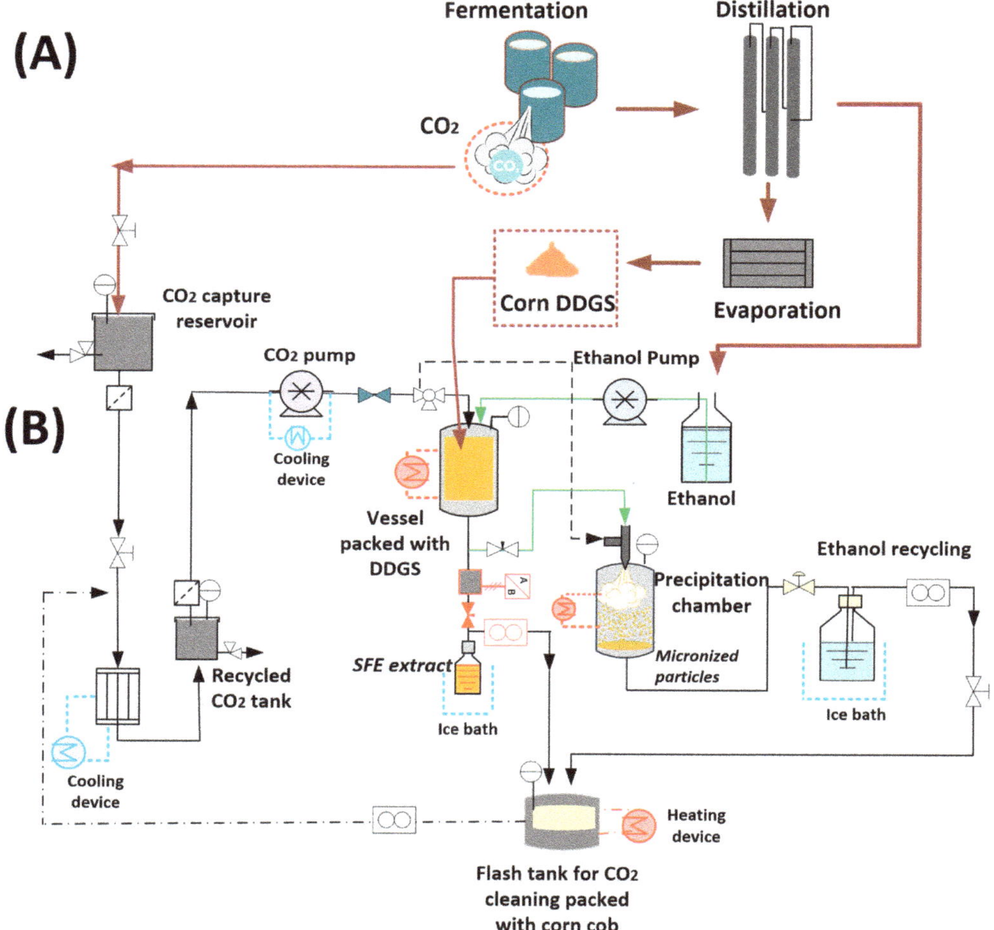

Figure 4. The future directions on the optimization of dry grinding (**A**) via integration with supercritical technology (**B**) for the reutilization of corn byproducts.

The CO_2 flows to the pump coupled to a cooling bath at $-5\,^\circ C$ to maintain a liquid state. Liquid CO_2 is pumped to the heated extraction vessel packed with corn DDGS to reach the supercritical state. In this first stage, supercritical CO_2 flows to the extraction vessel to obtain an extract enriched with oil, carotenoids, phytosterols, and tocols.

In the second stage, the valve that allows the CO_2 flow is closed, and pressurized liquid extraction (PLE) starts with a stream of ethanol to select phenolic acids from semi-defatted corn DDGS (Figure 4B).

Afterward, the CO_2 valve is opened, CO_2 is pumped and interacts with the ethanolic extract that is transferred to a nozzle installed before the precipitation chamber that is used for micronization, i.e., particle formation via size reduction of extracts resulting in the concentration of target compounds.

Once all of the ethanolic extract is injected, only CO_2 is injected at constant flow to remove all traces of ethanol in the extract. Ethanol leaves the precipitation vessel and is recycled in a solvent reservoir, whereas the CO_2 is transferred to a flash tank packed with an adsorbent to remove all possible impurities carried by CO_2 and is subsequently coupled to a second storage tank with recycled CO_2. Figure 4 shows the schematic diagram of process intensification proposed in this review.

Torres and coworkers [16], who included a storage tank reservoir for SFE of annatto seeds, showed that the recycling of CO_2 reduces the cost of manufacturing from USD 349.50/kg to USD 122.67/kg via reduced CO_2 consumption by 85.24%. The semi-defatted and depigmented corn DDGS may be used as feedstock to synthesize bioactive peptides.

7. Conclusions

Milling transforms corn into multiple products and supports the biorefinery concept to valorize the underutilized fractions, including stover, condensed distillers' grains, and DDGS. In this work, we reviewed the opportunities offered by supercritical technology to valorize corn milling byproducts by discussing the patents deposited, the processes used and their economic feasibility, and the possibilities of being considered for future works.

For instance, the dry grinding integrated with supercritical technology was suggested in this review to valorize DDGS via the production of oil enriched with carotenoids, particles enriched with phenolic compounds, and a solid fraction to be reused as protein feedstock for the generation of peptides of high biological value.

The authors expect this review to support future research with milling optimization via integration with supercritical technology, by process optimization at bench scale, followed by economic evaluation.

Author Contributions: Conceptualization: Á.L.S. and M.A.A.M.; Methodology: Á.L.S.; Formal Analysis, Á.L.S.; Investigation: Á.L.S. and M.A.A.M.; Writing—Original Draft Preparation, Á.L.S.; Writing—review and editing: Á.L.S. and M.A.A.M.; Validation: Á.L.S. and M.A.A.M.; Visualization: Á.L.S. and M.A.A.M. All authors have read and agreed to the published version of the manuscript.

Funding: This research received no external funding.

Data Availability Statement: The data availability statement was included. Data analyzed in this review were a re-analysis of existing data, which are available at the works cited in the reference section.

Conflicts of Interest: The authors declare no conflict of interest.

Nomenclature

CDS	Condensed Distillers' Solubles
DDGS	Dried Distillers' Grains with Solubles
PLE	Pressurized liquid extraction
SC-CO_2	Supercritical carbon dioxide
SFE	Supercritical Fluid Extraction
SWE	Subcritical Water Extraction
SWG	Supercritical Water Gasification
TS	Thin Stillage
WDG	Wet Distillers' Grains

References

1. USDA World Agricultural Production. Available online: https://apps.fas.usda.gov/psdonline/circulars/production.pdf (accessed on 18 December 2022).
2. Khanna, M.; Paulson, N. To Harvest Stover or Not: Is It Worth It? Available online: https://farmdocdaily.illinois.edu (accessed on 10 December 2022).
3. Ruan, Z.; Wang, X.; Liu, Y.; Liao, W. Corn; *Integrated Processing Technologies for Food and Agricultural By-Products*; Pan, Z., Zhang, R., Zicari, S., Eds.; Academic Press: Cambridge, MA, USA, 2019; pp. 59–72.
4. Stanford, J.P.; Keener, K.M. *Cedar Rapids Food and Bioprocessors Manufacturing Report*; Iowa State University: Ames, IA, USA, 2018.
5. Anderson, B.; Almeida, H. Corn Dry Milling: Processes, Products, and Applications. In *Corn*; Serna-Saldivar, S.O., Ed.; AACC International Press: Oxford, UK, 2019; pp. 405–433.
6. Rausch, K.D.; Hummel, D.; Johnson, L.A.; May, J.B. Wet Milling: The Basis for Corn Biorefineries. In *Corn*; Serna-Saldivar, S.O., Ed.; AACC International Press: Oxford, UK, 2019; pp. 501–535.
7. Malumba, P.; Boudry, C.; Roiseux, O.; Bindelle, J.; Beckers, Y.; Béra, F. Chemical characterisation and in vitro assessment of the nutritive value of co-products yield from the corn wet-milling process. *Food Chem.* **2015**, *166*, 143–149. [CrossRef] [PubMed]
8. Cargill Corn—Advances Sustainability Across Corn Supply Chains. Available online: https://www.cargill.com/sustainability/corn/sustainable-corn (accessed on 10 December 2022).
9. Bunge Milling. Available online: https://www.bunge.com/our-businesses/milling (accessed on 10 December 2022).
10. Gavilon Gavilon. Available online: https://www.gavilon.com/ (accessed on 10 December 2022).
11. J-Six Dry Corn Milling. Available online: https://www.jsixenterprises.com/services/dry-corn-milling/ (accessed on 10 December 2022).
12. Lagerkvist, A.; Dahlén, L. Solid Wastesolid wasteGenerationsolid wastegenerationand Characterizationsolid wastecharacterization. In *Encyclopedia of Sustainability Science and Technology*; Meyers, R.A., Ed.; Springer: New York, NY, USA, 2012; pp. 10000–10013. ISBN 978-1-4419-0851-3.
13. Qi, Y.Y.; Zhang, K.Y.; Tian, G.; Bai, S.P.; Ding, X.M.; Wang, J.P.; Peng, H.W.; LV, L.; Xuan, Y.; Zeng, Q.F. Effects of dietary corn germ meal levels on growth performance, serum biochemical parameters, meat quality, and standardized ileal digestibility of amino acids in Pekin ducks. *Poult. Sci.* **2022**, *101*, 101779. [CrossRef]
14. Shin, E.-C.; Shurson, G.C.; Gallaher, D.D. Antioxidant capacity and phytochemical content of 16 sources of corn distillers dried grains with solubles (DDGS). *Anim. Nutr.* **2018**, *4*, 435–441. [CrossRef]
15. Chañi-Paucar, L.O.; Santana, Á.L.; Albarelli, J.Q.; Meireles, M.A.A. Chapter 6—Extraction of polyphenols by sub/supercritical based technologies. In *Technologies to Recover Polyphenols from AgroFood By-Products and Wastes*; Pintado, M.E., Saraiva, J.M.A., da Cruz Alexandre, E.M., Eds.; Academic Press: Cambridge, MA, USA, 2022; pp. 137–168. ISBN 978-0-323-85273-9.
16. Torres, R.A.C.; Santana, Á.L.; Santos, D.T.; Albarelli, J.Q.; Meireles, M.A.A. A novel process for CO_2 purification and recycling based on subcritical adsorption in oat bran. *J. CO_2 Util.* **2019**, *34*, 362–374. [CrossRef]
17. Attard, T.M.; McElroy, C.R.; Hunt, A.J. Economic Assessment of Supercritical CO_2 Extraction of Waxes as Part of a Maize Stover Biorefinery. *Int. J. Mol. Sci.* **2015**, *16*, 17546–17564. [CrossRef] [PubMed]
18. Rosa, M.T.M.G.; Alvarez, V.H.; Albarelli, J.Q.; Santos, D.T.; Meireles, M.A.A.; Saldaña, M.D.A. Supercritical anti-solvent process as an alternative technology for vitamin complex encapsulation using zein as wall material: Technical-economic evaluation. *J. Supercrit. Fluids* **2020**, *159*, 104499. [CrossRef]
19. Knez, Ž. Enzymatic reactions in dense gases. *J. Supercrit. Fluidsupercrit. Fluids* **2009**, *47*, 357–372. [CrossRef]
20. Fărcaș, A.C.; Socaci, S.A.; Nemeș, S.A.; Salanță, L.C.; Chiș, M.S.; Pop, C.R.; Borșa, A.; Diaconeasa, Z.; Vodnar, D.C. Cereal Waste Valorization through Conventional and Current Extraction Techniques—An Up-to-Date Overview. *Foods* **2022**, *11*, 2454. [CrossRef]
21. Marinho, C.; Lemos, C.; Arvelos, S.; Barrozo, M.; Hori, C.; Watanabe, E. Extraction of corn germ oil with supercritical CO_2 and cosolvents. *J. Food Sci. Technol.* **2019**, *56*, 4448–4456. [CrossRef]
22. Yadav, G.; Fabiano, L.A.; Soh, L.; Zimmerman, J.; Sen, R.; Seider, W.D. Supercritical CO_2 Transesterification of Triolein to Methyl-Oleate in a Batch Reactor: Experimental and Simulation Results. *Processes* **2019**, *7*, 16. [CrossRef]
23. Li, H.; Hu, Y.; Wang, H.; Han, X.; El-Sayed, H.; Zeng, Y.; Charles Xu, C. Supercritical water gasification of lignocellulosic biomass: Development of a general kinetic model for prediction of gas yield. *Chem. Eng. J.* **2022**, *433*, 133618. [CrossRef]
24. Jayasinghe, S.; Miller, D.U.S. Corn Usage for Ethanol, Dry Mill Ethanol Co-Products Production, and Ethanol Yields Update. Available online: https://www.agmrc.org/ (accessed on 2 November 2022).
25. Deshwal, G.; Alam, T.; Panjagari, N.R.; Bhardwaj, A. Utilization of Cereal Crop Residues, Cereal Milling, Sugarcane and Dairy Processing By-Products for Sustainable Packaging Solutions. *J. Polym. Environ.* **2021**, *29*, 2046–2061. [CrossRef]
26. Vohra, M.; Manwar, J.; Manmode, R.; Padgilwar, S.; Patil, S. Bioethanol production: Feedstock and current technologies. *J. Environ. Chem. Eng.* **2014**, *2*, 573–584. [CrossRef]
27. Reis, C.E.R.; Rajendran, A.; Hu, B. New technologies in value addition to the thin stillage from corn-to-ethanol process. *Rev. Environ. Sci. Bio/Technol.* **2017**, *16*, 175–206. [CrossRef]
28. US Grains Council. Guide to Distiller's Dried Grains with Solubles (DDGS). Available online: https://www.canr.msu.edu/uploads/236/58572/cfans_asset_417244.pdf (accessed on 12 December 2022).

29. Shurson, J.; Spiehs, M.; Wilson, J.; Whitney, M. Value and Use of 'New Generation' Distiller's Dried Grains with Solubles in Swine Diets. Available online: https://en.engormix.com/pig-industry/articles/dried-grains-solubles-in-swine-diets-t33444.htm (accessed on 9 December 2022).
30. USDA Grain Crushings and CoProducts Production 2019 Summary. In *USDA, National Agricultural Statistics Service*; USDA: Washington, DC, USA, 2020; pp. 1–7. Available online: https://www.nass.usda.gov/Publications/Todays_Reports/reports/cagcan20.pdf (accessed on 14 November 2022).
31. Li, Q.; Singh, V.; de Mejia, E.G.; Somavat, P. Effect of sulfur dioxide and lactic acid in steeping water on the extraction of anthocyanins and bioactives from purple corn pericarp. *Cereal Chem.* **2019**, *96*, 575–589. [CrossRef]
32. Rausch, K.D.; Eckhoff, S.R. Maize: Wet Milling. In *Encyclopedia of Food Grain*; Wrigley, C., Corke, H., Seetharaman, K., Faubion, J., Eds.; Academic Press: Oxford, UK, 2016; pp. 467–481.
33. Sharma, S.; Pradhan, R.; Manickavasagan, A.; Thimmanagari, M.; Saha, D.; Singh, S.S.; Dutta, A. Production of antioxidative protein hydrolysates from corn distillers solubles: Process optimization, antioxidant activity evaluation, and peptide analysis. *Ind. Crops Prod.* **2022**, *184*, 115107. [CrossRef]
34. Sharma, S.; Pradhan, R.; Manickavasagan, A.; Tsopmo, A.; Thimmanagari, M.; Dutta, A. Corn distillers solubles by two-step proteolytic hydrolysis as a new source of plant-based protein hydrolysates with ACE and DPP4 inhibition activities. *Food Chem.* **2023**, *401*, 134120. [CrossRef]
35. Purdum, S.; Hanford, K.; Kreifels, B. Short-term effects of lower oil dried distillers grains with solubles in laying hen rations. *Poult. Sci.* **2014**, *93*, 2592–2595. [CrossRef]
36. Heuzé, V.; Tran, G.; Sauvant, D.; Noblet, J.; Renaudeau, D.; Bastianelli, D.; Lessire, M.; Lebas, F. Corn Distillers Grain. Available online: https://www.feedipedia.org/node/71 (accessed on 10 December 2022).
37. Langemeier, M. Explaining Fluctuations in DDG Prices. *Farmdoc Dly.* **2022**, *12*, 82.
38. Calendula Flower Powder Organic. Available online: https://www.starwest-botanicals.com/product/calendula-flower-powder-organic/ (accessed on 17 December 2022).
39. Luthria, D.L.; Liu, K.; Memon, A.A. Phenolic Acids and Antioxidant Capacity of Distillers Dried Grains with Solubles (DDGS) as Compared with Corn. *J. Am. Oil Chem. Soc.* **2012**, *89*, 1297–1304. [CrossRef]
40. Lorigooini, Z.; Jamshidi-kia, F.; Hosseini, Z. Chapter 4—Analysis of aromatic acids (phenolic acids and hydroxycinnamic acids). In *Recent Advances in Natural Products Analysis*; Sanches Silva, A., Nabavi, S.F., Saeedi, M., Nabavi, S.M., Eds.; Elsevier: Amsterdam, The Netherlands, 2020; pp. 199–219. ISBN 978-0-12-816455-6.
41. Impact of Oil Extraction on the Nutritional Value of DDGS. Available online: https://www.wengerfeeds.com/impact-of-oil-extraction-on-the-nutritional-value-of-ddgs/ (accessed on 18 December 2022).
42. Masisi, K.; Diehl-Jones, W.L.; Gordon, J.; Chapman, D.; Moghadasian, M.H.; Beta, T. Carotenoids of aleurone, germ, and endosperm fractions of barley, corn and wheat differentially inhibit oxidative stress. *J. Agric. Food Chem.* **2015**, *63*, 2715–2724. [CrossRef] [PubMed]
43. Espinosa-Pardo, F.A.; Savoire, R.; Subra-Paternault, P.; Harscoat-Schiavo, C. Oil and protein recovery from corn germ: Extraction yield, composition and protein functionality. *Food Bioprod. Process.* **2020**, *120*, 131–142. [CrossRef]
44. Whisner, C.M.; Martin, B.R.; Nakatsu, C.H.; Story, J.A.; MacDonald-Clarke, C.J.; McCabe, L.D.; McCabe, G.P.; Weaver, C.M. Soluble Corn Fiber Increases Calcium Absorption Associated with Shifts in the Gut Microbiome: A Randomized Dose-Response Trial in Free-Living Pubertal Females. *J. Nutr.* **2016**, *146*, 1298–1306. [CrossRef] [PubMed]
45. Costabile, A.; Bergillos-Meca, T.; Rasinkangas, P.; Korpela, K.; De Vos, W.M.; Gibson, G.R. Effects of soluble corn fiber alone or in synbiotic combination with lactobacillus rhamnosus GG and the pilus-deficient derivative GG-PB12 on fecal microbiota, metabolism, and markers of immune function: A randomized, double-blind, placebo-controlled, cro. *Front. Immunol.* **2017**, *8*, 1443. [CrossRef]
46. Fernandes, A.C.; Vieira, N.C.; de Santana, Á.L.; de Pádua Gandra, R.L.; Rubia, C.; Castro-Gamboa, I.; Macedo, J.A.; Macedo, G.A. Peanut skin polyphenols inhibit toxicity induced by advanced glycation end-products in RAW264.7 macrophages. *Food Chem. Toxicol.* **2020**, *145*, 111619. [CrossRef]
47. Zuo, G.; Song, X.; Cheng, F.; Shen, Z. Physical and structural characterization of edible bilayer films made with zein and corn-wheat starch. *J. Saudi Soc. Agric. Sci.* **2019**, *18*, 324–331. [CrossRef]
48. Liu, W.; Fang, L.; Feng, X.; Li, G.; Gu, R. In vitro antioxidant and angiotensin I-converting enzyme inhibitory properties of peptides derived from corn gluten meal. *Eur. Food Res. Technol.* **2020**, *246*, 2017–2027. [CrossRef]
49. Wang, L.; Ding, L.; Yu, Z.; Zhang, T.; Ma, S.; Liu, J. Intracellular ROS scavenging and antioxidant enzyme regulating capacities of corn gluten meal-derived antioxidant peptides in HepG2 cells. *Food Res. Int.* **2016**, *90*, 33–41. [CrossRef]
50. Heng, L.; Zhang, H.; Xiao, J.; Xiao, R. Life Cycle Assessment of Polyol Fuel from Corn Stover via Fast Pyrolysis and Upgrading. *ACS Sustain. Chem. Eng.* **2018**, *6*, 2733–2740. [CrossRef]
51. Saha, B.C.; Kennedy, G.J.; Qureshi, N.; Cotta, M.A. Biological pretreatment of corn stover with Phlebia brevispora NRRL-13108 for enhanced enzymatic hydrolysis and efficient ethanol production. *Biotechnol. Prog.* **2017**, *33*, 365–374. [CrossRef]
52. Duan, X.-L.; Yuan, C.-G.; Jing, T.-T.; Yuan, X.-D. Removal of elemental mercury using large surface area micro-porous corn cob activated carbon by zinc chloride activation. *Fuel* **2019**, *239*, 830–840. [CrossRef]

53. Abirami, S.; Priyalakshmi, M.; Soundariya, A.; Samrot, A.V.; Saigeetha, S.; Emilin, R.R.; Dhiva, S.; Inbathamizh, L. Antimicrobial activity, antiproliferative activity, amylase inhibitory activity and phytochemical analysis of ethanol extract of corn (*Zea mays* L.) silk. *Curr. Res. Green Sustain. Chem.* **2021**, *4*, 100089. [CrossRef]
54. Žilić, S.; Janković, M.; Basić, Z.; Vančetović, J.; Maksimović, V. Antioxidant activity, phenolic profile, chlorophyll and mineral matter content of corn silk (*Zea mays* L): Comparison with medicinal herbs. *J. Cereal Sci.* **2016**, *69*, 363–370. [CrossRef]
55. Lee, C.W.; Seo, J.Y.; Kim, S.-L.; Lee, J.; Choi, J.W.; Park, Y. Il Corn silk maysin ameliorates obesity in vitro and in vivo via suppression of lipogenesis, differentiation, and function of adipocytes. *Biomed. Pharmacother.* **2017**, *93*, 267–275. [CrossRef] [PubMed]
56. Rodríguez-López, L.; Vecino, X.; Barbosa-Pereira, L.; Moldes, A.B.; Cruz, J.M. A multifunctional extract from corn steep liquor: Antioxidant and surfactant activities. *Food Funct.* **2016**, *7*, 3724–3732. [CrossRef] [PubMed]
57. Costa, A.F.S.L.; Almeida, F.C.G.; Vinhas, G.M.; Sarubbo, L.A. Production of Bacterial Cellulose by Gluconacetobacter hansenii Using Corn Steep Liquor As Nutrient Sources. *Front. Microbiol.* **2017**, *8*, 2027. [CrossRef] [PubMed]
58. Zhu, M.-M.; Liu, E.-Q.; Bao, Y.; Duan, S.-L.; She, J.; Liu, H.; Wu, T.-T.; Cao, X.-Q.; Zhang, J.; Li, B.; et al. Low concentration of corn steep liquor promotes seed germination, plant growth, biomass production and flowering in soybean. *Plant Growth Regul.* **2019**, *87*, 29–37. [CrossRef]
59. Loy, D.D.; Lundy, E.L. Chapter 23—Nutritional Properties and Feeding Value of Corn and Its Coproducts. In *Corn*, 3rd ed.; Serna-Saldivar, S.O., Ed.; AACC International Press: Oxford, UK, 2019; pp. 633–659. ISBN 978-0-12-811971-6.
60. Kim, Y.; Mosier, N.S.; Hendrickson, R.; Ezeji, T.; Blaschek, H.; Dien, B.; Cotta, M.; Dale, B.; Ladisch, M.R. Composition of corn dry-grind ethanol by-products: DDGS, wet cake, and thin stillage. *Bioresour. Technol.* **2008**, *99*, 5165–5176. [CrossRef]
61. Di Lena, G.; Ondrejíčková, P.; del Pulgar, J.S.; Cyprichová, V.; Ježovič, T.; Lucarini, M.; Lombardi Boccia, G.; Ferrari Nicoli, S.; Gabrielli, P.; Aguzzi, A.; et al. Towards a Valorization of Corn Bioethanol Side Streams: Chemical Characterization of Post Fermentation Corn Oil and Thin Stillage. *Molecules* **2020**, *25*, 3549. [CrossRef]
62. Yang, X.; Nath, C.; Doering, A.; Goihl, J.; Baidoo, S.K. Effects of liquid feeding of corn condensed distiller's solubles and whole stillage on growth performance, carcass characteristics, and sensory traits of pigs. *J. Anim. Sci. Biotechnol.* **2017**, *8*, 9. [CrossRef]
63. Gao, J.L.; Wang, P.; Zhou, C.H.; Li, P.; Tang, H.Y.; Zhang, J.B.; Cai, Y. Chemical composition and in vitro digestibility of corn stover during field exposure and their fermentation characteristics of silage prepared with microbial additives. *Asian-Australas. J Anim. Sci.* **2019**, *32*, 1854–1863. [CrossRef]
64. Ma, Z.; Kasipandi, S.; Wen, Z.; Yu, L.; Cui, K.; Chen, H.; Li, Y. Highly efficient fractionation of corn stover into lignin monomers and cellulose-rich pulp over H2WO4. *Appl. Catal. B Environ.* **2021**, *284*, 119731. [CrossRef]
65. Liu, Z.H.; Qin, L.; Zhu, J.Q.; Li, B.Z.; Yuan, Y.J. Simultaneous saccharification and fermentation of steam-exploded corn stover at high glucan loading and high temperature. *Biotechnol. Biofuels* **2014**, *7*, 167. [CrossRef] [PubMed]
66. Chen, L.; Chen, W.; Zheng, B.; Yu, W.; Zheng, L.; Qu, Z.; Yan, X.; Wei, B.; Zhao, Z. Fermentation of NaHCO3-treated corn germ meal by Bacillus velezensis CL-4 promotes lignocellulose degradation and nutrient utilization. *Appl. Microbiol. Biotechnol.* **2022**, *106*, 6077–6094. [CrossRef] [PubMed]
67. Azizi-Shotorkhoft, A.; Sharifi, A.; Mirmohammadi, D.; Baluch-Gharaei, H.; Rezaei, J. Effects of feeding different levels of corn steep liquor on the performance of fattening lambs. *J. Anim. Physiol. Anim. Nutr.* **2016**, *100*, 109–117. [CrossRef]
68. Winkler-Moser, J.K.; Hwang, H.-S.; Byars, J.A.; Vaughn, S.F.; Aurandt-Pilgrim, J.; Kern, O. Variations in phytochemical content and composition in distillers corn oil from 30 U.S. ethanol plants. *Ind. Crops Prod.* **2023**, *193*, 116108. [CrossRef]
69. Deepak, T.S.; Jayadeep, P.A. Nutraceutical potential of maize (*Zea mays* L.) (corn) lab-scale wet milling by-products in terms of β-sitosterol in fiber, gamma-tocopherol in germ, and lutein & zeaxanthin in gluten. Research Square Preprint. Available online: https://www.researchsquare.com/article/rs-1518914/v1 (accessed on 12 December 2022).
70. Jiao, Y.; Li, D.; Chang, Y.; Xiao, Y. Effect of Freeze-Thaw Pretreatment on Extraction Yield and Antioxidant Bioactivity of Corn Carotenoids (Lutein and Zeaxanthin). *J. Food Qual.* **2018**, *2018*, 9843503. [CrossRef]
71. Burlini, I.; Grandini, A.; Tacchini, M.; Maresca, I.; Guerrini, A.; Sacchetti, G. Different Strategies to Obtain Corn (Zea mays L.) Germ Extracts with Enhanced Antioxidant Properties. *Nat. Prod. Commun.* **2020**, *15*, 1–9. [CrossRef]
72. Yadav, M.P.; Moreau, R.A.; Hicks, K.B. Phenolic Acids, Lipids, and Proteins Associated with Purified Corn Fiber Arabinoxylans. *J. Agric. Food Chem.* **2007**, *55*, 943–947. [CrossRef]
73. Vazquez-Olivo, G.; López-Martínez, L.X.; Contreras-Angulo, L.; Heredia, J.B. Antioxidant Capacity of Lignin and Phenolic Compounds from Corn Stover. *Waste Biomass Valor.* **2019**, *10*, 95–102. [CrossRef]
74. Johner, J.C.F.; de Meireles, M.A.A. Construction of a supercritical fluid extraction (SFE) equipment: Validation using annatto and fennel and extract analysis by thin layer chromatography coupled to image. *Food Sci. Technol.* **2016**, *36*, 210–247. [CrossRef]
75. Santos, D.T.; Santana, Á.L.; Meireles, M.A.A.; Petenate, A.J.; Silva, E.K.; Albarelli, J.Q.; Johner, J.C.F.; Gomes, M.T.; Torres, R.A.D.C.; Hatami, T. (Eds.) A Detailed Design and Construction of a Supercritical Antisolvent Precipitation Equipment. In *Supercritical Antisolvent Precipitation Process: Funtamentals, Applications and Perspectives*; Springer Nature: Cham, Switzerland, 2019; Chapter 1.
76. Santana, Á.L.; Zabot, G.L.; Osorio-Tobón, J.F.; Johner, J.C.F.; Coelho, A.S.; Schmiele, M.; Steel, C.J.; Meireles, M.A.A. Starch recovery from turmeric wastes using supercritical technology. *J. Food Eng.* **2017**, *214*, 266–276. [CrossRef]
77. Carvalho, P.I.N.; Osorio-Tobón, J.F.; Rostagno, M.A.; Petenate, A.J.; Meireles, M.A.A. Techno-economic evaluation of the extraction of turmeric (*Curcuma longa* L.) oil and ar-turmerone using supercritical carbon dioxide. *J. Supercrit. Fluids* **2015**, *105*, 44–54. [CrossRef]

78. Santana, Á.L.; Santos, D.T.; Meireles, M.A.A. Perspectives on small-scale integrated biorefineries using supercritical CO_2 as a green solvent. *Curr. Opin. Green Sustain. Chem.* **2019**, *18*, 1–12. [CrossRef]

79. Liu, J.; Lin, S.; Wang, Z.; Wang, C.; Wang, E.; Zhang, Y.; Liu, J. Supercritical fluid extraction of flavonoids from Maydis stigma and its nitrite-scavenging ability. *Food Bioprod. Process.* **2011**, *89*, 333–339. [CrossRef]

80. Sun, J.; Ding, R.; Yin, J. Pretreatment corn ingredient biomass with high pressure CO_2 for conversion to fermentable sugars via enzymatic hydrolysis of cellulose. *Ind. Crops Prod.* **2022**, *177*, 114518. [CrossRef]

81. Yin, J.; Hao, L.; Yu, W.; Wang, E.; Zhao, M.; Xu, Q.; Liu, Y. Enzymatic hydrolysis enhancement of corn lignocellulose by supercritical CO_2 combined with ultrasound pretreatment. *Chin. J. Catal.* **2014**, *35*, 763–769. [CrossRef]

82. Cardenas-Toro, F.P.; Forster-Carneiro, T.; Rostagno, M.A.; Petenate, A.J.; Maugeri Filho, F.; Meireles, M.A.A. Integrated supercritical fluid extraction and subcritical water hydrolysis for the recovery of bioactive compounds from pressed palm fiber. *J. Supercrit. Fluids* **2014**, *93*, 42–48. [CrossRef]

83. Zhang, J.; Wen, C.; Zhang, H.; Zandile, M.; Luo, X.; Duan, Y.; Ma, H. Structure of the zein protein as treated with subcritical water. *Int. J. Food Prop.* **2018**, *21*, 128–138. [CrossRef]

84. Moraes, M.N.; Zabot, G.L.; Meireles, M.A.A. Extraction of tocotrienols from annatto seeds by a pseudo continuously operated SFE process integrated with low-pressure solvent extraction for bixin production. *J. Supercrit. Fluids* **2015**, *96*, 262–271. [CrossRef]

85. Colombo, T.S.; Mazutti, M.A.; Di Luccio, M.; de Oliveira, D.; Oliveira, J.V. Enzymatic synthesis of soybean biodiesel using supercritical carbon dioxide as solvent in a continuous expanded-bed reactor. *J. Supercrit. Fluids* **2015**, *97*, 16–21. [CrossRef]

86. Casademont, P.; García-Jarana, M.B.; Sánchez-Oneto, J.; Portela, J.R.; de la Ossa, E.J.M. Supercritical water gasification: A patents review. *Rev. Chem. Eng.* **2017**, *33*, 237–261. [CrossRef]

87. Torres, A.R.C.; Santana, Á.L.; Santos, D.T.; Meireles, M.A.A. Perspectives on the application of supercritical antisolvent fractionation process for the purification of plant extracts: Effects of operating parameters and patent survey. *Recent Pat. Eng.* **2016**, *10*, 88–97. [CrossRef]

88. Santana, Á.L.; Meireles, M.A.A. Coprecipitation of turmeric extracts and polyethylene glycol with compressed carbon dioxide. *J. Supercrit. Fluids* **2017**, *125*, 31–41. [CrossRef]

89. Palazzo, I.; Campardelli, R.; Scognamiglio, M.; Reverchon, E. Zein/luteolin microparticles formation using a supercritical fluids assisted technique. *Powder Technol.* **2019**, *356*, 899–908. [CrossRef]

90. Liu, X.; Jia, J.; Duan, S.; Zhou, X.; Xiang, A.; Lian, Z.; Ge, F. Zein/MCM-41 Nanocomposite Film Incorporated with Cinnamon Essential Oil Loaded by Modified Supercritical CO_2 Impregnation for Long-Term Antibacterial Packaging. *Pharmaceutics* **2020**, *12*, 169. [CrossRef]

91. Dias, A.L.B.; Hatami, T.; Viganó, J.; Santos de Araújo, E.J.; Mei, L.H.I.; Rezende, C.A.; Martínez, J. Role of supercritical CO_2 impregnation variables on β-carotene loading into corn starch aerogel particles. *J. CO_2 Util.* **2022**, *63*, 102125. [CrossRef]

92. Chauvet, M.; Sauceau, M.; Fages, J. Extrusion assisted by supercritical CO_2: A review on its application to biopolymers. *J. Supercrit. Fluids* **2017**, *120*, 408–420. [CrossRef]

93. Wang, Y.-Y.; Ryu, G.-H. Physicochemical and antioxidant properties of extruded corn grits with corn fiber by CO_2 injection extrusion process. *J. Cereal Sci.* **2013**, *58*, 110–116. [CrossRef]

94. Turton, R.B.; Wallace, B.; Whiting, J.S.; Bhattacharyya, D. *Analysis, Synthesis and Design of Chemical Processes*, 3rd ed.; Prentice Hall: Upper Saddle River, NJ, USA, 2009.

95. Baldino, L.; De Luca, R.; Reverchon, E. Scale-up and economic analysis of a supercritical CO_2 plant for antimalarial active compounds extraction. *Chem. Eng. Trans.* **2018**, *64*, 31–36.

96. Chañi-Paucar, L.O.; Osorio-Tobón, J.F.; Johner, J.C.F.; Meireles, M.A.A. A comparative and economic study of the extraction of oil from Baru (Dipteryx alata) seeds by supercritical CO_2 with and without mechanical pressing. *Heliyon* **2021**, *7*, e05971. [CrossRef] [PubMed]

97. Koxholt, M.; Altieri, P.A.; Marentis, R.T.; Trzasko, P.T. Process for Purifying Starches. U.S. Patent 8216628B, 10 July 2012.

98. DeLine, K.E.; Claycamp, D.L.; Fetherston, D.; Marentis, R.T. Carbon Dioxide Corn Germ Oil Extraction System. U.S. Patent US8603328B2, 3 November 2009.

99. Zhao, F.; Liu, D.; Cai, S.; Wang, H.; Fu, H. Method for Extracting Corn Embryo Oil by Supercritical Carbon Dioxide. CN Patent 101077990A, 4 July 2007.

100. Kilambi, S.; Kadam, K.L. Solvo-Thermal Fractionation of Biomass. U.S. Patent 8282738B2, 16 July 2009.

101. Xu, J.; Gu, S.; Wu, H.; Zuo, Z.; Wei, W.; Hu, S.; Zhang, K.; Xu, X. Method for Improving Enzymolysis and Xylose Conversion Efficiency of Corn Straws by Supercritical Carbon Dioxide Coupled NaOH Pretreatment. CN Patent 112708647A, 24 December 2020.

102. Yu, D.; Xu, J.; Zhang, Q.; Zhang, S.; Wang, J.; Jiang, L.; Hu, L.; Wang, L.; Qi, Y.; Zhang, J.; et al. Method for Synthesizing Corn Oil Rich in Caprylin Enzymatically under Supercritical Carbon Dioxide (CO_2) State. CN Patent 102210356B, 1 April 2011.

MDPI AG

Grosspeteranlage 5

4052 Basel

Switzerland

Tel.: +41 61 683 77 34

www.mdpi.com

Processes Editorial Office

E-mail: processes@mdpi.com

www.mdpi.com/journal/processes